TETRAHEDRON ORGANIC CHEMISTR

VOLUME 23

Organolithiums:

Selectivity for Synthesis

Related Pergamon Titles of Interest

BOOKS

Tetrahedron Organic Chemistry Series:
CARRUTHERS: Cycloaddition Reactions in Organic Synthesis
CLARIDGE: High-Resolution NMR Techniques in Organic Chemistry
FINET: Ligand Coupling Reactions with Heteroatomic Compounds
GAWLEY & AUBÉ: Principles of Asymmetric Synthesis
HASSNER & STUMER: Organic Syntheses Based on Name Reactions and Unnamed Reactions
McKILLOP: Advanced Problems in Organic Reaction Mechanisms
OBRECHT & VILLALGORDO: Solid-Supported Combinatorial and Parallel Synthesis of Small-Molecular-Weight Compound Libraries
PERLMUTTER: Conjugate Addition Reactions in Organic Synthesis
SESSLER & WEGHORN: Expanded, Contracted & Isomeric Porphyrins
TANG & LEVY: Chemistry of *C*-Glycosides
WONG & WHITESIDES: Enzymes in Synthetic Organic Chemistry
LI & GRIBBLE: Palladium in Heterocyclic Chemistry
PIETRA: Biodiversity and Natural Product Diversity
HASSNER & STUMER: Organic Synthesis based on Name Reactions (2nd Edition)

RAHMAN: Studies in Natural Products Chemistry *(series)*

JOURNALS

BIOORGANIC & MEDICINAL CHEMISTRY
BIOORGANIC & MEDICINAL CHEMISTRY LETTERS
TETRAHEDRON
TETRAHEDRON LETTERS
TETRAHEDRON: ASYMMETRY

Full details of all Elsevier Science publications are available on www.elsevier.com or from your nearest Elsevier Science office

Organolithiums: Selectivity for Synthesis

JONATHAN CLAYDEN

Department of Chemistry,

University of Manchester,

Manchester, UK

2002

PERGAMON

An imprint of Elsevier Science

Amsterdam - Boston - London - New York - Oxford - Paris - San Diego
San Francisco - Singapore - Sydney - Tokyo

ELSEVIER SCIENCE Ltd
The Boulevard, Langford Lane
Kidlington, Oxford OX5 1GB, UK

First edition 2002

Library of Congress Cataloging in Publication Data
A catalog record from the Library of Congress has been applied for.

British Library Cataloguing in Publication Data
A catalogue record from the British Library has been applied for.

ISBN: 0 08 043262 X (hardbound)
ISBN: 0 08 043261 1 (paperback)

♾ The paper used in this publication meets the requirements of ANSI/NISO Z39.48-1992 (Permanence of Paper).
Printed in The Netherlands.

Table of Contents

Foreword

The importance of organolithium chemistry to organic synthesis is undisputed: functionalisation by lithiation and electrophilic quench is among the most fundamental of synthetic transformations. Chemists will often turn to organolithium methods because of their simplicity, and with standard laboratory equipment and careful technique organolithiums are straightforward reagents to handle. Not for some years, however, have the themes of organolithium chemistry been brought together in a single book. My aim has been to do this, reviewing established procedures such as directed metallation and reductive lithiation and summarising recent discoveries in developing areas such as the use of (–)-sparteine as a ligand. I have aimed to provide clear mechanistic explanations for the control of selectivity which organolithiums (often uniquely) allow, and I have aimed to provide a valuable resource for all chemists carrying out synthesis, whether graduate students or professional chemists in industry.

Jonathan Clayden, Manchester, May 2002

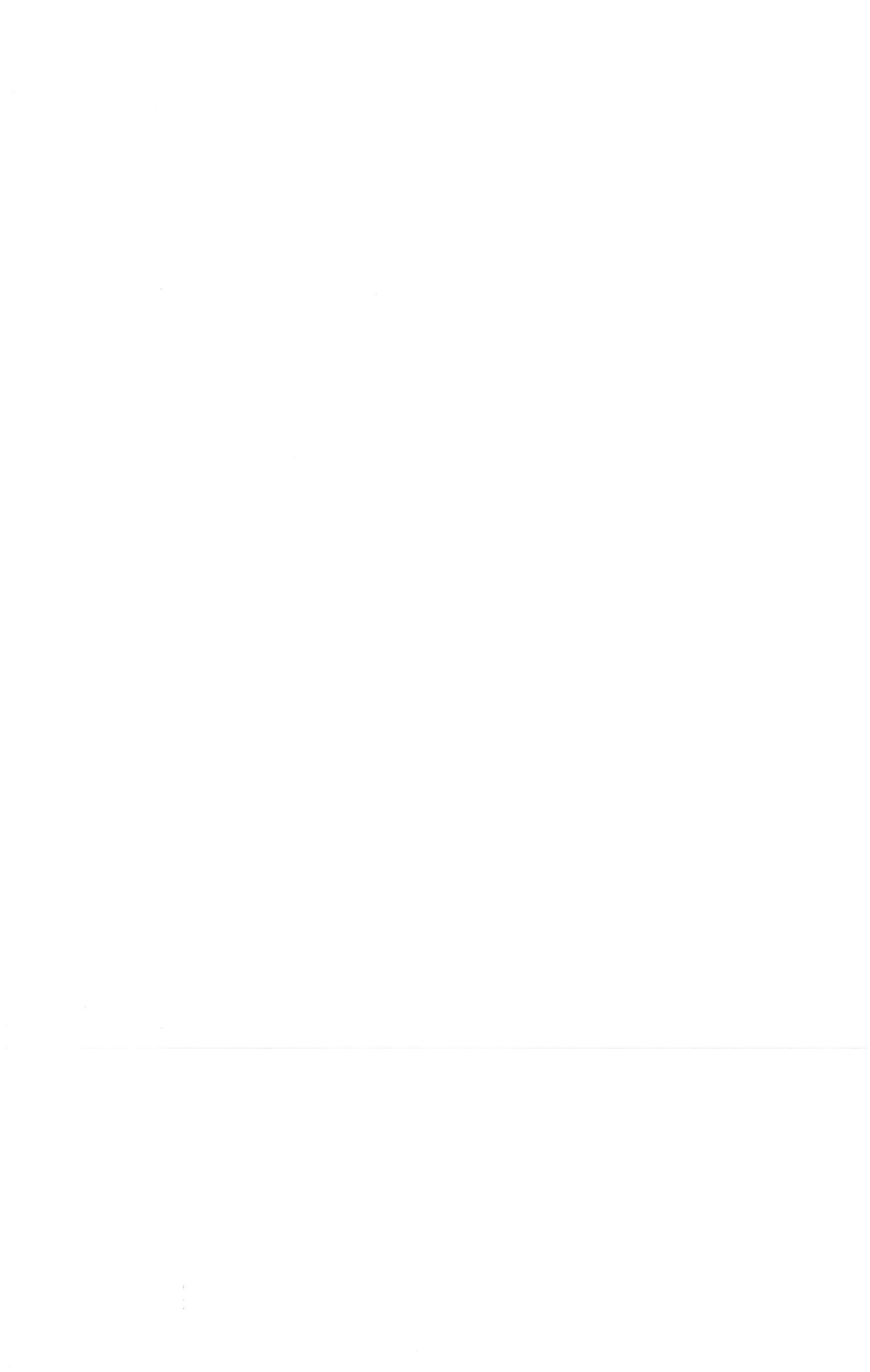

Acknowledgements

This book was written during the period 1998-2002 with the forbearance of my family and friends, to whom I am indebted. Parts were written during stays in Rouen while I was a visiting professor at the *Institut National des Sciences Appliquées*, and I am grateful to Jean-Charles Quirion and to Guy Quéguiner for the opportunity to spend time in Normandy. I would also like to thank the members of my research group (Katherine Hebditch, Martin Kenworthy, Przemek Kubinski, Andrew Lund, Christel Menet, Savroop Purewal and David Watson) who proof-read the manuscript and made many helpful suggestions, and Julie Clayden, who did the final editing.

Abbreviations

BuLi	*n*-butyllithium
DABCO	1,4-diazabicyclo[2.2.2]octane
DBB	4,4'-di-*tert*-butylbiphenyl
DMAN	1-(*N,N*-dimethylamino)naphthalene
DME	1,2-dimethoxyethane
DMPU	N,N'-dimethylpropylideneurea
ee	enantiomeric excess
ESR	electron spin resonance
Et_2O	diethyl ether
ether	diethyl ether
EVL	1-ethoxyvinyllithium
HMPA	*N,N,N',N',N'',N''*-hexamethylphosphoramide
i-PrLi	*iso*-propyllithium
KIE	kinetic isotope effect
LDA	*N*-lithio diisopropylamine
LiCKOR	BuLi–KO*t*-Bu mixtures
LiDMAE	lithiated 2-(*N,N*-dimethylamino)ethanol
LiTMP	*N*-lithio 2,2,6,6-tetramethylpiperidine
MEM	methoxyethoxymethyl
MOM	methoxymethyl
MTPA	2-methoxy-2-trifuloromethyl-2-phenylacetic acid
MTBE	methyl *tert*-butyl ether
NMO	*N*-methylmorpholine-*N*-oxide
NMR	nuclear magnetic resonance
Np	naphthalene
PMDTA	*N,N',N'',N''',N'''*-pentamethylethylenetriamine
s-BuLi	*sec*-butyllithium
S_E1	electrophilic substitution, first order
S_E2inv	electrophilic substitution, second order, with inversion of stereochemistry
S_E2ret	electrophilic substitution, second order, with retention of stereochemistry
SOMO	singly occupied molecular orbital

(–)-sp.	(–)-sparteine
TBAF	tetra-*n*-butylammonium fluoride
t-BuLi	*tert*-butyllithium
THF	tetrahydrofuran
THP	tetrahydropyran
TMEDA	*N,N,N',N'*-tetramethylethylenediamine
TPAP	tetra-*n*-propylammonium perruthenate

CHAPTER 1

Introduction

1.1 Scope and Overview

Organolithiums are central to so many aspects of synthetic organic chemistry that a book on organolithium chemistry must be a book on synthesis. Hardly a molecule is made without a bottle of BuLi: evaporation as butane is the destiny of at least one proton of the starting material in almost any synthetic sequence.

A book on organolithiums is also a book on mechanism. The observation of selectivity in an organolithium reaction has often led to mechanistic insights that later turn out to be general mechanistic features of organic reactions. Directed metallation, for example, started with the ortholithiation of anisole, and led to directed reactions of zinc and palladium. A decade of close investigation of configurational stability in the organolithium series preceded similar studies on organozincs and other organometallics. "Dynamic thermodynamic resolution" is a mechanistic feature first identified in the reactions of organolithiums in the presence of sparteine, and later exploited right outside of the organometallic sphere, in the synthesis of atropisomers.

Given their ubiquity, this book concentrates on one feature of the reactions of organolithiums (pushed to a wider sense in some areas than others) *selectivity*. This feature is the result of the civilisation of organolithiums from the savage beasts of 40 years ago (BuLi, benzene, reflux) to the tamed, well-trained species we use to coax out one proton at a time or to nudge a starting material over the energetic barrier of a spectacular cascade reaction.

To explain selectivity I have discussed mechanism in detail but not depth – the mechanistic discussions are intended not as a full account of current understanding of organolithium structure and reactivity, but as a tool for use in predicting likely outcomes of reactions and in accounting for unlikely ones. Similarly, structure is dealt with where necessary to explain a point, but detailed discussions of organolithium structure is outside the scope of a book primarily about organolithium *reactions*. I have also limited the definition of "organolithium" to those compounds in which there is a clear C–Li bond: compounds with any degree of enolate structure, and lithiated sulfones, sulfoxides, phosphonates, phosphine oxides etc. have been excluded. Inclusion or exclusion of a compound should not be taken to imply anything about its structure – a limit had to be drawn somewhere, and in some discussions the limit is stretched further tan in others.

General points about organolithiums in solution are considered briefly first, followed by chapters addressing the synthesis of functionalised organolithiums, and in particular the various methods for achieving regioselectivity. Some organolithiums have stereochemistry, and an account of the stereoselective synthesis of organolithiums follows an explanation of which and why. Discussions of reactions follow, but rather than give a thin account of a wide range of well-known additions and substitutions, I have limited the coverage to some important and developing areas: stereospecific and selective reactions, particularly those involving (–)-sparteine and other chiral additives, inter- and intramolecular additions to π systems, and rearrangements.

1.2 Organolithiums in solution

Organolithiums (with the exceptions of methyllithium and phenyllithium) are remarkably soluble even in hydrocarbon solvents,[1,2] and simple organolithium starting materials are available as stable hydrocarbon solutions (Table 1.2.1). Methyllithium and phenyllithium are indefinitely stable at ambient temperatures in the presence of ethers, and are solubilised by the addition of ether or THF.

organolithium	abbreviation[a]	solvent	concentration
methyllithium	MeLi	Et_2O	1.4 M
		cumene/THF	1.0 M
n-butyllithium	BuLi	cyclohexane	2.0 M
		hexanes	1.6, 2.5, 10 M
		pentane	2.0 M
sec-butyllithium	*s*-BuLi	cyclohexane	1.3 M
		(or cyclohexane/hexane)	
tert-butyllithium	*t*-BuLi	pentane	1.5, 1.7 M
phenyllithium	PhLi	cyclohexane/Et_2O	1.8 M

[a]*Abbreviation used in this book*

Table 1.2.1 Commercially available organolithiums in solution

Hydrocarbon solutions of *n*-, *s*- and *t*-BuLi are the ultimate source of most organolithiums, but a number of other bases are widely used to generate organolithiums from more acidic substrates. Among these are LDA, LiTMP and other more hindered lithium amide bases, and hindered aryllithiums such as mesityllithium and triisopropylphenyllithium.

The electron-deficient lithium atom of an organolithium compound requires greater stabilisation than can be provided by a single carbanionic ligand, and freezing-point measurements indicate that in hydrocarbon solution organolithiums are invariably aggregated as hexamers, tetramers or dimers.[3] The structure of these aggregates in solution can be deduced to a certain extent from the organolithiums' crystal structures[4] or by calculation:[5] the tetramers approximate to tetrahedra of lithium atoms bridged by the organic ligands; the hexamers approximate to octahedra of lithium atoms unsymmetrically bridged by the organic ligands.[6,7] The aggregation state of simple, unfunctionalised organolithiums depends primarily on steric hindrance. Primary organolithiums are hexamers in hydrocarbons, except when they are branched β to the lithium atom, when they are tetramers. Secondary and tertiary organolithiums are tetramers, while benzyllithium and very bulky alkyllithiums (menthyllithium) are dimers.[1,3]

Hexameric	Tetrameric	Dimeric	Monomeric
EtLi	*i*-$PrCH_2Li$	$PhCH_2Li$	–
BuLi	*i*-PrLi		
	t-BuLi		

Table 1.2.2 Typical aggregation state in hydrocarbon solution

Coordinating ligands – such as ethers or amines, or even metal alkoxides (see section 2.6) – can provide an alternative source of electron density for the electron-deficient lithium atoms. These ligands can first of all stabilise the aggregates by coordinating to the lithium atoms at their vertices, and then allow the organolithiums to shift to an entropically favoured lower degree of aggregation. As shown in Table 1.2.3, the presence of ether or THF typically causes a shift down in aggregation state, but only occasionally results in complete deaggregation to the monomer.[1] Methyllithium, ethyllithium and butyllithium remain tetramers in Et_2O, THF,[8] or dimethoxyethane (DME),[9] with some dimer forming at low temperatures;[10] *t*-BuLi becomes dimeric in Et_2O and monomeric in THF at low temperatures.[11] In the presence of TMEDA alone, however, *s*-BuLi remains a tetramer.[12]

Hexameric	Tetrameric	Dimeric	Monomeric
–	MeLi	*i*-PrLi	$PhCH_2Li$
	EtLi	*s*-BuLi	(ArLi)[a]
	BuLi	*t*-BuLi	(*t*-BuLi)[a]
	(*s*-BuLi)[b]	ArLi	

[a]*In THF <–100 °C or in TMEDA* [b]*In cyclohexane–TMEDA*

Table 1.2.3 Typical aggregation state in the presence of Et_2O or THF

Coordinating solvents also greatly increase the reactivity of the organolithiums, and an ether or amine solvent is indispensable in almost all organolithium reactions. The most important coordinating solvents commonly employed in organolithium reactions, in an approximate empirical ordering of decreasing activating power, are:

HMPA — $(Me_2N)_3P{=}O$ — *N,N,N',N',N'',N''*-hexamethylphosphoramide

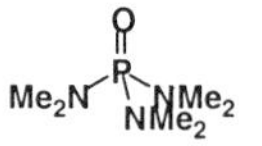

PMDTA — $Me_2N(CH_2)_2N(Me)(CH_2)_2NMe_2$ — *N,N,N',N'',N''*-pentamethyldiethylenetriamine

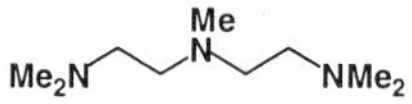

(–)-sparteine

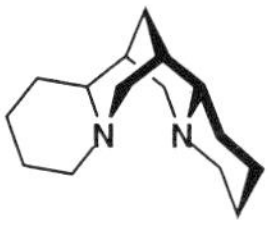

DMPU 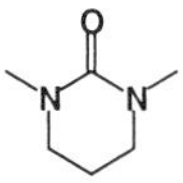 — *N,N'*-dimethylpropylideneurea

DME — MeO⌒OMe — 1,2-dimethoxyethane

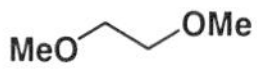

TMEDA — Me_2N⌒NMe_2 — *N,N,N',N'*-tetramethylethylenediamine

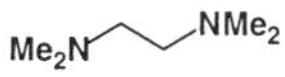

THF — tetrahydrofuran

t-BuOMe	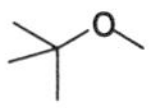	*t*-butyl methyl ether
Et_2O	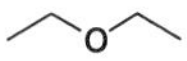	diethyl ether

Part, but not all, of the effect of these solvents is due to deaggregation of the organolithium. The importance of aggregation in determining reactivity is illustrated by methyllithium, which, as a monomer, should be more basic than phenyllithium by about 10 pK_a units. However, a 0.01 M solution of MeLi in THF is only three times more reactive than PhLi, and at 0.5 M concentration in THF PhLi is more reactive than MeLi.[13]

While deaggregation explains the difference in reactivity between an organolithium in hydrocarbon solution and one in ether or THF, there is little sound evidence that reactivity differences between organolithiums in different coordinating solvents are primarily due to their ability to deaggregate the organolithiums.[14] For example, while it is clear that TMEDA has a beneficial effect on certain lithiation reactions with *s*-BuLi (ortholithiation of *N,N*-diethylbenzamides for example, where it prevents self-condensation reactions) it is not at all clear that this is due to deaggregation of the *s*-BuLi (it could be, for example, due to stabilisation of the aryllithium product). Indeed, there is plenty of evidence that THF is in fact a better ligand for lithium than TMEDA, and that alkyllithiums are deaggregated just as much by THF alone as by TMEDA in the presence of THF.[14]

(–)-Sparteine is the most important chiral ligand for lithium, and will be discussed in greater detail in chapters 5 and 6. However, irrespective of its chirality, it is included in this list as it often leads to important enhancements of reactivity even when enantioselectivity is not required.[15] (–)-Sparteine can also make a very suitable replacement for toxic HMPA.[16] Another class of ligand yet to be exploited are terminal alkenes, which appear to be able to affect the outcome and selectivity of certain organolithium reactions,[17] perhaps by forming a Li–|| complex.[18]

All of the solvents in this list suffer to some extent from a tendency to react with organolithiums, precluding their use at temperatures above ambient, and in some cases limiting them to 0 °C or below. Et_2O is stable over a period of days in the presence of BuLi at room temperature,[19] but THF **1** is readily decomposed by organolithiums at this temperature[20,21] by reverse cycloaddition of the anion derived from **2**.[22] The products of this decomposition – ethylene and the lithium enolate of acetaldehyde **3** – can be trapped.[22,23] For example, acylation with phenylthiochloroformate of a solution of BuLi in THF gives good yields of the thiocarbonate **4**.[24] Carbolithiation of ethylene can be the source of mysterious doubly homologated products **5** arising from secondary and tertiary organolithiums,[25-27] and reactions of isopropyllithium in THF often leads to products arising from incorporation of 3-methylbutyllithium.[28]

Usual decomposition pathway for THF

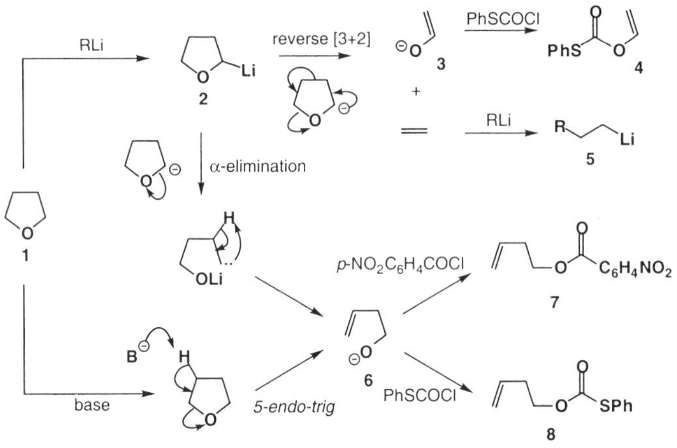

Alternative decomposition pathway for THF

THF occasionally decomposes by an alternative mechanism to give but-3-en-1-oxide **6**, first observed in solution by Fleming,[29] who isolated **7** after quenching the reaction of a basic organolithium in THF with *p*-nitrobenzoyl chloride. We later found but-3-en-1-oxide **6** to be the major decomposition product with RLi–HMPA mixtures.[30] Addition of phenylthiochloroformate to a *t*-BuLi–HMPA–THF mixture gives **8** in up to 61% yield. It is not yet clear whether this alternative elimination (which appears to occur only with very basic organolithiums) is a Baldwin-disfavoured *5-endo-trig* reaction[31] or the result of α-elimination of a carbene derived from **2**.

s-BuLi and *t*-BuLi decompose ethereal solvents much more rapidly than *n*-BuLi,[32] and it is wise to avoid THF for extended reactions of *t*-BuLi or for reactions conducted above 0 °C. Table 1.2.4[1,32] gives approximate half lives of a range of commonly used lithiating mixtures (RLi, solvent) at various temperatures. DME is particularly susceptible to attack by organolithiums,[33] and in general Et_2O is ten to fifty times as stable to an organolithium as THF at any given temperature. As a general rule, a temperature increase of 20 °C shortens the lifetime of an organolithium by a factor of 10. Extrapolating in this manner, BuLi turns out to be stable in THF for 24 h at about 0 °C (–15 °C if TMEDA is present). *t*-BuLi is stable in THF for 24 h only at –50 °C or below. Concentrated solutions of *n*-BuLi undergo slow b-elimination to give 1-butene and LiH.

t-Butyl methyl ether is a less widely used alternative to Et_2O offering better solubility for some compounds. The common cosolvents TMEDA, DMPU and HMPA are all susceptible to decomposition on extended exposure to organolithiums.[34-37] HMPA freezes at +7 °C and is a suspected carcinogen, but as a cosolvent in THF at low temperature it can have profound effects on organolithium reactivity. Benzene too is a carcinogenic, high freezing point solvent offering good solubility and low reactivity towards organolithiums. Toluene is liquid to a much lower temperature, but is slowly lithiated by alkyllithiums. Cumene is a more resistant replacement. For very low temperature work (less than –105 °C), the "Trapp mixture"[38] of 4:4:1 THF:Et_2O:pentane avoids the problems of viscosity and solidification of THF or Et_2O alone.

organolithium	solvent	temperature / °C	half-life
MeLi	Et_2O	25	3 months
PhLi	Et_2O	35	12 days
BuLi	Bu_2O	150	35 min
	DME	0	<5 min
	Et_2O	35	31 h
	Et_2O	25	153 h
	Et_2O + TMEDA	20	10 h
	THF	35	10 min
	THF + TMEDA	20	30 min
	THF + TMEDA	0	5 h
	THF + TMEDA	–20	50 h
t-BuLi	Et_2O	0	1 h
	Et_2O	–20	8 h
	THF	–20	45 min

Table 1.2.4 Stabilities of organolithiums in common solvents[1,32]

References

1. Wakefield, B. J. *The Chemistry of Organolithium Compounds*; Pergamon: Oxford, 1974.

2. Wakefield, B. J. *Organolithium Methods*; Academic Press: London, 1988.
3. Brown, T. L. *Acc. Chem. Res.* **1968,** *1*, 23.
4. Williard, P. G. In *Comprehensive Organic Synthesis;* Trost, B. M.; Fleming, I. Eds.; Pergamon: Oxford, 1990; Vol. 1; pp. 1.
5. Schleyer, P. v. R. *Pure Appl. Chem.* **1984,** *56*, 151.
6. Schlosser, M. In *Organometallics in Synthesis;* Schlosser, M. Ed.; Wiley: New York, 1994; pp. 1-166.
7. Sapse, A.-M.; Jain, D. C.; Raghavachari, K. In *Lithium Chemistry;* Sapse, A.-M.; Schleyer, P. v. R. Eds.; New York: Wiley, 1995.
8. West, P.; Waack, R. *J. Am. Chem. Soc.* **1967,** *69*, 4395.
9. Bergander, K.; He, R.; Chandrakumar, N.; Eppers, O.; Günther, H. *Tetrahedron* **1994,** *51*, 5861.
10. Nájera, C.; Yus, M.; Seebach, D. *Helv. Chim. Acta* **1984,** *67*, 289.
11. Hoffmann, R. W.; Kemper, B. *Tetrahedron Lett.* **1981,** *22*, 5263.
12. Hay, D. R.; Song, Z.; Smith, S. G.; Beak, P. *J. Am. Chem. Soc.* **1988,** *110*, 8145.
13. Screttas, C. G.; Smonou, I. C. *J. Organomet. Chem.* **1988,** *342*, 143.
14. Collum, D. B. *Acc. Chem. Res.* **1992,** *25*, 448.
15. Hodgson, D. M.; Norsikian, S. L. M. *Org. Lett.* **2001,** *3*, 461.
16. Dieter, K. *Synthesis* **1997**, 1114.
17. Ahmed, A.; Clayden, J. *unpublished results.*
18. Rölle, T.; Hoffmann, R. W. *J. Chem. Soc., Perkin Trans. 2* **1995**, 1953.
19. Gilman, H.; Haubein, A. H.; Harzfield, H. *J. Org. Chem.* **1954,** *19*, 1034.
20. Gilman, H.; Gaj, B. J. *J. Org. Chem.* **1957,** *22*, 1165.
21. Honeycutt, S. *J. Organomet. Chem.* **1971,** *29*, 1.
22. Bates, R. E.; Kroposki, L. M.; Potter, D. E. *J. Org. Chem.* **1972,** *37*, 560.
23. Blum, R. B.; Jung, M. E. *Tetrahedron Lett.* **1977**, 3791.
24. Duggan, A. J.; Roberts, F. E. *Tetrahedron Lett.* **1979**, 595.
25. Maercker, A.; Theysohn, W. *Liebigs Ann. Chem.* **1971,** *747*, 70.
26. Spialter, L.; Harris, C. W. *J. Org. Chem.* **1966,** *31*, 4263.
27. Krief, A.; Kenda, B.; Barbeaux, P.; Guittet, E. *Tetrahedron* **1994,** *50*, 7177.
28. Bartlett, P. D.; Friedman, S.; Stiles, M. *J. Am. Chem. Soc.* **1953,** *75*, 1771.
29. Fleming, I.; Mack, S. R.; Clark, B. P. *J. Chem. Soc., Chem. Commun.* **1998**, 713.
30. Clayden, J.; Yasin, S. A. *New. J. Chem.* in press.
31. Baldwin, J. E. *J. Chem. Soc., Chem. Commun.* **1976**, 734.
32. Stanetty, P.; Koller, H.; Mihovilovic, M. *J. Org. Chem.* **1992,** *57*, 6833.
33. Ellison, R. A.; Griffin, R.; Kotsonis, F. N. *J. Organomet. Chem.* **1972,** *36*, 209.
34. Peterson, D. J. *J. Organomet. Chem.* **1967,** *9*, 373.
35. Kaiser, E. M.; Petty, J. D.; Solter, L. E. *J. Organomet. Chem.* **1971,** *61*, C1.
36. Peterson, D. J.; Hays, H. R. *J. Org. Chem.* **1965,** *30*, 1939.
37. Peterson, D. J. *J. Organomet. Chem.* **1970,** *21*, P63.
38. Köbrich, G.; Trapp, H. *Chem. Ber.* **1966,** *99*, 670.

CHAPTER 2

Regioselective Synthesis of Organolithiums by Deprotonation

2.1 General points

Organolithiums may be formed most simply by one of three distinct methods, each discussed in one of the next three chapters. The formation of new organolithiums by addition to π systems is discussed in Chapter 7.

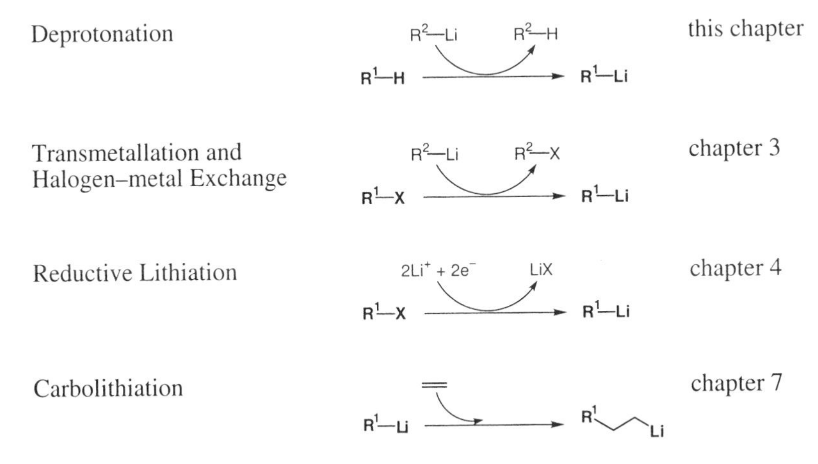

As a general guide, lithiation by *deprotonation* using commercially available alkyllithiums is feasible if the new organolithium has at least three stars in the following cumulative scoring system:

☆☆☆	Li bonded to digonal (sp-hybridised) carbon atom
☆☆☆	C–Li bond adjacent to carbonyl sulfoxide, sulfone, phosphonate or similar strongly electron-withdrawing substituent
☆☆	Li bonded to trigonal (sp^2-hybridised) carbon atom
☆☆	C–Li bond in a small ring (three or four members)
☆☆	C–Li bond α to S, P or other second or third row element
☆☆	C–Li bond in allylic or benzylic location
☆	C–Li bond adjacent to O or N
☆	Helpfully positioned heteroatom to stabilise C–Li by coordination
☆	More remote acidifying effect (electron-withdrawing group *ortho* on aromatic ring, for example)

Lithiation by deprotonation of a C–H bond takes place at a reasonable rate only if the organolithium product displays two features: intramolecular coordination of the electron-deficient lithium atom to a heteroatom (hydrocarbons are extremely slow to lithiate under most conditions, even at aromatic or vinylic sites) and stabilisation of the electron-rich C–Li bond by a nearby empty orbital or electron-withdrawing group. These two factors controlling lithiation are of varying importance according to the reaction in question, and the balance between them is a recurring theme of this chapter.

2.2 Lithiation α to heteroatoms

Lithiation adjacent to a heteroatom – α-lithiation – is favoured by strongly acidifying groups such as arylsulfonyl, arylsulfinyl or dialkylphosphonyl groups, but can also take place adjacent to oxygen- or nitrogen-based functional groups which apparently even decrease the acidity of the adjacent protons,[1] provided these groups coordinate strongly to the organolithium. Acidity is increased if the proton to be removed is benzylic, allylic, vinylic, or attached to an aromatic or small-ring saturated heterocycle, and many successful α-lithiations make use of allylic or benzylic stabilisation as well as the effect of the α-heteroatom.[1]

Rather more details about the mechanism of α-lithiation are known than about many other lithiations, and the sequence of events in the lithiation of **1** have been used as a model for directed lithiations in general. Stopped flow infra-red studies show that a complex between the amide and the organolithium forms much faster than the amide is lithiated.[2,3] This complex, presumably approximating to **2**, is believed to represent the first step in the lithiation sequence. The kinetics of the reaction are consistent with a deprotonation within the complex to give the organolithium **3**.[2,3]

ν(C=O) = 1650 cm^{-1} ν(C=O) = 1625 cm^{-1} ν(C=O) = 1588 cm^{-1} very slow rotation about this bond

(BuLi)$_n$ (BuLi) Li E$^+$ E

1 2 3 4

The rate of rotation about the C–N bond in 2,4,6-triisopropylbenzamides **4** is so slow that the groups *cis* and *trans* to oxygen about the C–N bond do not exchange, and compounds **4** may exhibit geometrical isomerism. Lithiation can therefore be shown always to occur *syn* to oxygen, consistent with delivery of the base by coordination to the amide.[4,5] Calculations show that in the absence of a lithium atom a *trans* anion would be more stable, as it would benefit from a bonding interaction with the C–O σ* orbital.[6]

Empirical demonstration of the importance of the lithium ion in deprotonation *syn* to C=O is provided by **5**, which is deprotonated to give **6** only with lithium bases (LiTMP). Sodium or potassium equivalents do not deprotonate **5**.[7]

Ph
Ph
LiTMP
N
O
5
Ph
Ph
N
O
Li
6

Meyers showed that the deprotonation of **7** displays no kinetic isotope effect – an observation consistent with rate-limiting formation of a formamidine-BuLi complex followed by deprotonation.[8] A "pre-lithiation complex" analogous to **2** has furthermore been isolated in the formamidine series.[9]

BuLi
N
H H/D NR*
7
N
H Li NR*
$k_H/k_D = 1$

More evidence that deprotonation involves more than one mechanistic step is provided by kinetic isotope effect studies on the stereochemically simpler amide **8**.[10,11] Two experiments – comparing firstly the deuterium content of **10** and secondly the relative amounts of lithiation of h_2-**8** and d_2-**8** – determined the intramolecular and intermolecular kinetic isotope effects for deprotonation by *s*-BuLi at –78 °C. The intramolecular KIE is the isotope effect displayed by the deprotonation step itself; the intermolecular isotope effect however is affected by other mechanistic steps preceding the deprotonation. For example, a rate-determining complexation with *s*-BuLi would be expected to show a very small isotope effect, and would lead to a low intermolecular KIE but leave the intramolecular KIE unaffected. In the experiment, the intramolecular kinetic isotope effect had a value of >20, while the intermolecular kinetic isotope effect had a value of 5-6. A difference such as this can only be accounted for by a mechanistic pathway consisting of more than one step. The value of the intermolecular isotope effect gives further information: since it is neither 1 (which would be consistent with rate-determining complexation) nor the same as the intramolecular effect (which would be consistent with rate determining deprotonation) both complexation and deprotonation appear to be occurring with competitive rates.

H D O
N NHMe
h,d-**8**
2 x *s*-BuLi, TMEDA
–78 °C
Li D OLi
N NMe
d-**9**
CO_2
HO_2C D O
N NHMe
d-**10**
Li H OLi
N NMe
h-**9**
CO_2
HO_2C H O
N NHMe
h-**10**
$k_H/k_D > 20$

2.2.1 Lithiation α to oxygen

In the general sense, deprotonation α to oxygen is unfavourable: the antibonding interaction of the oxygen's lone pairs with the C–Li bond overcomes the electron-withdrawing effect of the oxygen atom itself. The repulsion is lessened if the lone pairs are lowered in energy by delocalisation, especially into a carbonyl group, and the most versatile lithiations α to oxygen are the deprotonations of hindered carbamates under the influence of (–)-sparteine, discussed in section 5.4.[12]

Formation of vinylic (trigonal C–Li), allylic or benzylic organolithiums α to O is more favourable. The useful acyl anion equivalent methoxyvinyllithium **11**, for example, is formed on treatment of methyl vinyl ether with *t*-BuLi at –65 °C,[13] and furan is lithiated by BuLi in Et_2O at 0 °C to give **12**.[14] More complex carbohydrate-derived vinyl ethers such as **13** also lithiate readily.[15]

It is possible, but not always easy, to lithiate allyl ethers – *n*-BuLi deprotonates allyl phenyl ether **14**, though the organolithium product **15** reacts with poor regioselectivity.[16] Lithiated allyl triethylsilyl ethers react more regioselectively: **16** gives high (but very electrophile-dependent) yields of the γ-quenched isomer **18**.[16] This is a valuable stereoselective synthesis of a silyl enol ether: the *endo* configuration of the allyl lithium **17** (which is general, and not confined to O-substituted allyl systems[17-19]) ensures the stereoselectivity.

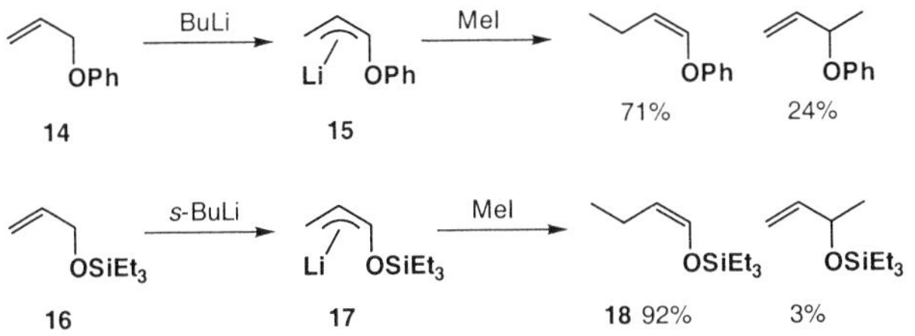

Allyl carbamates **19** are even more versatile, and the lithio derivatives **20** of allyl carbamates are the most important class of homoenolate equivalents.[17] Lithiated allyl carbamates react reliably at the γ-position with aldehydes and ketones but less regioselectively with alkylating and silylating agents. *O*-Benzyl carbamates **21** are readily deprotonated and can be quenched with electrophiles.[17,20]

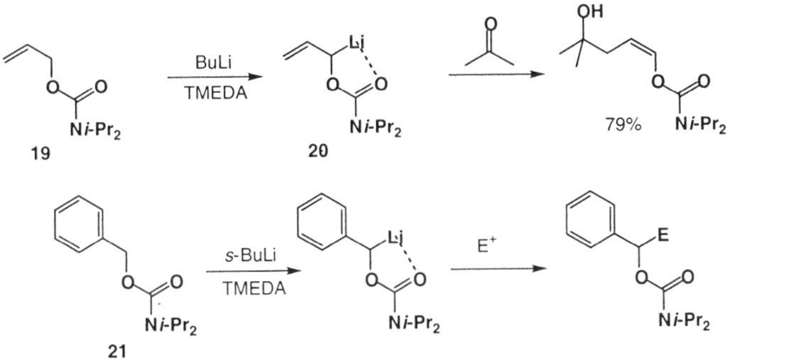

Allyl acetals,[21] and carbamates,[17] can be lithiated and quenched with metallic electrophiles at the α-position to provide allylmetals, for example, the allyl boronate **22**.[21] Allyl and benzyl esters have also been lithiated α to O.[22]

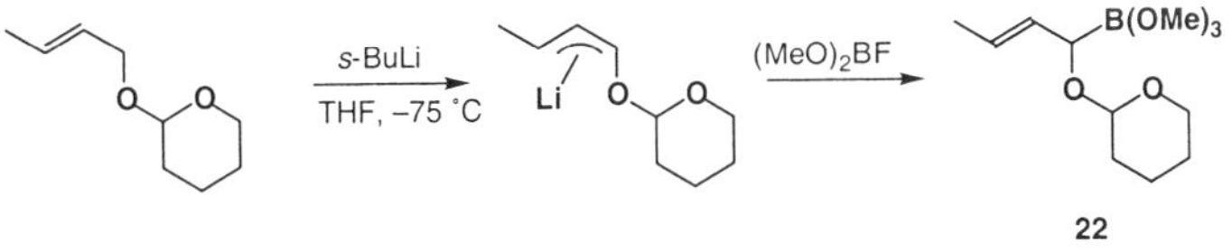

Lithiation and alkylation of the doubly allylic ether **23** with alkyl halides provides a useful synthesis of 1,3-butadienes. Alkyl triflates and tosylates, and carbonyl electrophiles, react with α-selectivity and give skipped dienes **24**.[23]

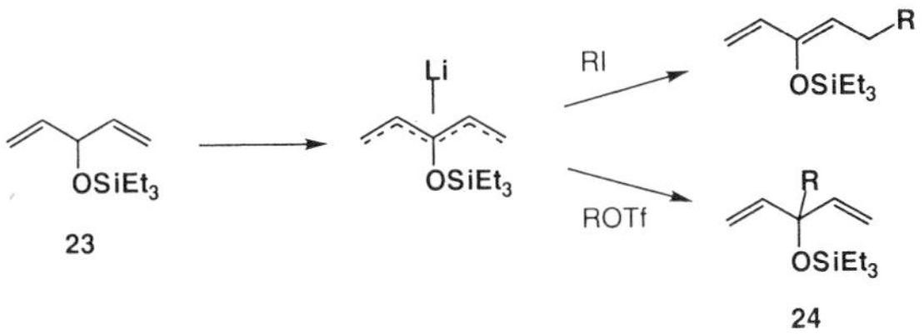

Hoppe has shown[12] that lithiation α to O of a simple alkoxy group is best achieved with carabamates and *s*-BuLi-(–)-sparteine (see section 5.4) but provided attack at C=O is prevented by steric hindrance, even alkyl esters can be lithiated α to O. **25** and its *O*-methyl analogue can be lithiated and functionalised.[24]

s-BuLi, TMEDA, –95 °C, THF → ; D_2O →

25

The relatively easy lithiation of cyclic ethers is at the origin of the susceptibility of THF to attack by organolithiums (section 1.2). Until recently lithiated cyclic ethers had never been trapped before they decomposed. However, Hodgson has shown that *s*-BuLi in the presence of (–)-sparteine in hexane at –90 °C is able to deprotonate epoxides to yield organolithiums **26** which may be deuterated to give **27** in 70% yield. An internal Me_3SiCl quench (in other words, Me_3SiCl present in the reaction mixture as the deprotonation is carried out) yielded silyl epoxides **28**,[25] but quenches with external electrophiles (that is, electrophiles added after the lithiation step is complete) other than D_2O failed.[26]

s-BuLi, (–)-sparteine, –90 °C, hexane → **26** ; D_2O → **27** (70%; 75% D) ; **28**

2.2.2 Lithiation α to nitrogen[27-29]

While nitrogen lone pairs can coordinate strongly to an incoming organolithium reagent, they destabilise an adjacent C–Li bond by an antibonding interaction even worse than that of lone pairs at oxygen. Direct deprotonation α to nitrogen is therefore usually impossible[30] (except with superbasic reagents[31]) unless the nitrogen's lone pair is involved in conjugation with a carbonyl group or delocalised around an aromatic ring. Not only does conjugation with a carbonyl group make the nitrogen-containing group more electron-withdrawing, but it also provides a new, electron-rich oxygen atom which can itself stabilise the organolithium by coordination. Organolithiums stabilised by the $N(\delta^+)$–$C{=}O(\delta^-)$ dipole of an amide (or the dipoles of similar functional groups) form an important group of compounds which have been termed "dipole-stabilised carbanions".[32]

destabilising effect   stabilising effect

2.2.2.1 Amides[29]

Lithiation α to amide nitrogen atoms is commonly possible, but the product organolithiums are stable only with hindered amides such as **1** described above and **29** and **33** shown below. BuLi would rather attack the α-protons of **29** and **22** than add to their C=O groups, and the α-lithiated species **30** and **34** react with non-enolisable aldehydes, ketones[33,34] and some alkylating agents.[4] Removal of the hindered acyl group from **31** requires MeLi or Birch reduction; **35** can be hydrolysed via rearrangement to the ester **36** in acid. At temperatures above 0 °C, **30** undergoes "anion translocation"[35] to an ortholithiated compound **32** which undergoes further rearrangements. Less hindered amides, including even the amides of pivalic acid, undergo extensive side-reactions under these conditions.[28]

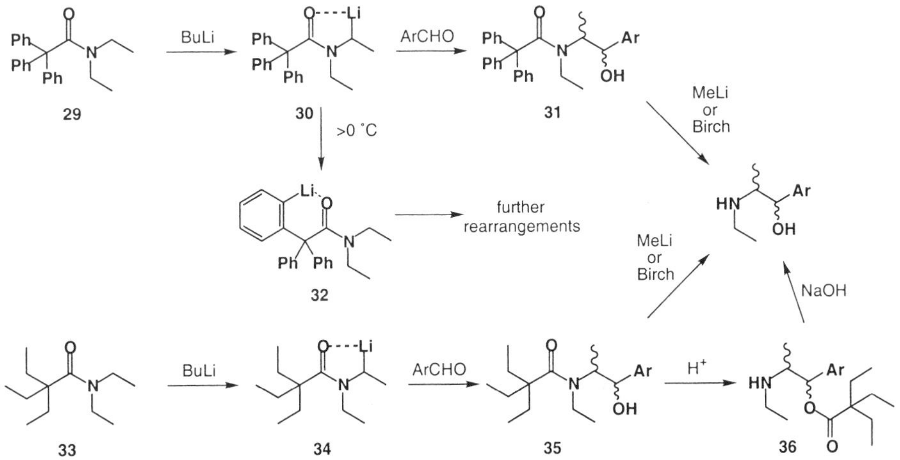

Though the destabilisation is much less severe than with a nitrogen lone pair, an amide-stabilised organolithium prefers to minimise the interaction (shown below) between the C–Li bond and the π-system of the amide. Hindered amides of piperidines such as **37** are therefore deprotonated to give an equatorial organolithium such as **38**.[5,4] The equatorially substituted piperidine amide products **39** and **40** are less stable than their axially substituted products owing to the interaction between the new substituent and the pseudoequatorial amide group. Oxidation and equilibration leads to the axial substituted ketone **41**.[4]

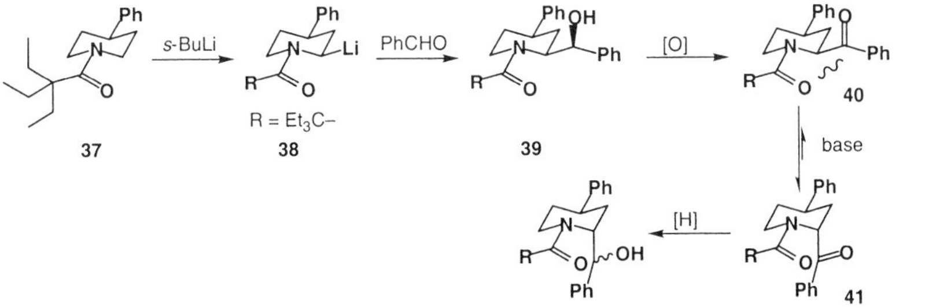

Equatorial preference in ***38***:

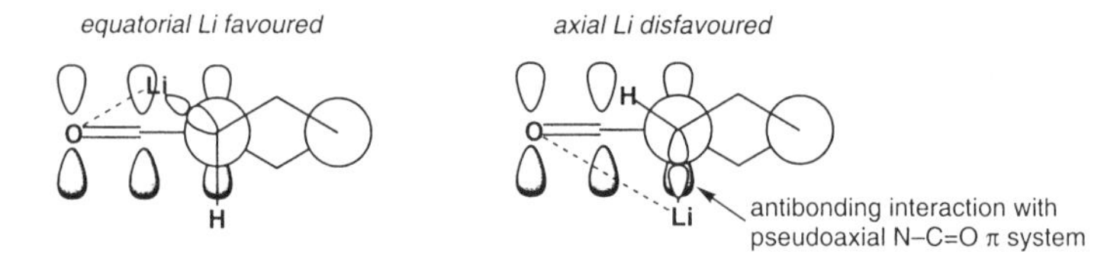

Other saturated nitrogen heterocycles can be lithiated in the same way to give, for example, **42**, **43** and **44**.[33,34,36]

42 **43** **44**

More versatile are the lithiations of carbamates (*N*-Boc amines) – as with lithiation α to oxygen, the lack of electrophilicity in a carbamate C=O group and its particularly electron-rich oxygen lone pairs conspire to facilitate lithiation. In their Boc protected form, even simple amines such as **45** may be lithiated.[37]

45 — *s*-BuLi, TMEDA, −78 °C → — E^+ →

These amides are α-lithiated because no other deprotonation can compete. With less hindered aromatic *N,N*-dimethylamides, α-lithiation still takes precedence over ortholithiation but the organolithium is immediately acylated by another molecule of amide. For example, LiTMP α-lithiates **46** even though ortholithiation could compete, but the product rapidly forms **47**.[7] With less readily lithiated groups on nitrogen (such as ethyl or isopropyl) ortholithiation (see section 2.3) takes over as the main reaction pathway.

46 — LiTMP → → **47** 53%

With *N*-benzyl or *N*-allyl amides, the α-lithiated products are more stable. *N,N*-dibenzylbenzamide **48**, for example, is α-lithiated with LDA to give **49** and hence **50**,[38] while two equivalents of BuLi or LDA are sufficient to α-lithiate the secondary amides **51** and **54** to give **52** or **55** which react with electrophiles to produce **53** and **56**.[39] The lithiated *N*-Boc allyl amides **57** react with selectivity which is highly electrophile dependent, but transmetallation of **58** to the allylzinc reagent leads to good α-selectivity.[40]

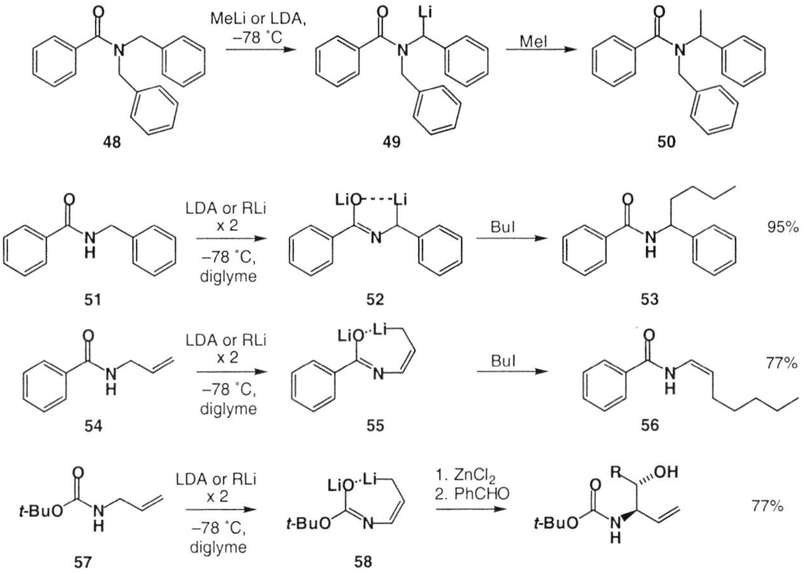

It is interesting to contrast these double lithiations of secondary *N*-allyl amides with the double lithiation of a secondary *N*-silyl amine **59**, which leads to vinylic γ-lithiation, rather than allylic α-lithiation.[41,42] The product **60** cyclises onto carbonyl electrophiles to yield pyrroles such as **61**.

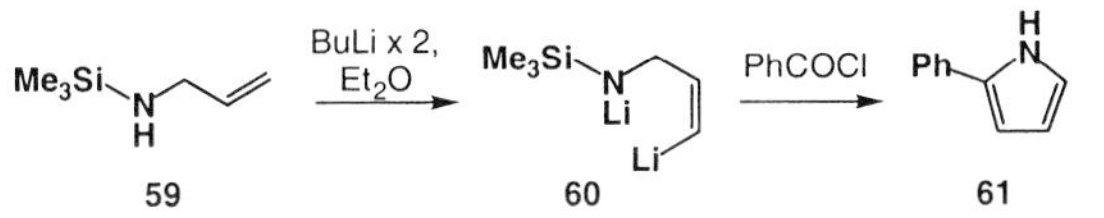

The choice between *ortho* and benzylic α-lithiation is less clear cut when one of the substituents on nitrogen is bulky, lying *cis* to the amide carbonyl and preventing Li–O coordination, and mixtures of regioisomers may be obtained. For example, lithiating **62** gives organolithiums **63** and **64** whose quench leads to different regioselectivities with different electrophiles suggesting they are in rapid equilibrium.[35]

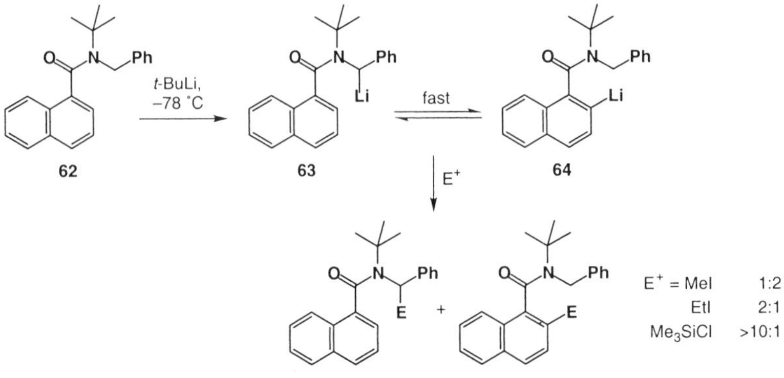

Each of the two isomeric organolithiums **63** and **64** could be formed selectively by using deuterium to block lithiation at the other site via the kinetic isotope effect (see below).[43] The less stabilised α-lithio species **63** have a tendency to undergo dearomatising cyclisations,[44,45] as described in section 7.2.4.6.

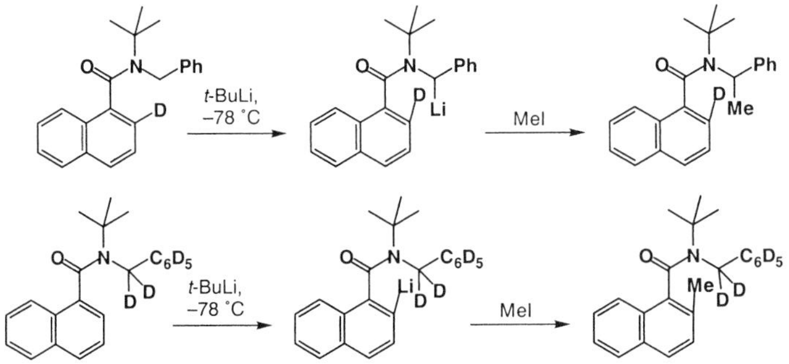

α-Lithiation at benzylic sites even outpaces enolisation of an amide, as in the reaction of *N*-benzyl lactam **65** with BuLi.[46] The α-lithiation of *N*-benzyl-β-lactams **66** and **68** is the first step in the formation of azepinone **67** and pyrrolidinone **69** by [1,2]- or [2,3]-aza-Wittig rearrangement (section 8.4.4).[47]

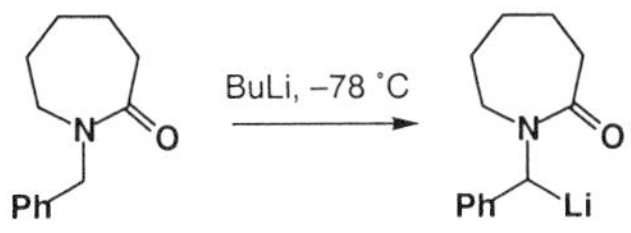

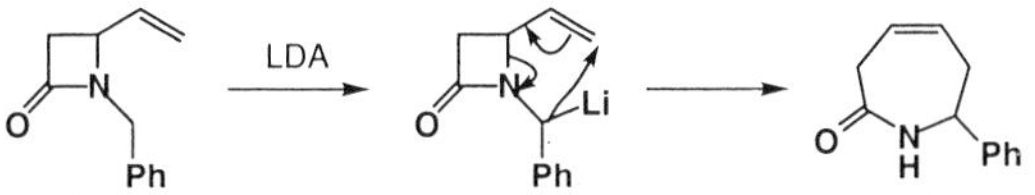

The isoquinolines **70** are an important subclass of the tertiary *N*-benzylamides, and their lithiation and benzylation is an important way of synthesising a range of benzylisoquinoline alkaloids.[48,49]

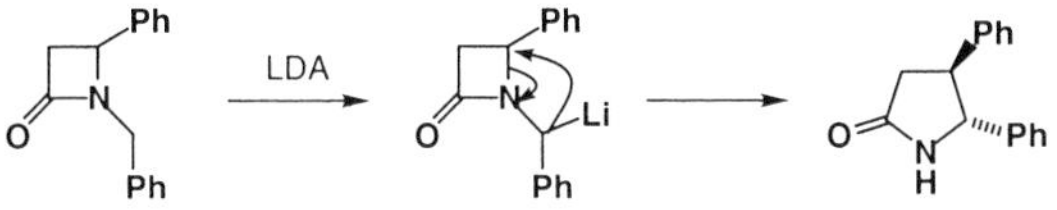

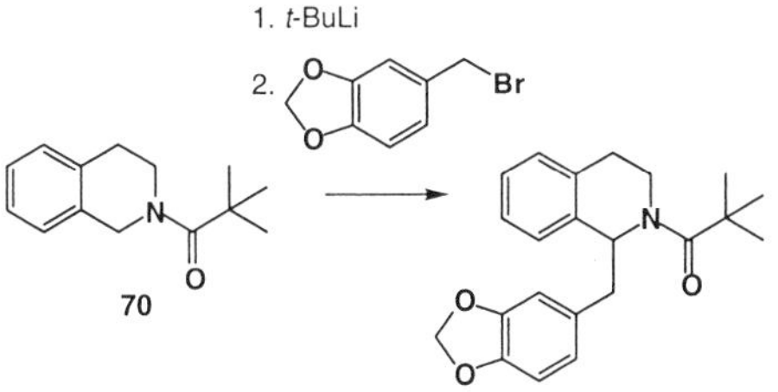

α-Lithiation of a heterocyclic ring adjacent to heterocyclic nitrogen is discussed in section 2.3.3 as a form of ortholithiation, but α-lithiation of *N*-benzyl heterocycles is also possible outside the ring. *N*-Benzylpyridones (**71**) and pyrimidines, for example, can be lithiated:[50,51,28]

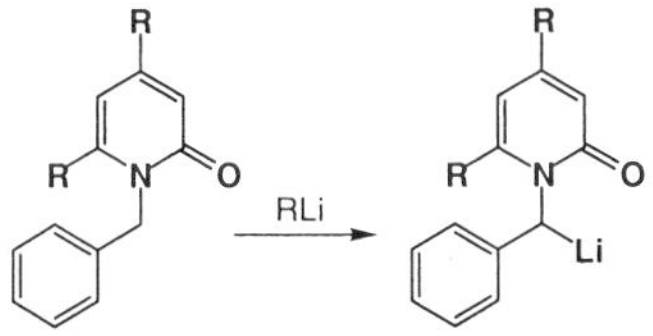

N-Allyl tertiary amides **72** can be lithiated with *t*-BuLi and, like the secondary amides **54** and **57** they react γ to nitrogen to give *cis* acylenamine products **74**, with the intermediate organolithium adopting a structure approximating to **72**.[1,52] The stereochemistry of the reactions of lithiated allyl amides and carbamates in the presence of (–)-sparteine[53] is discussed further in chapter 6.

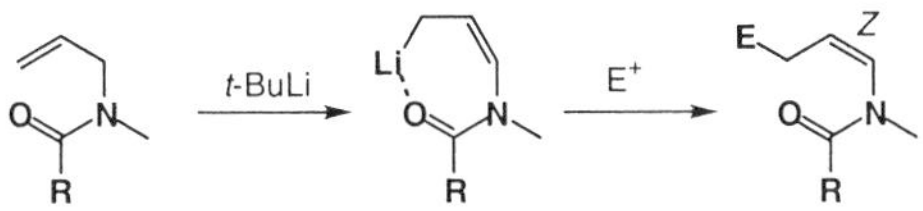

Much of the α-deprotonation chemistry of the amides is mirrored by hindered thioamides, imides, ureas, carbamates and phosphonamides,[28] and the important asymmetric versions of these reactions are discussed in chapters 5 and 6. Difficulties removing the heavily substituted groups required for "protection" of the carbonyl group in these compounds have been overcome in such cases as the urea **75**, which is resistant to strong base, but which undergoes acid-catalysed hydrolysis and retro-Michael reaction to reveal the simpler derivative **76**.[54]

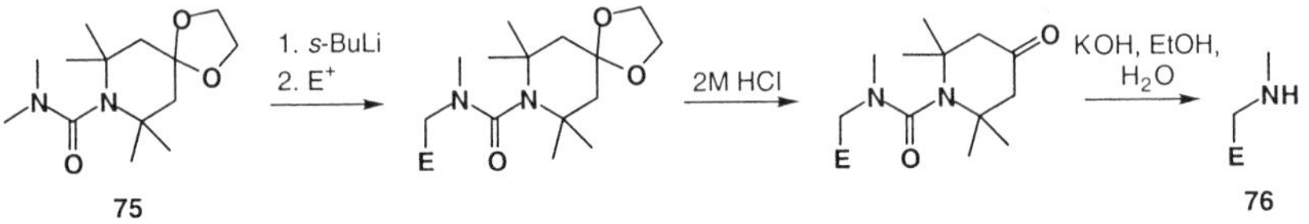

1. *s*-BuLi 2. E$^+$ — 2M HCl — KOH, EtOH, H_2O

75 **76**

2.2.2.2 Benzotriazoles and other 5-membered nitrogen heterocycles

Metallation of the *N*-alkyl groups of 5-membered heterocycles is frequently possible – the pyrazole **77** for example gives **78** in good yield.

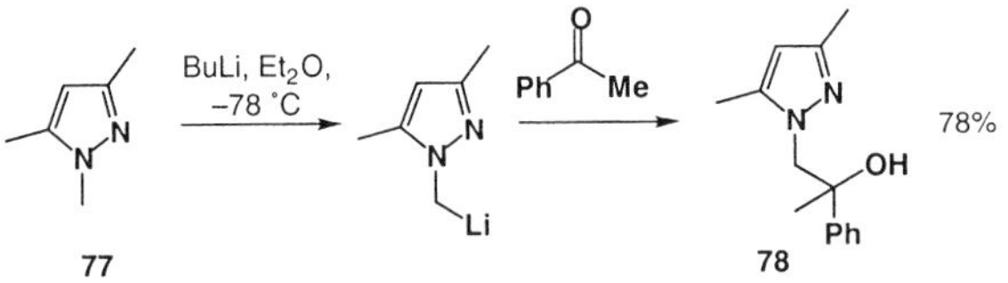

BuLi, Et_2O, −78 °C — PhC(O)Me — 78%

77 **78**

Benzotriazole is the most versatile of these heterocycles,[55] because it is easily introduced and easily removed. It promotes α-lithiation even at hindered tertiary centres, and the lithiation of α-benzotriazole ethers **79** gives useful acyl anion equivalents **80**.[56]

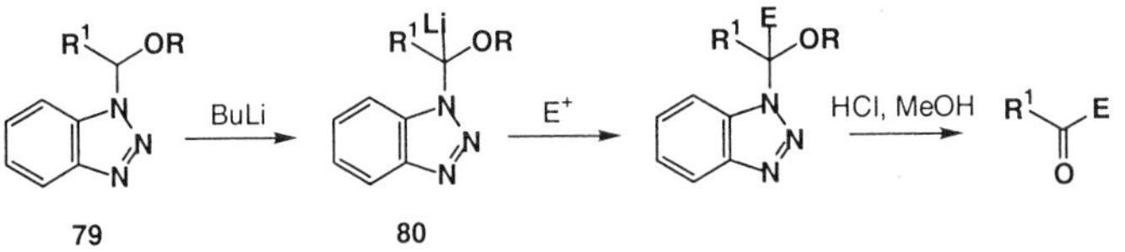

BuLi — E$^+$ — HCl, MeOH

79 **80**

2.2.2.3 Formamidines

Formamidines such as **81** may be lithiated with *s*-BuLi or *t*-BuLi and give stabilised organolithiums **82** which react with a wide range of electrophiles. Cleavage is much easier than cleavages in the amide series; acidic methanolysis gives a secondary amine **83** while hydride reduction gives a tertiary amine **84**.[57]

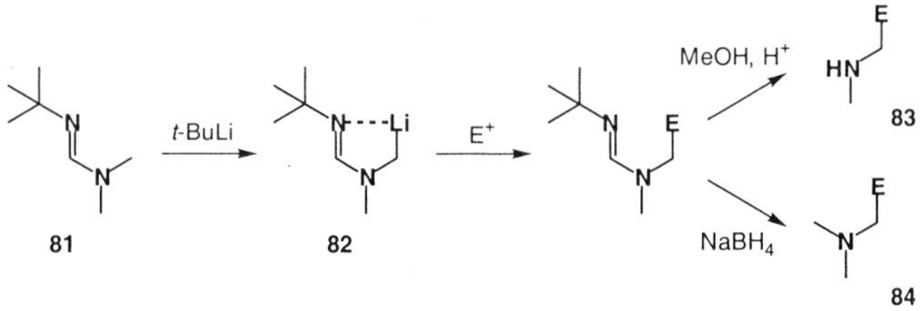

Formamidines also activate secondary centres towards lithiation, and have been used extensively in the synthesis of the benzylisoquinolines, where the lithiation takes place at a benzylic position. Formamidines bearing chiral substituents (for example the serine-derived **85**) allow the introduction of asymmetry at such centres.[58]

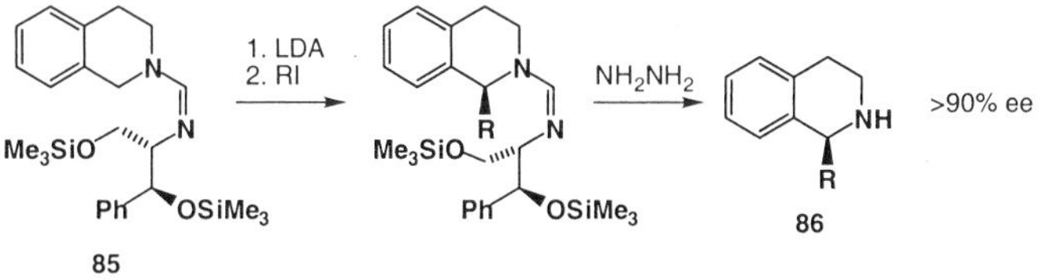

Formamidines can also be used to promote the lithiation of vinylic centres α to nitrogen. **87** gives the organolithium **88** on treatment with *t*-BuLi. This is an acyl anion equivalent which adds to aldehydes to give, after hydrolysis, hydroxyketones **89**.[59]

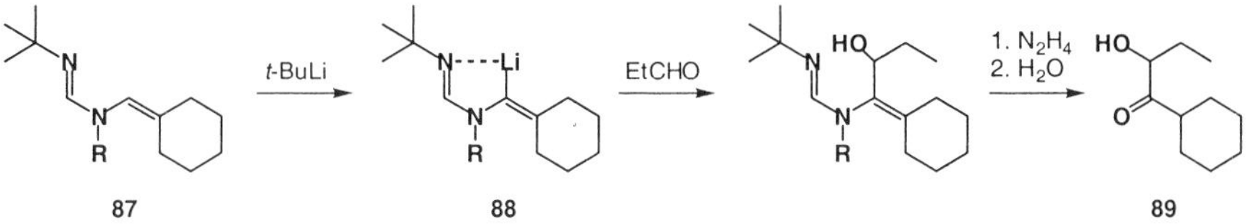

2.2.2.4 Nitrosoamines[60,29]

N-Nitrosoamines are highly versatile substrates for α-lithiation, because while the N=O group has activating and coordinating properties similar to the carbonyl group, it lacks the carbonyl group's electrophilicity. The nitroso group will direct lithiation even to a tertiary position (**90** → **91**), and can be removed by reduction with Raney nickel.[60] The drawback is the toxicity of *N*-nitrosoamines.

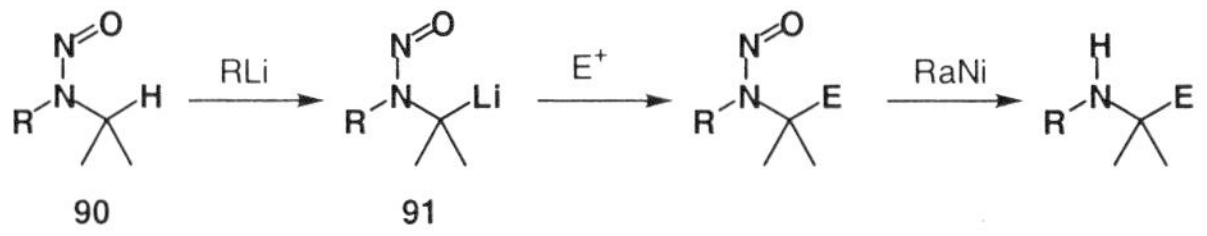

N-Nitroso pyrrolidine **92** can be lithiated just with LDA. Double benzylation gives principally the *trans* isomer of **93** after reduction.[61]

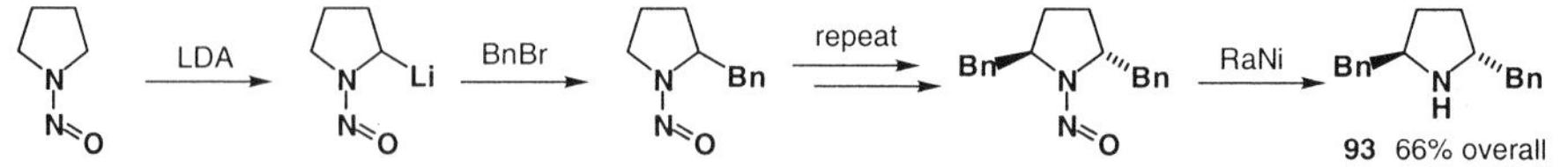

LDA BnBr repeat RaNi

92

93 66% overall
2:1-4:1 *trans:cis*

An important difference between the lithiated *N*-nitrosoamines and lithiated amides highlights differences in the way in which the X=O group stabilises the organolithium (X = C or N). While *N*-acylpiperidines react to give equatorially substituted products (the organolithium orients its C–Li bond orthogonal to the N–C=O π system), the *N*-nitrosopiperidines such as **94** give axially substituted products[60] because the organolithium is stabilised by interaction with the LUMO of the adjacent π-system more than it is destabilised by the HOMO.[62-64] The fact that delocalisation is an important factor in the stabilisation of lithiated *N*-nitrosamines is supported by the unreactivity of **97** towards lithiation.[65]

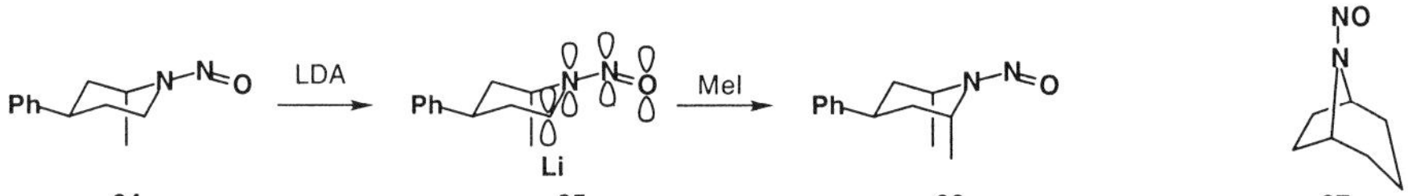

LDA MeI

94 95 96 97

2.2.2.5 *Imines*

Imines (**98**) may be lithiated[66] if (a) they have no other acidic protons α to C=N and (b) they are *N*-methyl imines or they have other activation towards α-lithiation. The *N*-allyl amidines **99**, for example, give interesting chiral organolithiums **100** with BuLi.[67]

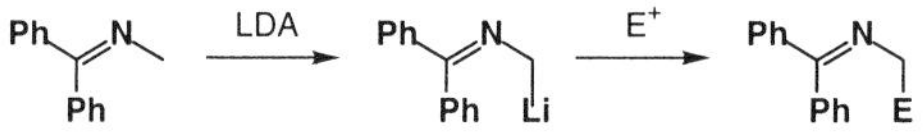

LDA E^+

98

BuLi

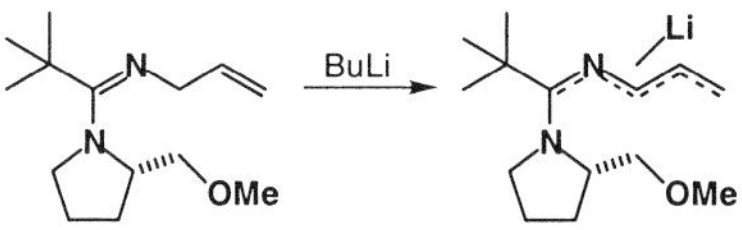

99 100

Remarkably, some hydrazones may be doubly lithiated with the second lithiation occurring α to N at the trigonal carbon atom (**101**).[68] Doubly lithiated hydroazones of a different kind are intermediates in the Shapiro reaction (section 8.1).

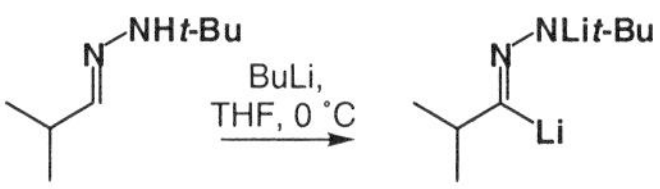

BuLi, THF, 0 °C

101

2.2.2.6 Isocyanides

Methyl and primary isocyanides are easily lithiated and react with carbonyl compounds to form heterocycles such as **102**. Secondary isocyanides (with the exception of small rings) cannot be lithiated.[69]

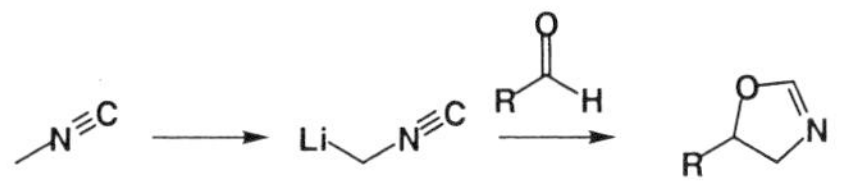

2.2.2.7 Lithiation of trigonal C–H α to nitrogen

Lithiation of nitrogen heterocycles α to N is discussed in section 2.3.3. Enamines may be lithiated with *t*-BuLi.[70] Already lithiated nitroalkanes **103** can be further lithiated α to N to yield versatile species **104**.[71]

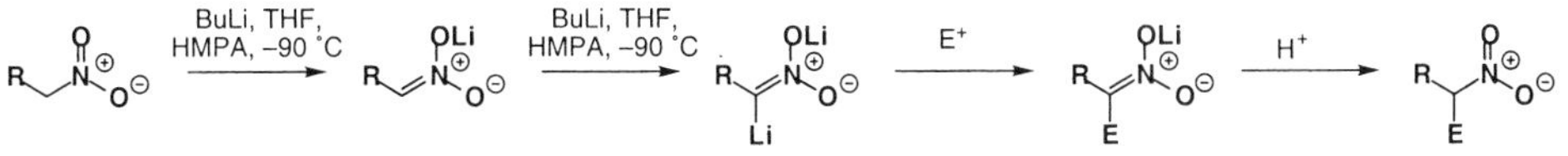

2.2.2.8 N-oxides and amine-borane complexes

Pyridine *N*-oxide can be α-lithiated in a reaction which is essentially an ortholithiation.[29] The reaction works in the saturated series too – quinuclidine-*N*-oxide **105** may be lithiated and functionalised as shown below.[72]

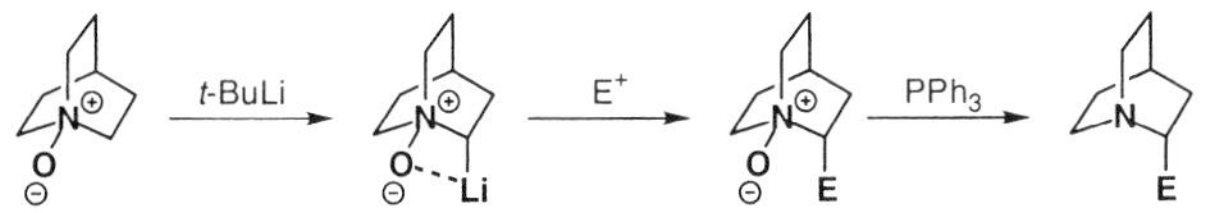

Even without the directing effect of coordination to O, incorporation of the amine lone pairs into an amine-borane complex **106** allows the easy deprotonation at the benzylic site to give **107**.[73] Refluxing ethanol deprotects the nitrogen.

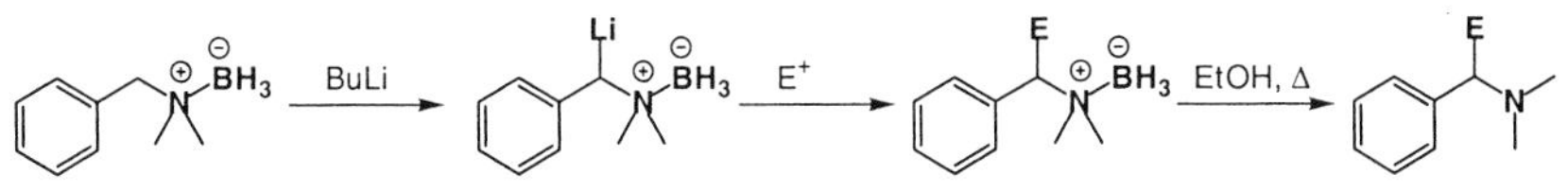

2.2.3 Lithiation α to sulfur

The lithiation of sulfoxides and sulfones gives stabilised species which are arguably not organolithiums but anions with O–Li bonds.[74-76] Lithiation α to phosphorus likewise gives anions which are outside the scope of this book.

Organolithiums α to sulfide S can be produced with difficulty – thioanisole **108** requires BuLi and DABCO in THF at 0 °C to produce the organolithium **109**[77] (without DABCO only a low yield is formed[78]).

BuLi, DABCO, THF, 0 °C; 108 → 109, 97%; 110: interaction with S–C σ*

The effect of the sulfur atom is principally one of acidification – a phenylsulfanyl (PhS) substituent increases the acidity of α-protons by about 5-10 pK_a units.[79] This effect is now generally considered not to be a result of interaction with sulfur's d-orbitals, but rather of sulfur's polarisability[80] and of hyperconjugation with the antiperiplanar S–C σ* orbital (**110**).[81]

α-Thio organolithiums form much more readily when there is additional stabilisation in the form of conjugation or coordination, or, in the case of the dithianes, a second S atom. 1,3-Dithiane itself has a pK_a of 31.1,[82] and is readily lithiated with *n*-BuLi.

Benzylsulfides may be lithiated with PhLi,[83] and the acetal group of **111** both assists lithiation and imparts stereoselectivity on the reaction (the configurational stability of organolithiums such as **112** is discussed in section 5.1).[84]

111 —BuLi→ 112 —E⁺→ (SPh, E)

Allyl sulfides are readily lithiated, and give species **113** which are configurationally stable about the allyl system at <0 °C.[1]

PhS–CH₂–CH=C(R¹)(R²) —RLi→ 113 —E⁺→ PhS–CH(E)–CH=C(R¹)(R²); 113 does not interconvert with the isomer bearing R¹ and R² exchanged

Lithiated allyl sulfides (in common with allyl sulfoxides and selenides, but in contrast with allyl ethers and allyl amine derivatives) tend to react with alkyl halides at the α position, adjacent to S.[1] The best α-selectivities are obtained with lithium-coordinating S-substituents such as pyridyl (**114**),[85] imidazolyl (**115**), and dimethylaminocarbonyl (**116**).[1]

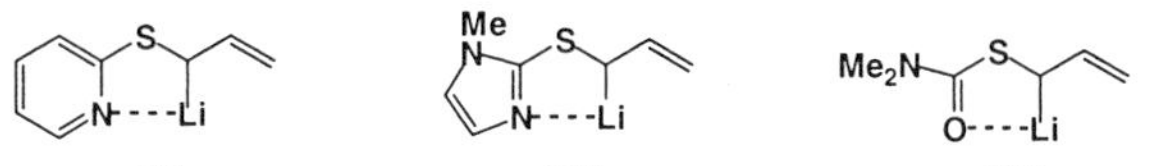

Hindered aromatic thioesters may be lithiated in a similar way to hindered aromatic esters **25**.[29] Thioimidate **117** is lithiated to give an organolithium **118** which reacts with aldehydes and ketones to form thiiranes **119**.[86,87]

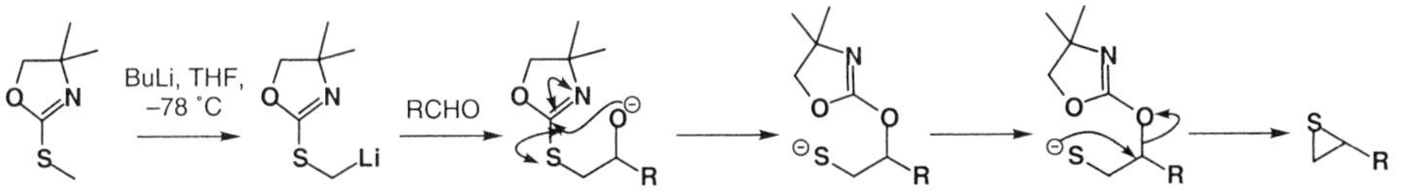

In the six-membered series, α-lithiation of **120** is achieved under kinetic lithiation conditions; with LDA the more stable azaenolate **121** forms.[88]

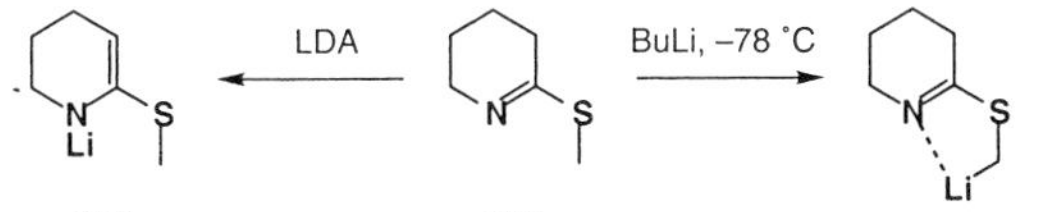

Thiophenes are readily lithiated next to sulfur: **122** gives **123** with BuLi in THF at 0 °C.[89] Interestingly, lithiation α to oxygen (**124**) occurs in a non-coordinating solvent – presumably coordination to oxygen becomes more important than acidification by sulfur under these conditions. A similar contrast, probably for similar reasons, is evident in the lithiation of **125** with LDA and with BuLi.[90]

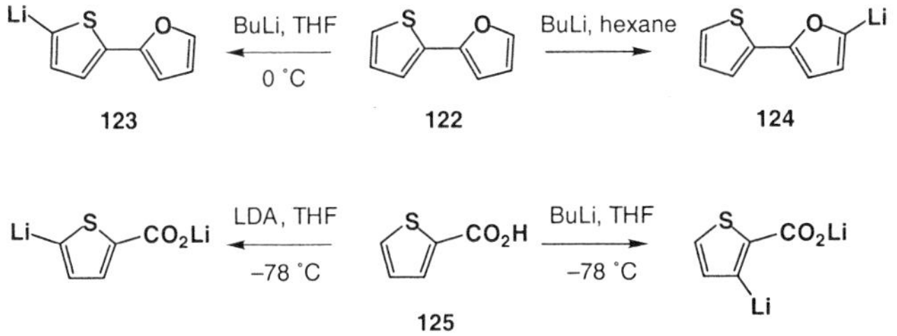

The lithiation of thiophene and its addition to ethylene oxide is a key step in the synthesis of the platelet aggregation inhibitor ticlopidine.[91]

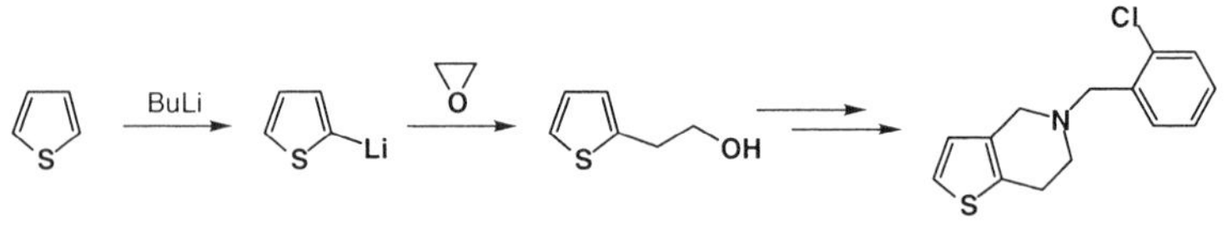

ticlopidine

Lithiation next to selenium is by contrast impossible: organolithiums attack the Se atom directly.[92] α-Lithio selenides are best formed by selenium-lithium exchange (section 3.3.1)

2.2.4 Lithiation α to silicon

Allyltrimethylsilane **126** and benzyltrimethylsilane **127** are lithiated with BuLi in the presence of TMEDA or HMPA. Methyllithium, however, attacks the silicon atom.[93-95]

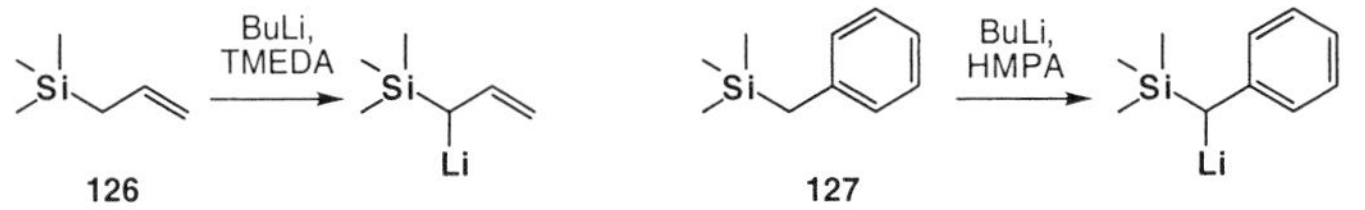

126 127

Without additional stabilisation of the organolithium, deprotonation α to silicon is difficult. It can occur however when there is a nearby coordinating heteroatom, as in the formation of **129** from **128**.[96] The organolithium **129** is unstable and cyclises to give **130** which hydrolyses to **131**. Treatment with BF_3 gives the stable flurorsilyl compound **132**.

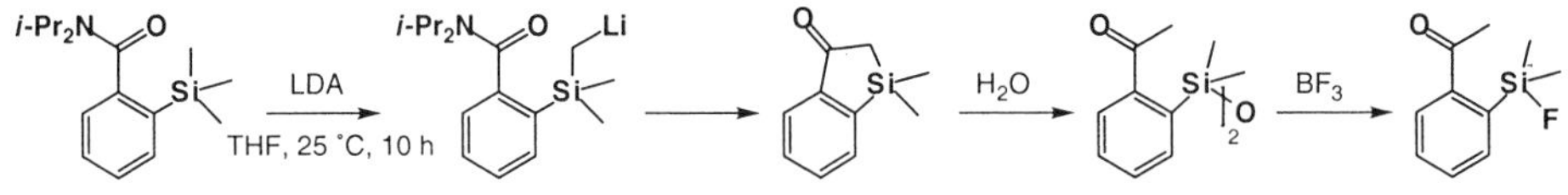

128 129 130 131 **132** 85%

Attempted double silylation of 1,8-dilithionaphthalene **133** was frustrated by a deprotonation α to silicon in the intermediate **134** which rearranged to **135** faster than it reacted with Me_3SiCl.[97]

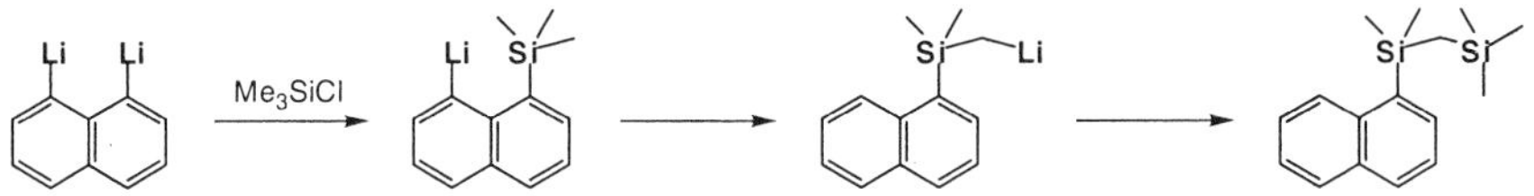

133 134 135 136

2.2.5 Lithiation at unfunctionalised allylic positions[98]

Although allylic lithiation by deprotonation of non-heterosubstituted compounds is possible using superbases (see section 2.6), in most cases allylic lithiation requires a directing heteroatom. (Non-heterosubstituted allyllithiums are best produced by reductive lithiation of allyl ethers or allyl sulfides - see section 4.4.) One of the few cases where this heteroatom is not α to the new organolithium is shown below: the β-lithiation of a homoallylic amide **137**. The reaction is particularly remarkable because of the possibility of competing deprotonation at the position α to the amide group. The proton at that position is some 10 pK_a units more acidic, yet the kinetically favoured process is removal of the β proton – no enolate **139** is formed, even as an intermediate.[99] Coordination of Li to the amide is confirmed by the *cis* stereochemistry of the product **138**.

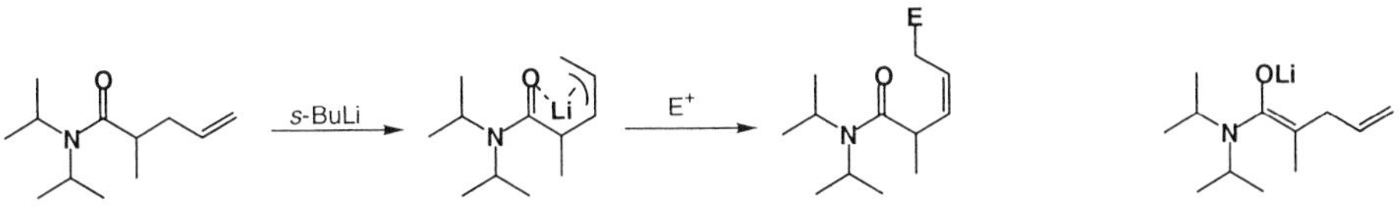

s-BuLi E+

137 138 139

The lithiation of allylic positions directed by remote carbonyl groups is observable even when enolates *are* formed – labelling shows that only the *cis* methyl group of **140** is deprotonated during the formation of the extended enolate **141**.[46]

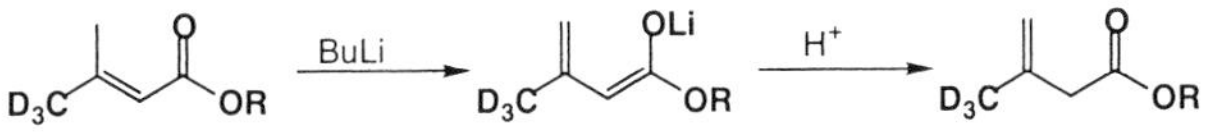

BuLi H+

140 141

Extended allylic systems can be formed by deprotonation of dienes such as **142**, **144** and **147** with *s*-BuLi. The dienyllithiums **143**, **146** and **148** adopt an extended "W" conformation, and react to give 1,3-butadienes **146** and **149** with retention of double bond geometry.[19] The equivalent species **150** formed by deprotonation with "LiCKOR" superbases (see section 2.6) adopt a "U"-shaped configuration.

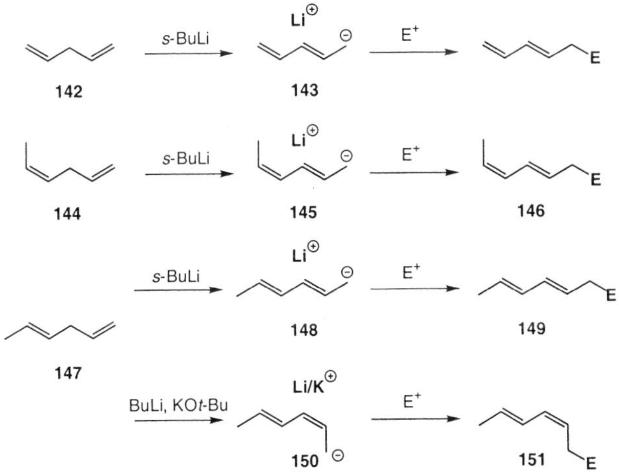

s-BuLi E+

142 143

144 145 146

147 148 149

BuLi, KOt-Bu

150 151

2.3 Ortholithiation

2.3.1 Introduction: mechanism

Ortholithiation – the directed metallation of an aromatic ring adjacent to a heteroatom-containing functional group – has arguably overtaken classical electrophilic aromatic substitution as the principal means of making regiospecifically substituted aromatic rings. Landmarks in the development of ortholithiation since the first metallations of anisole by Gilman[100] and by Wittig[101] have included the publication of the extensive and seminal review in the area by Gschwend and Rodriguez,[102] and more recently the introduction and development of amide and oxazoline-based directing groups by Beak,[103] Meyers[104] and Snieckus.[105,106] Given the detailed reviews[107,102,106] which exist in this area, my intention in this section is not to provide comprehensive coverage of ortholithiation, but instead to discuss the efficiency and therefore the value of the ortholithiation reaction according to directing group – and some groups direct far more efficiently in terms of rate of lithiation than others – and to highlight the selectivities possible in molecules bearing more than one directing group.

Ortholithiations typically involve the deprotonation of a substituted aromatic ring by an organolithium – usually *n*-, *s*-, or *t*-butyllithium or (for more electron-deficient aromatic rings) LDA. Given that benzene is some ten orders of magnitude more acidic than butane, thermodynamics pose no barrier to the removal of any aromatic proton by butyllithium. However, the kinetics of most such reactions impede their usefulness: *n*-butyllithium deprotonates benzene in hexane negligibly after 3 h at room temperature.[107,108] The problem is that reactions of hexameric *n*-BuLi (its structure in hydrocarbons)[109] are extremely slow, and only after a lithium-coordinating reagent such as THF or TMEDA has broken up the BuLi aggregates can lithiation proceed at a reasonable rate. Addition of TMEDA to a BuLi/benzene/hydrocarbon mixture is sufficient to do this – the less aggregated BuLi-TMEDA complexes deprotonate benzene at room temperature almost quantitatively.[108]

BuLi, TMEDA, 25 °C → PhLi 92%

Ortholithiation is considered to work by providing the alkyllithium with a point of coordination, increasing reactivity specifically in the locality of the coordination site (typically a basic heteroatom) of the substrate, and hence directing the regioselectivity.[110] Although there is evidence that this is not the case with weaker ortholithiation directors (see below), it can generally be assumed that for a strong director coordination between substrate and the alkyllithium, perhaps entailing deaggregation of the alkyllithium, marks the first step towards lithiation, whether ortholithiation or not. IR and kinetic studies have shown for example, that a substrate-alkyllithium complex is formed en route to α-lithiation (see Section 2.2). The functional groups involved in these α-lithiations are essentially the same as those which direct ortholithiation, and we might suppose that ortholithiation's deprotonation step, forming the more basic phenyl anion, is even more in need of assistance than the benzylic α-lithiations.

Attempts to use intermolecular and intramolecular kinetic isotope effects (see section 2.2) to identify a complexation step during ortholithiation have so far been inconclusive. Both intramolecular and intermolecular KIE's for the deprotonation of **152** and **153** by *s*-BuLi at –78 °C have values too high to measure, perhaps because complexation is fast and reversible but deprotonation is slow.[11]

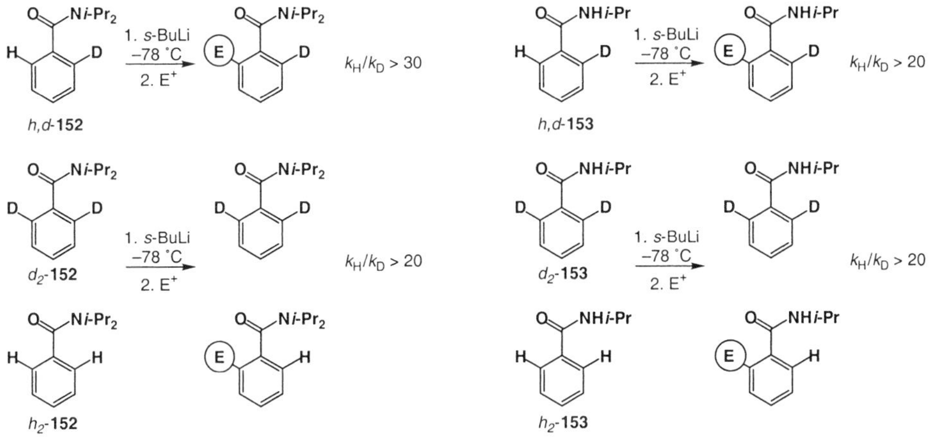

1. *s*-BuLi, –78 °C; 2. E^+ — *h,d*-**152**: $k_H/k_D > 30$; *h,d*-**153**: $k_H/k_D > 20$

1. *s*-BuLi, –78 °C; 2. E^+ — *d*$_2$-**152** / *h*$_2$-**152**: $k_H/k_D > 20$; *d*$_2$-**153** / *h*$_2$-**153**: $k_H/k_D > 20$

Stratakis[111] showed that a similar situation exists for the deprotonation of anisole by BuLi in Et_2O or TMEDA. The kinetic isotope effects are much lower, but give no evidence regarding the existence of an anisole-BuLi complex prior to deprotonation.

Given that BuLi–TMEDA will deprotonate benzene,[108] substrate coordination alone should clearly be sufficient to allow the deprotonation of substituted aromatics. And indeed, amine **154**, whose aromatic protons are no more acidic than those of benzene, is deprotonated rapidly (much more so than benzene)[112,113] and regioselectively (at the 2-position, closest to the directing group).[114]

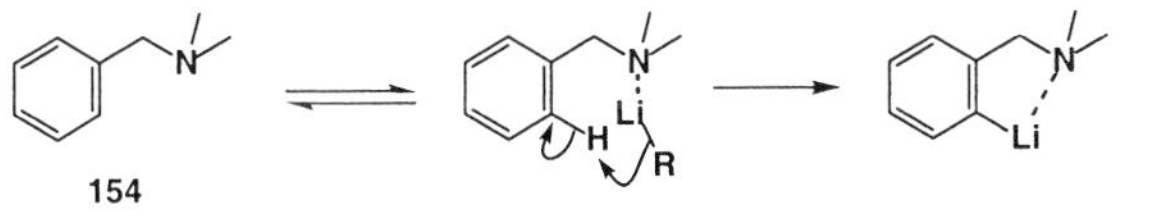

154

Detailed NMR and theoretical studies have identified and characterised a number of the complexes along this proposed reaction pathway for anisole, 1,2-dimethoxybenzene and *N,N*-dimethylaniline.[115,116] For example, anisole deaggreagates the BuLi hexamer (see chapter 1) to form a tetrameric BuLi-anisole complex **155**. Adding TMEDA displaces the anisole from the tetramer and breaks it down further to give a BuLi-TMEDA dimer **156**, which deprotonates anisole at > 0 °C yielding **157**.

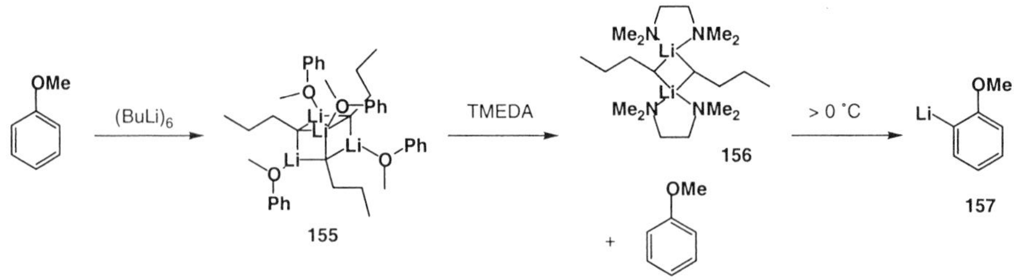

It has usually been assumed that the lithiation step involves loss of TMEDA and re-formation of a BuLi-anisole complex prior to the deprotonation itself. However, the kinetics of deprotonaton step are inconsistent with this proposition: both TMEDA molecules remain part of the complex during the deprotonation,[117] which may therefore involve no O–Li coordination and be directed purely by the acidifying effect on nearby protons of the σ-electron-withdrawing MeO substituent.[118]

Acidity is evidently the only factor directing lithiation when coordination to the heteroatom is electronically or geometrically impossible. For example, fluorobenzene is slowly lithiated by BuLi–TMEDA at –50 °C, despite the unlikelihood of a strong F–Li complex forming (in contrast with anisole, no PhF–BuLi complex is discernible by NMR[115]).

As far as synthetic utility goes, it is clear that rings bearing electron-withdrawing substituents which acidify nearby protons (via the inductive effect) are usually (not always) lithiated much more rapidly than those which acidify nearby protons only weakly or not at all. Although steric effects are undoubtedly involved as well, the differing regioselectivities in the series of peri- and ortholithiations of 1-substituted naphthalenes **158** shown in Table 2.3.1[119,120] serves to illustrate the increasing importance of acidity as substituents become more inductively withdrawing. While a group X directing lithiation by coordination can promote lithiation at either the *ortho* or the *peri* position of **158**, a group whose directing effect derives mainly from its ability to acidify nearby protons can direct lithiation only to the *ortho* position.

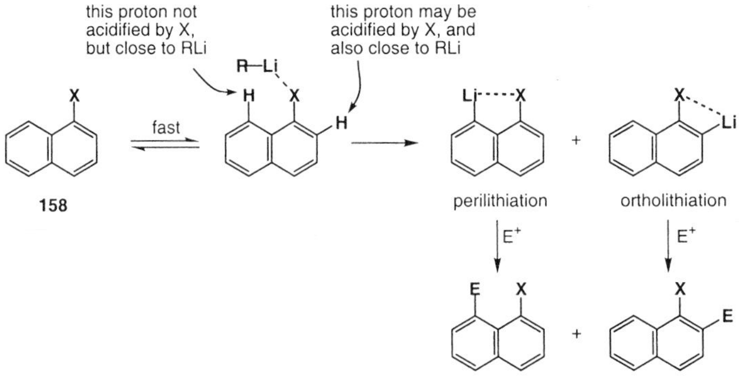

The aminoalkyl substituted compound analogous to **154** (entry 1) is lithiated solely in the *peri* position, presumably because this is a purely coordination-driven lithiation, and the geometry for perilithiation by the amine-complexed BuLi is more favourable than the geometry for ortholithiation. The same is true for other amino-substituted naphthalenes (entries 2-4). With alkyl groups themselves bearing electron withdrawing substituents, ortholithiation begins to make an appearance (entries 5 and 6). Acidity seems to become more important when the naphthalene bears the more inductively withdrawing oxygen substituents (entries 7-10) – they give either peri- or ortholithiation (or both) depending on conditions. More strongly withdrawing and coordinating acetal (entry 11) and carbamate (entry 12) groups give even higher levels of ortholithiation, as does the strongly electron-withdrawing amide substituent (entry 13). For these last two or three entries, the geometry for coordination to a *peri*-lithium atom may also be very unfavourable. Interestingly, secondary sulfonamides (entry 14) also direct lithiation to the *peri*-position. The sulfone (entry 15) on the other hand directs *ortho*.

entry[ref]	X =	conditions	E+ =	yield *peri*-**159**	yield *ortho*-**159**
1[121]	CH_2NMe_2	BuLi, Et_2O, hexane, 20 °C	Ph_2CO	58 (91)[a]	0 (9)[a]
2[122]	NH_2	BuLi x 3, Et_2O, Δ	CO_2	20	0
3[123,124]	NLiR	*t*-BuLi, Et_2O, 20 °C	D_2O	(100)[a]	(0)[a]
4[125,126]	NMe_2	BuLi, Et_2O, 20 °C	DMF	76	0
5[127]	$CH(NEt_2)OLi$	BuLi, PhH, Δ	DMF	32	2
6[127]	$CH(OMe)_2$	*t*-BuLi, Et_2O	MeI	27	13
7[128]	OH	*t*-BuLi, TMEDA, 20 °C	various	(100)[a]	(0)[a]
8[129]	OH	*n*-BuLi, THP, 50 °C	Me_2S_2	50	19
9[130]	OMe	*t*-BuLi, cyclohexane, 20 °C	CO_2	35	0
10[130]	OMe	*n*-BuLi, TMEDA, 20 °C	CO_2	0	59
11[131]	OMOM	*n*-BuLi, TMEDA	RCHO	0	73
12[106]	$OCONR_2$	*s*-BuLi, TMEDA, THF, –78 °C	MeI	0	90
13[132,119]	$CONR_2$	*s*-BuLi, TMEDA, THF, –78 °C	various	0	76-93
14[133]	SONHt-Bu	BuLi, THF, 25 °C	CO_2	14	0
15[134]	SO_2t-Bu	BuLi, –70 °C	CO_2	0	47

[a]*Ratio in crude product*

Table 2.3.1 Ortho- and perilithiation of 1-substituted naphthalenes[119]

The relative importance of coordination and acidity can furthermore depend on the base employed in the lithiation. Once coordinated to a basic solvent (TMEDA or THF, for example) alkyllithiums become less Lewis-acidic: they have a somewhat decreased tendency to be directed by coordination and acidity can become the dominant factor. Deprotonation of **160** and **161** illustrates this effect: in the absence of TMEDA, deprotonation occurs *ortho* to the more Lewis-basic amino group, while in the presence of TMEDA, the deprotonation occurs *ortho* the more electronegative, and therefore more acidifying, MeO group. [135]

BuLi/TMEDA deprotonates here

BuLi alone deprotonates here

OMe, NMe_2 **160**; OMe, Me_2N **161**

Coordination to strongly *ortho*-directing groups is responsible for the regiochemistry of some other reactions which do not involve ortholithiation. For example, while the electron-withdrawing nature of the oxazoline would be expected to direct the addition of the organolithium nucleophile to benzyne **162** towards the meta position, the major product that arises is the result of addition at the *ortho* position to give **163**.[136]

RLi, −78 - −15 °C

162 **163**

By and large, the correlation between ring acidity and the rate of deprotonation is relatively loose: under the kinetically-controlled conditions usually used for an ortholithiation reaction it is anyway rather difficult to quantify "acidity". Fraser has published a quantified scale of acidifying effect for a range of functional groups by measuring the extent of lithiation by lithium tetramethylpiperidide (pK_a = 37.8) using NMR spectroscopy.[137] The results are shown in Table 2.3.2. There is a moderately good correlation between pK_a and the *peri*- vs. *ortho*-directing ability summarised in Table 2.3.1. With lithium amide bases, lithiation becomes reversible and a pK_a scale of acidity is useful for determining lithiation regioselectivity, as we shall see.

complexation favoured by basicity of heteroatom

deprotonation favoured by acidity of proton

To summarise, ortholithiation is a reaction with two steps (complex-formation and deprotonation) in which two features (rate and regioselectivity of lithiation) are controlled by two factors (coordination between organolithium and a heteroatom and acidity of the proton to be removed). In some cases, some of these points are less important (acidity, for example, or the coordination step), but in attempting a new ortholithiation reaction it is worthwhile giving consideration to all six. Indeed, the best directing groups tend to have a mixture of the basic properties required for good coordination to lithium and the acidic properties required for rapid and efficient deprotonation – we might call them "amphoteric".

164

entry	X =	pK_a
1	CH_2NMe_2	>40.3[a]
2	NHCO*t*-Bu	>40.3[a]
3	NMe_2	>40.3[a]
4	OLi	>40.3[a]
5	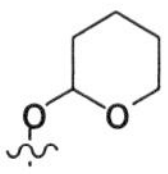 O-(tetrahydropyran-2-yl)	40.0
6	OMe	39.0
7	OPh	38.5
8	SO_2NEt_2	38.2
9	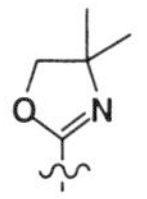 4,4-dimethyloxazolin-2-yl	38.1
10	CN	38.1
11	CON*i*-Pr_2	37.8
12	$OCONEt_2$	37.2

[a]*Not lithiated by LiTMP*

Table 2.3.2 pK_a and ortholithiation[135]

2.3.2 Classes of directing group

In his review of 1990,[104] Snieckus divided the typical ortholithiation-directing groups into classes, according to the ease with which they may be lithiated and the practicalities of their application in synthesis. It is clear that in the meantime certain metallation-directing groups have remained little more than curiosities, while others have become widely used. In this section, we shall survey metallation-directing groups in broad assemblies which share certain features.

As far as their simple ability to direct metallation goes, functional groups may broadly be placed in the sequence shown below. Several studies of relative directing ability have been carried out,[101,133,136-139] sometimes with conflicting results (though these may often be ascribed to differences in conditions, of whether the competition was inter- or intramolecular), and these studies are taken into account here. Also in the scheme is a guide to the conditions typically required for lithiation of a certain functional group. This is a broad-brush picture, and is not intended to provide detailed information, but aims to give a general impression of the approach required for each class of functional group.

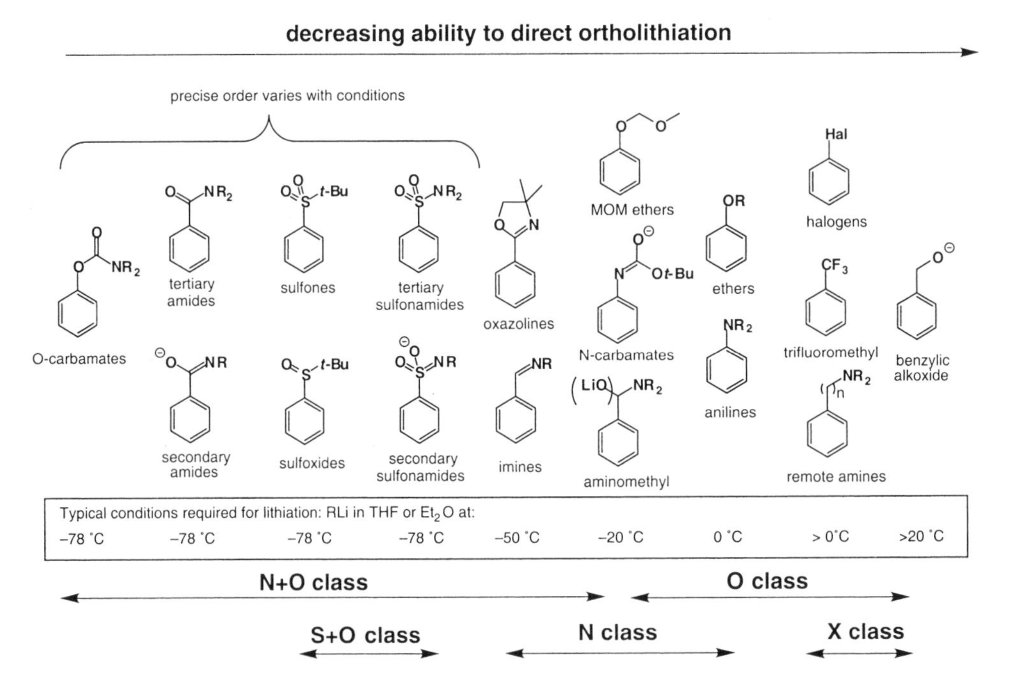

The most powerful classes, for several reasons, comprise carboxylic acid and carbonate-derived functions containing both nitrogen and oxygen: secondary and tertiary amides,[104] oxazolines,[140,102] and carbamates.[104] We have grouped these functional groups into the "N+O class". Their importance stems firstly from their "amphoteric" nature: they are all functional groups which have a highly basic heteroatom (the oxygen atom in the carbonyl group of an amide is among the most basic of neutral oxygen atoms in organic chemistry) and many are strongly electron-withdrawing groups which acidify the protons of the ring. Secondly, they are functional groups which are transformable to useful targets. They do nonetheless have a drawback: they all contain an electrophilic carbonyl group (or equivalent) which may itself suffer attack by the lithiating agent. Typically, therefore, they need to be highly sterically encumbered to prevent this, and this may make their subsequent removal or transformation difficult. Steric hindrance is avoidable if the functional group can be deactivated by deprotonation, and such amide and carbamate anions (and in fact many other anionic functional groups) are surprisingly powerful directors, considering they have little power to lower the acidity of the ring protons.

Sulfones and sulfonamides – the "S+O class" are similarly powerful directors,[141] and do not suffer electrophilic attack at sulfur (though occasionally suffer nucleophilic attack on the aromatic ring)[134,143] but are less useful because of their more limited synthetic applications.[144,143] Aryl *t*-butylsulfoxides are also powerful *ortho*-directors,[145] but less hindered diaryl sulfoxides are susceptible to attack by organolithiums at sulfur[146] (see section 3.3.3 for discussion of this "sulfur-lithium exchange").

Less powerful with regard to directing ability (they are less basic and less acidic) but of high synthetic value are functional groups containing oxygen only (ethers, acetals, carboxylates... the "O class") and nitrogen only (amines, imines, nitriles... the "N class"). They typically direct lithiation more slowly or at higher temperatures, but in many cases have the advantage that they cannot be attacked by the organolithium reagent. No special structural features are therefore necessary, and for example the simple methyl ethers of MOM acetals occurring in normal synthetic routes or even target molecules may be used as lithiation substrates. The most powerful members of the N class are those which can use a nitrogen lone pair to coordinate to the organolithium – in other words, those other than the anilines. In the O-class, the best directors are on the other hand those in which the oxygen is attached to the ring, inductively acidifying nearby protons, and those which carry a second oxygen atom (an acetal, for example) for lithium coordination.

The halogens, the "X class", are gaining importance as lithiation directors, particularly for the lithiation of heteroaromatics. Halogens direct by an inductive, acidifying effect alone, and work best when there are additional factors favouring deprotonation – if the aromatic ring is an electron-deficient heterocycle for example. Halogens are susceptible to attack by organolithiums, and ortholithiated haloarenes are prone to elimination to give benzynes, so conditions must be carefully controlled: typically these directing groups are used in conjunction not with organolithiums but with LDA.

It is interesting to note the surprising fact that even a lithio substituent activates an aromatic ring towards lithiation: the deprotonation of phenyllithium is easier than the deprotonation of benzene![147] Double ortholithiation (to give a dilithiated ring) is usually feasible when two separate directing groups are involved, but using one group to direct simultaneously to both *ortho* positions usually fails.[148] Simultaneous triple lithiation has never been achieved even with three separate groups.[148]

2.3.2.1 N+O Class

2.3.2.1.1 Secondary and Tertiary Amides

Häuser reported in 1973[149] that *N,N*-diethylbenzamide **165** is attacked at the carbonyl group by *n*-BuLi to give aryl ketone **166**. However, by using the hindered, non-nucleophilic base LiTMP Beak was able to avoid attack at C=O and to ortholithiate **165**.[7] The product **167** immediately attacks another molecule of starting material (a recurring problem in $-CONEt_2$ ortholithiation chemistry) to yield 2-benzoyl benzamide **168**. A similar reaction occurs with the diethylamide and *s*-BuLi, but not if TMEDA is added[150] – presumably the starting material's deprotonation then outpaces its attack by the newly formed ortholithiated species. For this reason, the addition of TMEDA is essential for the lithiation of diethyl amides in THF. *N,N*-Diisopropyl amides **169**, on the other hand, are almost completely resistant to attack at C=O, and can be ortholithiated without difficulty with *s*-BuLi, *t*-BuLi or even *n*-BuLi in the presence or absence of TMEDA. For practicality's sake, the most appropriate conditions with $-CONi\text{-}Pr_2$ amides are *s*-BuLi or *n*-BuLi in THF at –78 °C (no TMEDA required).[132]

without TMEDA, reaction of **165** with product **167** competes with lithiation

with TMEDA, lithiation is fast and all of **165** is lithiated before **167** can attack it – **167** then reacts cleanly with an electrophile

diisopropyl amides are so hindered that they ortholithiate cleanly even in the absence of TMEDA

The ortholithiated products **167** and **170** will then react with a wide range of electrophiles: the only reported important exceptions are enolisable aldehydes and allylic halides. Products requiring these electrophiles are best made by first transmetallating the organolithium to a Grignard reagent with $MgCl_2$[106] or copper.[151]

N,N-Dimethyl amides are susceptible to attack at C=O, but can be successfully ortholithiated if kept cold. Keck used successive ortholithiations of **171** in a route towards pancratistatin.[152] (Danishefsky's similar route using *N,N*-diethylamides[153] suffered from difficulties removing the amide group – see below.)

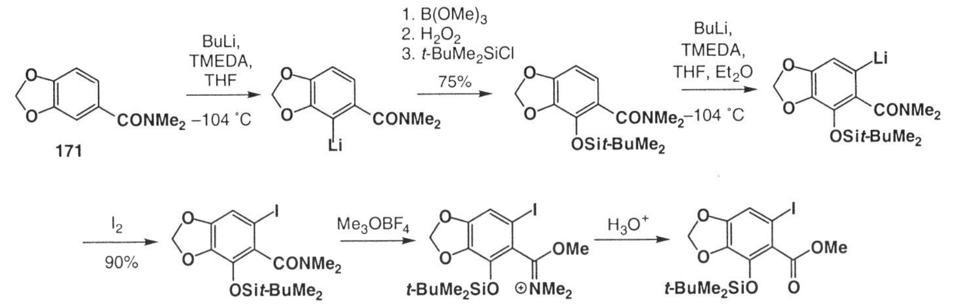

In more heavily substituted amides, the amide group is forced to lie perpendicular to the aromatic ring[132,154] – even in **169** the amide and the ring are not coplanar.[155] Clearly this poses greater difficulties for ortholithiation, and Beak has shown that the rate of lithiation is dependent on the angle between the amide plane and the plane of the ring.[155] With ring and amide perpendicular, a single lithium atom would be unable to bridge between O and C, and X-ray crystal structures of **170** and **173**[156] show that in fact ortholithiation of **169** and **172** gives organolithiums which are (at least in the solid state) bridged dimers, with one lithium atom more or less in the plane of the amide and the other more or less in the plane of the ring.

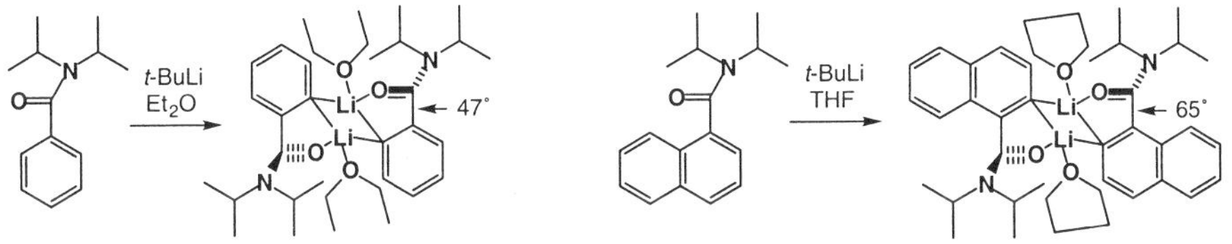

For the same reason that they resist attack at C=O by alkyllithiums, tertiary amides can be extremely difficult to hydrolyse – almost impossible in the case of –CON*i*-Pr_2, and even –$CONEt_2$ amides are stable to 6 M HCl for 72 h. For reactions in which an amide is not required in the product therefore, it is preferable to use –$CONEt_2$, and to remove the amide from the product by reduction, as in the scheme below (note the cooperative effect of the amide and methoxy group in the first step).[157] Hydrolysis can also be achieved via an imidate (see the sequence above).

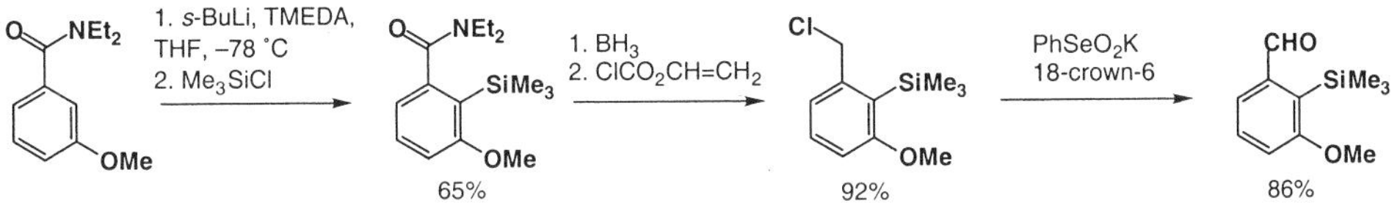

Removal of the amide function is much easier if the reaction is intramolecular, and $-CONEt_2$ amides (sometimes even $-CONi\text{-}Pr_2$ amides) may be converted to lactones, lactams and other heterocycles in this way.[120,158] Addition of an aldehyde or ketone as an electrophile generates a hydroxyl group (in some cases, atroposelectively, as it happens[159,132] – though this is usually irrelevant to the stereochemistry of the product) which cyclises to give a lactone via a benzylic cation in acid. This reaction has found wide use in the synthesis of polycyclic aromatics, particularly alkaloids.

Lactonisation of alcohols derived by organometallic addition to 2-formyl amides, lactol formation from 2-formyl amides themselves[106] and lactam formation from their nitrogen counterparts[160] are equally useful synthetically. Further examples are presented in section 2.4.

Rather than fighting against the lack of electrophilicity at the amide carbonyl group, it can be more fruitful to exploit its electron-rich nature in cyclisations onto carbocations. For example, the allyl-substituted amides **174** cyclise to the 6-membered lactones **175** in refluxing 6M HCl.[161]

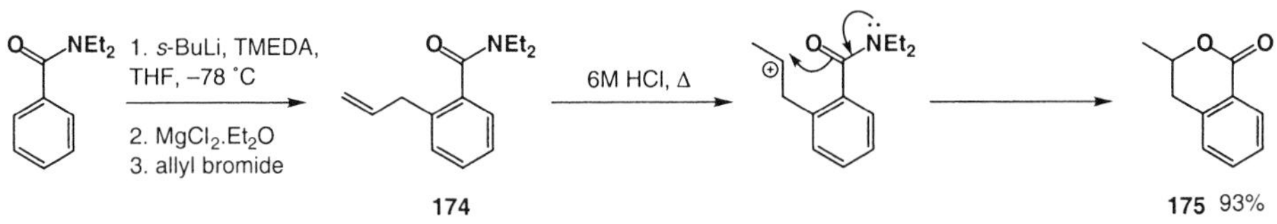

1. *s*-BuLi, TMEDA, THF, –78 °C; 2. $MgCl_2.Et_2O$; 3. allyl bromide → **174**; 6M HCl, Δ → **175** 93%

Although there is no shortage of applications for the lithiation of diethylamides, their potential would clearly be widened if the limitations of the amides' reactivity could be overcome. Alternative *N*-substituents have been devised with the aim of increasing the number of transformations available at the amide group, and a few of the more recent ones are shown below. Clearly, the *N*-substituents must be highly resistant to strong base, but can be sensitive to acid or, say, fluoride. One solution is to replace the methyl groups of one isopropyl with silyl groups, as in **176**.[162] After functionalisation, fluoride treatment gives **177** which is then much more easily hydrolysed or reduced.

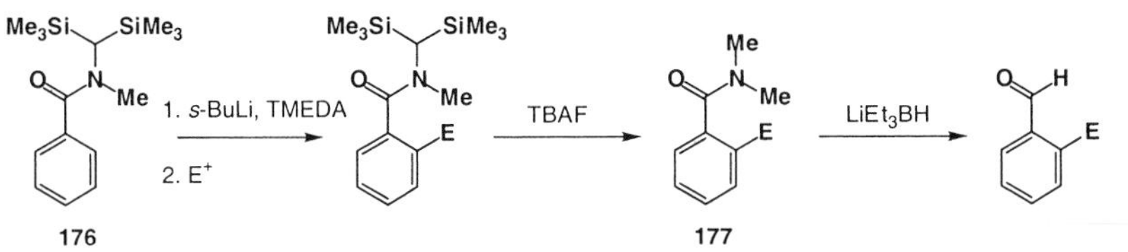

176 → 1. *s*-BuLi, TMEDA; 2. E^+ → TBAF → **177** → $LiEt_3BH$

More practical from the point of view of starting material availability is the use of the *N*-*t*-butyl-*N*-methyl amides **178**, whose *t*-butyl group may be removed with acid.[163] De-*t*-butylation of amides can be capricious however, though it seems much more successful with acyclic than cyclic amides.

1. *s*-BuLi
2. MeI

1. CF_3CO_2H, Δ
2. $NaNO_2$, HCl
3. KOH, EtOH

178

We have introduced an amide based on the acid-sensitive oxazolidine employed by Hoppe[164] in his base-resistant, acid-labile alkylcarbamates. Our aim was to retain a high degree of steric hindrance around the nitrogen atom for purposes of stereoselectivity. The amides based on **179** seem to have many of the properties of *N,N*-diisopropylamides, but are readily hydrolysed by treatment with $MeSO_3H$ for 5 min in refluxing MeOH.[165] One disadvantage of these acyl oxazolidines is their existence as a mixture of slowly interconverting conformers (sometimes separable geometrical isomers) which complicates their characterisation by NMR.

1. *s*-BuLi, −78 °C
2. RCHO

$MeSO_3H$, MeOH, 5 min, Δ

179

Other more readily cleavable tertiary amides include **180**[166] and **181**.[167]

180 **181**

Secondary amides (but not primary amides, which resist ortholithiation under all conditions) are ortholithiated with two equivalents of organolithium – the first-formed amide anion prevents nucleophilic attack by the second equivalent of organolithium.[168] Many anionic groups are surprisingly powerful directors of ortholithiation, given that the acidity of the ring protons can hardly be enhanced by the presence of a negatively charged heteroatom. Any loss of acidity must be more than outweighed by the highly favourable complexation of the organolithium to this charged intermediate. Reaction with simple electophiles leads to ortho-substituted amides **182**; with diketones, isoquinolones **183** are formed.[169]

Secondary amides have the advantage over tertiary amides that they are relatively easy to remove. It is quite difficult to stop the addition products from aldehydes, ketones, amides, epoxides and nitriles cyclising directly to give a variety of lactone derivatives (by attack of OH on the secondary amide) or lactam derivatives (by attack of the secondary amide on the new electrophilic centre).[102] Thioamides behave similarly.[170]

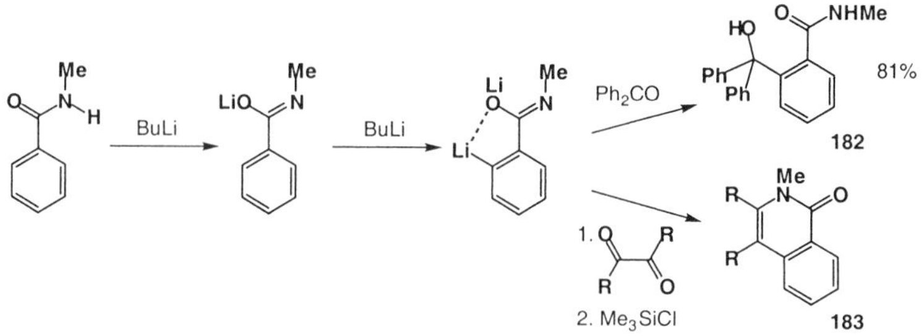

Secondary amides **184** based on cumylamine[171] or dialkylhydrazines[172,173] are particularly useful. The cumyl protecting group is removed by E1 elimination in strong acid (**185**) or by formation of a nitrile **186**, while hydrazines can be cleaved oxidatively.

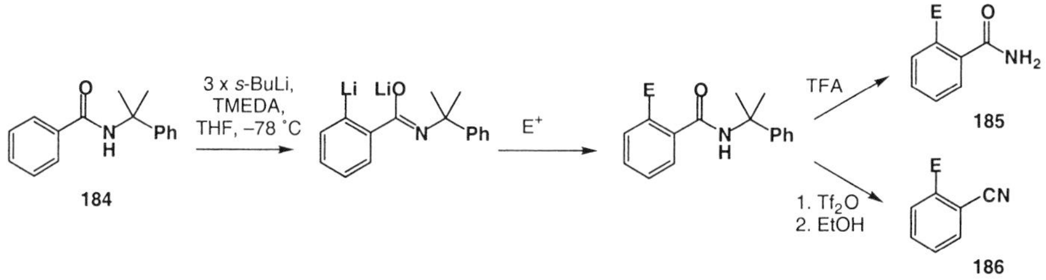

Attachment to a solid support via a secondary amide linkage allows ortholithiations to be carried out in the solid phase. After a reaction with an aldehyde or ketone, refluxing in toluene returns the amino-substituted polymer **187**.[174]

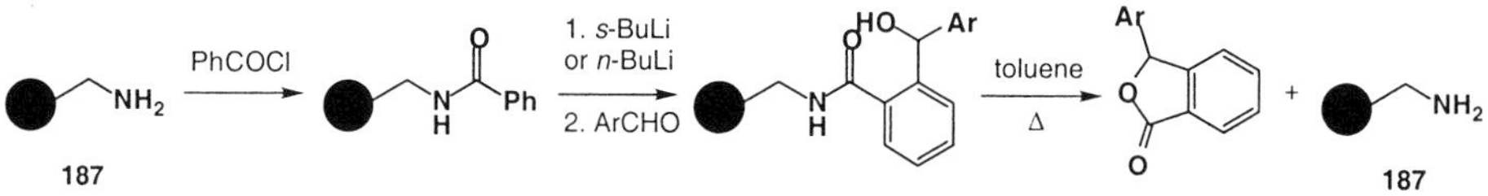

Secondary and tertiary amides are particularly effective for directing the metallation of heterocycles.[106,175-177,90,178-180,173,181] Lithiation α to the heteroatom can be a competing reaction with the 5-membered heterocycles.[178] Lithiation of furan-2-carboxamides **188** can lead to ring cleavage to give **190**,[171] though the intermediate **189** is stable with a chelating amide group.[181] Thiophene-2-carboxamides lithiate preferentially α to sulfur. Lithiation in the 3-position is achieved by double deprotonation to give **192** or by silyl protection: **193** gives **194**.[106] The strategy of silyl protection is discussed in more detail in section 2.7.

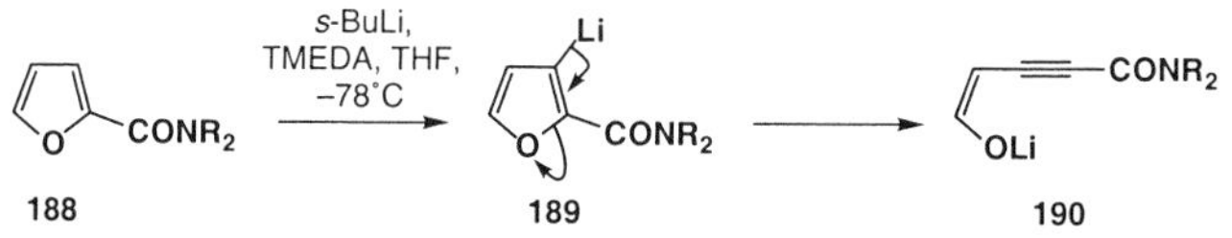

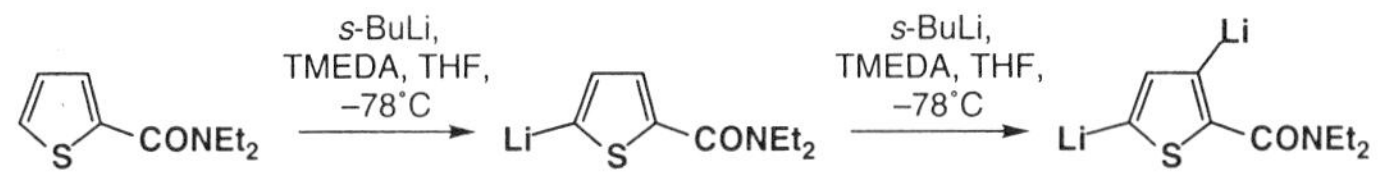

Pyridines bearing secondary amides may be lithiated with *n*-BuLi at –78 °C; with tertiary amides the optimum conditions are LiTMP, DME, 5-15 min.[182] Lithiated tertiary amidopyridines react well with carbonyl electrophiles but poorly with alkylating agents. Lithiation of the bromopyridine **195** with LDA is a key step in the synthesis of eupoluramine.[183]

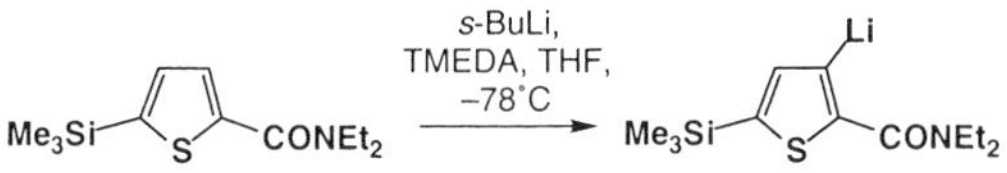

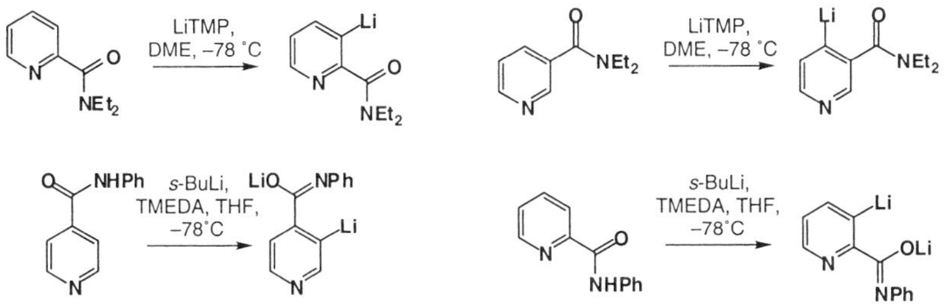

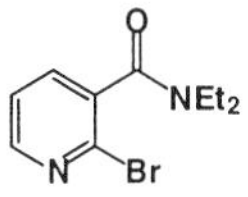

The example below shows the synthesis of a quinone by a second lithiation α to the sulfur of a thiophene.[184]

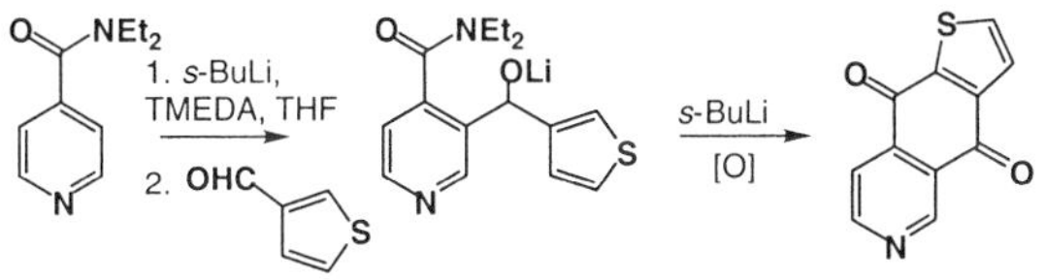

To summarise: the amides are most suitable for the formation, by ortholithiation, of condensed heterocycles and polycyclic aromatics (in which subsequent rings are formed by intramolecular attack on the amide group). In other cases the removal of the amide group may be problematic, though if carboxylic acids, aldehydes, or hydroxymethyl-substituted compounds are required, alternative amide substituents may be used.

2.3.2.1.2 α-Amino alkoxides

N,N-Dimethylbenzamides can be used as lithiation directors even under conditions that lead to alkyllithium attack at the carbonyl group to give ketones. The tetrahedral intermediate in this sequence **196** is a good director of lithiation (note its similarity to a deprotonated secondary amide) and an excess of organolithium intercepts this intermediate and ortholithiates it to give **197**. The overall synthetic result is effectively that of an ortholithiation-directing ketone or aldehyde.

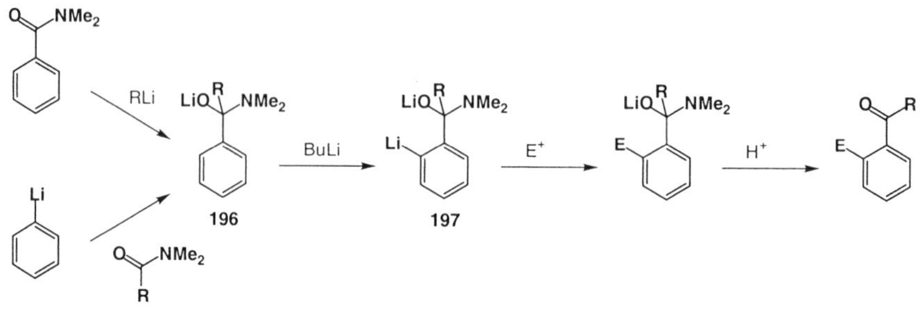

The same intermediate **196** is formed when an aryllithium attacks dimethylformamide *en route* to an aldehyde. Intercepting this intermediate can provide a useful way of synthesising *ortho*-substituted aromatic aldehydes such as **198**.[185]

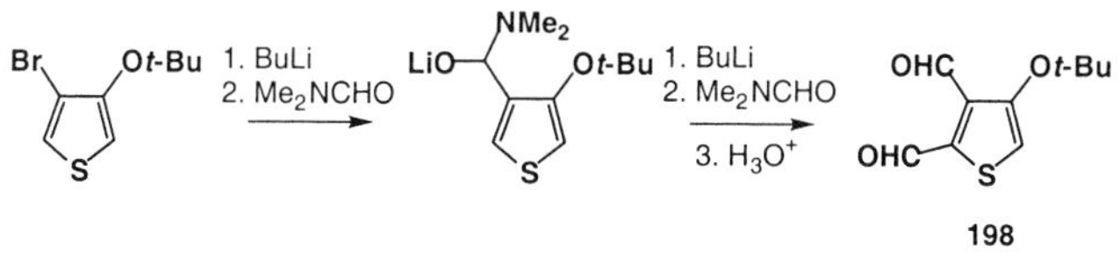

Addition of a lithiated secondary amine to an aldehyde both protects the aldehyde from attack by RLi and turns it into an ortholithiation directing group. Comins has shown that the best lithioamines for this purpose are *N*-lithio-*N,N,N'*-trimethylethylenediamine **202** and *N*-lithio-*N*-methylpiperazine **199**, which optimise the opportunity for coordination of BuLi to the intermediate alkoxide **200**.[127,186,187]

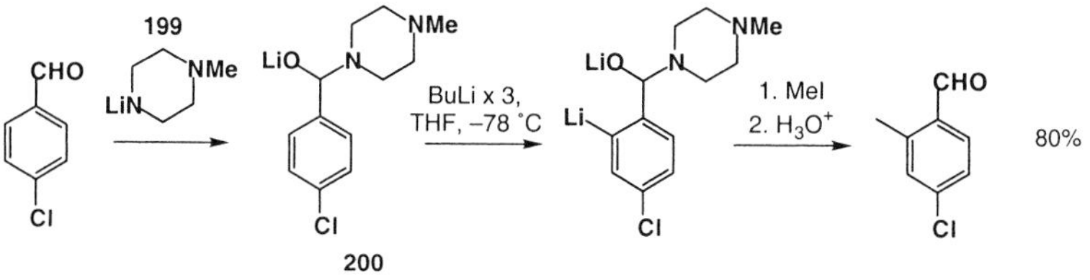

The more conformationally flexible lithioamine derived from *N,N,N'*-trimethylethylenediamine enhances the directing group ability of the aminoalkoxide: while both **199** and **202** can be used for the protection of **201** during the lithiation step, **202** leads to a directing group stronger than OMe, while **199** leads to a directing group weaker than OMe.[188]

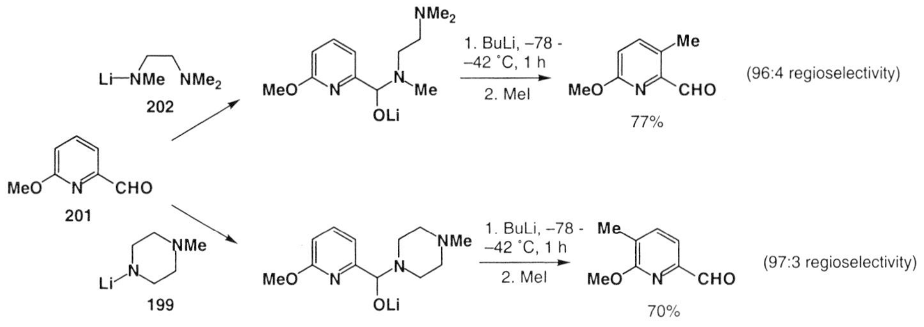

2.3.2.1.3 Oxazolines

Aryl oxazolines exhibit resistance to nucleophilic attack and can be lithiated with *n*-BuLi. A range of methods are available for activation of the oxazoline group of the products to allow constructive conversion to other functional groups.[142,104] A sequence of simple lithiations of **203** illustrate the potential of this method at the start of this synthesis of a precursor to a steroid.[189]

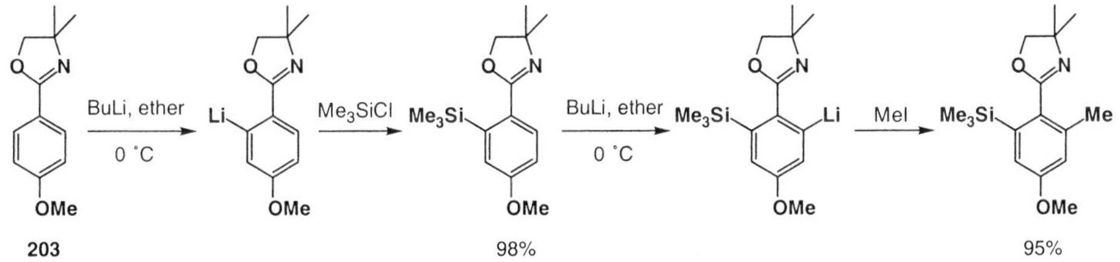

The cooperative effect of MeO and the oxazoline directs the lithiation of **204** in this synthesis of a lipoxygenase inhibitor AC-5-1 **205**.[190]

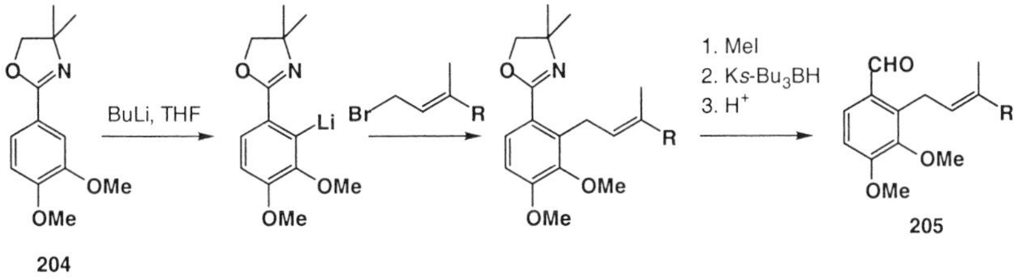

Ortholithiated oxazoline **206** is best transmetallated to its magnesium analogue before reaction with aldehydes. As with the equivalent amide reaction, treatment with 4.5 M HCl then cyclises the products to lactones.[191]

BuLi, THF, 0 °C → 206 (Li) → MgBr₂ → MgBr → MeCHO → OH → 4.5 M HCl →

2.3.2.1.4 *O*-Carbamates

The *O*-aryl *N,N*-diethyl carbamates **207** are the most readily lithiated of the oxygen-based metallation directing groups, and have chemistry which is closely associated with that of tertiary amides.[106] Like the amides, they are hard to hydrolyse, requiring vigorous basic conditions.[192] Hydrolysis is easier with nearby functional groups, which must play an assisting role; alternatively, the carbamates may be reduced to phenols by $LiAlH_4$.

OH → $ClCONEt_2$ → **207** → 1. *s*-BuLi, −78 °C 2. E^+ → $LiAlH_4$, THF, Δ or NaOH, MeOH (or ethylene glycol), Δ → OH

N,N-Diisopropylcarbamates generally offer little advantage over their *N,N*-diethyl analogues, and their removal is even more difficult. *O*-Aryl *N*-cumyl carbamates do however offer some advantages, and can be hydrolysed to phenols under acid conditions.[171]

N,N-Dimethylcarbamates are unstable once ortholithiated, and rearrange rapidly by a carbamoyl transfer mechanism known as the "anionic *ortho*-Fries rearrangement".[106] With *N,N*-diethylcarbamates, this rearrangement can be controlled: the ortholithiated carbamate **208** is stable at −78 °C, but on warming to room temperature, rearranges to give a 2-hydroxyamide **209**.[192]

207 → *s*-BuLi, −78 °C → **208** → −78 → +20 °C, *anionic ortho-Fries rearrangement* → NEt_2 OLi → H^+ → **209**

This chemistry can be very powerful, since the amide product itself offers further possibilities for functionalisation by lithiation. The synthesis of the natural product ochratoxin A (section 9.1) illustrates this point. Two successive ortholithiations of carbamate **210** are used first to introduce one amide group and then a second, by anionic *ortho*-Fries rearrangement. The symmetrical diamide **211** can be allylated and then cyclised in acid, with concomitant hydrolysis of the second amide and deprotection of the phenol to yield a known intermediate

212 in the synthesis of ochratoxin A **213** (notwithstanding any question of stereochemistry).[193]

1. *s*-BuLi, TMEDA –78 °C 2. ClCONEt2

1. *s*-BuLi, TMEDA –78 °C 2. –78 → +20 °C 3. K_2CO_3, MeI

1. *s*-BuLi, TMEDA –78 °C 2. $MgCl_2$ 3. allyl bromide

6M HCl, Δ

ochratoxin A

210 **211** **212** **213**

An anionic *ortho*-Fries rearrangement has been used to make BINOL derivatives.[194]

2.3.2.1.5 Anilides and *N*-aryl carbamates

Anilines are poor directors of lithiation:[195] the nitrogen's lone pair is less basic than an amine's, and the nitrogen has a weaker acidifying effect than a corresponding oxygen substituent. *N,N*-Dimethylaniline can nonetheless be lithiated by BuLi in refluxing hexane,[196,197] and the lithiation of *N,N*-dialkylanilines is made more efficient by the addition of TMEDA.[198]

By far the best solution to the problem of lithiating an amino-substituted aromatic ring is to acylate an aniline to give a pivalanilide **214**[199] or an *N*-Boc carbamate **217**.[200] Two equivalents of base (BuLi) deprotonate first the nitrogen and then cleanly ortholithiate the anion to give **216**. As with the secondary amides described above, the lack of acidifying effect in the intermediate anion (which this time *is* conjugated with the ring) is more than outweighed by the powerful lithium-coordinating ability of the anionic intermediate **215**.

BuLi BuLi Me_2S_2 78%

214 **215** **216**

With *N*-Boc anilides **217**, the best conditions for lithiation are *t*-BuLi (2.2 equivalents) in Et_2O at –10 °C.

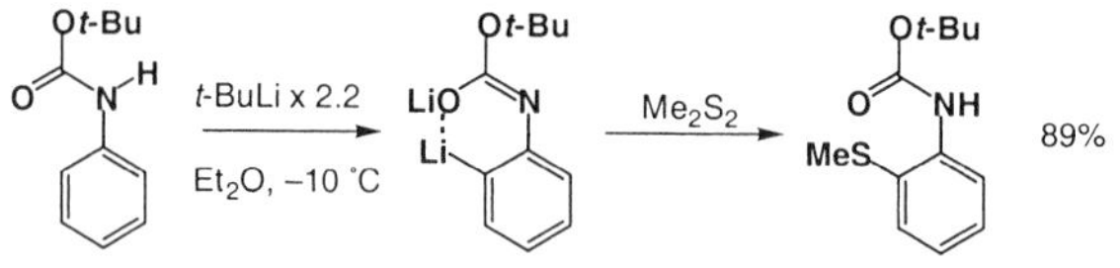

217

Amides of aminopyridines have also been widely used to direct lithiation, and are most effective when lithiated with BuLi in the absence of TMEDA.[175] The lithiation of **218** can be used as a key step in the synthesis of naphthyridines and other condensed polycyclic heterocycles.[201]

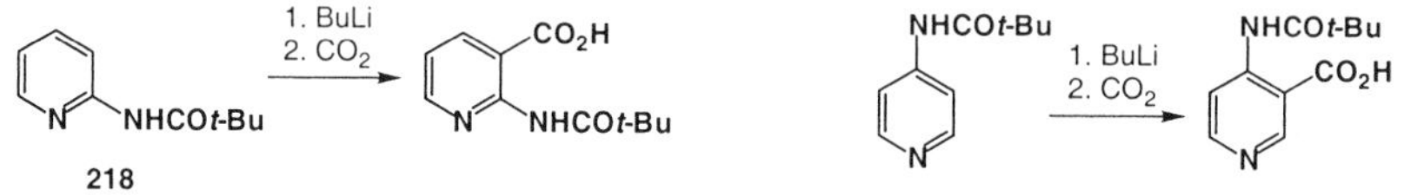

218

Ortholithiation of thioanilides **219** has been used to construct benzothiazole rings **221** via the benzyne **220**.[202]

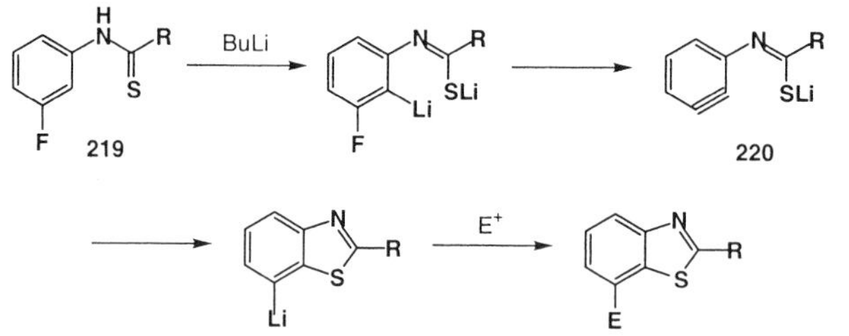

219 220 221

Ureas of anilines **222** can also be lithiated.[203,204] The products are generally very hard to cleave, but quenching the intermediate organolithiums **223** with carbon monoxide generates acyllithiums **224** which cyclise to give isatins **225**.[205]

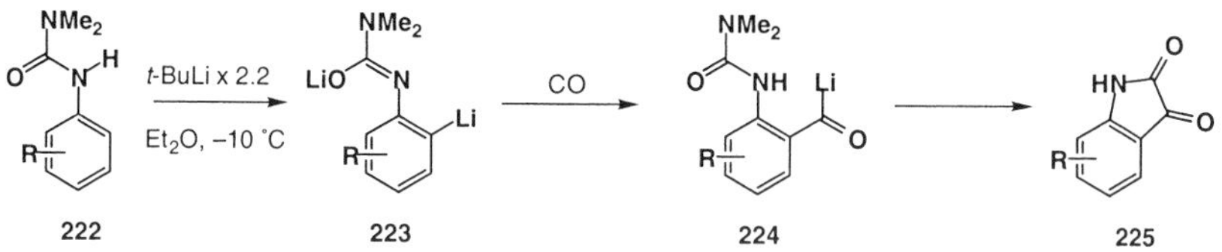

222 223 224 225

2.3.2.2 S+O Class

Sulfides are weak orthodirectors, and the lithiation of thioanisole **226** with BuLi leads to a mixture of α- and ortholithiated compounds **227** and **228**.[206] The ortholithiated compound forms about one third of the kinetic product mixture, but slow isomerisation to the α-lithiated sulfide follows. The isomerisation is much faster (and therefore the yield of α-lithiated sulfide much higher) if BuLi is used in the presence of DABCO.[77] With two equivalents of

BuLi, clean *ortho*+α double lithiation occurs, giving **229**: the SCH_2Li group is itself an *ortho*-director,[207] though a weaker one than OMe, since doubly lithiated **230** has the regiochemistry **231**.[208] A methylthio substituent in conjunction with a methoxy substituent directs to their mutual *ortho* position (**232** → **233**).

228 ⇌ 227 ← BuLi — 226 — BuLi x 2 → 229

37:63 (initial ratio)
4:96 (equilibrated ratio)

230 — BuLi x 2 → 231

232 — BuLi → 233

Ethyl phenyl sulfide is lithiated mainly in the *ortho* position, but with significant amounts of *meta* and *para* lithiation and substitution products. Superbases, on the other hand, prefer to deprotonate alkylthio benzenes at benzylic or α-positions, rather than on the ring.[209,210]

Diarylsulfides are lithiated *ortho* to sulfur, but less efficiently than diaryl ethers: dibenzothiophene **234** for example lithiates to give **235** and hence **236**. Highlighting preferential lithiation at the more acidic positions *ortho* to O, **237** gives **238** rather than **239**.[107]

234 — *n*-BuLi → 235 — Br_2 → 236 77%

237 — *n*-BuLi → 238 — CO_2 → (CO_2H) 35%

239

In sulfoxides, sulfur's weak acidifying effect is enhanced, and the oxygen atom introduces a powerful coordination effect: in contrast with sulfides, sulfoxides are very powerful directors of both ortholithiation and α-lithiation. Ortholithiation is possible with aryl sulfoxides lacking an α-proton, but since sulfoxides suffer from the disadvantage of being electrophilic at sulfur, diaryl sulfoxides must be lithiated with lithium amide bases rather than alkyllithiums. Pyridyl sulfoxides **240** are ortholithiated by LDA, and have the added feature that the sulfinyl group can subsequently be displaced by Grignard reagents, leading for example to the bipyridyl **241**.[211] When the electrophile is an aldehyde, the chiral sulfoxide can control up to 7:1 stereoselectivity at the new stereogenic centre.[212]

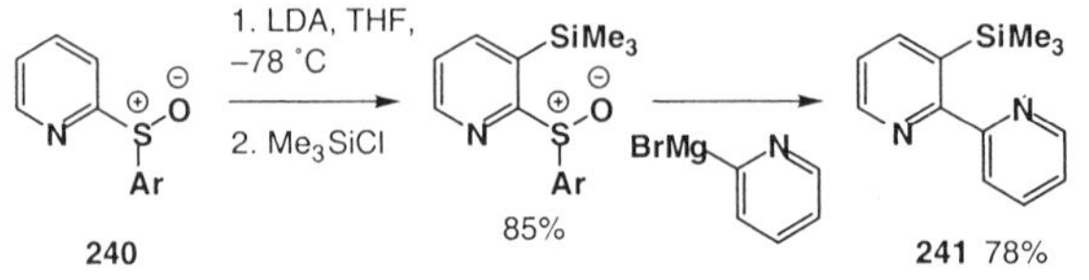
1. LDA, THF, −78 °C
2. Me_3SiCl
$SiMe_3$
Ar
85%
BrMg
$SiMe_3$
240
241 78%

Sulfoxide removal using sulfoxide-lithium exchange (see section 3.3.3) is also effective.[213] It was employed in tandem with a sulfoxide-directed stereoselective ortholithiation of the ferrocene **242**[214,215] in this synthesis of the phosphine ligand **243**.[216] Ferrocene lithiation is discussed further in section 2.3.4.2.

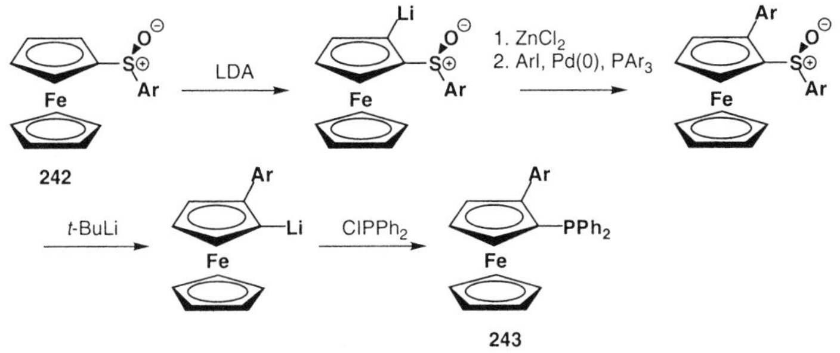
LDA
Li
1. $ZnCl_2$
2. ArI, Pd(0), PAr_3
Fe
Ar
242
t-BuLi
ClPPh₂
PPh_2
243

Aryl *t*-butylsulfoxides are sufficiently hindered for attack at sulfur to be less of a problem.[145] The chiral sulfoxide offers no control over the new stereogenic centre in the reaction of **244** to give **245**.

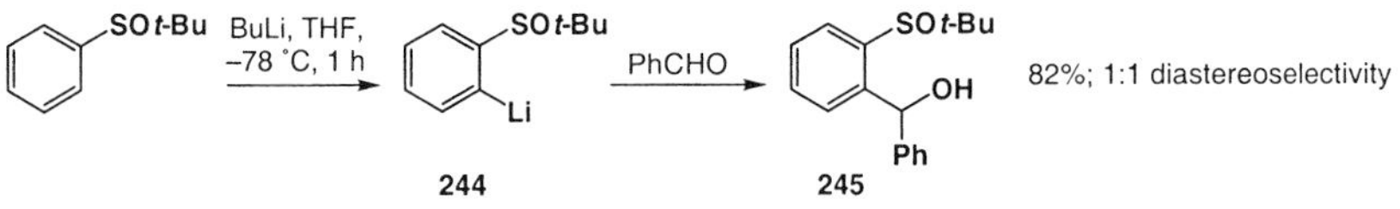
SO*t*-Bu
BuLi, THF, −78 °C, 1 h
Li
PhCHO
OH
Ph
82%; 1:1 diastereoselectivity
244
245

Sulfones are similar in some ways: even more acidifying, and with a powerful ability to coordinate, but less likely to be attacked at S. As with sulfoxides, lithiation α to S competes, and ortholithiation is useful only with sulfones lacking α-protons.[141] After lithiation, the removal of sulfones can sometimes be accomplished by transition metal-catalysed reduction or substitution:[144,143]

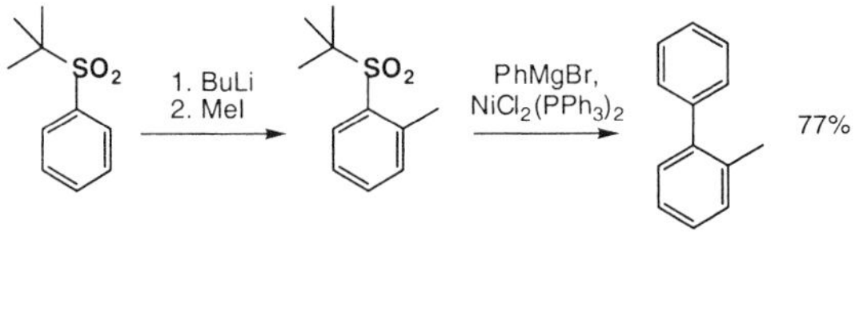
SO_2
1. BuLi
2. MeI
PhMgBr, $NiCl_2(PPh_3)_2$
77%

Even sulfonate esters **246** are powerful directing groups, competing well with tertiary amides. No substitution accompanies ortholithiation of ethyl or isopropyl benzenesulfonate by BuLi. Hydrolysis and chlorination of the products **247** gives functionalised sulfonyl chlorides **248**.[217]

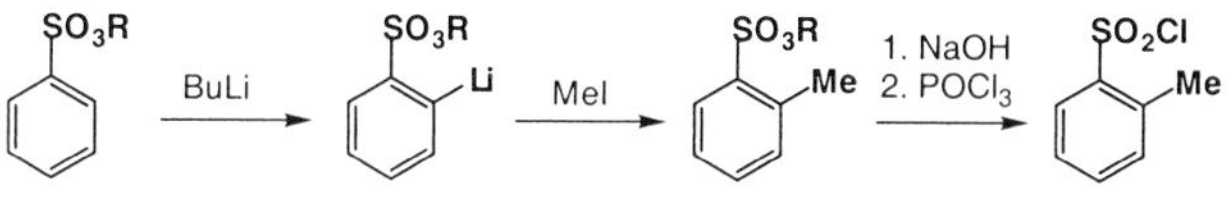

246 247 95% 248

Secondary and tertiary sulfonamides are among the most powerful ortho-directing groups known. There is no danger of attack at S, and *N,N*-dimethylsulfonamides **249** may be lithiated with *n*-BuLi alone.[218] Secondary *N-t*-butyl sulfonamides **250** are particularly useful because the *t*-butyl group is readily removed in polyphosphoric acid. Carbonation of **251** and cyclisation in acid gives the saccharin analogues **252**.[133]

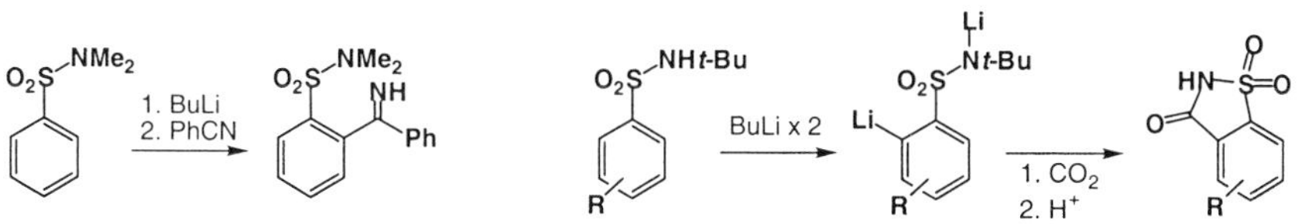

249 79% 250 251 252

Naphthyl sulfones **253** are susceptible to nucleophilic attack at the 2-position of the naphthalene ring by the alkyllithium;[134] secondary sulfonamides **254**, on the other hand, surprisingly undergo perilithiation with alkyllithiums.[133]

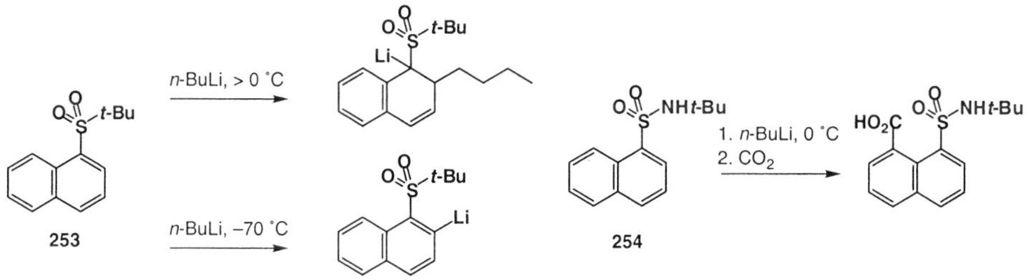

253 254

N-Aryl sulfonamides have not been investigated in detail, but "anion translocation" from the *S*-substituted ring to the *N*-substituted ring takes place in diaryl sulfonamides **255**, ultimately leading to a rearrangement as shown below.[219,220]

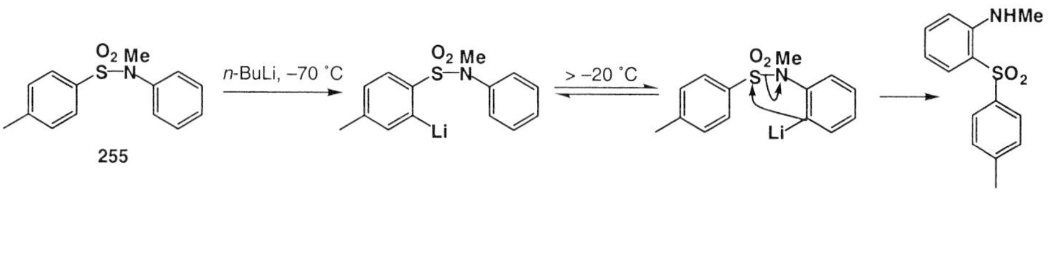

255

The ortholithiation of benzenesulfinamides **256** is of use in the regioselective synthesis of *meta*-substituted compounds **257**: the sulfinamide group is used to set up two *ortho* relationships and then removed reductively.[221] The same concept has been explored with other sulfur-containing *ortho*-directors. [141,143]

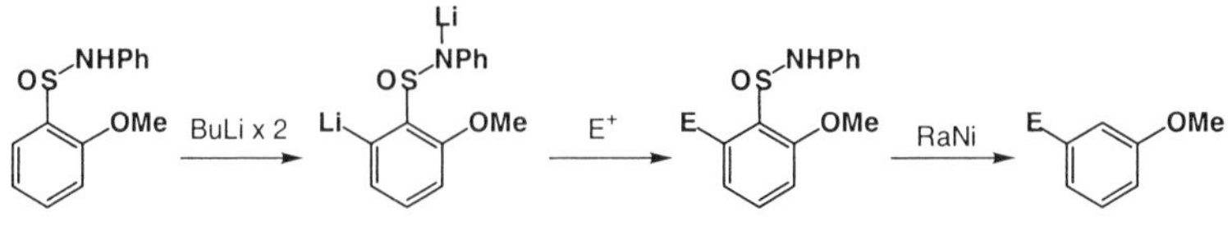

Phosphine oxides are similar: excellent orthodirectors which have seen only limited use so far. The iodide **260**, for example – a precursor of a new class of bisphosphine ligands – can be made by cooperatively-directed ortholithiation of the phosphine oxide **259**, itself derived by halogen-metal exchange from **258**.[222]

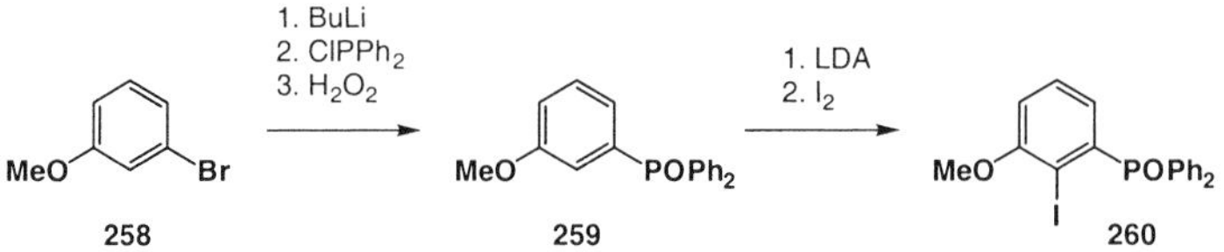

2.3.2.3 N Class

2.3.2.3.1 Aminomethyl groups

Dimethylaminomethyl groups are among the most powerful of the non-acidifying functional groups with regard to directing ability, and reliably direct ortholithiation with *n*-BuLi.[114,223,224] Other dialkylaminomethyl (**261**),[102] and even monalkylaminoalkyl groups (**262**, R = H), which are doubly deprotonated by BuLi–TMEDA,[225] are similarly powerful directors. By contrast, the related amides, carbamates and ureas (**262**, R = COAr, $CONR_2$, CO_2R) usually undergo benzylic α-lithiation (see section 2.2.2). The bias can be shifted towards ortholithiation by additional electron withdrawing substituents on the ring.[226]

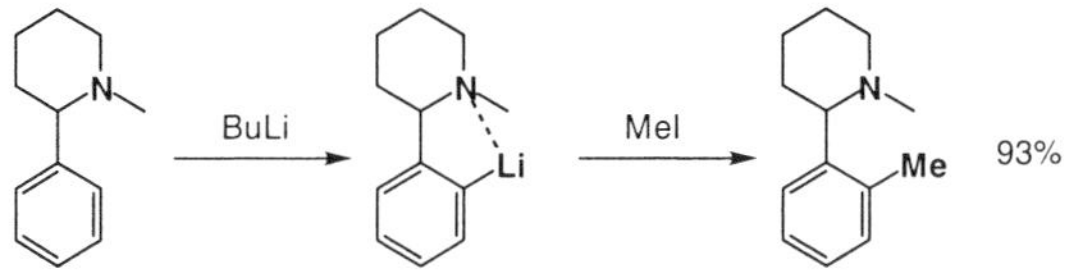

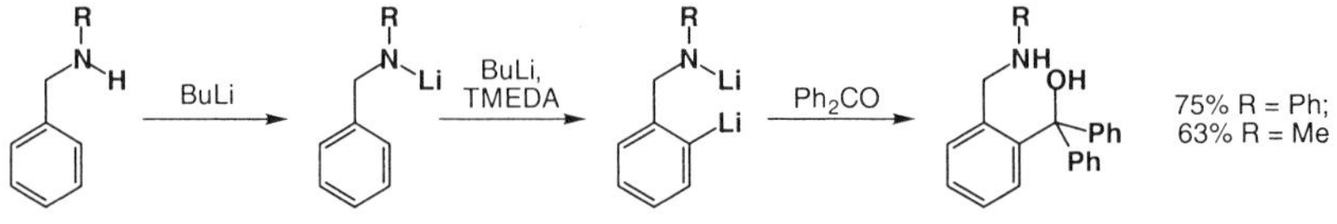

Groups bearing amino substituents even further from the aromatic ring are very weak *ortho*-directors.[102]

2.3.2.3.2 Anilines and isocyanides

Anilines are poor directors of lithiation because the delocalisation of the lone pair into the ring, which lessens the nitrogen's coordinating ability, is not counterbalanced by an increase in acidity of the ring protons (N is not sufficiently electronegative). However, the theme emerging in the N+O class – that anionic substituents are suprisingly good lithiation directors – is echoed for the lithium salts of anilines as shown below. The neutral heterocycle **263** gains a weak acidifying effect from both N and from S, and lithiates poorly *ortho* to both. In the anion **265** derived from heterocycle **264**, the negative charge is delocalised around the rings, and the nitrogen now bears a lone pair perpendicular to the π-system. Even though the entire system must have lessened acidity, ortholithiation is much more efficient because of coordination to this lone pair.[227]

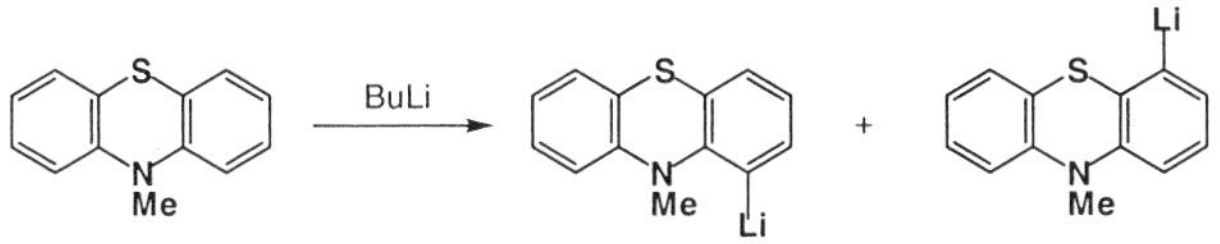

263 BuLi 14% + 16%

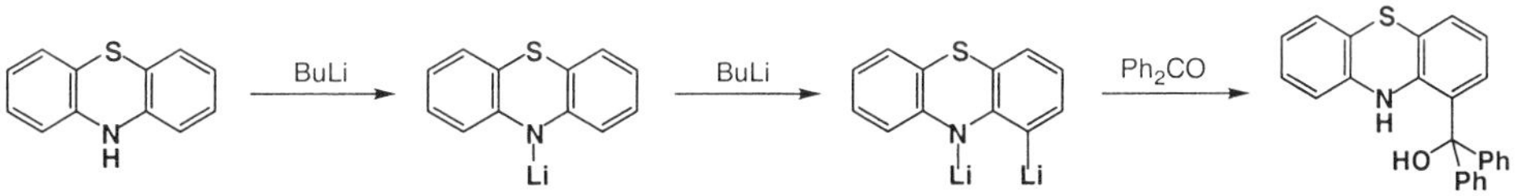

264 BuLi 265 BuLi Ph_2CO 70%

Isocyanides are converted, on addition of *t*-BuLi at the electrophilic carbon atom, to lithioimines – another class of anionic, nitrogen containing function which turn out to have good *ortho*-directing ability.[228] The electrophile reacts at both lithium bearing centres of **266**.

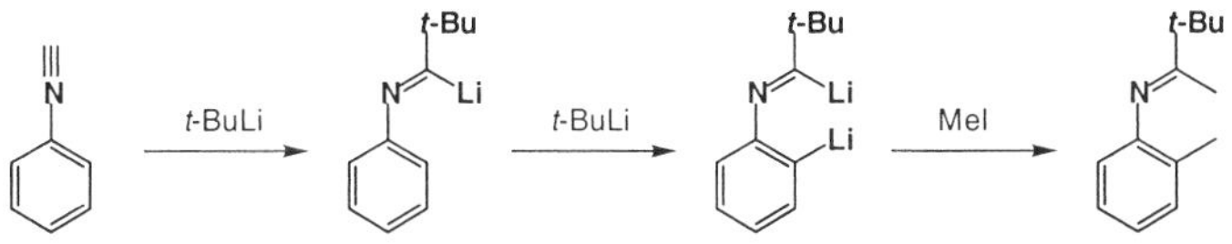

t-BuLi *t*-BuLi 266 MeI

2.3.2.3.3 Imines, nitriles, hydrazones and nitrogen heterocycles

The powerful ability of trigonal (sp^2) nitrogen to coordinate to lithium[229] means compounds of general structure **267** may be ortholithiated, provided addition to C=N can be avoided.

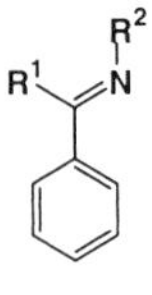

R^1 = H, alkyl, aryl
R^2 = NR_2, alkyl, aryl
or R^1, R^2 = completion of heteroaromatic ring

267

It is just about possible to ortholithiate imines, but in nearly all cases, the side reaction of nucleophilic attack at C=N is at least as important. Ortholithiation can overcome this side-reaction if it is assisted by an additional directing group:[230,231]

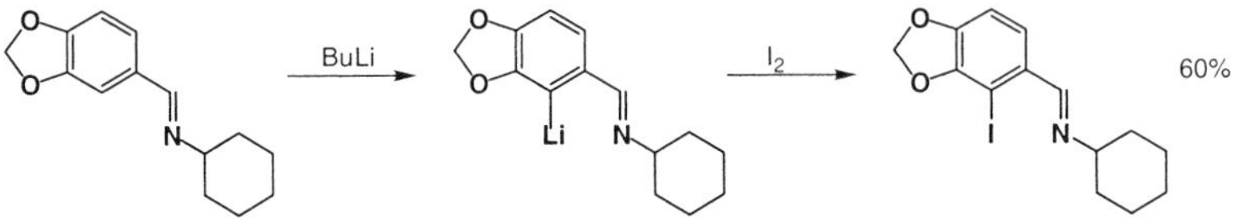

Hydrazones may also direct lithiation, and are particularly effective after deprotonation to an azaenolate **268**.[102]

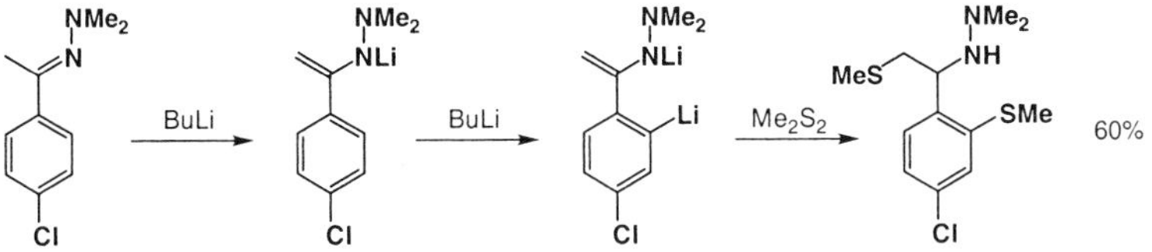

Nitriles will direct lithiation with non-nucleophilic bases such as LiTMP, particularly in conjuction with another nitrile group.[232] The nitriles presumably act by an acidifying effect alone – no intramolecular N–Li coordination is possible in the intermediate.

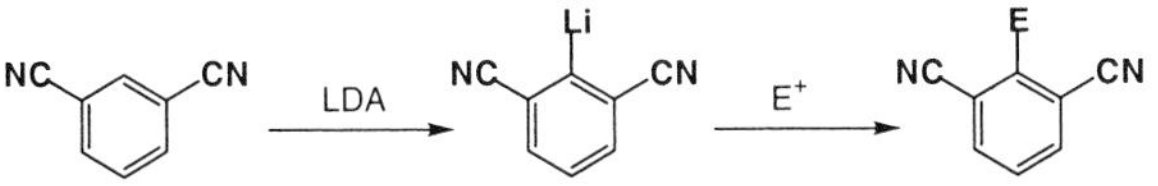

No heterocycle containing a C=N bond is as powerful a director as the oxazolines described above, but their imidazoline analogues **269** direct well if deprotonated to the amidine equivalent **270** of a secondary amide anion.[102] Pyrazoles **271** also direct lithiation, but need protecting with a bulky *N*-substituent to prevent nucleophilic attack by the base.[102]

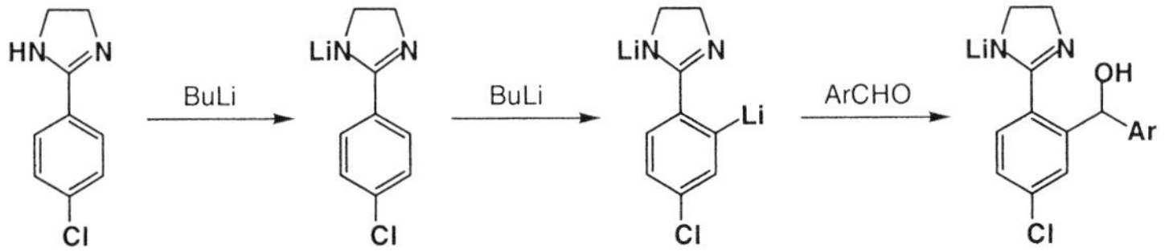

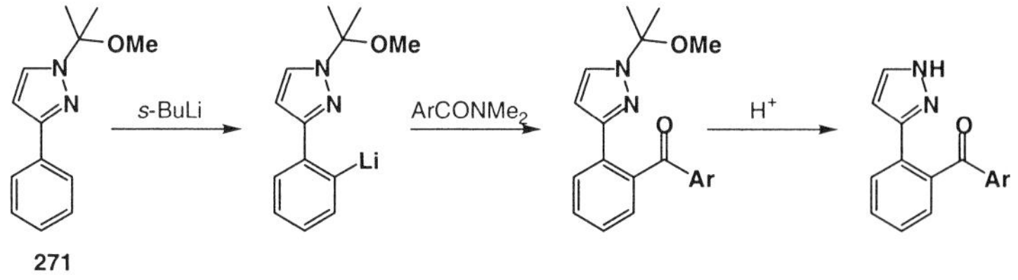

OMe OMe OMe NH N N N N N N N s-BuLi ArCONMe2 H+ Li O O Ar Ar 271

Pyridines are yet more susceptible to nucleophilic attack, but may just about direct lithiation if ring addition can be avoided. Organolithium **272** is unstable and slowly isomerises to **273**.[102]

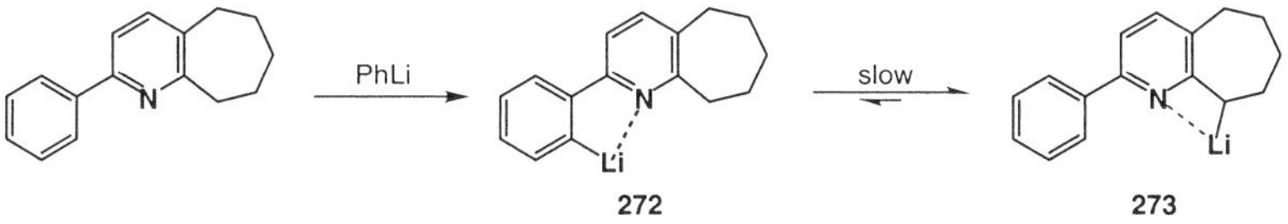

N PhLi N slow N Li Li 272 273

2.3.2.4 O class

2.3.2.4.1 Ethers and alkoxides

Alkyl aryl ethers have a long history in lithiation reactions, and there are detailed reports on the mechanism (see section 2.3.1) and relative efficiency[135] of the lithiation of anisole and its derivatives. Methyl aryl ethers are often lithiated in Et_2O, but THF can be just as effective. The stabilisation afforded to an ortholithiated anisole relative to a *para*-lithiated anisole can be judged from their relative heats of protonation by *s*-BuOH.[233]

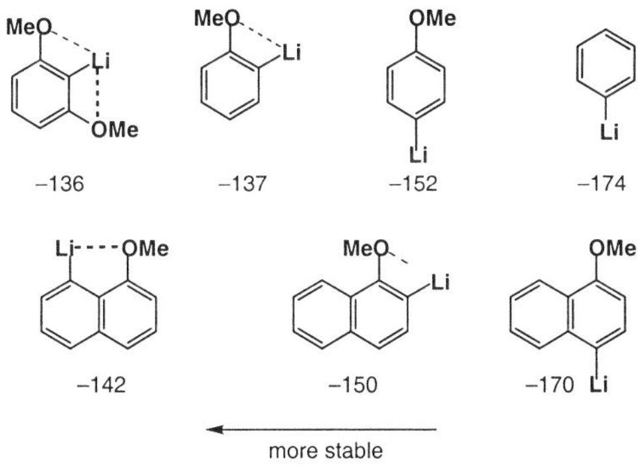

MeO MeO OMe Li Li OMe Li Li –136 –137 –152 –174 Li OMe MeO OMe Li –142 –150 –170 Li more stable

Relative heats of protonation (kJ mol^{-1})[233]

Particularly powerful synthetically is the ability of two *meta*-disposed methoxy groups jointly to direct lithiation to the position between them.[234,235] Lithiation and carboxylation of 1,3-dimethoxybenzene, for example, is a key step in the synthesis of methicillin.[91]

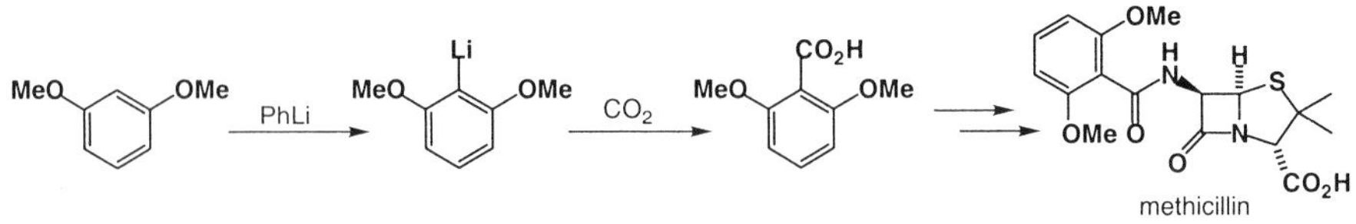

1,3,5-Trimethoxybenzene **274** can be lithiated twice (but not three times) by excess BuLi.[236]

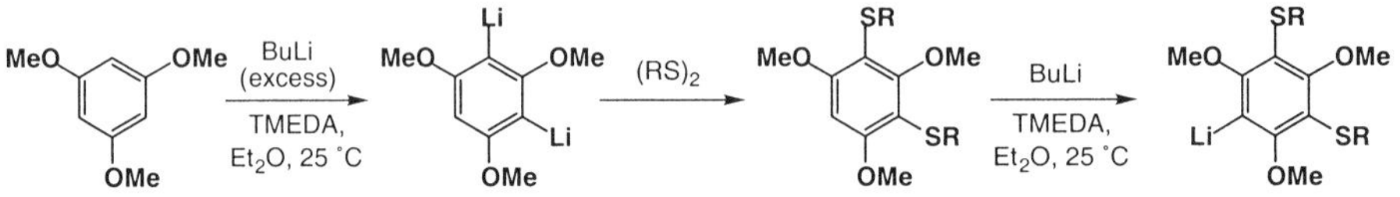

A single methoxy group alone is a relatively weak director, but their small size means methoxy groups are extremely good at biasing lithiation to the position in between themselves and another, better director, leading to 1,2,3-substituted aromatics: there are several examples in this chapter, for example in section 2.3.2.1.3.

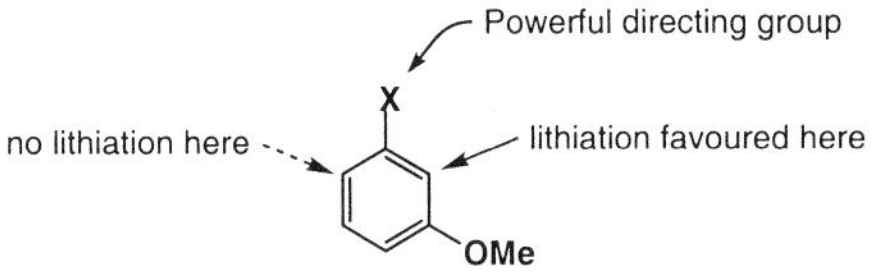

Other alkyl aryl ethers lithiate cleanly - for instance, benzoxepine **275**.[237] Lithiation of a fluoroacetal forms the first step in a route to the drug fludioxinil.

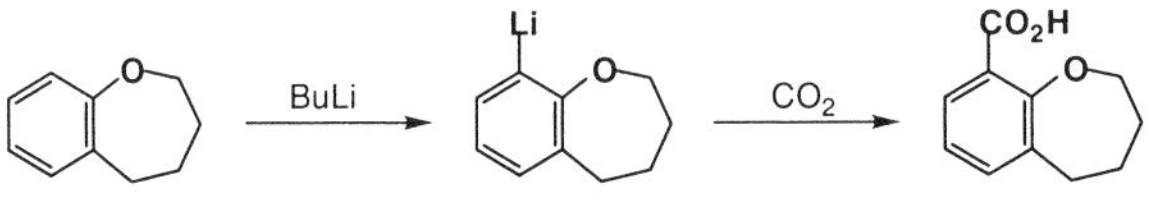

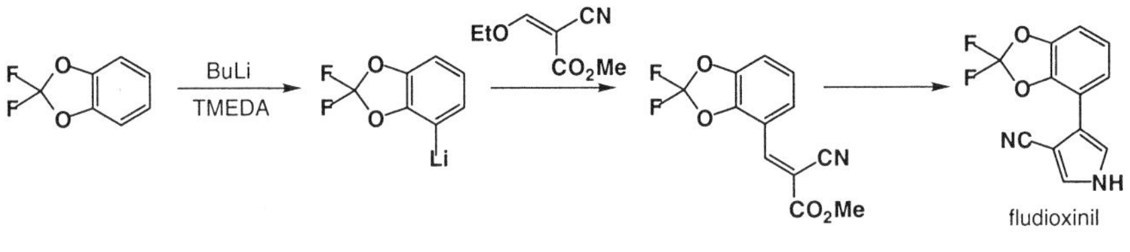

The lithiation of phenols protected as acetals – methoxymethyl acetals like **276** in particular – is especially valuable: the second oxygen supplies a powerful coordination component to their directing effect.[238,239]

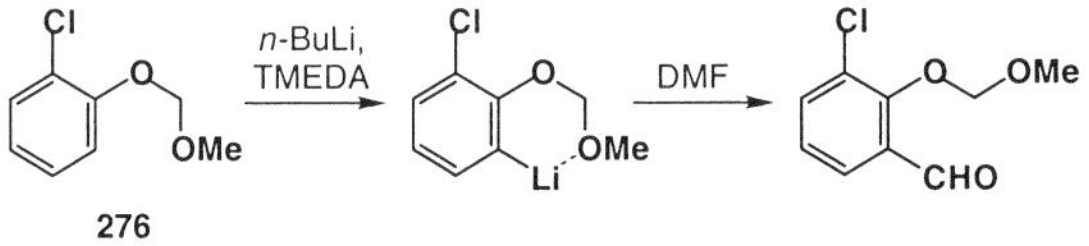

The regioselective lithiation of **277** was used in the synthesis of the pterocarpans 4'-deoxycabenegrins A-I.

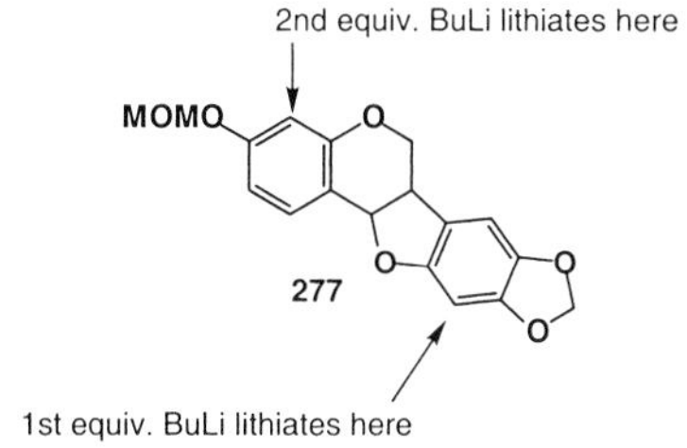

2nd equiv. BuLi lithiates here

MOMO

277

1st equiv. BuLi lithiates here

The reactions below[240,241] illustrate the fact that –MOM acetals lie between –OMe and $-OCON$*i*-Pr_2 groups in their directing ability. Orthogonal deprotection conditions (acid for MOM, base for the carbamate) makes MOM and OCON*i*-Pr_2 a useful pair of directors for the regioselective synthesis of substituted phenols and aryl ethers.

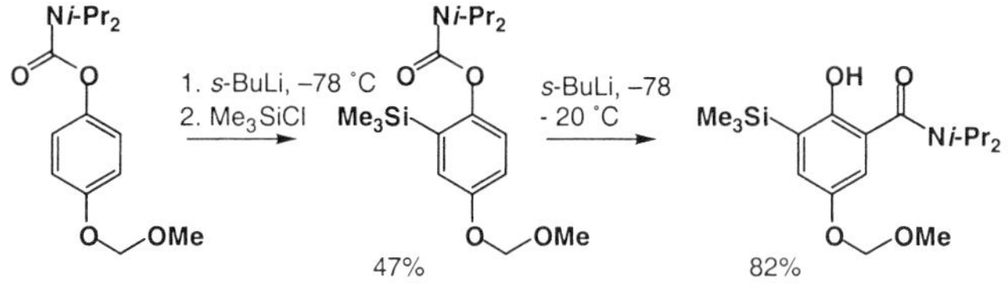

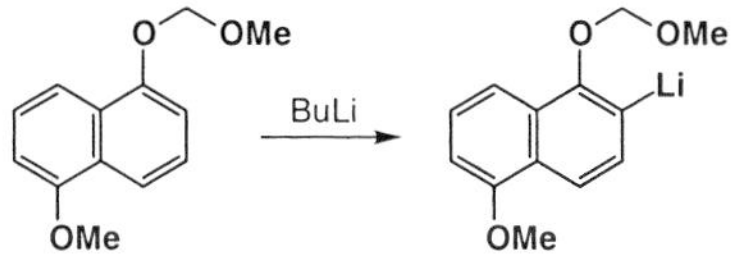

1. *s*-BuLi, –78 °C
2. Me_3SiCl

47%

s-BuLi, –78 - 20 °C

82%

BuLi

A note of warning: both MOM acetals and methyl ethers *ortho* to electron-withdrawing groups – particularly oxazolines, aldehydes, imines and amides – are susceptible to nucleophilic aromatic substitution reactions involving loss of the alkoxy substituent.[104,242,173]

Aryl ethers are less powerful directing groups than alkyl ethers: **278** gives only **279**.[243] Diaryl ethers can usually be lithiated on one (the more acidic) or both rings, according to the amount of alkyllithium employed. Diphenyl ether **280** will give, for example, **281** on quench with a dichlorosilane.

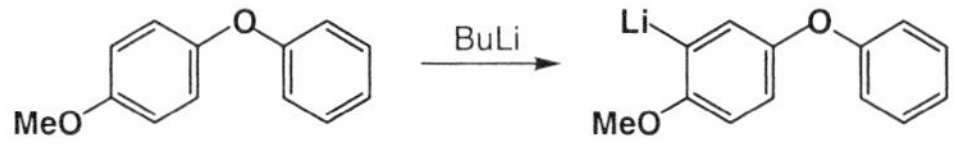

BuLi

278 **279**

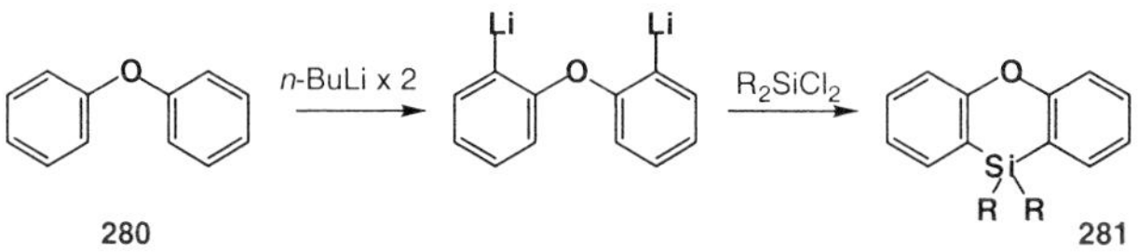

n-BuLi x 2

R_2SiCl_2

280 **281**

Double lithiation of the xanthene derivative **282**[244] is the key step in an important synthesis of the ligand xantphos **283**.[245,246] Benzofuran **284** is also doubly lithiated by excess BuLi/TMEDA.[247,248]

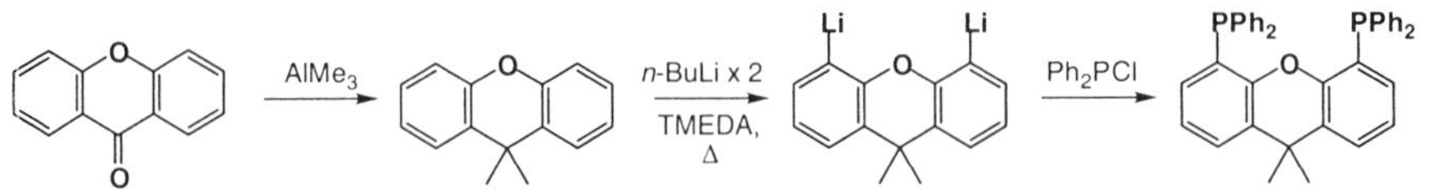

282 xantphos **283**

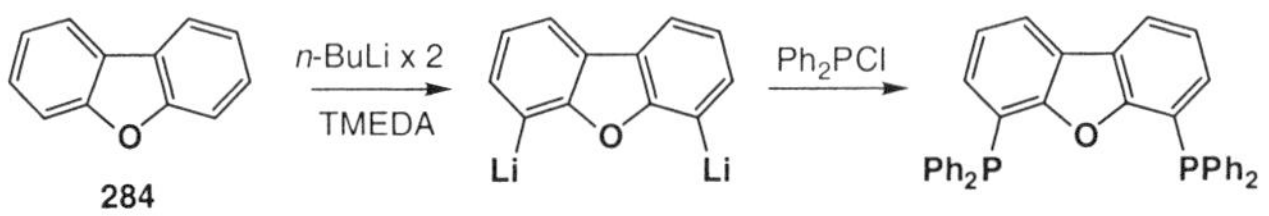

284

Unlike benzylic amines, benzylic ethers undergo benzylic lithiation and 1,2-Wittig rearrangement (section 8.3) too readily to be directors of lithiation.[249] In the absence of benzylic protons, ortholithiation of **285** can occur, for example to give **286**, which undergoes a curious addition to a second phenyl ring, ultimately eliminating methoxide to give 9-phenylfluorenyllithium **287**.[250]

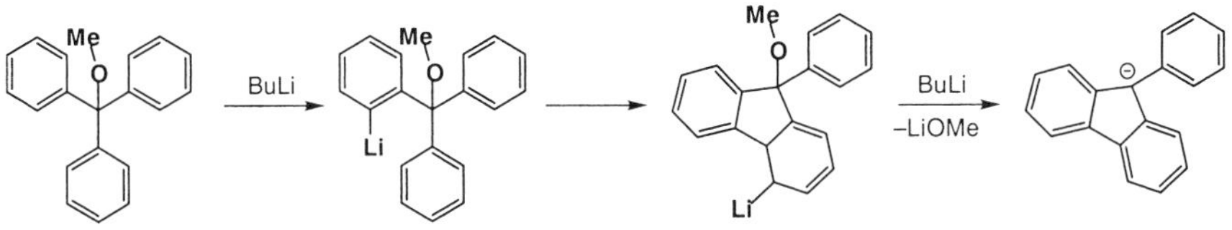

285 **286** **287**

Free phenols (as their phenoxides) have a weak, and synthetically useless, directing effect,[251,129,128] but the alkoxides of benzylic alcohols can be useful as lithiation directors (see below),[252,107] especially when assisted by a second director:[253]

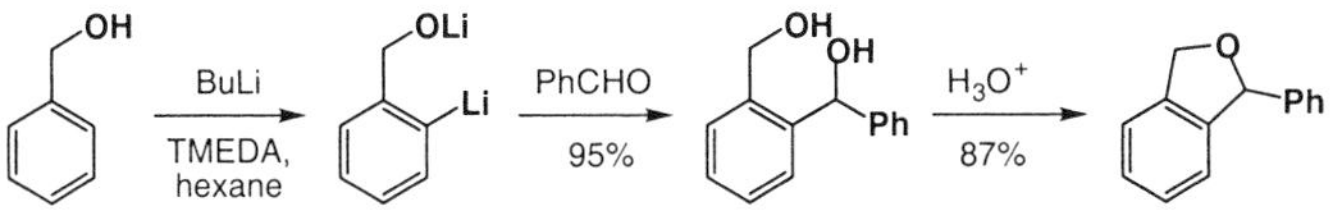

2.3.2.4.2 Ketones, Esters and Carboxylates

Lithium amides (LiTMP) can give good yields of products resulting from ortholithiation of both ketones and lithium carboxylates. Benzophenone **288**, for example, with LiTMP gives a good yield of the ortholithiation–dimerisation product **289**.[103] In general, however, ketones and aldehydes are best lithiated by the method developed by Comins described in section 2.3.2.1.2.

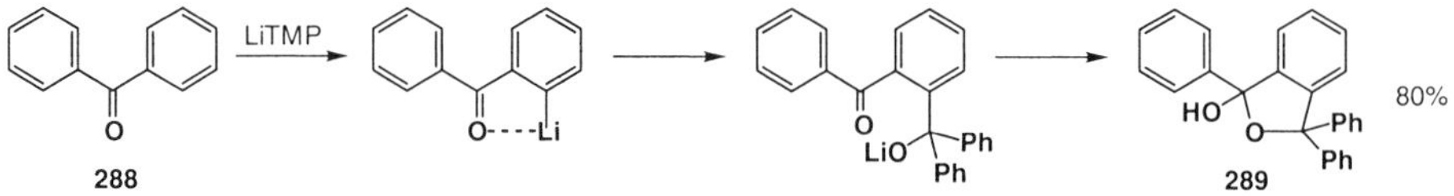

The lithiation of lithium carboxylates was believed to be impossible by Gschwend and Rodriguez[102] because of competing attack at C=O by an organolithium. Carboxylates are insufficiently strong orthodirectors to be lithiated by LDA, but provided temperatures are kept low, BuLi lithiation without nucleophilic addition is possible. Treatment of **290** with 2.2 equivalents of BuLi at –90 °C in the presence of TMEDA leads to good yields of ortholithiated product **291**.[254,255] In terms of their directing power, $-CO_2Li$ is weaker than those in the N+O class, but among the strongest of the O class.[256]

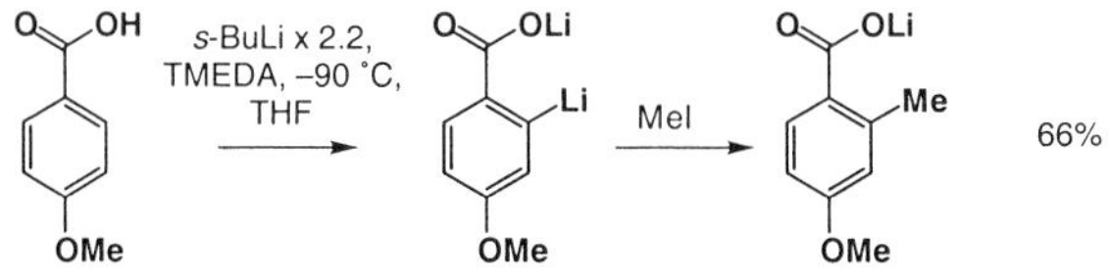

In the pyridinecarboxylic acid series, the greater acidity of the ring protons means that LDA or LiTMP can be used for metallation; all three pyridinecarboxylic acids **292-294** are lithiated in good yield.[257]

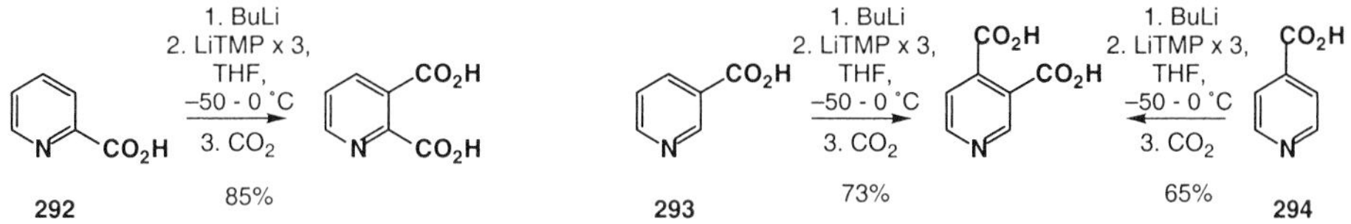

The thiophenecarboxylic acid **295** is lithiated *ortho* to the carboxylate group by BuLi,[90] but α to the ring sulfur by LDA[258] – presumably a case of regioselectivity determined kinetically (with BuLi – better coordination to O than S) or thermodynamically (with LDA – acidification of positions α to S).

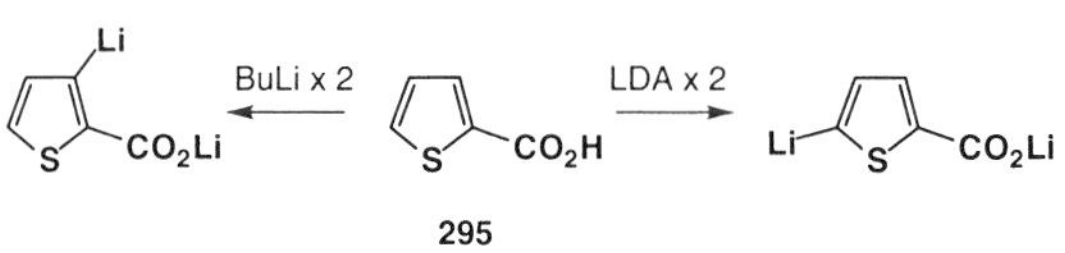

Even hindered esters can be ortholithiated, provided the electrophile is present *in situ* during the lithiation. Neopentyl ester **296** gives, on treatment with LDA in the presence of triisopropylborate, the boron derivative **297**, which can be deprotected and used in Suzuki coupling reactions.[259]

LDA, $B(O\textit{i}\text{-}Pr)_3$, –78 °C

296 → 297 ($B(O\textit{i}\text{-}Pr)_2$) 90%

2.3.2.5 X class

Fluorine acidifies[260] an *ortho* proton rather more than chlorine or bromine, whose weak directing effects are more or less comparable.[261] Deprotonation of fluorobenzene is feasible in THF, and at temperatures below –50 °C the lithiated species **298** is stable and does not collapse to a benzyne.[262,147]

n-BuLi, –60 °C → 298 → CO_2 → 60%

The trifluoromethyl group behaves in a similar way,[235] but it is now clear that deprotonation of sites *ortho* to such acidifying but non-coordinating and non-electrophilic substituents is best carried out with BuLi–KO*t*-Bu superbases (see section 2.6). A combination of BuLi metallation and superbase metallation of fluoroarenes has been used in the synthesis of components **299** and **300** for fluorinated liquid crystals:[263]

KO*t*-Bu, BuLi THF, –100 °C → Li/K → I_2 → 299 51%

BuLi, THF, –70 °C → Li → 1. $B(OMe)_3$ 2. H_2O_2 → 300 50-60%

Chlorobenzene is lithiated more slowly[261] and cannot be lithiated completely at temperatures where benzyne formation is slow.[102] With 1,2,3,4-tetrachlorobenzene **301**, MeLi leads to ortholithiation; *t*-BuLi on the other hand leads to halogen-metal exchange.[264]

MeLi ← 301 → *t*-BuLi

Successful lithiation of aryl halides – carbocyclic or heterocyclic – with alkyllithiums is however the exception rather than the rule. The instability of ortholithiated carbocyclic aryl halides towards benzyne formation is always a limiting feature of their use, and aryl bromides and iodides undergo halogen-metal exchange in preference to deprotonation. Lithium amide bases avoid the second of these problems, but work well only with aryl halides benefitting from some additional acidifying feature. Chlorobenzene and bromobenzene can be lithiated with moderate yield and selectivity by LDA or LiTMP at –75 or –100 °C.[261]

Meyers has exploited both the lithiation-directing and benzyne-forming abilities of a chloro substituent to form the benzyne **304** from oxazoline **302**. Excess organolithium adds regioselectively to the 2-position of the benzyne (probably directed by coordination to the oxazoline) to give **305**.[265,266]

RLi, RLi, E^+

302 303 304 305

2.3.3 Ortholithiation of Aromatic Heterocycles

2.3.3.1 Electron-deficient heterocycles

Electron-deficient six-membered aromatic heterocycles are distinctly more acidic than their carbocyclic analogues, and not only are LDA and other lithium amides capable of deprotonating them in good yield,[175] but they are in many cases also a necessity since alkyllithiums prefer addition to the electron-deficient ring over deprotonation. Even when heterocycles can be successfully lithiated, the addition of the organolithium product to remaining starting material may itself pose a problem – as in the attempted lithiation of pyridine with LDA at –70 °C, which gives a 50% yield of bipyridine **308** via **307**.[267]

LDA, HMPA, Et_2O, –70 °C; H_2O, [O]

307 308

Lithiation of pyridine itself (and quinoline) is best carried out using a combination of BuLi and the lithiated amino alcohol LiDMAE **309**.[268]

2 equiv. BuLi, LiDMAE, hexane, –78 °C, 1 h; E^+; 25-80%

306

LiO NMe$_2$

LiDMAE **309**

Bipyridines can be lithiated with LiTMP (if only in moderate yield), with one nitrogen atom directing the lithiation of the other ring.[269] Both 2,2'- and 2,4-bipyridines **310** and **311** can be lithiated in this way:

LiTMP, THF, –45 °C

310

LiTMP, THF, –45 °C

311

Halogen substituents are of course easy to introduce to heteroaromatic rings, and they also enhance the acidity of the ring protons. *n*-BuLi will for example lithiate the tetrafluoropyridine **312** at –60 °C in ether,[270,271] but with pyridine itself it leads to addition/reoxidation products.[272] Addition to the ring is the major product with 2-fluoropyridine **313**, though some metallation can be detected; selectivity in favour of metallation is complete with LDA in THF at –75 °C[273,274] or with phenyllithium and catalytic *i*-Pr_2NH at –50 °C.[275] Similar results are obtained with quinolines.[276]

n-BuLi, –60 °C

CO_2

312

LDA, THF, –75 °C or PhLi, cat. *i*-Pr_2NH, THF, –50 °C

Me_3SiCl

313

n-BuLi, –60 °C

E^+

With even more electrophilic heterocycles, addition of the lithiated species itself to the starting material can become a problem – for example, LDA will lithiate pyrimidine **314** at –10 °C, but the product, after work up, is the biaryl **315** resulting from ortholithiation and readdition. By lithiating in the presence of benzaldehyde, a moderate yield of the alcohol **316** is obtainable.[277] Strategies for the lithiation of pyrimidines and other very electrophilic heterocycles are discussed below.[278]

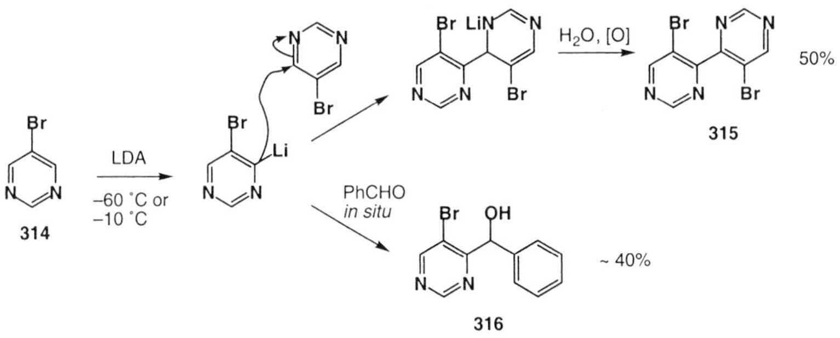

Because lithiation with lithium amides leads to the thermodynamically favoured product, regioselectivity in the lithiation of electron-deficient heterocycles is often determined more by anion stability than by the rate of deprotonation – in other words, the coordination aspect of lithiation may take second place to the acidifying effect of the substituents.[279] This has an important consequence for the regiochemistry of lithiation, because while the nitrogen lone pair of pyridine may on occasions act as a directing group by coordinating to Li, it also has a destabilising effect on an anion formed at the adjacent carbon because of repulsion between the lone pair and the C–Li bond. The relative stabilities of organolithiums at the 4-, 3- and 2-positions on a pyridine ring are indicated by the relative acidities of the protons: 700:72:1.[280]

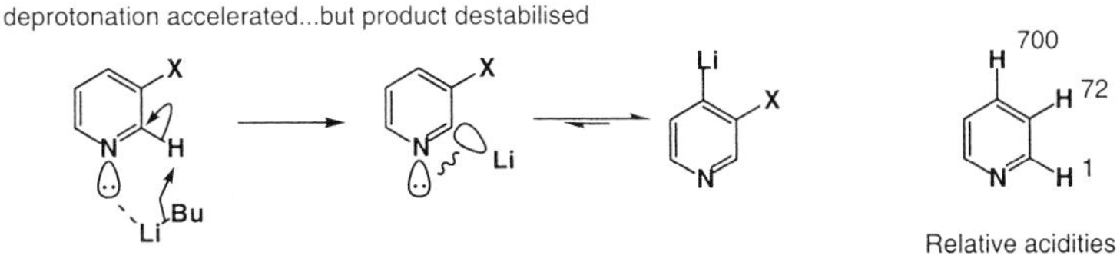

Overall, the most stable pyridinyllithiums are those bearing Li at the 3- or 4-positions; 2-pyridinyllithiums may on occasion be formed faster, but given a proton source they will readily isomerise to 4-pyridinyllithiums. The deprotonation of 3-fluoropyridine **317** at the 4-position with Schlosser's BuLi–KO*t*-Bu ("LiC-KOR") superbase[281] illustrates this.[282] In contrast, BuLi–TMEDA in Et_2O leads mainly to 2-lithiation. BuLi–TMEDA in THF, on the other hand, gives 4-metallation, showing how closely balanced these two factors are.

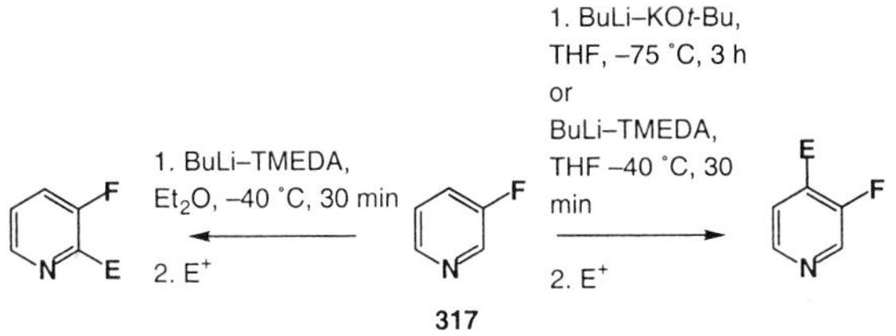

With 3-trifluoromethylpyridine **318**, BuLi in ether also lithiates cleanly at the 2-position.[283]

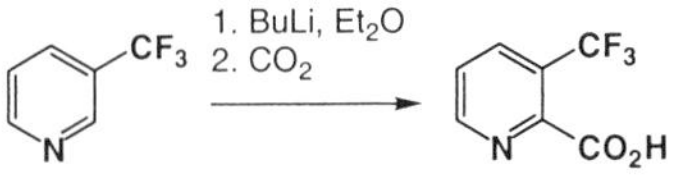

318

Blocking either the 2- or 4-position allows lithiation at the other: LDA or BuLi can be used to 4-lithiate the 2-blocked 3-fluoropyridines **319**,[284] **320**[285] and **321**,[286,287] and BuLi 2-lithiates the 4-blocked 3-fluoropyridine **322**.[284,288,289] Note the selectivity for removal of the acidified pyridine protons in **321** and **322** over the coordination-activated ones adjacent to the pivalanilide group. A double orthodirecting effect ensures 4-lithiation of **323**.[290]

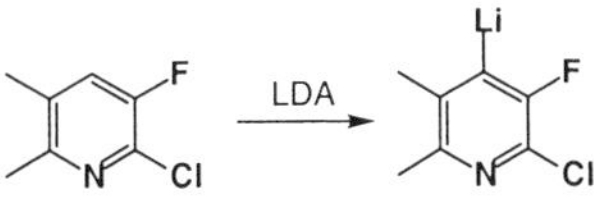

319

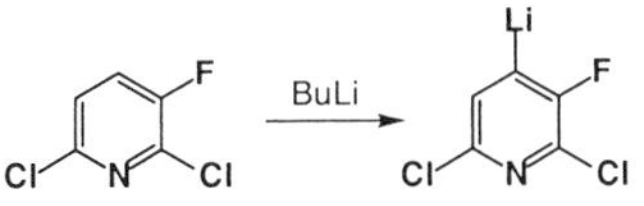

320

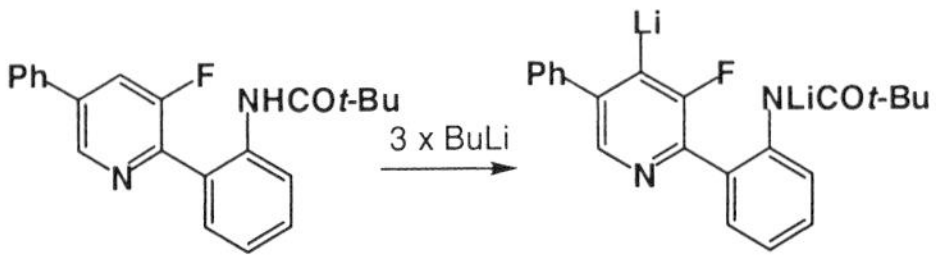

321

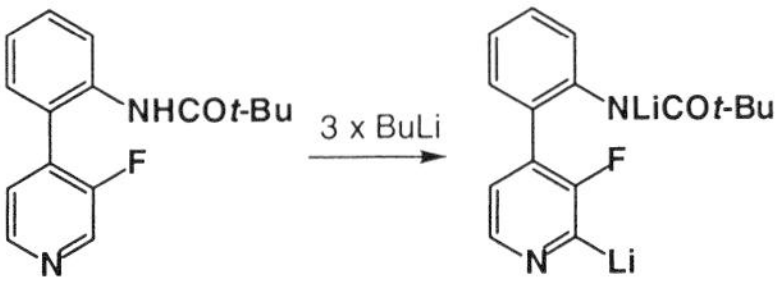

322

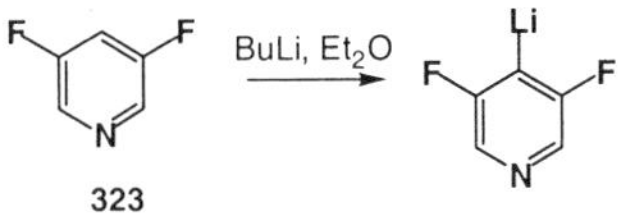

323

Fluoropyridines form valuable starting materials for a range of disubstituted pyridines because after lithiation, nucleophilic substitution of fluoride[274] (sometimes via the *N*-oxide) can be used to introduce, say, N or O substituents as in **324** and **325**. Subsequent annelations can allow complex polycyclic heteroaromatics to be constructed.[175]

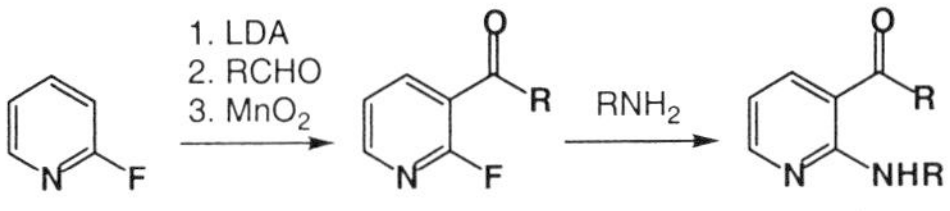

324

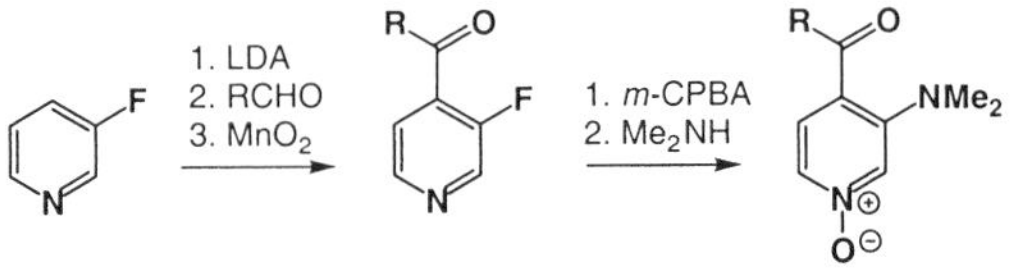

325

Chloropyridines behave largely similarly to fluoropyridines: they are lithiated by LDA (or in some cases BuLi, though addition to the pyridine ring may pose problems).[175,279] 3-Chloropyridines such as **326** are readily 4-lithiated,[291,292] and 2-chloropyridines **327-329** can be 3-lithiated with LDA or PhLi/cat. i-Pr_2NH.[275,293,260] Chloropyridines **328** and **329** are lithiated *ortho* to Cl rather than CF_3 or even the usually more strongly directing MeO.

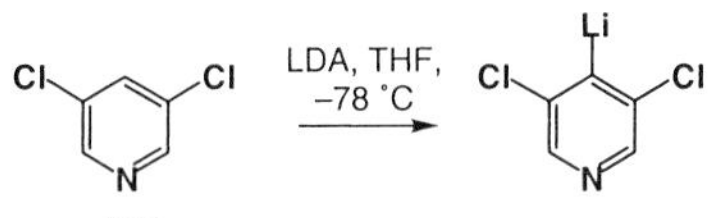

Since chlorine is less electronegative than fluorine, the role of the inherent relative acidities of the positions on the pyridine ring become more important in chloropyridine lithiations.[294] Not surprisingly, **330** is lithiated cleanly in the 4-position rather than the less acidic 3-position.[295] When **331** is lithiated with BuLi, the preference for 4-lithiation wins out even though the intermediate lithium lacks an adjacent acidifying Cl atom. By contrast, LDA lithiation of this compound under thermodynamic control gives the expected regioselectivity.[293] A similar effect is observed in the lithiation of **329** with BuLi and with LDA.[260]

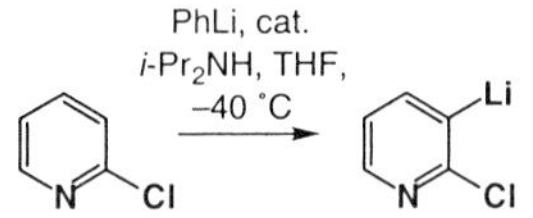

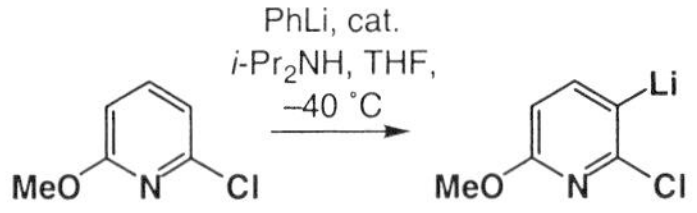

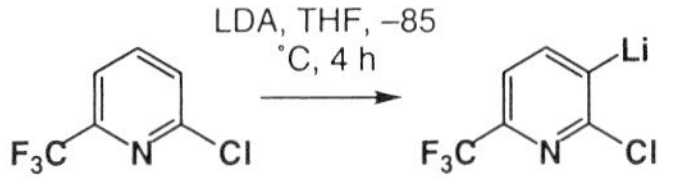

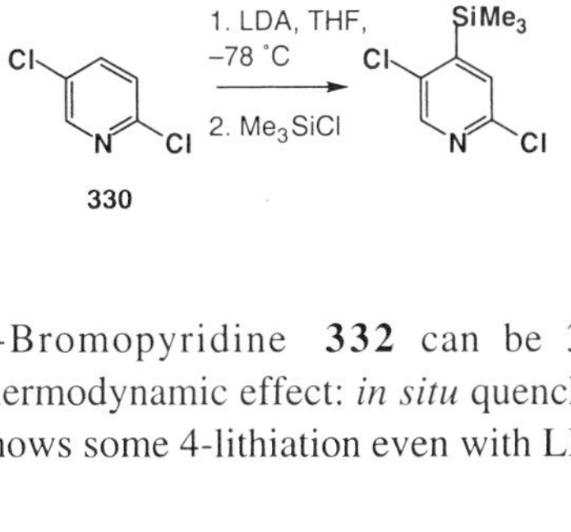

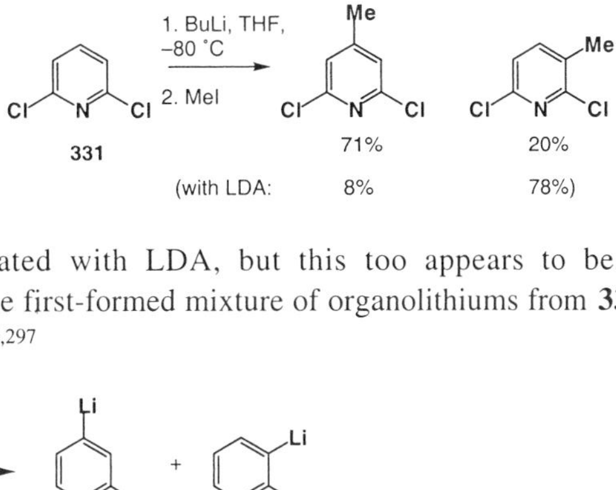

2-Bromopyridine **332** can be 3-lithiated with LDA, but this too appears to be a thermodynamic effect: *in situ* quench of the first-formed mixture of organolithiums from **332** shows some 4-lithiation even with LDA.[296,297]

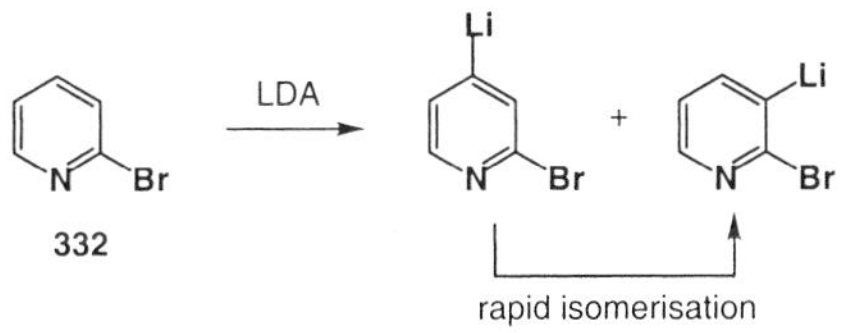

4-Bromopyridines **333**[298] and 3,5-dibromopyridines **334**[299] lithiate quite normally with LDA, as does 3-bromopyridine **335** provided the temperature is kept below –78 °C.[300] At –100 °C, or with *in situ* Me_3SiCl, evidence that the 2-lithiopyridine **337** is an intermediate in the formation of **336** can be obtained.[296,297]

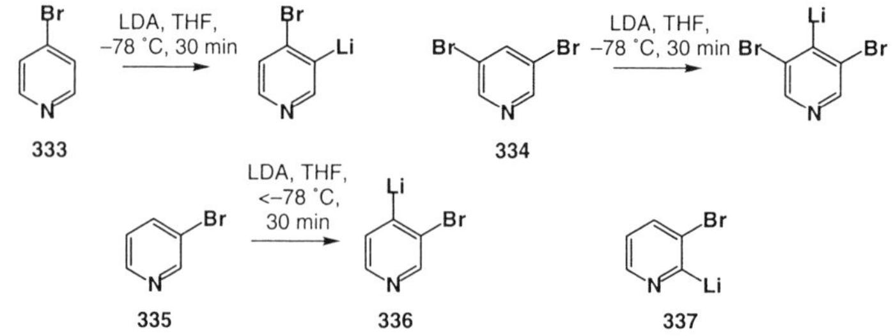

Lithiation of 3-bromopyridine at higher temperatures leads to a mixture of products of which the major (**342**) has the bromine at the 4-position and is clearly formed by rearrangement of **338** to **341** via halogen-metal exchange, probably through formation of a catalytic amount of 3,4-dibromopyridine **340**.[301]

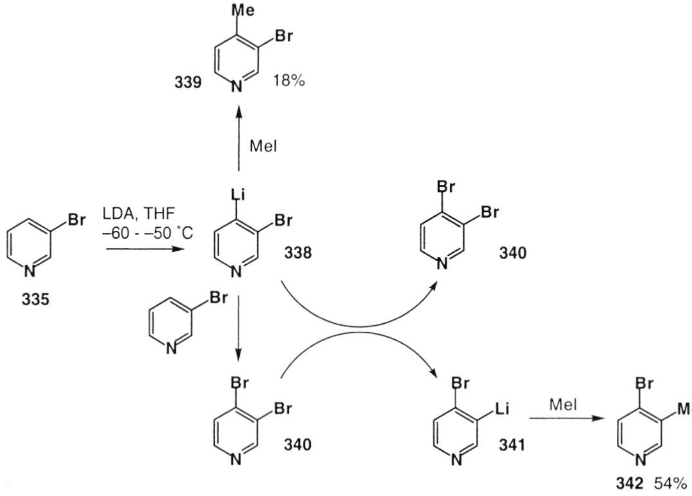

Halogen-metal exchange processes can also isomerise a less stable 2-lithiopyridine to a more stable 4-lithiopyridine. In some cases, this leads to complex mixtures of regioisomers.[175] However, these rearrangements (known as "halogen-dance" reactions) can be put to good use in the synthesis of unusual substitution patterns of bromo and iodopyridines.[302] For example, kinetic lithiation of **343** takes place at C-2, but rapid isomerisation by reversible iodine-lithium exchange ensues, and products are isolated resulting from electrophilic quench of **344**.[302]

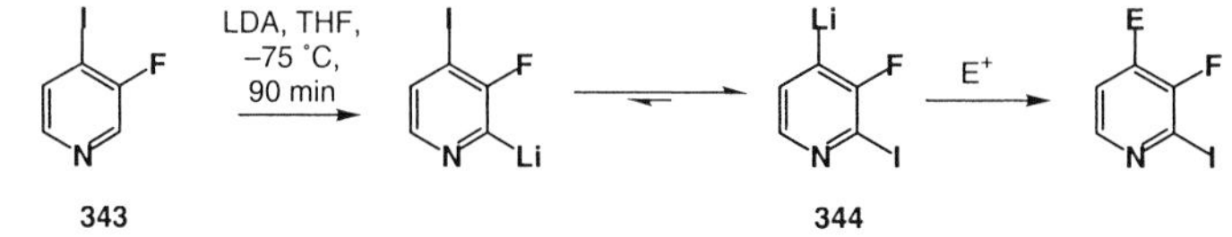

Metallation of simple iodopyridines fails,[300] but iodopyridines containing additional halogens metallate and rearrange to place the lithium between the two directing groups, as shown in the scheme below.[302]

LDA, THF, –75 °C ; E^+

2-Fluoro-3-iodopyridine **345** was used by Comins as the starting material in a synthesis of mappicine.[303] The first formed organolithium **346** rearranges to **347** and a quench gives **348**. Iodine-lithium exchange allows introducton of the C-4 hydroxypropyl substituent of **349**.

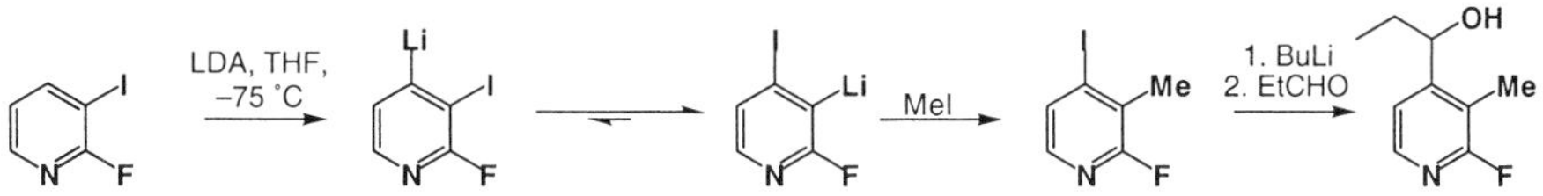

LDA, THF, –75 °C ; MeI ; 1. BuLi 2. EtCHO

345 346 347 348 349

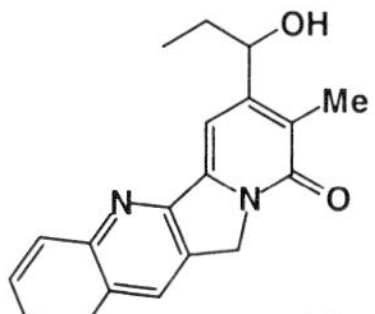

mappicine

Iodopyridines bearing other lithium-stabilising groups behave similarly,[304] and in the case of the amide **350** the intermediate 4-lithio species **351** was detectable by *in situ* silylation.[305]

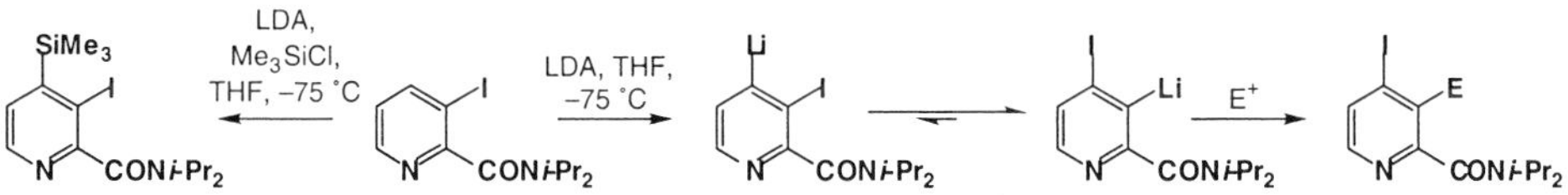

LDA, Me_3SiCl, THF, –75 °C ; LDA, THF, –75 °C ; E^+

350 351

A similar bromine-lithium exchange mediated isomerisation of 2-lithiopyridine **352**[306] was used in a synthesis of atpenin B (see chapter 9).[307]

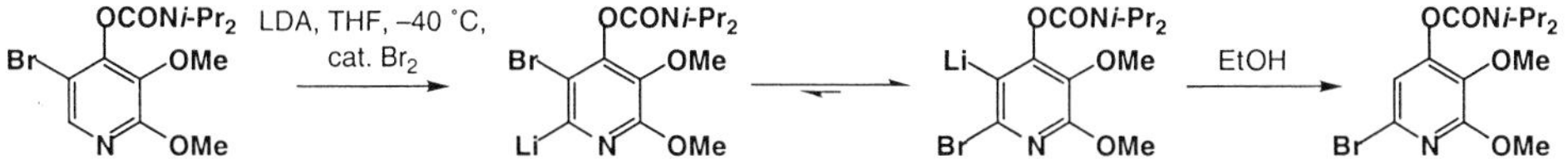

LDA, THF, –40 °C, cat. Br_2 ; EtOH

352

Methoxypyridines are much more resistant than halopyridines to addition to the heterocyclic ring and they may be lithiated with alkyllithiums.[308,309] They generally show unremarkable selectivities:[310] in the case of **354** the 2-lithio species – presumably the kinetic product – is preferred, and a similar reaction was used in the synthesis of the natural product UK-2A.[311] The lithiation of **353** with PhLi was a key step in the synthesis of caerulomycin C.[306]

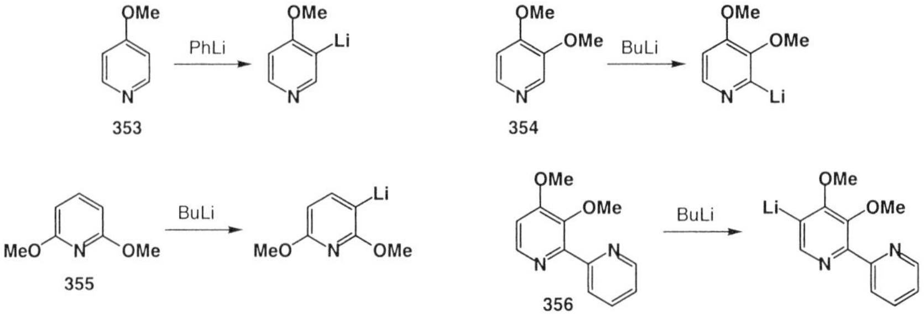

The delicate balance between 2- and 4-lithiation of 3-substituted pyridines is again illustrated clearly by the way that MeLi lithiates **357** at C-4, while BuLi–TMEDA lithiates it at C-2.[175] More strongly coordinating directing groups (such as OMOM) direct even more strongly to C-4 because they rely less on coordination to the nitrogen atom.[175] Base-dependent selectivities are evident when OR and Cl are in competition as directing groups on a pyridine ring.[279,312,313]

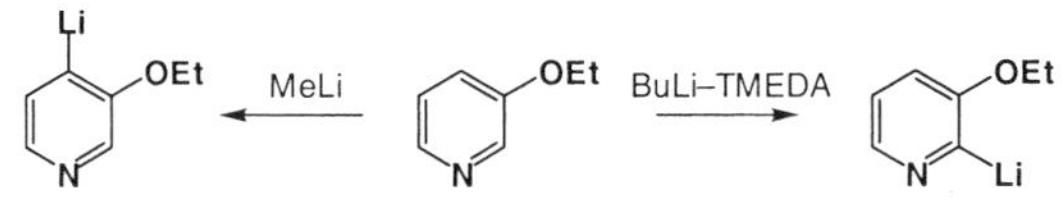

On deprotonation, pyridones and quinolones carry an oxyanion substituent which turns out to be a good director of lithiation. Two equivalents of LDA lead to the compounds **358-360**.[297]

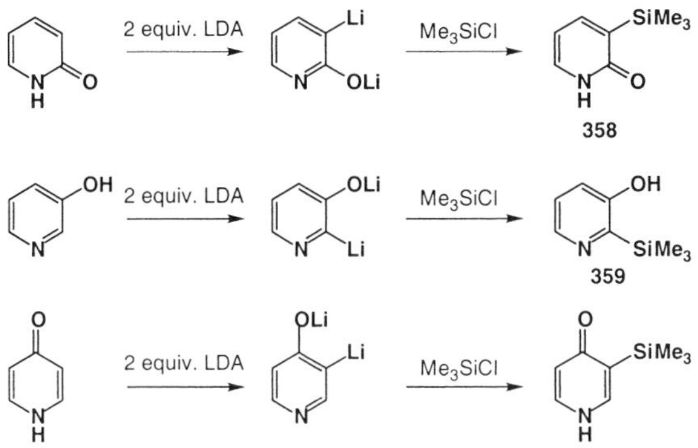

2-Quinolones behave similarly, and unlike quinolines are not attacked by BuLi at the ring.[314] *N*-substituted pyridones and quinolones are lithiated, sometimes on the *N*-substituent, sometimes *ortho* to N and sometimes *ortho* to O.[279] By way of example, the 4-pyridone **361**

lithiates α to N in the ring, while in the 2-pyridone **362** the combined lateral directing effect of the O and the α-directing effect of N lead to lithiation on the side chain.[315]

BuLi, –78 °C
361
BuLi, –78 °C
362

Closely related are the pyridine-*N*-oxides, which are readily lithiated in the 2-position,[175,32] even in examples such as **363** where there is a powerful competing directing group.[316] Lithiation of the bipyridine-*N*-oxide **364** was used in a synthesis of the caerulomycins and collismycins.[317]

LDA, THF, –75 °C
363
LDA, THF, –7 °C
364

N-activation by BF_3 can be used to promote ortholithiation of pyridine itself,[318,319] though not quinoline and isoquinoline. Temporary formation of a pyridine-hexafluoroacetone adduct **366** achieves the same result.[320] Complexation of pyridine by chromium tricarbonyl also makes the ring protons sufficiently acidic to be removed by LDA.[321]

BF_3 → **365** LiTMP →

$(CF_3)_2C=O$ → **366** LiTMP →

More powerful directing groups such as those based on amides and sulfonamides are successful with pyridines as with carboxylic rings, and will not be discussed separately. The enhanced acidity of pyridine ring protons makes the simple carboxylate substituent an ideal director of lithiation in pyridine systems.[257] The pyridinecarboxylic acids **367-369** are deprotonated with BuLi and then lithiated with an excess of LiTMP: all the substitution patterns are lithiated; nicotinic acid **368** is lithiated in the 2-position. The method provides a valuable way of introducing substituents into the picolinic, nicotinic and isonicotinic acid series.

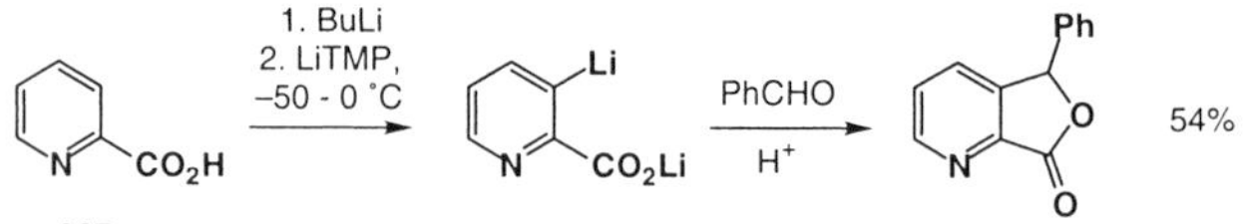

Another directing group whose value and effectiveness is greatly enhanced in the pyridine series is the α-aminoalkoxide group derived by addition of a lithioamine to an aldehyde[186] (see section 2.3.2.1.2). Aldehyde **370** was converted to **371**, and thence to an intermediate **372** in the synthesis of the alkaloid maxonine, by this method.[322]

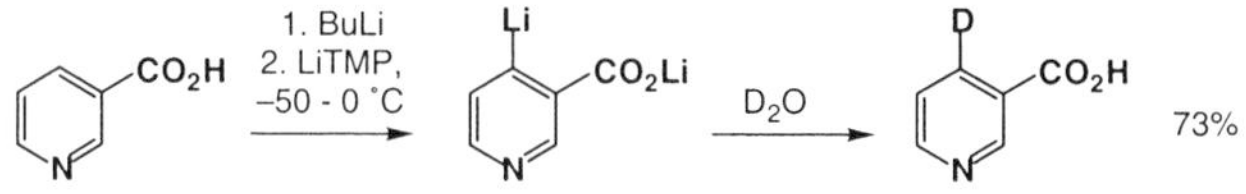

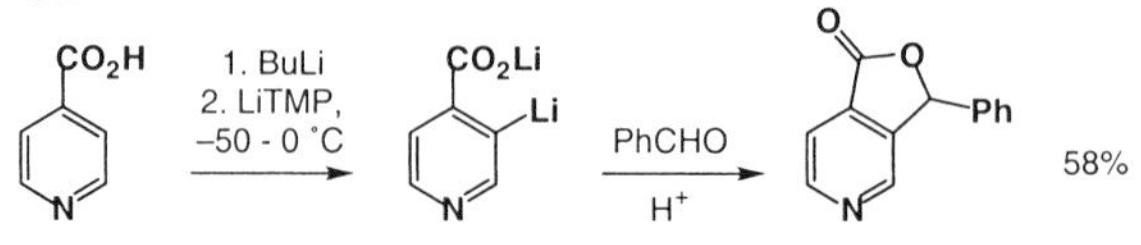

Halogenated *quinolines* very often undergo nucleophilic addition with BuLi, so must be lithiated with LDA.[175,279] The regioselectivity of their reactions parallels that of the pyridines, and similar halogen migrations are observed:[323,324]

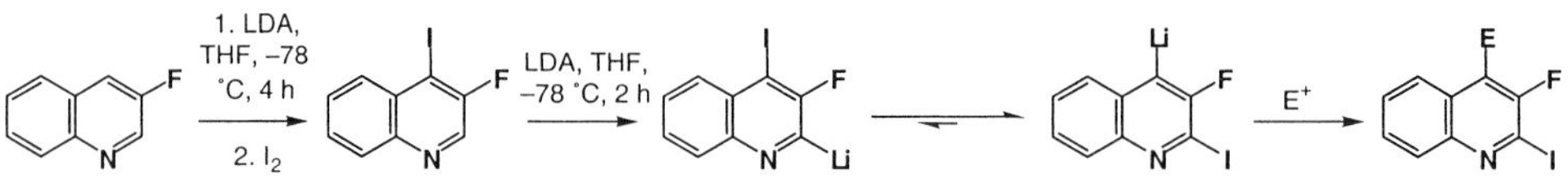

Fluoropyrimidines[325,326] and fluoropyrazines[327] must also be metallated with LDA or LiTMP to avoid ring addition. Replacing F with CF_3 also allows metallation, but side-reactions are more of a problem:

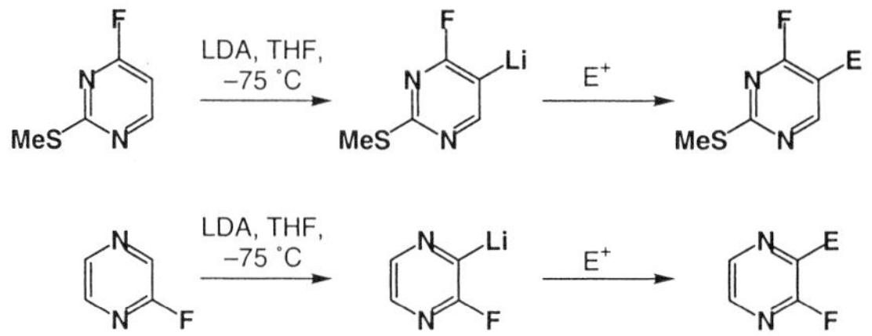

Certain chloropyrimidines can be metallated even with BuLi,[325,278] and the metallation of chloropyrazines[328] is useful in the synthesis of flavouring compounds[329] and the antiarrhythmia drug arglecin.[330]

The protons of pyrimidines, pyrazines and pyridazines are relatively acidic[278] even without halogen activation, and the three simple heterocycles **373-375** have been lithiated (with varying success) with LiTMP.[331]

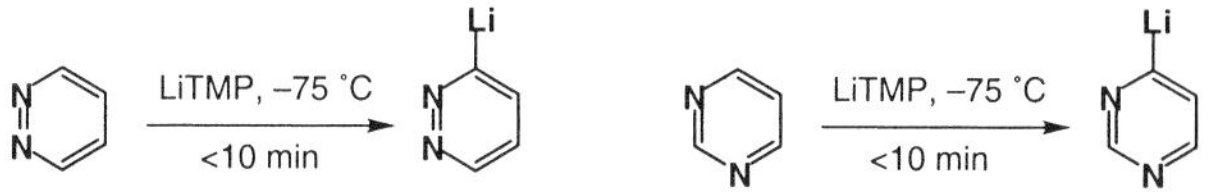

LiTMP, −75 °C, <10 min

373 **374** 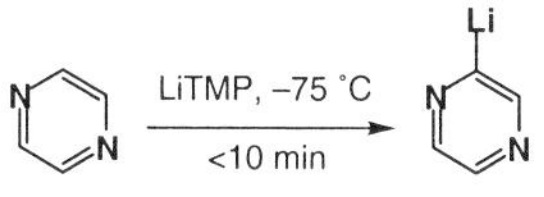 **375**

This acidity means that even iodopyrimidines and iodopyrazines may be lithiated because hindered, non-nucleophilic lithium amide bases will deprotonate them. For example, the base **377**, which is easily made by BuLi attack on the imine, deprotonates **376** α to N rather than *ortho* to I,[332] and the lithiation of **378** with LiTMP is also successful.[332]

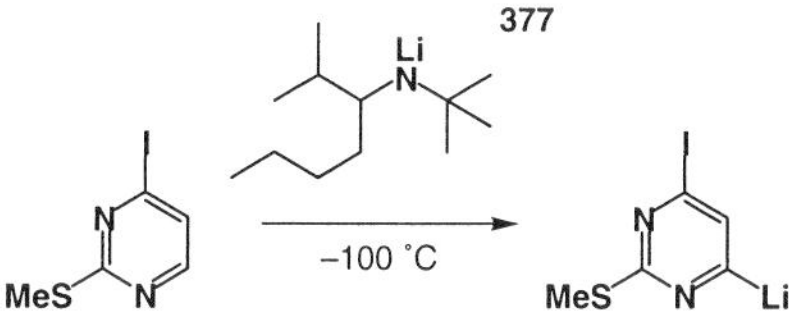

377 −100 °C **376**

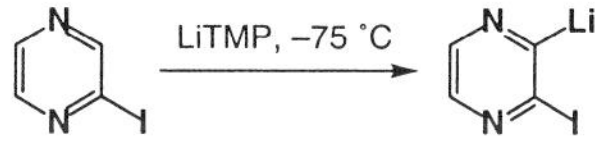

LiTMP, −75 °C **378**

Comparing the ortholithiation of **376** with the ortholithiation of the methoxypyrimidine **379** – the key step in a synthesis of an analogue **380** of bacimethrin[333] – contrasts the directing effects of I and MeO. This substitution pattern reacts well,[334] as does the symmetrical **381**.[335]

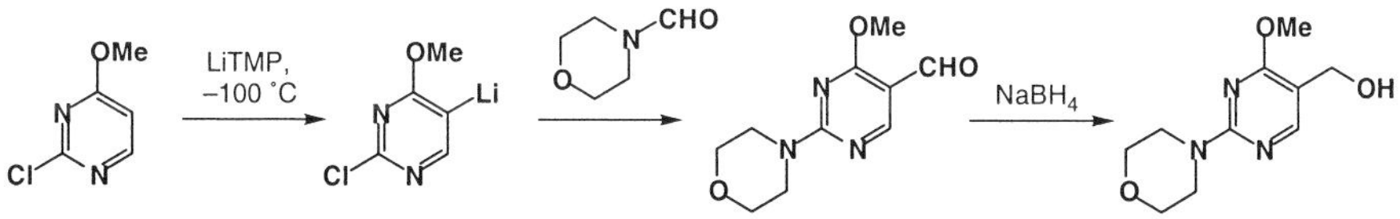

LiTMP, −100 °C; NaBH$_4$ **379** **380**

LiTMP, −75 °C; TsN$_3$; PPh$_3$ **381**

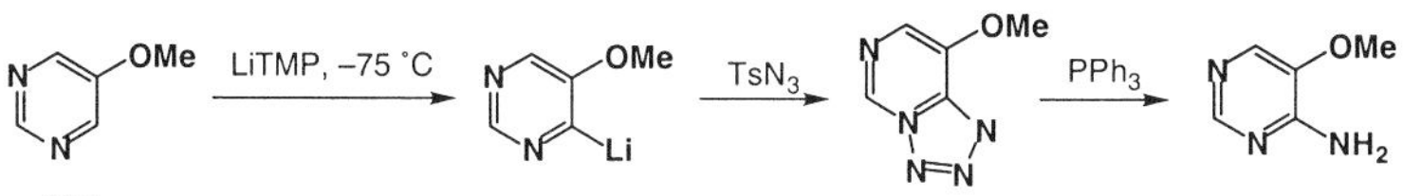

Methoxypyrazines **382** and **383** are lithiated with LiTMP as shown in the two syntheses below,[336] the first being part of a route to kelfizine and sulfalene.[335]

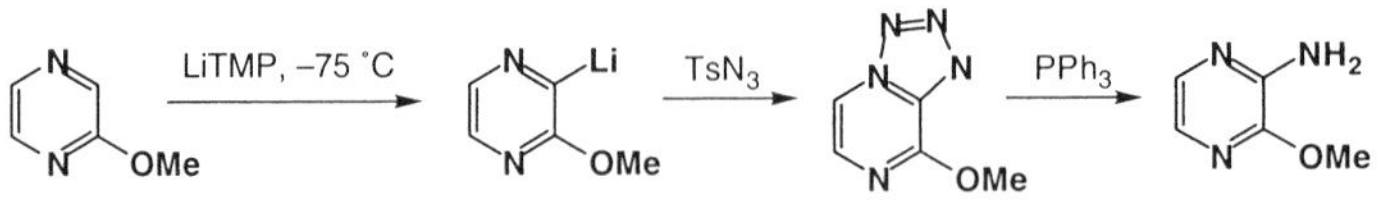

The lithiation of methoxypyridazines (which, for **384**, works even with BuLi) has been used to develop new routes to heterocyclic biaryls such as **385**.[337] In the pyridazine series MeO turns out to be a stronger director than Cl, as expected.[330] It also out-directs an alkylsulfanyl (RS) substituent, but sulfinyl and sulfonyl groups direct more strongly than MeO.[338]

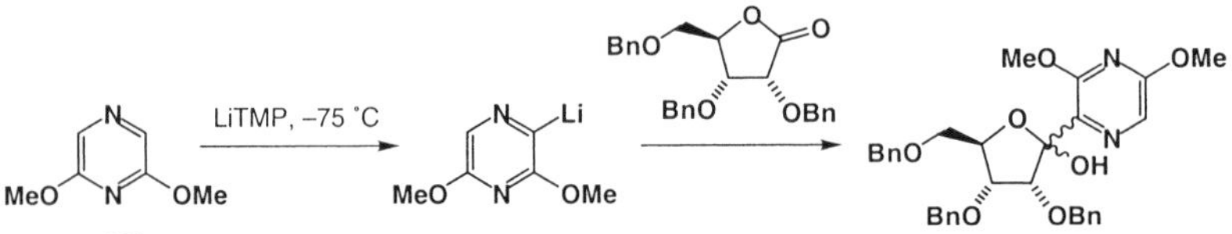

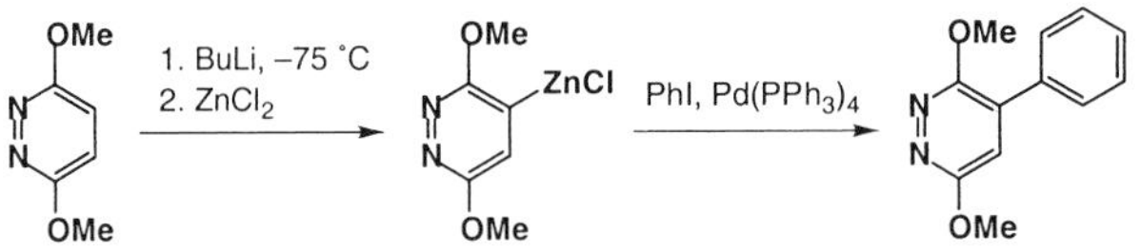

The usually very powerfully orthodirecting groups such as secondary carboxamide surprisingly do not always lead to ortholithiation on pyrazine and pyridazine rings: lithiation of **386** and **387**, for example, takes place principally (at least kinetically) at the "*meta*" and "*para*" positions.[339,340]

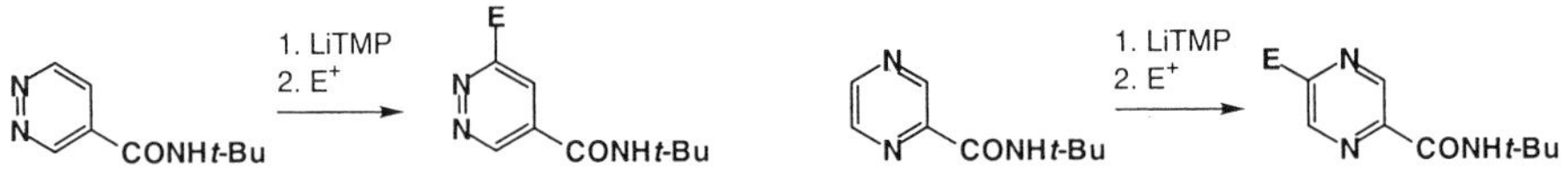

2.3.3.2 Electron-rich heterocycles

In many ways, the electron-rich five-membered aromatic heterocycles behave very much like carbocyclic aromatic compounds when it comes to lithiation. Lithiation α to O or S of furan and thiophene is straightforward, and was dealt with in sections 2.2.1 and 2.2.2. The usual selection of *ortho*-directing groups allow lithiation at other positions,[106,175-177,90,178-180,173,341,181] and some examples were given in section 2.3.2.

Pyrroles are lithiated at the 2-position whether or not the *N*-protecting group has the ability to coordinate:[14]

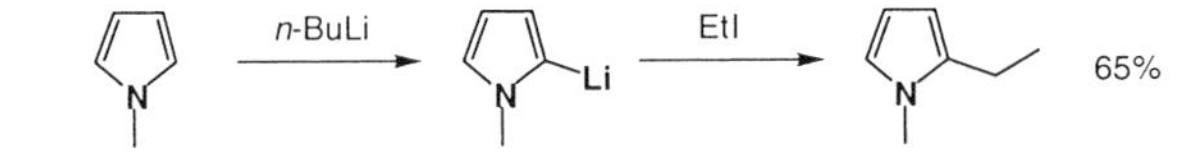

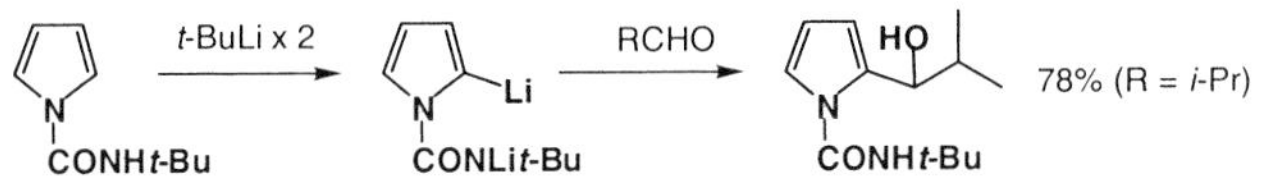

Interestingly, indoles are lithiated at either the 2- or the 3-position, depending on protectoin at N.[342] For 2-lithiation, probably the best technique is *in situ* protection of nitrogen by CO_2. The lithiocarbamate group of **388** acts as a good *ortho*-director.[343]

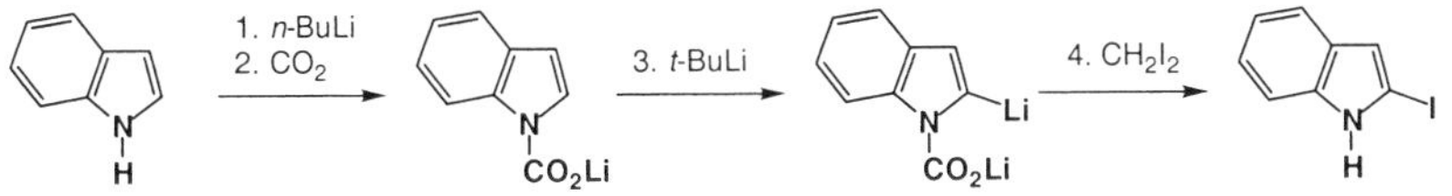

With more bulky, non-coordinating *N*-protecting groups such as triisopropylsilyl (**389**), lithiation occurs at C-3.[344]

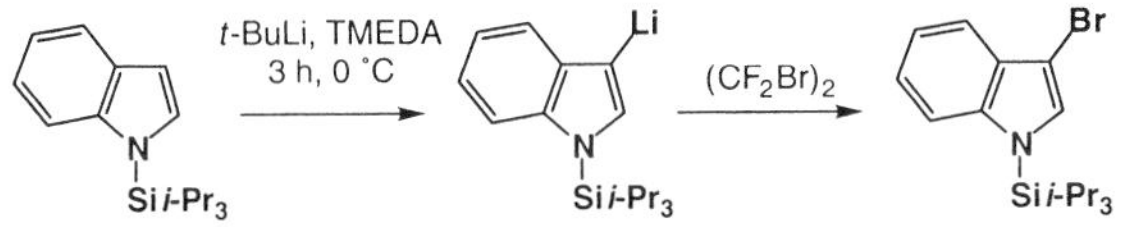

2.3.4 Lithiation of metal arene complexes

By coordinating to arenes, transition metals can facilitate ring lithiation by decreasing the electron density in the ring and acidifying the ring protons. We shall consider briefly the two most important metal-arene complexes in this regard – arenechromium tricarbonyls and ferrocenes.

2.3.4.1 Chromium-arene complexes[345]

Benzenechromium tricarbonyl **390** is deprotonated by BuLi in Et_2O–THF at –40 °C in a reaction that needs careful control for good yields.[346] The product **391** can be silylated to give **392** in 60% yield. Toluenechromium tricarbonyl lithiates non-regioselectively on the ring (but at the benzylic position with Na or K bases). Excess base can lead to polylithiation.[347]

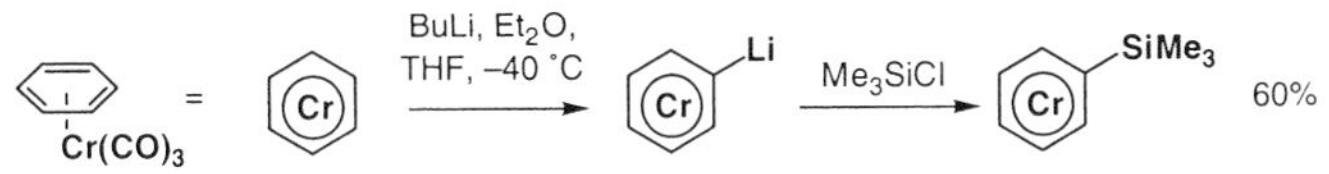

Anisolechromium tricarbonyl, however, is lithiated cleanly *ortho* to the methoxy group, giving products such as **393**.[348,349] Coordination of **394** to $Cr(CO)_3$ changes the regioselectivity from the expective cooperatively directed **395** to **396**.

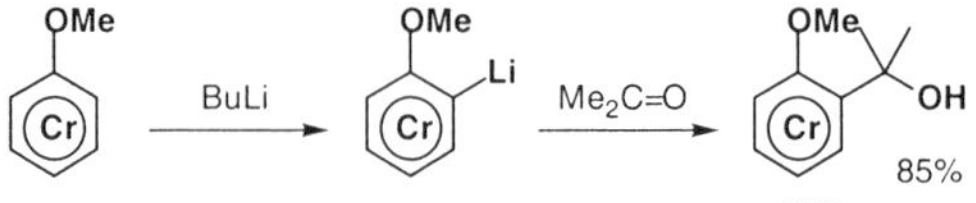

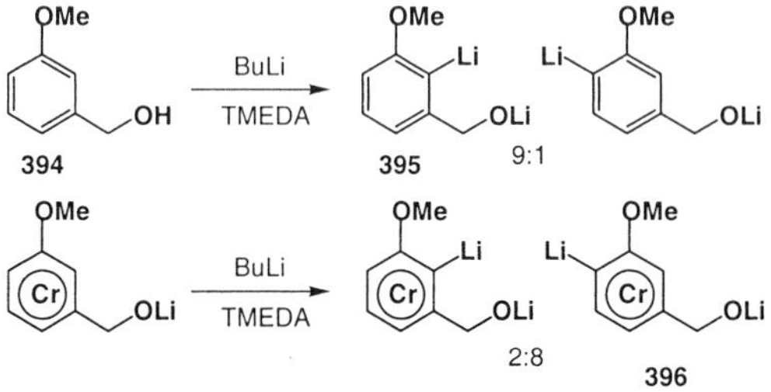

Multiple lithiation of anisolechromium tricarbonyl is easy to achieve: three equivalents of LiTMP give **397**.[347]

The chirality of lithiated arenechromium tricarbonyls offers the opportunity for asymmetric lithiation. This has been achieved either with auxiliaries[350-353] or using chiral bases.[354-359]

2.3.4.2 Ferrocenes

Ferrocene is best deprotonated by *t*-BuLi/*t*-BuOK in THF at 0 °C,[360] since BuLi alone will not lithiate ferrocene in the absence of TMEDA and leads to multiple lithiation in the presence of TMEDA. In the example below,[216] a sulfur electrophile and a Kagan-Sharpless epoxidation lead to the enantiomerically pure sulfinyl ferrocene **398**. The sulfinyl group directs stereoselective ortholithiation (see section 2.3.2.2), allowing the formation of products such as **399**. Nucleophilic attack at sulfur is avoided by using triisopropylphenyllithium for this lithiation.

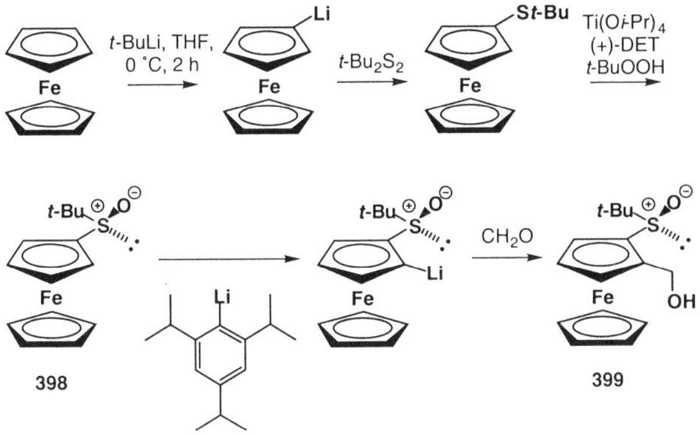

Apart from the use of enantiomerically pure sulfinyl groups,[361] ferrocenes have been lithiated enantioselectively using chiral bases.[362-364]

2.4 Lateral Lithiation

Lateral lithiation is the lithiation of benzylic alkyl groups which are themselves ortho to a directing group.[365] A general scheme for a lateral lithiation directed by a group G is shown below.

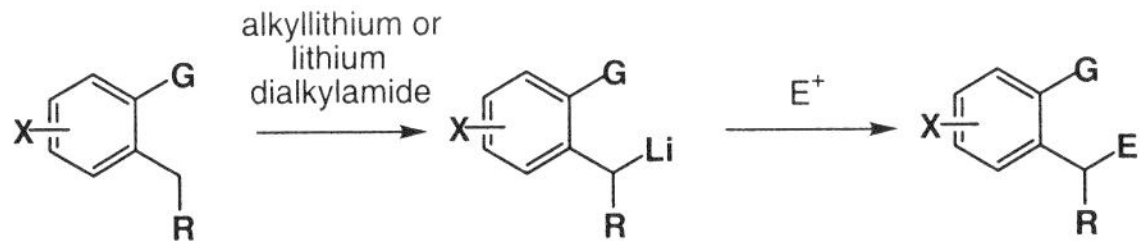

The organolithium deriving from a lateral lithiation is benzylic, and therefore often of significantly greater thermodynamic stability than the equivalent ortholithiated species. In general, ortho- and lateral lithiation strategies have developed in parallel with one another, and since the starting materials for a lateral lithiation may often be made by ortholithiation there are many links between the two classes of reaction.

2.4.1 Mechanism and regioselectivity

Benzylic lithiation requires an activating group if it is to be an efficient and synthetically valuable process. Both the ability to coordinate and the ability to acidify are important in a lateral lithiation-directing group, and most of the classes of directing group which will promote ortholithiation will also promote lateral lithiation. However, it is important to note one fundamental difference. In ortholithiation acidifying groups must operate by an inductive effect since the C-Li bond is in the plane of the aromatic ring. For lateral lithiation, the benzylic C-Li bond means that acidifying groups work best by conjugation.

Factors favouring ortholithiation

Li

X

Coordination

Inductive electron withdrawal

Factors favouring lateral lithiation

Li

Z

Coordination

Conjugation

Toluene itself can be lithiated by *n*-BuLi–TMEDA at or above room temperature, and deprotonation occurs almost exclusively at the methyl group – about 10% ring metallation (mainly in the *meta* position) is observed with *n*-BuLi–TMEDA.[366,367] At lower temperatures deprotonation is very slow,[368] and the best conditions for achieving the metallation of toluene are the Lochmann-Schlosser superbases (see section 2.6).[369]

n-BuLi–TMEDA; Li + Li

400; NMe_2; *n*-BuLi; Et_2O, hexane; NMe_2 Li

By contrast, the lithiation of **400** with *n*-BuLi, which is assisted by coordination to the NMe_2 group, is faster, reaching completion in less than 6 h at 25 °C, and completely regioselective.[370,371] The mesitylmethyl amine **401** is lithiated only at the methyl group *ortho* to the aminomethyl substituent.

NMe_2; BuLi; Et_2O, hexane; NMe_2; Li

401

The dimethylaminomethyl group must operate solely by coordination to Li, and it is assumed, as for ortholithiation, that the deprotonation takes place after the initial equilibrium formation of a BuLi–amine complex.

Similar is the lateral lithiation of amide **402**,[149] which is presumably assisted both by coordination of the amide to the lithium and by conjugation of the aromatic ring with the amide. Both **402** and **405** may be lithiated with LDA,[103] suggesting that conjugation is more important than coordination in stabilising the benzylic organolithiums with such electron-withdrawing groups. Given LDA's tendency to deprotonate its substrates reversibly, the formation of **406** is probably a result of thermodynamic control (though the methyl group may nonetheless be the site of kinetic lithiation with LDA): kinetically controlled lithiation with *s*-BuLi–TMEDA gives the ortholithiated **404**.[103] As noted before, coordination to lithium is a less important factor in LDA-promoted deprotonation, though it does ensure that mesitamides such as **407** are deprotonated only at the 2-methyl group.[372]

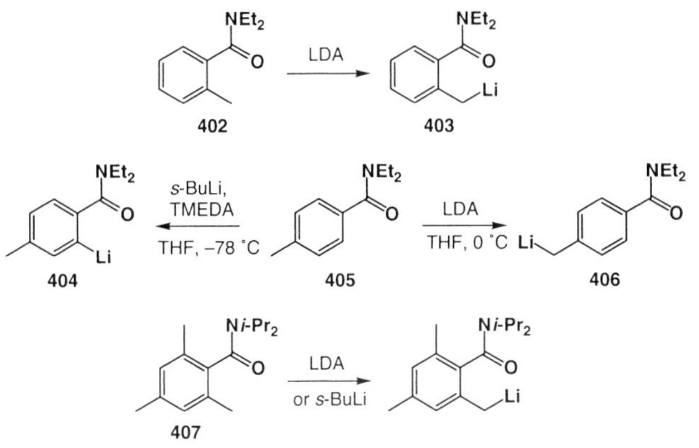

Organolithium **403** has been represented by the extended enolate structure **408**, though the chirality of some analogues of **403** proves the localised structure **403**.[373]

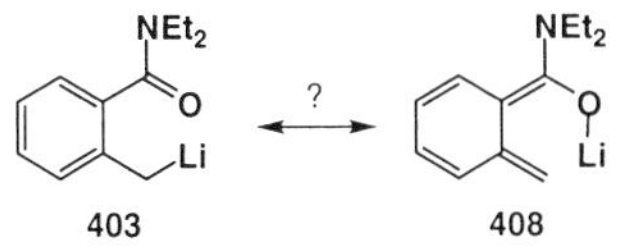

2.4.2 Classes of directing group

2.4.2.1 Secondary and tertiary amides

Resistance of the deprotonated amide to attack by alkyllithium means that two equivalents of BuLi, even at 0 °C, will give the dilithiated **409** in good yield, and various electrophiles can be used to introduce benzylic functionality.[374]

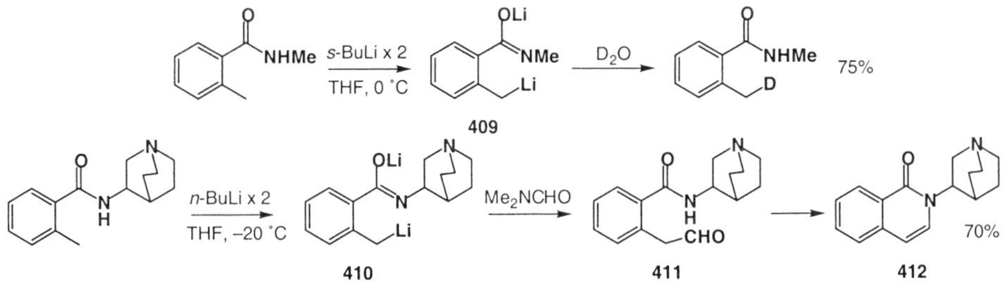

In the case of **410**, the product **411** cyclises to the isoquinolone **412**, and the amide substituent is a required part of the target molecule.[375] However, it frequently occurs that the amide substituent is not required in the final product, and the acid sensitive alkenyl substituent of **413** has been used as a solution to the problem of cleaving a C–N bond in the product.[376,377] Weinreb-type amides **414** can also be laterally lithiated, and the methoxy group removed from **415** by $TiCl_4$.[378] Hydrazones similarly can be laterally lithiated and oxidatively deprotected.[173]

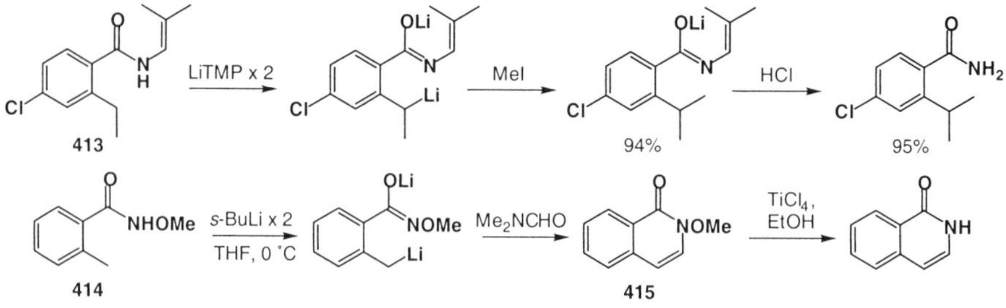

Laterally lithiated *tertiary* amides are more prone to self-condensation than the anions of secondary amides, so they are best lithiated at low temperature (–78 °C). *N,N*-Dimethyl, diethyl (**416**) and diisopropyl amides have all been laterally lithiated with alkyllithiums or LDA, but, as discussed in section 2.3.2.1.1, these functional groups are resistant to manipulation other than by intramolecular attack.[379] Clark has used the addition of a laterally lithiated tertiary amide **417** to an imine to generate an amino-amide **418** product whose cyclisation to lactams such as **419** is a useful (if rather low-yielding) way of building up isoquinoline portions of alkaloid structures.[380] The addition of laterally lithiated amines to imines needs careful control as it may be reversible at higher temperatures.[381]

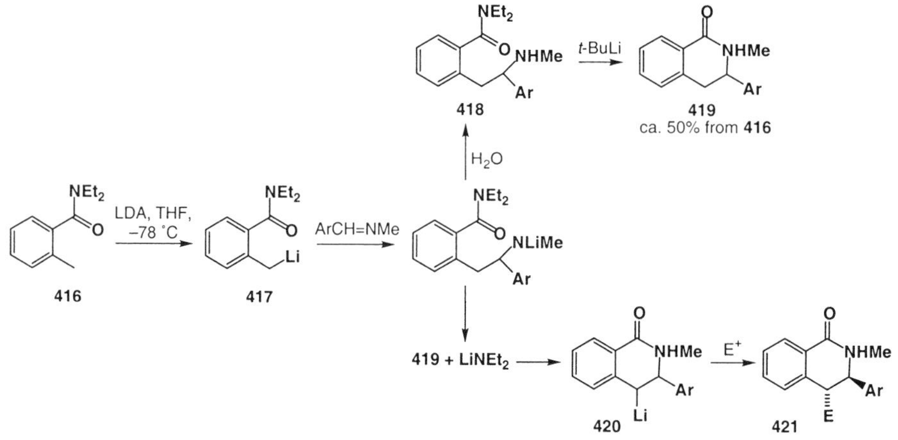

Under certain conditions, the cyclisation to lactam **419** occurs spontaneously, eliminating one equivalent of lithium diethylamide, which goes on to deprotonate the benzylic position a second time to yield **420** and then **421**.[380]

The labile tertiary amide groups described in section 2.3.2.1.1 are also applicable to lateral lithiations;[163] the piperazine-based amide **422** has been used to direct lateral lithiation before being methylated and cleaved to the acid **423**.[382]

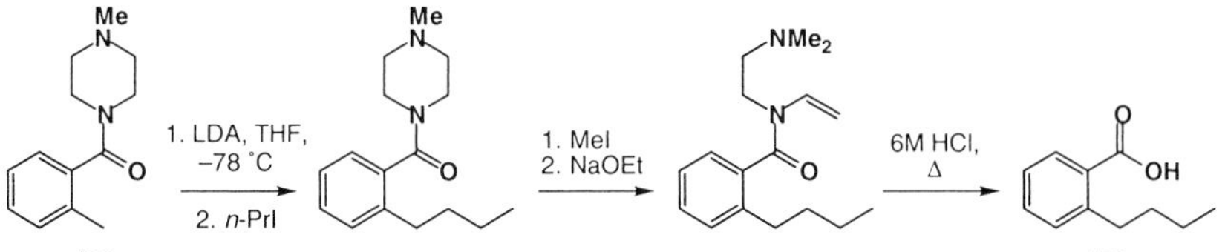

1. LDA, THF, −78 °C 2. n-PrI; 1. MeI 2. NaOEt; 6M HCl, Δ

422 423

2.4.2.2 Nitriles

Lateral lithiation of nitriles can be achieved – and self-condensation avoided – if LiTMP is used in THF at −78 °C.[383]

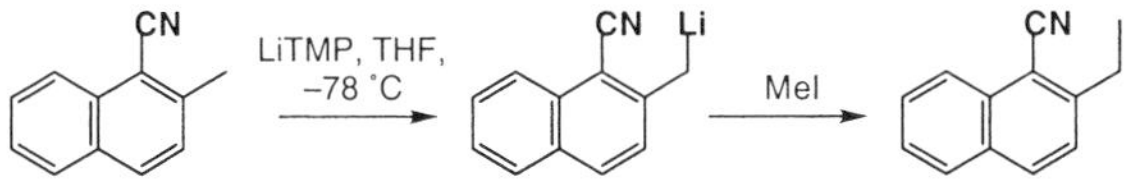

LiTMP, THF, −78 °C; MeI; 85%

2.4.2.3 Oxazolines, imidazolines and tetrazoles

Oxazolines,[384] imidazolines[385] and tetrazoles[386] can all be laterally lithiated. Oxazolines have been used in this regard rather less than for ortholithiation.

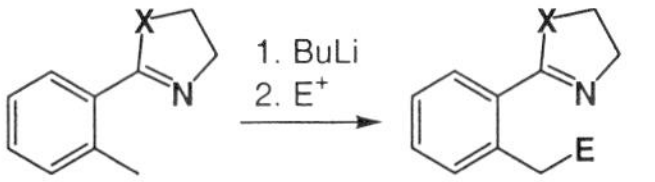

1. BuLi 2. E⁺; X = O, NH

2.4.2.4 Carboxylates

2-Methylbenzoic acid **424** can be laterally lithiated with two equivalents of lithium amide base (LDA[387] or LiTMP[388]) or alkyllithium,[389,390] provided the temperature is kept low to avoid addition to the carbonyl group. It is usually preferable to carry out the lithiation using alkyllithiums,[389] since with lithium amides the subsequent reaction of **425** with electrophiles is disrupted by the presence of the amine by-product (diisopropylamine, for example).[389,391] The dilithio species **425** is stable in THF even at room temperature, and (as with the amide **404**) since LDA will also dilithiate **426** stabilisation presumably comes principally from conjugation with the carboxylate.

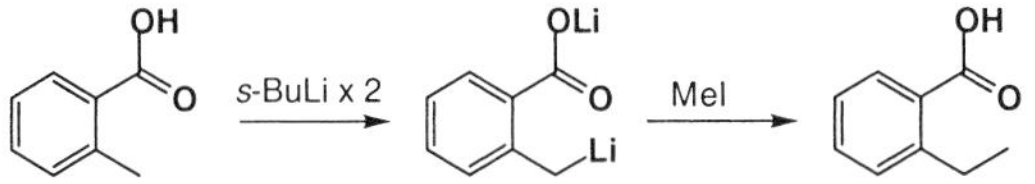

s-BuLi x 2; MeI

424 425

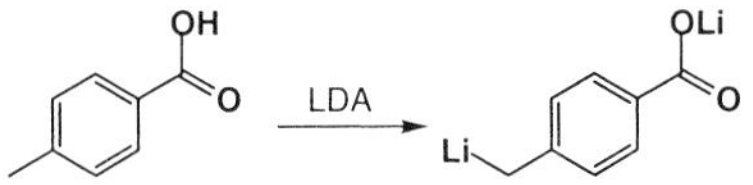

LDA

426

2.4.2.5 Carboxylic esters

The greater acidity of lateral protons means that LDA can usually be used to remove them and hence much more electrophilic directing groups can be used for lateral lithiation than ortholithiation. Ethyl 2-methylbenzoate **427** is deprotonated at –78 °C by LDA but as soon as the product organolithium forms it adds to unreacted starting material to give dimeric products **428**.[392]

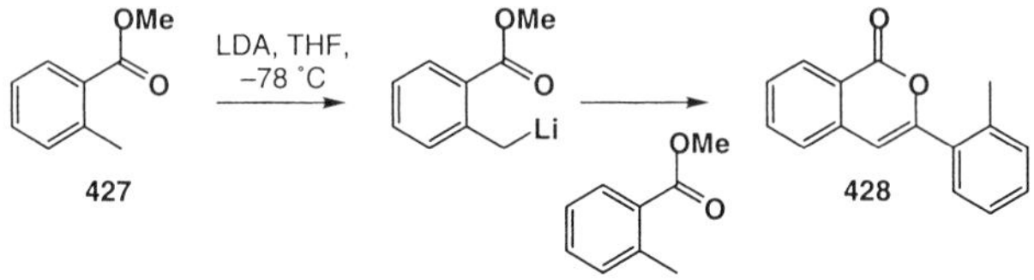

This addition reaction is much slower with 2,6-disubstituted esters (and phenoxides), and organolithiums **429** and **431** can be formed and reacted with external electrophiles to give compunds such as **430** and **432**.[392,393] An asymmetric version of this reaction has been developed.[394,395]

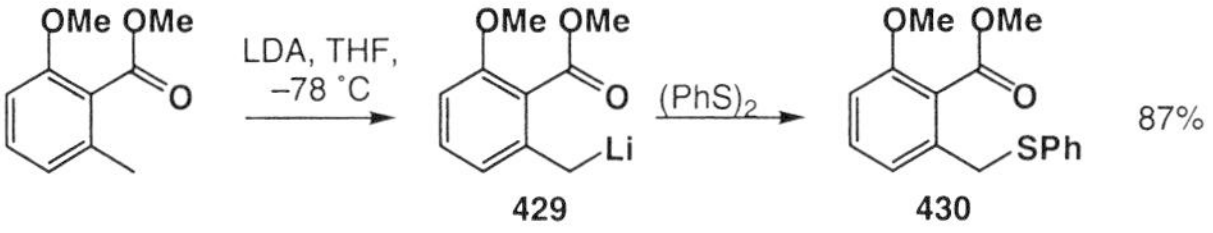

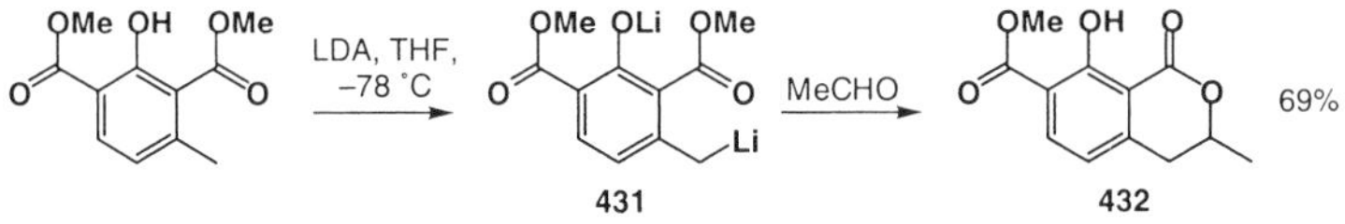

2.4.2.6 Ketones

Ketones may direct lateral lithiation even if the ketone itself is enolised: enolates appear to have moderate lateral-directing ability. Mesityl ketone **433**, for example, yields **434** after silylation – BuLi is successful here because of the extreme steric hindrance around the carbonyl group.[396] The lithium enolate can equally well be made from less hindered ketones by starting with a silyl enol ether.[396]

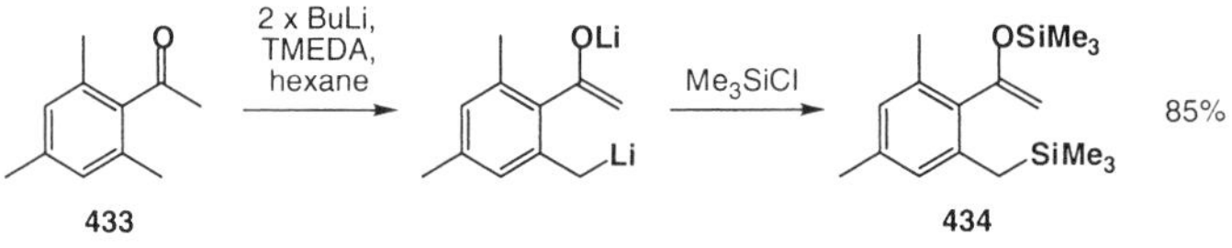

When the ketone cannot be enolised, lateral lithiation appears to be very easy; **435** is lithiated by LDA at –78 °C.[397]

LDA, THF, −78 °C; Me_2CO; 94%

435

2.4.2.7 Aldehydes protected as α-amino alkoxides

Temporary protection of an aldehyde by addition of a lithium amide can be used to facilitate lateral lithiation by *n*-BuLi. The best lithium amide for this purpose is **202**: interestingly, lithium piperidide promotes ortho-, rather than lateral, lithiation of **437**.[398]

202; CHO; NLiMe; OLi; BuLi; 1. E^+; 2. H^+; CHO; E

436 437

Aldehydes can also be laterally lithiated if protected as imidazolidines (**438**)[399] or as imines (**439**).[400] With imines, LiTMP must be used to prevent nucleophilic addition to C=N.

BuLi, TMEDA, Et_2O, 25 °C; Li; CHO; LiTMP, THF, −15 °C

438 439

2.4.2.8 Alcohols and phenols (cresols) and their derivatives

2-Methylbenzyl alcohol **440** can just about be lithiated by treatment with BuLi in Et_2O at room temperature, but the activation of the methyl group is very weak.[401] Lateral lithiation of cresol **441** is even harder to achieve, and the superbase conditions required are similar to those used to deprotonate toluene.[402] The coordinating effect of the oxyanion is more than outweighed by electron-donation into the ring.

OH; 2 x *n*-BuLi, Et_2O, 20 °C, 24 h; OLi; Li; EtBr; OH; 92%

440

OH; *n*-BuLi–*t*-BuOK, heptane, reflux, 3 h; OLi/K; Li/K; E^+; OH; E

441

The deactivating effect of a phenoxide oxyanion is removed in the ether series, but in cases such as **442** where ortholithiation can compete with lateral lithiation mixtures of products are frequently obtained.[403] The MOM acetal **443** is fully *ortho*-selective in its reaction with *t*-BuLi.[404,405]

without TMEDA 2:1
with TMEDA 1:4

>99:1

The prospects for lateral lithiation are slightly improved if ortholithiation is blocked, though even then yields are moderate at best.

48%; R = cyclohexyl

Better for the lateral lithiation of phenols are the *N,N*-dialkylcarbamate derivatives **444**. These may be lithiated with LDA, allowing complete selectivity for the lateral position, presumably because this is the thermodynamic product.[192] With *s*-BuLi ortholithiation is the predominant reaction pathway. If the lateral organolithium **445** is warmed to room temperature, an acyl transfer from O to C takes place, analogous to the anionic *ortho*-Fries (see section 2.3.2.1.4), giving amide **446**.[365]

LDA, THF — sole product
s-BuLi, TMEDA — 2:1

The best phenol-derived lateral director of all appears to be the diaminophosphoryl group. Lithiation of **447** at –105 °C with *s*-BuLi gives a highly reactive organolithium which adds to electrophiles even at –90 °C.[406]

2.4.2.9 Sulfur-based functional groups

The lithiation regioselectivity problem is particularly acute with arylsulfides since sulfur can promote α, lateral and ortholithiation to a broadly similar degree. Lithiating **448** in ether, for example, gives a 20:10:1 ratio of the α, lateral and orthofunctionalised products **449**, **450** and **451**.[407] If **448** is deprotonated twice, a good yield of the dilithio species **452**, and hence the doubly quenched compounds **453**, results.[408] The SCH_2Li group is clearly having some directing effect in this reaction, since no second deprotonation occurs if the methyl group is in the *meta* or *para* position.

Me SMe 448 BuLi x 1 449 450 451 20 : 10 : 1 BuLi x 2 452 MeI 453 63-68%

α-Lithiation is suppressed in aryl isopropylsulfides, but now ortholithiation wins out over lateral lithiation unless there are no deprotonatable *ortho* positions.[407]

Sulfoxides direct lateral lithiation in a reaction which is also highly stereoselective.[409] In common with other electrophiles, $ClCO_2Et$ produces as a single diastereoisomer of **455** from **454**, and Raney nickel can be used to remove the sulfinyl group from the product, making this a very versatile method for asymmetric functionalisation of a benzyl group.

LDA 454 $ClCO_2Et$ 455 75% RaNi

Remarkably, ethyl *o*-tolylsulfonate **456** can be laterally lithiated by BuLi at –78 °C.[410] With the methyl ester, the expected substitution at the methyl group takes place. The *p*-substituted ester also undergoes lithiation at the methyl group, suggesting that the sulfonate group acts substantially by conjugation.

SO_3Et 456 BuLi, THF, –78 °C $Ph_2C=O$ 91%

By contrast, while the *o*-methyl sulfonamide **457** can be laterally lithiated, its *p*-methyl isomer undergoes ortholithiation rather than benzylic lithiation.[411-413]

2.4.2.10 Aniline and aminoalkylbenzene derivatives

N,N-Dimethyl-*o*-toluidine **458** is reluctantly lithiated with *n*-BuLi–TMEDA at 25 °C; ortholithiation occurs to some extent in the reaction but the yield of laterally functionalised product is maximised after 3 h.[198] Without TMEDA, the extent of ortholithiation is increased.

Much more versatile than the simple anilines are their anilide derivatives. Pivalanilides, benzanilides and other non[199,414] (or scarcely[415]) enolisable amides **459** are laterally lithiated on treatment with two equivalents of BuLi, and may be quenched with electrophiles to give **461**. In the absence of an electrophile, the organolithiums **460** cyclise to indoles **462**.

A similar cyclisation can result from lithiation of an isonitrile; lithiation of **463** requires two equivalents of LDA and the organolithium **464** can either be trapped with other electrophiles at low temperature or warmed to give an indole **465**.[416,417] It is quite clear that isonitriles activate purely by conjugation, and indeed they promote deprotonation of methyl groups *para* to an isonitrile just as well as *ortho*. The ease with which isonitriles can be made from formamides suggests that these methods could be rather more widely used than they are.

The carbamates **459** (R = O*t*-Bu) behave similarly, though they must be lithiated with *s*-BuLi to avoid addition to the carbonyl group.[418] It is possible simply to use a lithium carbamate to protect an amino group during a lateral lithiation: an initial deprotonation and carbonation generates the lithium carbamate **466**, which is then deprotonated twice more by *t*-BuLi. After electrophilic quench, acid hydrolysis of the carbamic acid returns the unprotected aniline.[419,420] An alternative *in situ* protection sequence relies on temporary trimethylsilylation of the aniline.[421]

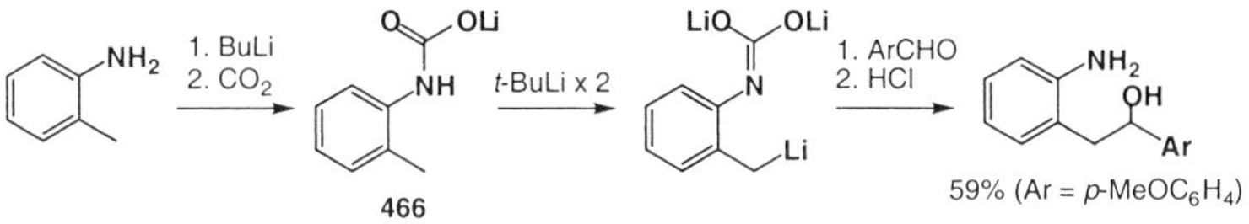

1. BuLi 2. CO_2 → **466** → *t*-BuLi x 2 → 1. ArCHO 2. HCl → 59% (Ar = *p*-$MeOC_6H_4$)

Amines in the benzylic position, and their carbamate derivatives, promote lateral lithiation through coordination to Li, and lack the disadvantage that their ArNHR isomers have of attenuated acidity due to the nitrogen lone pair. As described above, simple benzylamines are readily lithiated – more so than the equivalent anilines;[114,371] their carbamate derivatives such as **467** also lithiate well in the lateral position[377] (in the absence of a 2-methyl group, benzylic lithiation α to nitrogen is the major pathway[422]). As the amino group is moved further and further from the aromatic ring, benzylic lithiation of the aminoalkyl chain becomes a more and more important side reaction.[423]

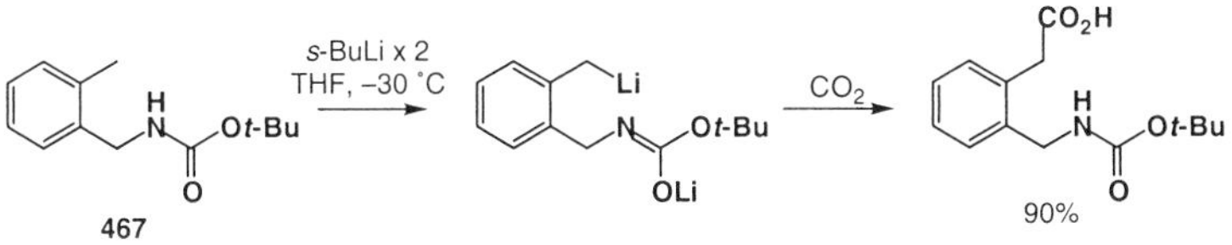

467 → *s*-BuLi x 2, THF, –30 °C → CO_2 → 90%

2.4.2.11 *Halogens*

2-Fluorotoluene can be deprotonated in the benzylic position by the superbases, but the halogen appears to offer little activation to the process. 2-Trifluoromethyltoluene is more readily deprotonated, but decomposes by elimination of fluoride even at –100 °C.[365]

2.4.2.12 *Lateral lithiation of heterocycles*[14]

Alkyl groups attached to electron-deficient heterocyclic rings can usually be lithiated even in the absence of a directing group. The pK_a values of methyl groups attached to some representative heterocycles are shown below.[424]

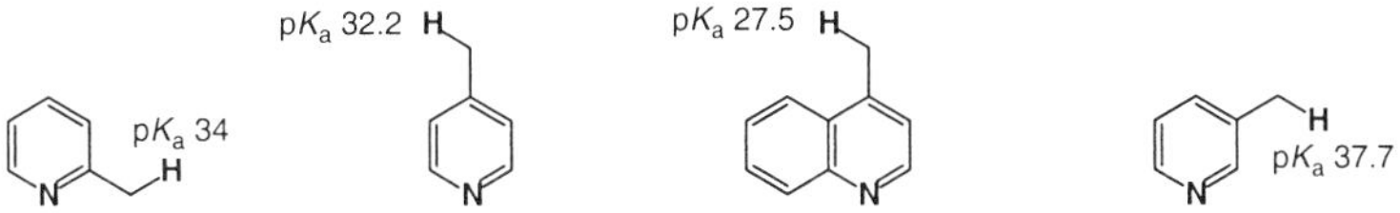

pK_a 34 · pK_a 32.2 · pK_a 27.5 · pK_a 37.7

Those of 2- and 4-substituted pyridines fall about half-way between those of toluene (ca. 40) and of a ketone (ca. 20) – the organolithiums have significant "azaenolate" character (as shown in **468** and **469**). A representative reaction is the synthesis of **470**.[425] Lithiation of 2-methylpyridine to give **468** has also been used in the synthesis of the anti-angina drug perhexiline.[91]

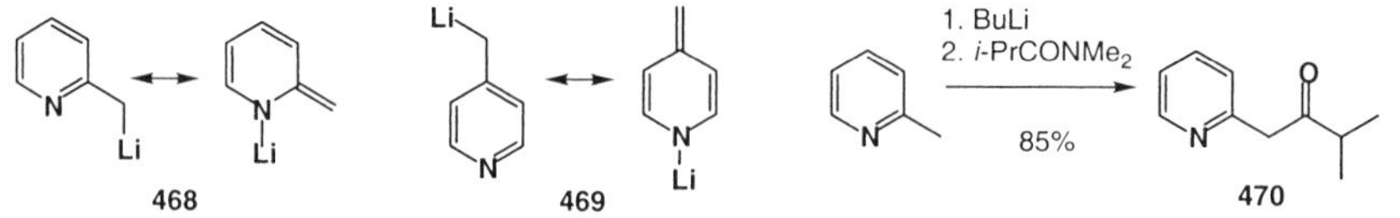

1. BuLi
2. *i*-PrCONMe$_2$
85%

468 469 470

The pK_a of 3-methylpyridine, whose anion cannot be delocalised onto N, is closer to that of toluene, and deprotonation gives only low yields with most bases. However, with a combination of BuLi and lithio-dimethylaminoethanol (LiDMAE) deprotonation is quantitative but yields products **471** arising from apparent lithiation α to N![426] Trying to force lateral lithiation of pyridines is generally doomed to failure, as ring lithiation or nucleophilic addition usually takes place first.[365]

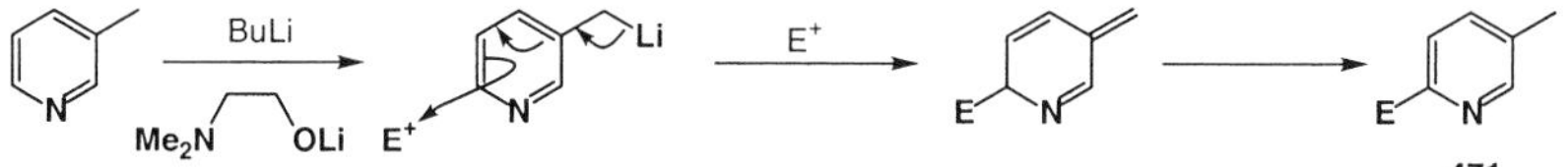

BuLi, Me$_2$N~OLi; E$^+$

471

Electron-rich heterocycles, such as pyrrole and furan, bear more resemblance to carbocyclic rings: their side chains are much less acidic, and undergo lateral lithiation much less readily. Without a second directing group, methyl groups only at the 2-position of furan, pyrrole or thiophene may be deprotonated.

Even with directing groups, lateral lithiations of furans and thiophenes especially are beset by uncertainty over regiocontrol.[365] The best lateral lithiations of electron-rich heterocycles are those of 2-alkyl indoles and pyrroles where the directing group is carried by the ring nitrogen atom and where there is no question of regioselectivity.[427] A remarkable example is shown below: the indole **472** is temporarily protected as its lithium carbamate, and the carbamate group is powerful enough to direct two successive lithiations of the methyl group, giving **473**.[427]

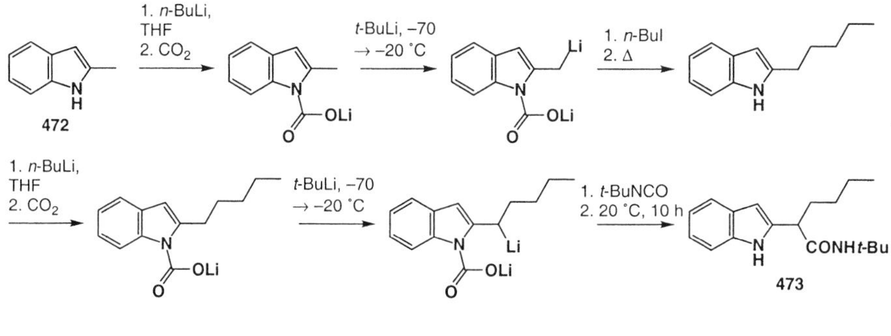

1. *n*-BuLi, THF 2. CO$_2$; *t*-BuLi, −70 → −20 °C; 1. *n*-BuI 2. Δ

472

1. *n*-BuLi, THF 2. CO$_2$; *t*-BuLi, −70 → −20 °C; 1. *t*-BuNCO 2. 20 °C, 10 h

CONH*t*-Bu

473

The usual directing groups such as secondary amides will also successfully direct lateral lithiation at the 2-methyl group of a pyrrole.[375]

The reduced electrophilicity of indole-3-carboxylic esters (they are vinylogous carbamates) means that they are much more versatile directors of lateral lithiation than the comparable benzoates, as illustrated by the synthesis of **474**.[428]

Pyrroles and indoles have one further unique mode of lateral lithiation – deprotonation of an *N*-methyl group (also an α lithiation). The reaction works particularly well with an aldehyde director, temporarily protected as the α-amino alkoxide **475**.[180,429]

With furans and thiophenes, ring lithiation α to the heteroatom is usually the preferred process. The only lateral directing group studied to any extent for furans and thiophenes is the lithium carboxylate group, and **476** and **477** are the only two compounds in this class to undergo lateral lithiation reliably. With the 2-position blocked (**478**), thiophene still lithiates at the 2-methyl group rather than the one lateral to CO_2Li.[365]

2.5 Remote lithiation, and β-lithiation of non-aromatic compounds

A number of reactions have close similarities to ortho- and lateral lithiation, even if they do not fall under the more rigid definition of the terms.[233] For example, vinylic protons with nearby directing groups can frequently be lithiated readily. Some examples are shown – **479-483** are all lithiated as though the double bond were part of an aromatic ring.[233] The importance of coordination in these reactions is shown by **483**, which lithiates at the more acidic position α to S if HMPA is added to disrupt Li–O interactions. Other similar lithiations are known in the cyclopropyl and cyclobutyl series.[233]

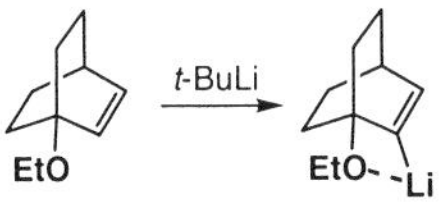

479

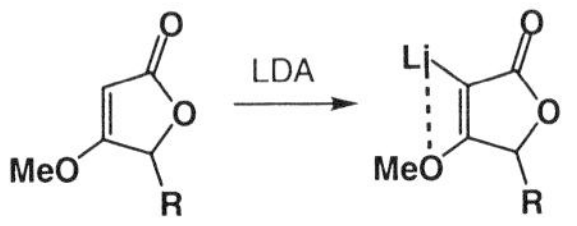

480

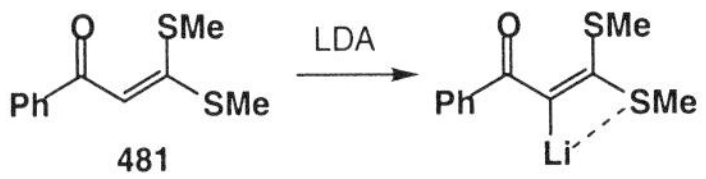

481

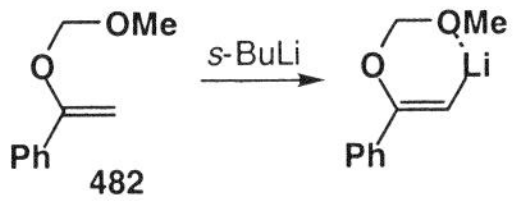

482

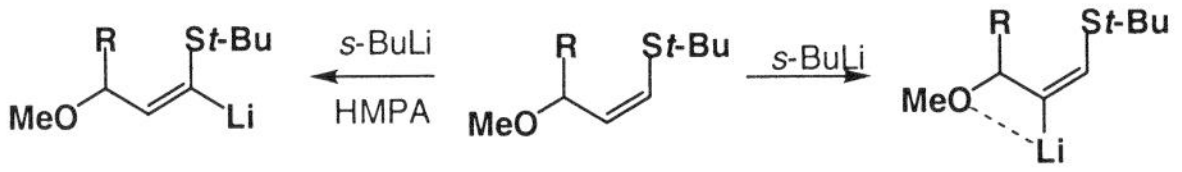

483

Tertiary amides can direct vinylic lithiations in the manner of ortholithiations as shown by the example of **484** and **485**.[430,431] Even methyl groups can be lithiated given an appropriate director and base: **486** forms the cyclopropane **488** on treatment with **487** in refluxing heptane.[432]

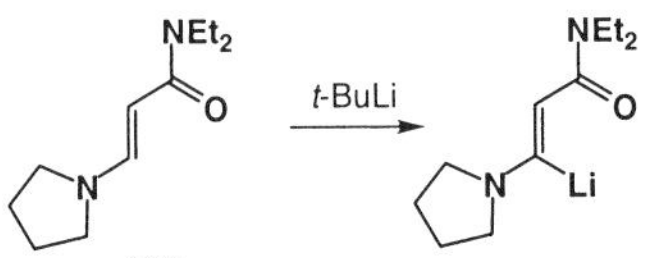

484

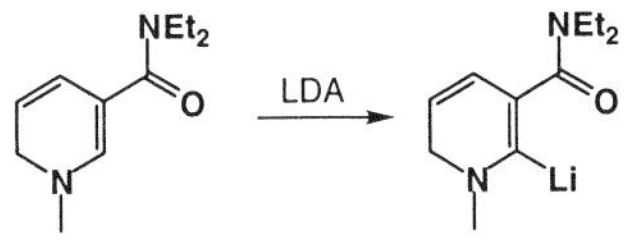

485

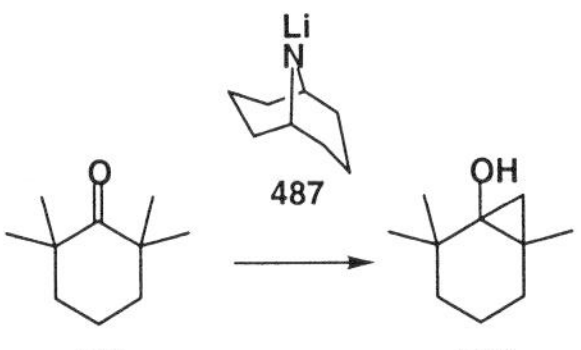

486 **488**

In aromatic compounds, potent but frustrated (by having their *ortho* positions blocked) directing groups may lead to lithiations at positions other than *ortho*. For example, when the carbamate **489** is treated with LDA in refluxing THF, lithiation occurs at a remote position (not *peri*, note) and an anionic Fries rearrangement ensues to give **490** (see section 2.3.2.1.4). Lactonisation gives **491**.[433]

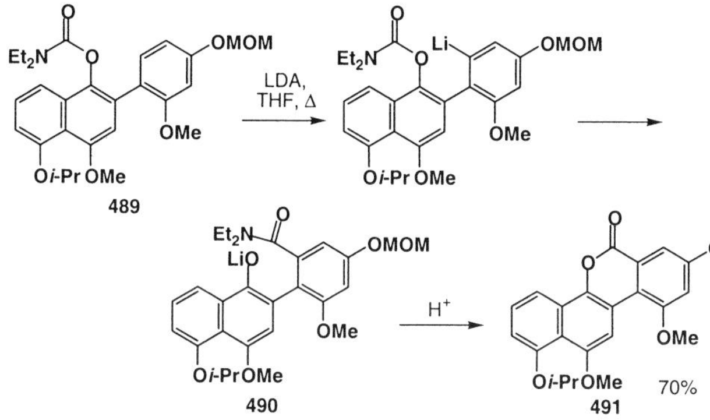

Carbamate **492** under the same conditions also undergoes remote lithiation and a remote anionic Fries rearrangement, and then the product amide **493** proceeds to direct a remote benzylic lithiation, even though it has a free *ortho* position. Finally, the benzylic organolithium **494** cyclises onto the amide to form **495** – all in one pot.[434]

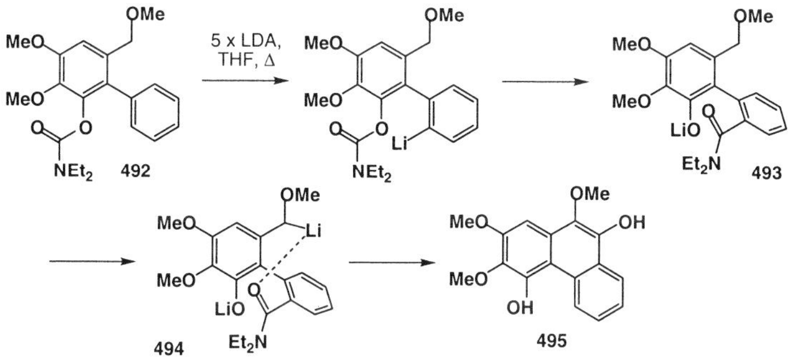

2.6 Superbases[435]

The combination of an alkyllithium with a metal alkoxide provides a marked increase in the basicity of the organolithium.[436,369,437] The most widely used of these "superbases" is the one obtained from BuLi and KO*t*-Bu, known as "LiCKOR" (Li–C + KOR).[281] The exact nature of the products obtained by superbase deprotonations – whether they are organolithiums, organopotassiums, or a mixture of both, is debatable, as is the precise nature of the superbase itself. For example, while prolonged mixing of alkyllithium and KO*t*-Bu in hexane gives a precipitate of butylpotassium,[438,439,436] the reactivity of a slurry of BuLi–KO*t*-Bu does not match that of BuK.[440] Simplistically, superbases can be considered to be organolithiums solvated by very electron-donating ligands (much more so than THF or TMEDA). As synthetic tools they provide a useful top end to the armoury of bases for regioselective functionalisation by deprotonation.

The violence of superbasic slurries towards functionalised organic molecules means that they are at their most effective with simple hydrocarbons; they also tolerate ethers and fluoro substituents. LiCKOR will deprotonate allylic, benzylic, vinylic, aromatic and cyclopropane C–H bonds with no additional assistance. From benzene, for example, it forms a mixture of mono and dimetallated compounds **496** and **497**.[441] ("Li/K" indicates metallation with a structurally ill-defined mixture of lithium and potassium.)

BuLi, KO*t*-Bu / hexane → Li/K **496** + Li/K, Li/K **497**

Typically, superbases care little for coordination effects, and simply remove the most acidic proton on offer; this provides useful alternative selectivities in the lithiation of aromatic rings, for example. With groups that direct principally by acidification, orthometallation occurs, and treatment with BuLi–KO*t*-Bu is the most efficient way of orthofunctionalising fluorobenzene or trifluoromethylbenzene.[442]

F: BuLi, KO*t*-Bu / hexane → F, Li/K

CF_3: BuLi, KO*t*-Bu / hexane → CF_3, Li/K

BuLi–*t*-BuOK metallates *ortho* to powerfully electron-withdrawing and acidifying fluoro even if much stronger directors (by the usual considerations) are present, such as methoxy (**498, 499**) or anilide (**500**) groups.[443]

OMe, F, Li ← BuLi–*t*-BuOK — OMe, F **498** — BuLi, THF → Li, OMe, F

Li, OMe, F ← BuLi–*t*-BuOK — OMe, F **499** — BuLi, THF → Li, OMe, F

Li, F, N, O, OLi ← *t*-BuLi–*t*-BuOK — H, N, O, O, F **500** — *t*-BuLi → Li, N, O, OLi, F

The deprotonation and electrophilic quench of unfunctionalised allylic compounds provides a useful way of making functionalised Z-alkenes, since the intermediate allylmetal species prefers the *endo* configuration.[281,19] For example, 1-dodecene **501** can be transformed in 57% yield to Z-dodecenol **502** on a 15 g scale.[281]

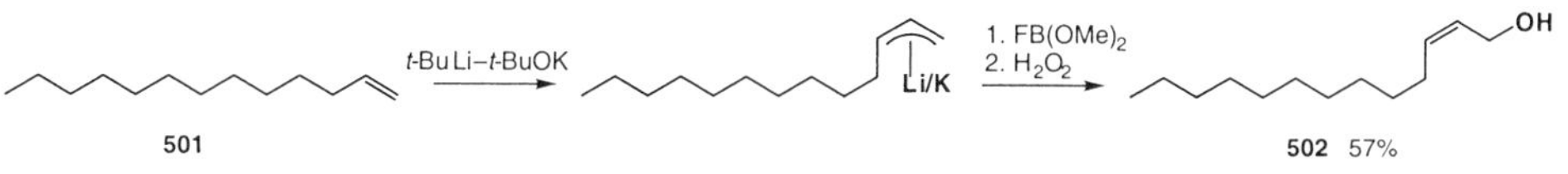

Similarly, cyclohexene can be used as a source of cyclohexenecarboxylic acid **503**.[281]

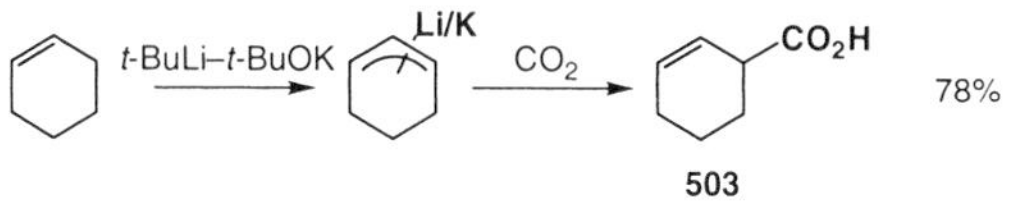

Successive metallations at benzylic sites have been used by Schlosser in an elegant and simple synthesis of ibuprofen using only superbase chemistry.[444] Starting with *para*-xylene **504**, two successive metallations and alkylations give **505** which is once more metallated at the less hindered benzylic site and carbonated to give ibuprofen **506**.

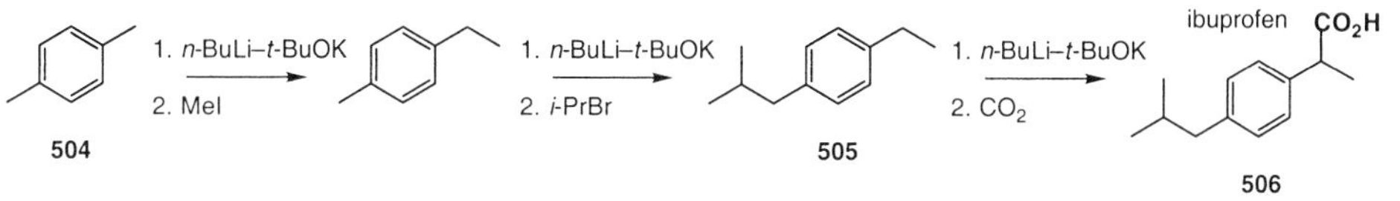

The BuLi-*t*-BuOK superbase will also deprotonate α to nitrogen[31] or halogens. The latter leads to carbenoids such as **508, 510** and **512**, which either decompose to carbenes (**509, 511**) or, if an appropriately placed silicon substituent is available, undergo Brook rearrangement (**513**).[445]

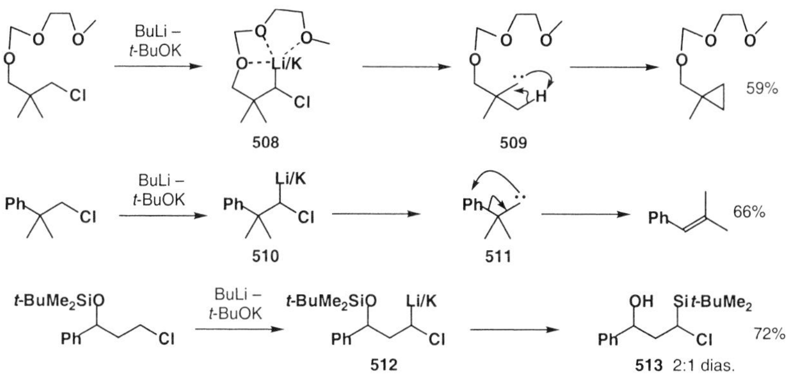

Other superbases based on LDA+BuLi,[281] LDA+*t*-BuOK[281] and BuLi + lithiodimethylamino-ethanol (LiDMAE)[426,446] are also widely used.

2.7 Cooperation, competition and regioselectivity

Numerous studies of the relative abilities of different ortholithiation directors to control regioselectivity have been published, and these are summarised in section 2.3.2. While the scheme in section 2.3.2 aims to place *ortho*-directors into an approximate order of potency, these studies show quite clearly that relative directing power depends on a variety of factors and varies from compound to compound.

What is clearer is that, if two lithiation-directing groups are placed *meta* to one another, lithiation nearly always occurs between them; there are examples in sections 2.3, 2.4 and 2.6. The only exceptions are when the resulting polysubstituted arene ring would be exceptionally hindered; lithiation then takes place *ortho* to the better of the two directing groups. For example, lithiation of **514** gives **515** by lithiation at the 5- rather than the 2-position.[447]

not here
BuHN NHBu
lithiates here
514
1. BuLi, Et_2O, THF
2. (epoxide)
BuHN NHBu OH
515
1. H^+
2. HO^-
HO_2C

Lithiation between two directing groups which would lead to a 1,2,3,4-tetrasubstituted ring is usually difficult.[448] For example, **516** undergoes kinetic lithiation at a site (**517**) which avoids the formation of a 1,2,3,4-tetrasubstituted ring. Only on standing does isomerisation to the thermodynamically more stable **518** occur.[449]

Me_2N MeO MeO
516
BuLi
Me_2N MeO MeO Li
517
Me_2 N Li MeO MeO
518

Change of regioselectivity with time is a feature of a number of lithiations where the site which is deprotonated fastest does not give the most stable possible organolithium. Dibenzothiophene **519** provides a nice example. Double lithiation occurs initially *ortho* to the sulfur atom to give **520**, but on heating, an isomerisation takes place to give the organolithium **521**, perhaps because of beneficial chelation of the two adjacent lithium atoms by TMEDA.[450] Similarly, BuLi at 0 °C ortholithiates the sulfonate **522**, but at 25 °C deprotonation occurs at the *para* methyl group where the benzylic anion can gain conjugation with the sulfonate.[451]

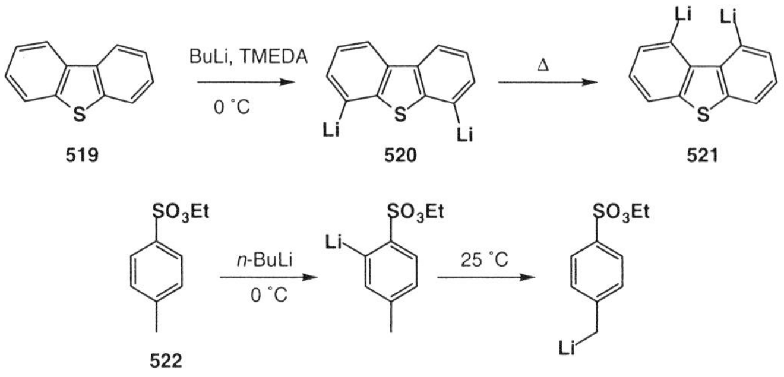

Formation of the most stable organolithium is much faster when a lithium amide is used as the base, since re-deprotonation of the amine by-product provides a mechanism for "anion translocation" from one site to another. Several examples were discussed in sections 2.3 and 2.4: the lithiations of **523** and of **524** with BuLi and with LDA illustrate the point well.

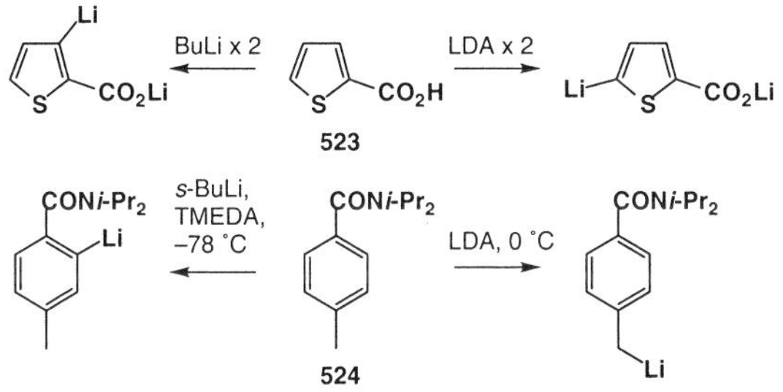

Organolithiums formed from lithium amide bases are presumably formed with a secondary amine ligand bound to lithium, and it is frequently noted that intramolecular coordination to a basic heteroatom is less important for their stability. The presence of even weak donor ligands such as TMEDA[135] can bias lithiation selectivity in favour of a more acidic site rather than a more readily coordinated site. The ethoxyvinyllithium (EVL)–HMPA combination discovered by Meyers,[452,453] demonstrates unusual regioselectivities more similar to those typically of the superbases (section 2.6). For example, it selectively deprotonates *ortho* to OMe rather than *ortho* to an amide **525** or oxazoline **526**, and chooses to deprotonate **527** *para* rather than *ortho* to the oxazoline. Once the benefits of coordination to a heteroatom have been removed (lithium bearing HMPA ligands is already well supplied with electron density) EVL-HMPA presumably prefers to deprotonate the site that is (a) next to the most inductively withdrawing group and (b) the least hindered.

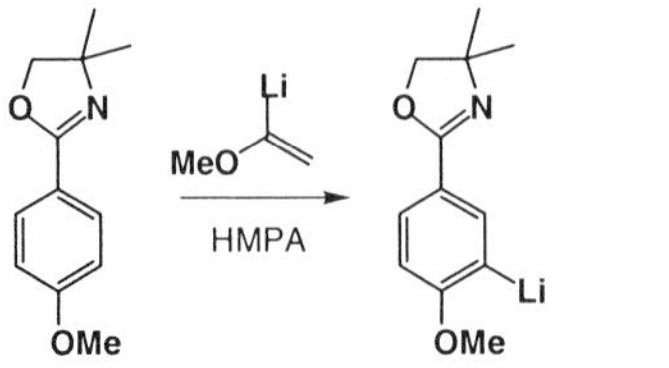

525 >99:1 regioselectivity

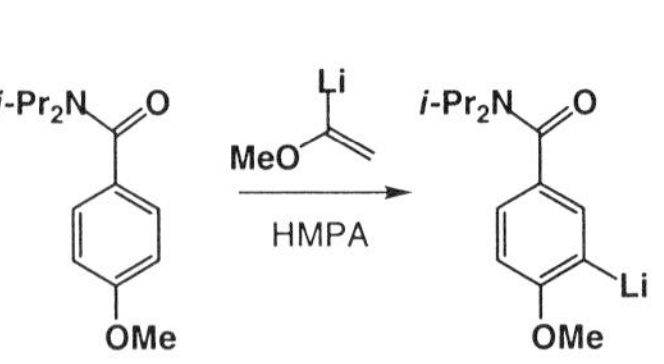

526 >98:2 regioselectivity

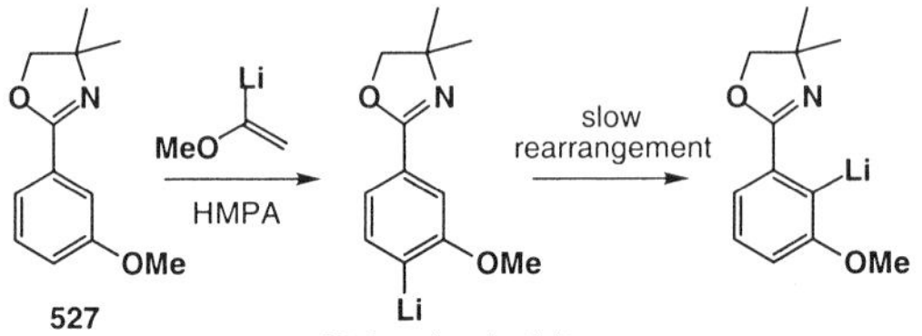

527 >98:2 regioselectivity

There are reports that other HMPA–organolithium complexes have a dramatic effect on lithiation regioselectivity away from coordination-favoured sites.[454]

The superbases (section 2.6) are similar in that they avoid hindered positions next to coordinating groups, and prefer to deprotonate *ortho* to small, powerfully inductively withdrawing OMe, or even at benzylic positions. Superbases prefer to deprotonate both **528** and **529** *ortho* to OMe rather than the usually more powerfully directing anilide group.[455]

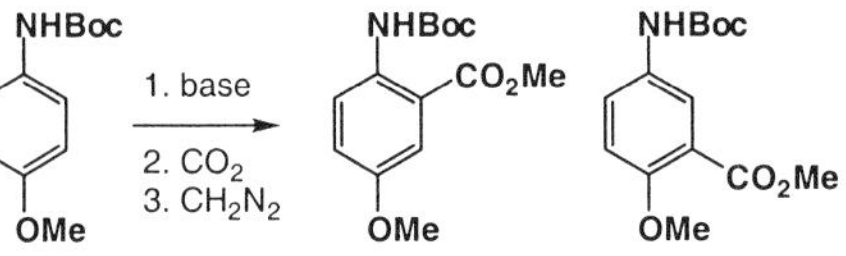

528

BuLi 88:12
BuLi–KO*t*-Bu 6:94

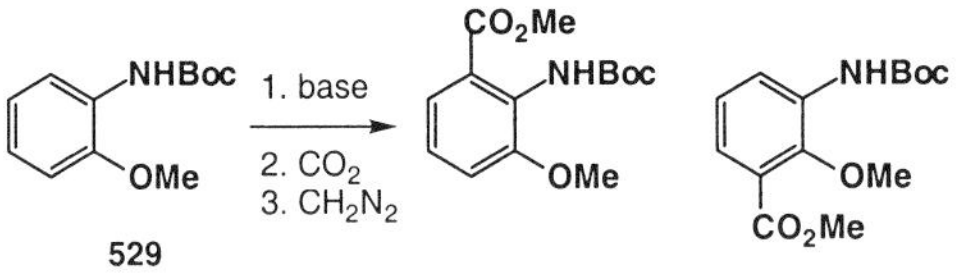

529

65% with BuLi 78% with BuLi–KO*t*-Bu

Steric hindrance controls a three-way divergence of regioselectivity in the metallation of **530**.[456] *n*-BuLi lithiates *ortho* to OMe at the site where coordination to the secondary amide can also be achieved. *t*-BuLi prefers to lithiate at the less hindered benzylic site, still presumably benefitting from amide coordination. The superbase, on the other hand, cares nothing for coordination to the amide and metallates at the less hindered site *ortho* to OMe.

t-BuLi metallates here

NHCOt-Bu

t-BuLi / KOt-Bu metallates here

OMe

n-BuLi metallates here

530

Other additives can have unexpected effects on regioselectivity: one equivalent of water, for example, changes the completely unselective lithiation of **531** into a synthetically useful reaction.[457]

3 x s-BuLi, 1 x H_2O, Et_2O, –78 °C, 1 h; s-BuLi, TMEDA; DMF

1:1:1 mixture — 531 — one regioisomer — 78%

Coordination to more than one heteroatom may turn off an otherwise favourable lithiation by preventing the organolithium reagent from approaching the substrate sufficiently closely. For example, the oxazoline **532** is not lithiated by BuLi even under forcing conditions, presumably because coordination to both N and O prevents approach of BuLi to the *ortho* position.[9]

532 — BuLi

There are a number of other reports of difficulties in ortholithiation when further coordination sites are present: molecules containing more than one strong coordinating substituent frequently require numerous equivalents of alkyllithium for lithiation.[244]

The competition between ortholithiation and alternative reactivity – α- or lateral lithiation for example, or halogen-metal exchange – has been quantified in only a few cases, and where such issues of chemoselectivity arise they are mentioned in the sections on these topics. Two general points are worth bearing in mind:

Ortholithiation vs. halogen-metal exchange. Provided the target organolithium is not too basic, ortholithiation can be favoured by using a more stable organolithium (PhLi for example,[448] or its more hindered congeners **537**[216]) or a lithium amide base. The possibilities for altering chemo- or regioselectivity with a change in base are well illustrated by the following series of reactions of bromothiophenes and bromofurans **533-536**.[458]

537 (R = Me or *i*-Pr)

α-lithiation	5-Li-2-Br-thiophene	← LDA	**533**	BuLi →	2-Li-thiophene	Br-Li exchange
ortholithiation	3-Li-2,5-Br₂-thiophene	← LDA	**534**	BuLi →	2-Br-5-Li-thiophene	Br-Li exchange
ortho/α-lithiation	3-Br-2-Li-furan	← LDA	**535**	BuLi →	3-Li-thiophene	Br-Li exchange
α-lithiation	5-Li-2,3-Br₂-furan	← LDA	**536**	BuLi →	3-Br-2-Li-thiophene	Br-Li exchange

Ortholithiation vs. benzylic or lateral lithiation. Usually, ortholithiated species are more basic than benzylically lithiated species, and lithium amide bases favour formation of the latter. Alkyllithiums give a better chance of achieving *ortho*-selectivity. The factors controlling *ortho* vs. lateral lithiation in tertiary amides have been studied in detail,[459] and show that increased substitution at an *ortho*-alkyl group disfavours lateral lithiation – ortho-isopropyl groups (**539**) cannot be lithiated. Increasing substitution at the *meta* position, which restricts the ability of the alkyl group to adopt a conformation suitable for deprotonation, also disfavours lateral lithiation. **537**, for example, gives only **538** with *s*-BuLi, while **541** gives a mixture of **542** and **543**. LDA gives only **542** from **541**.

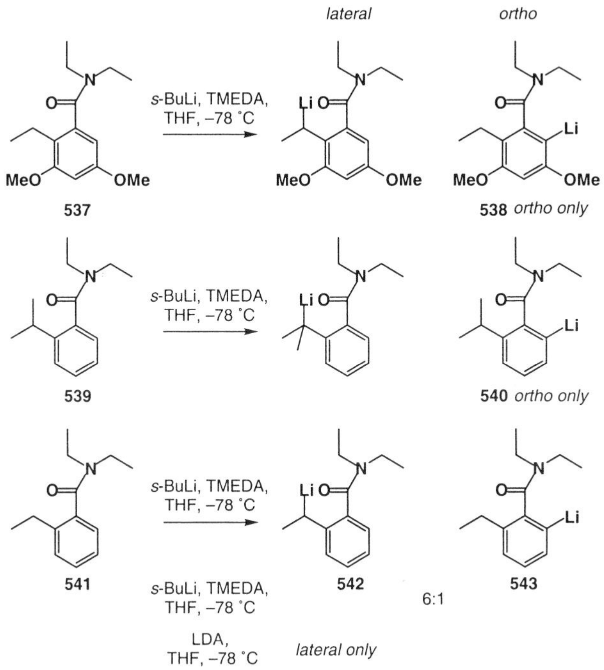

lateral *ortho*

MeO OMe s-BuLi, TMEDA, THF, –78 °C Li MeO OMe MeO Li OMe

537 **538** *ortho only*

s-BuLi, TMEDA, THF, –78 °C Li Li

539 **540** *ortho only*

s-BuLi, TMEDA, THF, –78 °C Li Li

541 **542** **543**

s-BuLi, TMEDA, THF, –78 °C 6:1

LDA, THF, –78 °C *lateral only*

For cases where the regioselectivity of lithiation cannot be biased in the required direction, silylation can be used to block acidic sites while further lithiations are carried out. The blocking silyl group is later readily removed with fluoride.[460] This method was used to avoid a competing lateral lithiation in the synthesis of the aldehyde **544**.[461]

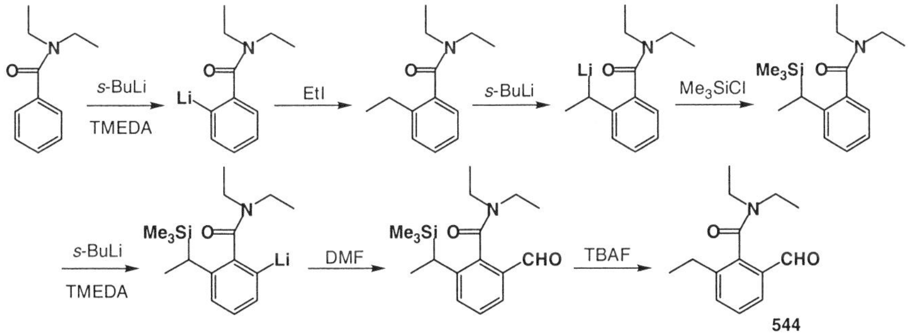

s-BuLi TMEDA Li EtI s-BuLi Li Me_3SiCl Me_3Si

s-BuLi TMEDA Me_3Si Li DMF Me_3Si CHO TBAF CHO

544

Silicon protection is also commonly used to direct lithiation chemistry in five-membered heterocycles. For example, oxazoles,[462] thiazoles[463] and *N*-alkylimidazoles[464,465] lithiate preferentially at C-2, where the inductive effect of the heteroatoms is greatest. If C-2 is blocked, lithiation occurs at C-5, where there is no adjacent lone pair to destabilise the organolithium. Functionalisation of these heterocycles at C-5 can therefore be achieved by first silylating C-2, reacting at C-5 and then removing the silyl group. The synthesis of **545** illustrates this sort of sequence.[466]

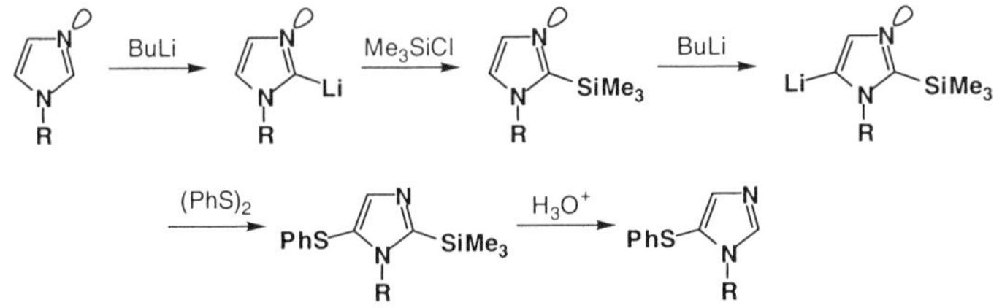

A less well developed, but potentially less intrusive alternative is to block acidic sites with deuterium and to use the kinetic isotope effect to prevent lithiation.[35,43] If the substrate can be polylithiated, the use of dianions (with the last formed anion being likely to be the most reactive) may give the required selectivity: this method works well in the halogen-metal exchange area (section 3.1).

References

1. Biellman, J.-F.; Ducep, J. B. *Org. Reac.* **1982,** *27*, 1.
2. Al-Aseer, M.; Beak, P.; Hay, D.; Kempf, D. J.; Mills, S.; Smith, S. G. *J. Am. Chem. Soc.* **1983,** *105*, 2080.
3. Hay, D. R.; Song, Z.; Smith, S. G.; Beak, P. *J. Am. Chem. Soc.* **1988,** *110*, 8145.
4. Beak, P.; Zajdel, W. J. *J. Am. Chem. Soc.* **1984,** *106*, 1010.
5. Seebach, D.; Wykypiel, W.; Lubosch, W.; Kalinowski, H. O. *Helv. Chim. Acta* **1978,** *61*, 3100.
6. Rondan, N. G.; Houk, K. N.; Beak, P.; Zajdel, W. J.; Chandrasekhar, J.; Schleyer, P. v. R. *J. Org. Chem.* **1981,** *46*, 4108.
7. Beak, P.; Brubaker, G. R.; Farney, R. *J. Am. Chem. Soc.* **1976,** *98*, 3621.
8. Meyers, A. I.; Dickman, D. I. *J. Am. Chem. Soc.* **1987,** *109*, 1263.
9. Meyers, A. I.; Ricker, W. F.; Fuentes, L. M. *J. Am. Chem. Soc.* **1983,** *105*, 2082.
10. Resek, J. E.; Beak, P. *J. Am. Chem. Soc.* **1994,** *116*, 405.
11. Anderson, D. R.; Faibish, N. C.; Beak, P. *J. Org. Chem.* **1999,** *121*, 7553.
12. Hoppe, D.; Hense, T. *Angew. Chem., Int. Ed. Engl.* **1997,** *36*, 2282.
13. Baldwin, J. E.; Höfle, G. A.; Lever, O. W. *J. Am. Chem. Soc.* **1974,** *96*, 7125.
14. Joule, J. A.; Mills, K.; Smith, G. F. *Heterocyclic Chemistry*, Third ed.; Chapman and Hall: London, 1995.
15. McClure, M. S.; Roschanger, F.; Hodson, S. J.; Millar, A.; Osterhout, M. A. *Synthesis* **2001**, 1681.
16. Still, W. C.; Macdonald, T. L. *J. Am. Chem. Soc.* **1974,** *96*, 5561.
17. Hoppe, D. *Angew. Chem., Int. Ed. Engl.* **1984,** *23*, 932.
18. Schlosser, M. In *Organometallics in Synthesis;* Schlosser, M. Ed.; Wiley: New York, 1994; pp. 1-166.
19. Schlosser, M.; Zellner, A.; Leroux, F. *Synthesis* **2001**, 1830.

20. Zhang, P.; Gawley, R. E. *J. Org. Chem.* **1993,** *58*, 3223.
21. Moret, E.; Schlosser, M. *Tetrahedron Lett.* **1984,** *25*, 4491.
22. Upton, C. J.; Beak, P. *J. Org. Chem.* **1975,** *40*, 1094.
23. Oppolzer, W.; Snowden, R. L. *Tetrahedron Lett.* **1976**, 4187.
24. Beak, P.; McKinnie, B. G. *J. Am. Chem. Soc.* **1977,** *99*, 5213.
25. Hodgson, D. M.; Norsikian, S. L. M. *Org. Lett.* **2001,** *3*, 461.
26. Yamauchi, Y.; Katagiri, T.; Uneyama, K. *Org. Lett.* **2002,** *4*, 173.
27. Gawley, R. E.; Rein, K. In *Comprehensive Organic Synthesis;* Trost, B. M.; Fleming, I. Eds.; Pergamon: Oxford, 1990; Vol. 1; pp. 459.
28. Beak, P.; Zajdel, W. J.; Reitz, D. B. *Chem. Rev.* **1984,** *84*, 471.
29. Beak, P.; Reitz, D. B. *Chem. Rev.* **1978,** *78*, 275.
30. Hillis, L. R.; Ronald, R. C. *J. Org. Chem.* **1981,** *46*, 3349.
31. Ahlbrecht, H.; Dollinger, H. *Tetrahedron Lett.* **1984,** *25*, 1353.
32. Beak, P.; Reitz, D. B. *Chem. Rev.* **1979,** *78*, 275.
33. Schlecker, R.; Seebach, D.; Lubosch, W. *Helv. Chim. Acta* **1978,** *61*, 512.
34. Wykypiel, W.; Lohmann, J.; Seebach, D. *Helv. Chim. Acta* **1981,** *64*, 1337.
35. Ahmed, A.; Clayden, J.; Rowley, M. *Tetrahedron Lett.* **1998,** *39*, 6103.
36. Kise, N.; Urai, T.; Yoshida, Y.-i. *Tetrahedron Asymmetry* **1998,** *9*, 3125.
37. Snieckus, V.; Rogers-Evans, M.; Beak, P.; Lee, W. K.; Yum, E. K.; Freskos, J. *Tetrahedron Lett.* **1994,** *35*, 4067.
38. Fraser, R. R.; Boussard, G.; Potescu, I. D.; Whiting, J. J.; Wigfield, Y. Y. *Can. J. Chem.* **1973,** *51*, 1109.
39. Tischler, A. N.; Tischler, M. H. *Tetrahedron Lett.* **1978**, 3407.
40. Resek, J. E.; Beak, P. *Tetrahedron Lett.* **1993,** *34*, 3043.
41. Burns, S. A.; Corriu, R. J. P.; Huynh, V.; Moreau, J. J. E. *J. Organomet. Chem.* **1987,** *333*, 281.
42. Jacobson, M. A.; Williard, P. G. *J. Org. Chem.* **2002,** *67*, 32.
43. Clayden, J.; Pink, J. H.; Westlund, N.; Wilson, F. X. *Tetrahedron Lett.* **1998,** *39*, 8377.
44. Ahmed, A.; Clayden, J.; Rowley, M. *J. Chem. Soc., Chem. Commun.* **1998**, 297.
45. Ahmed, A.; Clayden, J.; Yasin, S. A. *J. Chem. Soc., Chem. Commun.* **1999**, 231.
46. Beak, P.; Meyers, A. I. *Acc. Chem. Res.* **1986,** *19*, 356.
47. Durst, T.; van den Elzen, R.; LeBelle, M. J. *J. Am. Chem. Soc.* **1972,** *94*, 9261.
48. Lohmann, J.; Seebach, D.; Syfrig, M. A.; Yoshifuji, M. *Angew. Chem., Int. Ed. Engl.* **1981,** *20*, 128.
49. Seebach, D.; Lohmann, J.; Syfrig, M. A.; Yoshifuji, M. *Tetrahedron* **1983,** *39*, 1963.
50. Katritzky, A. R.; Grzeskowiak, N. E.; Salgado, H. J.; bin Zahari, Z. *Tetrahedron Lett.* **1980**, 4451.
51. Katritzky, A. R.; Arrowsmith, J.; bin Zahari, Z.; Jayaram, C.; Siddiqui, T.; Vassilatos, S. N. *J. Chem. Soc., Perkin Trans. 1* **1980**, 2851.
52. Beak, P.; Lee, B. *J. Org. Chem.* **1989,** *54*, 458.
53. Whisler, M. C.; Vaillancourt, L.; Beak, P. *Org. Lett.* **2000,** *2*, 2655.
54. Hassel, T.; Seebach, D. *Helv. Chim. Acta* **1978,** *61*, 2237.
55. Katritzky, A. R.; Jayaram, C.; Vassilatos, S. N. *Tetrahedron* **1983,** *39*, 2023.

56. Katritzky, A. R.; Huang, Z.; Fang, Y.; Prakash, I. *J. Org. Chem.* **1999,** *64*, 2124.
57. Meyers, A. I.; ten Hoeve, W. *J. Am. Chem. Soc.* **1980,** *102*, 7125.
58. Meyers, A. I.; Fuentes, L. M. *J. Am. Chem. Soc.* **1983,** *105*, 117.
59. Meyers, A. I.; Edwards, P. D.; Bailey, T. R.; Jagdmann, G. E. *J. Org. Chem.* **1985,** *50*, 1019.
60. Seebach, D.; Enders, D. *Angew. Chem., Int. Ed. Engl.* **1975,** *14*, 15.
61. Fraser, R. R.; Passannanti, S. *Synthesis* **1976**, 548.
62. Fraser, R. R.; Grindley, T. B.; Passannanti, S. *Can. J. Chem.* **1975,** *53*, 2473.
63. Lyle, R. E.; Saveedra, J. E.; Lyle, G. G.; Fribush, H. M.; Marshall, J. L. *Tetrahedron Lett.* **1976**, 4431.
64. Renger, B.; Kalinowski, H. O.; Seebach, D. *Chem. Ber.* **1977,** *110*, 1866.
65. Seebach, D.; Enders, D. *Chem. Ber.* **1975,** *108*, 1293.
66. Kauffmann, T.; Köppelmann, E.; Berg, H. *Angew. Chem., Int. Ed. Engl.* **1970,** *9*, 163.
67. Vossmann, P.; Hornig, K.; Fröhlich, R.; Würthwein, E.-U. *Synthesis* **2001**, 1415.
68. Adlington, R. M.; Baldwin, J. E.; Bottaro, J. C.; Perry, M. W. D. *J. Chem. Soc., Chem. Commun.* **1983**, 1040.
69. Schöllkopf, U. *Angew. Chem., Int. Ed. Engl.* **1977,** *16*, 339.
70. Ahlbrecht, H.; Eichler, M. *Synthesis* **1970**, 672.
71. Seebach, D.; Lehr, F. *Angew. Chem., Int. Ed. Engl.* **1976,** *15*, 505.
72. Barton, D. H. R.; Beugelmans, R.; Young, R. N. *Nouv. J. Chim.* **1978,** *2*, 363.
73. Ebden, M. R.; Simpkins, N. S. *Tetrahedron Lett.* **1995,** *36*, 8697.
74. Raabe, G.; Gais, H.-J.; Fleischhauer, J. *J. Am. Chem. Soc.* **1996,** *118*, 4622.
75. Simpkins, N. S. *Sulphones in Organic Synthesis*; Pergamon: Oxford, 1993.
76. Ogura, K. In *Comprehensive Organic Synthesis;* Trost, B. M.; Fleming, I. Eds.; Pergamon: Oxford, 1990; Vol. 1; pp. 505.
77. Corey, E. J.; Seebach, D. *J. Org. Chem.* **1966,** *31*, 4097.
78. Gilman, H.; Webb, F. J. *J. Am. Chem. Soc.* **1949,** *71*, 4062.
79. Bordwell, F. G.; Matthews, W. S.; Vanier, N. R. *J. Am. Chem. Soc.* **1975,** *97*, 442.
80. Borden, W. T.; Davidson, E. R.; Andersen, N. H.; Denniston, A. D.; Epiotis, N. D. *J. Am. Chem. Soc.* **1978,** *100*, 1604.
81. Epiotis, N. D.; Yates, R. L.; Bernardi, F.; Wolfe, S. *J. Am. Chem. Soc.* **1976,** *98*, 5435.
82. Streitwieser, A.; Ewing, S. P. *J. Am. Chem. Soc.* **1975,** *95*, 190.
83. Kingsbury, C. A. *J. Org. Chem.* **1972,** *37*, 102.
84. McDougal, P. G.; Condon, B. D.; Lafosse, M. D.; Lauro, A. M.; VanDerveer, D. *Tetrahedron Lett.* **1988,** *29*, 2547.
85. Mukaiyama, T.; Narasaka, K.; Maekawa, K.; Furusato, M. *Bull. Chem. Soc. Jap.* **1971,** *44*, 2285.
86. Meyers, A. I.; Ford, M. *J. Org. Chem.* **1976,** *41*, 1735.
87. Johnson, C.; Tanaka, K. *Synthesis* **1976**, 413.
88. Trost, B. M.; Vaultier, M.; Santiago, M. L. *J. Am. Chem. Soc.* **1980,** *102*, 7929.
89. Carpita, A.; Rossi, R.; Veracini, C. A. *Tetrahedron* **1985,** *41*, 1919.
90. Carpenter, A. J.; CHadwick, D. J. *Tetrahedron Lett.* **1985,** *26*, 1777.

91. Totter, F.; Rittmeyer, P. In *Organometallics in Synthesis;* Schlosser, M. Ed.; Wiley: New York, 1994; pp. 167-194.
92. Krief, A. In *Comprehensive Organic Chemistry;* Trost, B. M.; Fleming, I. Eds.; Pergamon: Oxford, 1990; Vol. 1; pp. 629.
93. Chan, T. H.; Chang, E.; Vinokur, E. *Tetrahedron Lett.* **1970**, 1137.
94. Corriu, R. J. P.; Masse, E. *J. Organomet. Chem.* **1973,** *57*, C5.
95. Peterson, D. J. *J. Org. Chem.* **1968,** *33*, 780.
96. Brough, P. A.; Fisher, S.; Zhao, B.; Thomas, R. C.; Snieckus, V. *Tetrahedron Lett.* **1996,** *37*, 2915.
97. Wroczynski, R. J.; Baum, M. W.; Kost, D.; Mislow, K.; Vick, S. L.; Seyferth, D. *J. Organomet. Chem.* **1979,** *170*, C29.
98. Fraenkel, G.; Qiu, F. *J. Am. Chem. Soc.* **2000,** *122*, 12806.
99. Beak, P.; Hunter, J. E.; Jun, Y. M. *J. Am. Chem. Soc.* **1983,** *105*, 6350.
100. Gilman, H.; Bebb, R. L. *J. Am. Chem. Soc.* **1939,** *61*, 109.
101. Wittig, G.; Fuhrman, G. *Chem. Ber.* **1940,** *73*, 1197.
102. Gschwend, H. W.; Rodriguez, H. R. *Org. Reac.* **1979,** *26*, 1.
103. Beak, P.; Brown, R. A. *J. Org. Chem.* **1982,** *47*, 34.
104. Gant, T. G.; Meyers, A. I. *Tetrahedron* **1994,** *50*, 2297.
105. Beak, P.; Snieckus, V. *Acc. Chem. Res.* **1982,** *15*, 306.
106. Snieckus, V. *Chem. Rev.* **1990,** *90*, 879.
107. Gilman, H.; Morton, J. W. *Org. Reac.* **1954,** *8*, 258.
108. Rausch, M. D.; Ciappenelli, D. J. *J. Organomet. Chem.* **1967,** *10*, 127.
109. Lewis, H. L.; Brown, T. L. *J. Am. Chem. Soc.* **1970,** *92*, 4664.
110. Roberts, J. D.; Curtin, D. Y. *J. Am. Chem. Soc.* **1946,** *68*, 1658.
111. Stratakis, M. *J. Org. Chem.* **1997,** *62*, 3024.
112. Shatenshtein, A. I. *Tetrahedron* **1962,** *18*, 95.
113. Huisgen, R.; Mack, W.; Herbig, K.; Ott, N.; Anneser, E. *Chem. Ber.* **1960,** *93*, 412.
114. Jones, F. N.; Zinn, M. F.; Hauser, C. R. *J. Org. Chem.* **1963,** *28*, 663.
115. Bauer, W.; Schleyer, P. v. R. *J. Am. Chem. Soc.* **1989,** *111*, 7191.
116. van Eikema Hommes, N. J. R.; Schleyer, P. v. R. *Tetrahedron* **1994,** *50*, 5903.
117. Rennels, R. A.; Maliakal, A. S.; Collum, D. B. *J. Am. Chem. Soc.* **1998,** *120*, 421.
118. Chadwick, S. T.; Rennels, R. A.; Rutherford, J. L.; Collum, D. B. *J. Am. Chem. Soc.* **2000,** *122*, 8640.
119. Clayden, J.; Frampton, C. S.; McCarthy, C.; Westlund, N. *Tetrahedron* **1999,** *55*, 14161.
120. Narasimhan, N. S.; Mali, R. S. *Synthesis* **1983**, 957.
121. Gay, R. L.; Hauser, C. R. *J. Am. Chem. Soc.* **1967,** *89*, 2297.
122. Eaborn, C.; Golborn, P.; Taylor, R. *J. Organomet. Chem.* **1967,** *10*, 171.
123. Barluenga, J.; González, R.; Fañanás, F. J.; Yus, M.; Foubelo, F. *J. Chem. Soc., Perkin Trans. 1* **1994**, 1069.
124. Narasimhan, N. S.; Ranade, A. C. *Indian J. Chem.* **1969,** *7*, 538.
125. Kirby, A. J.; Percy, J. M. *Tetrahedron* **1988,** *44*, 6903.
126. Kiefl, C.; Mannschreck, A. *Synthesis* **1995**, 1033.

127. Comins, D. L.; Brown, J. D. *J. Org. Chem.* **1984,** *49*, 1078.
128. Saá, J. M.; Morey, J.; Frontera, A.; Deyá, P. M. *J. Am. Chem. Soc.* **1995,** *117*, 1105.
129. Coll, G.; Morey, J.; Costa, A.; Saá, J. M. *J. Org. Chem.* **1988,** *53*, 5345.
130. Shirley, D. A.; Cheng, C. F. *J. Organomet. Chem.* **1969,** *20*, 251.
131. Kawikawa, T.; Kubo, I. *Synthesis* **1986**, 431.
132. Bowles, P.; Clayden, J.; Helliwell, M.; McCarthy, C.; Tomkinson, M.; Westlund, N. *J. Chem. Soc., Perkin Trans. 1* **1997**, 2607.
133. Lombardino, J. G. *J. Org. Chem.* **1971,** *36*, 1843.
134. Stoyanovich, F. M.; Karpenko, R. G.; Gol'dfarb, Y. L. *Tetrahedron* **1971,** *27*, 433.
135. Slocum, D. W.; Jennings, C. A. *J. Org. Chem.* **1976,** *41*, 3653.
136. Meyers, A. I.; Pansegrau, P. D. *Tetrahedron Lett.* **1983,** *24*, 4935.
137. Fraser, R. R.; Bresse, M.; Mansour, T. S. *J. Am. Chem. Soc.* **1983,** *105*, 7790.
138. Meyers, A. I.; Lutomski, K. *J. Org. Chem.* **1979,** *44*, 4464.
139. Beak, P.; Brown, R. A. *J. Org. Chem.* **1979,** *44*, 4463.
140. Beak, P.; Tse, A.; Hawkins, J.; Chen, C.-W.; Mills, S. *Tetrahedron* **1983,** *39*, 1983.
141. Iwao, M.; Iihama, T.; Mahalanabis, K. K.; Perrier, H.; Snieckus, V. *J. Org. Chem.* **1989,** *54*, 24.
142. Reuman, M.; Meyers, A. I. *Tetrahedron* **1985,** *41*, 837.
143. Clayden, J.; Cooney, J. J. A.; Julia, M. *J. Chem. Soc., Perkin Trans. 1* **1995**, 7.
144. Clayden, J.; Julia, M. *J. Chem. Soc., Chem. Commun.* **1993**, 1682.
145. Quesnelle, C.; Iihama, T.; Aubert, T.; Perrier, H.; Snieckus, V. *Tetrahedron Lett.* **1992,** *33*, 2625.
146. Lockard, J. P.; Schroeck, C. W.; Johnson, C. R. *Synthesis* **1973**, 485.
147. Schlosser, M.; Guio, L.; Leroux, F. *J. Am. Chem. Soc.* **2001,** *123*, 3822.
148. Mills, R. J.; Horvath, R. F.; Sibi, M. P.; Snieckus, V. *Tetrahedron Lett.* **1985,** *26*, 1145.
149. Ludt, R. E.; Griffiths, T. S.; McGrath, K. N.; Hauser, C. R. *J. Org. Chem.* **1973,** *38*, 1668.
150. Beak, P.; Brown, R. A. *J. Org. Chem.* **1977,** *42*, 1823.
151. Casas, R.; Cavé, C.; d'Angelo, J. *Tetrahedron Lett.* **1996,** *36*, 1039.
152. Keck, G. E.; Wager, T. T.; Rodriguez, J. F. D. *J. Am. Chem. Soc.* **1999,** *121*, 5176.
153. Park, T. K.; Danishefsky, S. J. *Tetrahedron Lett.* **1995,** *36*, 195.
154. Bond, A. D.; Clayden, J.; Wheatley, A. E. H. *Acta Cryst E* **2001,** *57*, 291.
155. Beak, P.; Kerrick, S. T.; Gallagher, D. J. *J. Am. Chem. Soc.* **1993,** *115*, 10628.
156. Clayden, J.; Davies, R. P.; Hendy, M. A.; Snaith, R.; Wheatley, A. E. H. *Angew. Chem., Int. Ed. Engl.* **2001,** *40*, 1238.
157. Sibi, M. P.; Shankaran, K.; Hahn, W. R.; Alo, B. I.; Snieckus, V. *Tetrahedron Lett.* **1987,** *28*, 2933.
158. Brimble, M. A.; Chan, S. H. *Aust. J. Chem.* **1998,** *51*, 275.
159. Bowles, P.; Clayden, J.; Tomkinson, M. *Tetrahedron Lett.* **1995,** *36*, 9219.
160. Clayden, J.; Westlund, N.; Wilson, F. X. *Tetrahedron Lett.* **1999,** *40*, 3329.
161. Sibi, M. P.; Miah, M. A. J.; Snieckus, V. *J. Org. Chem.* **1984,** *49*, 737.
162. Cuevas, J.-C.; Patil, P.; Snieckus, V. *Tetrahedron Lett.* **1989,** *30*, 5841.

163. Reitz, D. B.; Massey, S. M. *J. Org. Chem.* **1990,** *55*, 1375.
164. Hintze, F.; Hoppe, D. *Synthesis* **1992**, 1216.
165. Clayden, J.; Youssef, L. H. *unpublished work.*
166. Comins, D. L.; Brown, J. D. *J. Org. Chem.* **1986,** *51*, 3566.
167. Phillion, D. P.; Walker, D. M. *J. Org. Chem.* **1995,** *60*, 8417.
168. Puterbaugh, W. H.; Hauser, C. R. *J. Org. Chem.* **1964,** *29*, 853.
169. Kiselyov, A. S. *Tetrahedron Lett.* **1995,** *36*, 493.
170. Fitt, J. J.; Gschwend, H. W. *J. Org. Chem.* **1976,** *41*, 4029.
171. Metallinos, C.; Nerdinger, S.; Snieckus, V. *Org. Lett.* **1999,** *1*, 1183.
172. Pratt, S. A.; Goble, M. P.; Mulvaney, M. J.; Wuts, P. G. M. *Tetrahedron Lett.* **2000,** *41*, 3559.
173. McCombie, S. W.; Lin, S.-I.; Vice, S. F. *Tetrahedron Lett.* **1999,** *40*, 8767.
174. Garibay, P.; Toy, P. H.; Hoeg-Jensen, T.; Janda, K. D. *Synlett* **1999**, 1478.
175. Quéguiner, G.; Marsais, F.; Snieckus, V.; Epsztajn, J. *Adv. Heterocyclic Chem* **1991,** *52*, 187.
176. Vecchia, L. D.; Vlattas, I. *J. Org. Chem.* **1977,** *42*, 2649.
177. Chadwick, D. J.; McKnight, M. V.; Ngochindo, R. *J. Chem. Soc., Perkin Trans. 1* **1982**, 1343.
178. Carpenter, A. J.; Chadwick, D. J. *J. Org. Chem.* **1985,** *50*, 4362.
179. Doast, E. G.; Snieckus, V. *Tetrahedron Lett.* **1985,** *26*, 1149.
180. Comins, D. L.; Killpack, M. O. *J. Org. Chem.* **1987,** *52*, 104.
181. Donohoe, T. J.; Guillermin, J.-B.; Calabrese, A. A.; Walter, D. S. *Tetrahedron Lett.* **2001,** *42*, 5841.
182. Iwao, M.; Kuraishi, T. *Tetrahedron Lett.* **1983,** *24*, 2649.
183. Wang, X.; Snieckus, V. *Tetrahedron Lett.* **1991,** *32*, 4883.
184. Watanabe, M.; Snieckus, V. *J. Am. Chem. Soc.* **1980,** *102*, 1457.
185. Paulmier, C.; Morel, J.; Semard, D.; Pastour, P. *Bull. Soc. Chim. Fr.* **1973**, 2434.
186. Comins, D. L. *Synlett* **1992**, 615.
187. Comins, D. L.; Brown, J. D.; Mantlo, N. B. *Tetrahedron Lett.* **1982,** *23*, 3979.
188. Comins, D. L.; Killpack, M. O. *J. Org. Chem.* **1990,** *55*, 69.
189. Butenschon, H.; Winkler, M.; Vollhardt, K. P. C. *J. Chem. Soc., Chem. Commun.* **1986**, 388.
190. Nakano, J.; Uchida, K.; Fujimoto, Y. *Heterocycles* **1989,** *29*, 427.
191. Boulet, C. A.; Poulton, G. A. *Heterocycles* **1989,** *28*, 405.
192. Sibi, M. P.; Snieckus, V. *J. Org. Chem.* **1983,** *48*, 1935.
193. Sibi, M. P.; Chattopadhyay, S.; Dankwardt, J. W.; Snieckus, V. *J. Am. Chem. Soc.* **1985,** *107*, 6312.
194. Cunningham, A.; Dennis, M. R.; Woodward, S. *J. Chem. Soc., Perkin Trans. 1* **2000,** 4422.
195. Sisko, J.; Weinreb, S. M. *Synth. Commun.* **1988**, 1035.
196. Lepley, A. P.; Khan, W. A.; Giumanini, A. B.; Giumanini, A. G. *J. Org. Chem.* **1966,** *31*, 2047.
197. Slocum, D. W.; Bock, G.; Jennings, C. A. *Tetrahedron Lett.* **1970**, 3443.

198. Ludt, R. E.; Crowther, G. P.; Hauser, C. R. *J. Org. Chem.* **1970,** *35*, 1288.
199. Fuhrer, W.; Gschwend, H. W. *J. Org. Chem.* **1979,** *44*, 1133.
200. Stanetty, P.; Koller, H.; Mihovilovic, M. *J. Org. Chem.* **1992,** *57*, 6833.
201. Turner, J. A. *J. Org. Chem.* **1983,** *48*, 3401.
202. Stanetty, P.; Krumpak, B. *J. Org. Chem.* **1996,** *61*, 5130.
203. Smith, K.; El-Hiti, G. A.; Shukla, A. P. *J. Chem. Soc., Perkin Trans. 1* **1999**, 2305.
204. Meigh, J.-P.; Álvarez, M.; Joule, J. A. *J. Chem. Soc., Perkin Trans. 1* **2001**, 2012.
205. Smith, K.; El-Hiti, G. A.; Hawes, A. C. *Synlett* **1999**, 945.
206. Shirley, D. A.; Reeves, B. J. *J. Organomet. Chem.* **1969,** *16*, 1.
207. Cabiddu, S.; Floris, C.; Melis, S. *Tetrahedron Lett.* **1986,** *27*, 4625.
208. Cabiddu, S.; Fattuoni, C.; Floris, C.; Gelli, G.; Melis, S.; Sotgiu, F. *Tetrahedron* **1990,** *46*, 861.
209. Cabiddu, S.; Fattuoni, C.; Floris, C.; Gelli, G.; Melis, S. *Tetrahedron* **1993,** *49*, 4965.
210. Cabiddu, S.; Fattuoni, C.; Floris, C.; Melis, S.; Serci, A. *Tetrahedron* **1994,** *50*, 6037.
211. Furukawa, N.; Shibutani, T.; Fujihara, H. *Tetrahedron Lett.* **1989,** *30*, 7091.
212. Shibutani, T.; Fujihara, H.; Furukawa, N. *Tetrahedron Lett.* **1991,** *32*, 2943.
213. Argouarch, G.; Samuel, O.; Riant, O.; Daran, J.-C.; Kagan, H. B. *Eur. J. Org. Chem.* **2000**, 2893.
214. Riant, O.; Argouarch, G.; Guillaneux, D.; Samuel, O.; Kagan, H. B. *J. Org. Chem.* **1998,** *63*, 3511.
215. Hua, D. H.; Lagneau, N. M.; Chen, Y.; Robben, P. M.; Clapham, G.; Robinson, P. D. *J. Org. Chem.* **1996,** *61*, 4508.
216. Pedersen, H. L.; Johanssen, M. *J. Chem. Soc., Chem. Commun.* **1999**, 2517.
217. Spangler, L. A. *Tetrahedron Lett.* **1996,** *37*, 3639.
218. Watanabe, H.; Schwarz, R. A.; Hauser, C. R.; Lewis, J.; Slocum, D. W. *Can. J. Chem.* **2969,** *47*, 1543.
219. Shafer, S. J.; Closson, W. D. *J. Org. Chem.* **1975,** *40*, 889.
220. Hellwinkel, D.; Supp, M. *Tetrahedron Lett.* **1975**, 1499.
221. Katritzky, A. R.; Lue, R. *J. Org. Chem.* **1996,** *55*, 74.
222. Schmid, R.; Foricher, J.; Cereghetti, M.; Schönholzer, P. *Helv. Chim. Acta* **1988,** *74*, 370.
223. Jones, F. N.; Vaulx, R.; Hauser, C. R. *J. Org. Chem.* **1963,** *28*, 3461.
224. Klein, K. P.; Hauser, C. R. *J. Org. Chem.* **1967,** *32*, 1479.
225. Ludt, R. E.; Hauser, C. R. *J. Org. Chem.* **1971,** *36*, 1607.
226. Simig, G.; Schlosser, M. *Tetrahedron Lett.* **1988,** *29*, 4277.
227. Cauquil, G.; Casadevall, A.; Casadevall, E. *Bull. Soc. Chim. Fr.* **1960**, 1049.
228. Walborsky, H. M.; Ronman, P. *J. Org. Chem.* **1978,** *43*, 731.
229. Bruce, M. I. *Angew. Chem., Int. Ed. Engl.* **1977,** *16*, 73.
230. Ziegler, F. E.; Fowler, K. W. *J. Org. Chem.* **1976,** *41*, 1564.
231. Flippin, L. A.; Muchowski, J. M. *J. Org. Chem.* **1993,** *58*, 2631.
232. Krizan, T. D.; Martin, J. C. *J. Org. Chem.* **1982,** *47*, 2681.
233. Klumpp, G. W. *Rec. Trav. Chim. Pays-Bas* **1986,** *105*, 1.
234. Shirley, D. A.; Hendrix, J. P. *J. Organomet. Chem.* **1968,** *11*, 217.

235. Shirley, D. A.; Johnson, J. R.; Hendrix, J. P. *J. Organomet. Chem.* **1968,** *11*, 209.
236. Cabiddu, S.; Contini, L.; Fattuoni, C.; Floris, C.; Gelli, G. *Tetrahedron* **1991,** *47*, 9279.
237. Christensen, H. *Synth. Commun.* **1974,** *4*, 1.
238. Christensen, H. *Synth. Commun.* **1975,** *5*, 65.
239. Stern, R.; English, J.; Cassidy, H. G. *J. Am. Chem. Soc.* **1957,** *79*, 5797.
240. Clayden, J.; Helliwell, M.; McCarthy, C.; Westlund, N. *J. Chem. Soc., Perkin Trans. 1* **2000**, 3232.
241. Kamikawa, T.; Kubo, I. *Synthesis* **1986**, 431.
242. Pocci, M.; Bertini, V.; Lucchesini, F.; de Munno, A.; Picci, N.; Iemma, F.; Alfei, S. *Tetrahedron Lett.* **2001,** *42*, 1351.
243. Langham, W.; Brewster, R. Q.; Gilman, H. *J. Am. Chem. Soc.* **1941,** *63*, 545.
244. Clayden, J.; Kenworthy, M. N.; Youssef, L. H.; Helliwell, M. *Tetrahedron Lett.* **2000,** *41*, 5171.
245. Hillebrand, S.; Bruckmann, J.; Krüger, C.; Haenel, M. W. *Tetrahedron Lett.* **1995,** *36*, 75.
246. McWilliams, K.; Kelly, J. W. *J. Org. Chem.* **1996,** *61*, 7408.
247. Haenel, M. W.; Jakubik, D.; Rothenberger, E.; Schroth, G. *Chem. Ber.* **1991,** *124*, 1705.
248. Jean, F.; Melnyk, O.; Tartar, A. *Tetrahedron Lett.* **1995,** *36*, 7657.
249. Schöllkopf, U. *Angew. Chem., Int. Ed. Engl.* **1979,** *18*, 763.
250. Gilman, H.; Meikle, W. J.; Morton, J. W. *J. Am. Chem. Soc.* **1952,** *74*, 6282.
251. Santucci, L.; Gilman, H. *J. Am. Chem. Soc.* **1958,** *80*, 4537.
252. Meyer, N.; Seebach, D. *Angew. Chem., Int. Ed. Engl.* **1978,** *17*, 522.
253. Uemura, M.; Tokuyama, S.; Sakan, T. *Chem. Lett.* **1975**, 1195.
254. Mortier, J.; Moyroud, J.; Bennetau, B.; Cain, P. A. *J. Org. Chem.* **1994,** *59*, 4042.
255. Bennetau, B.; Mortier, J.; Moyroud, J.; Guesnet, J.-L. *J. Chem. Soc., Perkin Trans. 1* **1995**, 1265.
256. Ameline, G.; Vaultier, M.; Mortier, J. *Tetrahedron Lett.* **1996,** *37*, 8175.
257. Mongin, F.; Trécourt, F.; Quéguiner, G. *Tetrahedron Lett.* **1999,** *40*, 5483.
258. Knight, D. W.; Nott, A. P. *J. Chem. Soc., Perkin Trans. 1* **1983**, 1125.
259. Caron, S.; Hawkins, J. M. *J. Org. Chem.* **1998,** *63*, 2054.
260. Mongin, F.; Tognini, A.; Cottet, F.; Schlosser, M. *Tetrahedron Lett.* **1998,** *39*, 1749.
261. Mongin, F.; Schlosser, M. *Tetrahedron Lett.* **1997,** *38*, 1559.
262. Gilman, H.; Soddy, T. S. *J. Org. Chem.* **1957,** *22*, 1715.
263. Kirsch, P.; Bremer, M. *Angew. Chem., Int. Ed. Engl.* **2000,** *39*, 4217.
264. Haiduc, I.; Gilman, H. *Chem. and Ind. (London)* **1968**, 1278.
265. Meyers, A. I.; Pansegrau, P. D. *J. Chem. Soc., Chem. Commun.* **1985**, 690.
266. Pansegrau, P. D.; Rieker, W. F.; Meyers, A. I. *J. Am. Chem. Soc.* **1988,** *110*, 7178.
267. Clarke, A. J.; McNamara, S.; Meth-Cohn, O. *Tetrahedron Lett.* **1974**, 2373.
268. Gros, P.; Fort, Y.; Caubère, P. *J. Chem. Soc., Perkin Trans. 1* **1997**, 3597.
269. Zoltewicz, J. A.; Dill, C. D. *Tetrahedron* **1996,** *52*, 14469.
270. Chambers, R. D.; Drakesmith, F. G.; Musgrave, W. K. R. *J. Chem. Soc.* **1965**, 5045.

271. Chambers, R. D.; Heaton, C. A.; Musgrave, W. K. R. *J. Chem. Soc. (C)* **1969**, 1700.
272. Francis, R. F.; Crews, C. D.; Scott, B. S. *J. Org. Chem.* **1978,** *43*, 3227.
273. Marsais, F.; Granger, P.; Quéguiner, G. *J. Org. Chem.* **1981,** *46*, 4494.
274. Güngör, T.; Marsais, F.; Quéguiner, G. *J. Organomet. Chem.* **1981,** *215*, 139.
275. Mallet, M. *J. Organomet. Chem.* **1991,** *406*, 49.
276. Marsais, F.; Bouley, E.; Quéguiner, G. *J. Organomet. Chem.* **1979,** *171*, 273.
277. Kress, T. J. *J. Org. Chem.* **1979,** *44*, 2081.
278. Turck, A.; Plé, N.; Mongin, F.; Quéguiner, G. *Tetrahedron* **2001,** *57*, 4489.
279. Mongin, F.; Quéguiner, G. *Tetrahedron* **2001,** *57*, 4059.
280. Marsais, F.; Quéguiner, G. *Tetrahedron* **1983,** *39*, 2009.
281. Schlosser, M. *Mod. Synth. Methods* **1992,** *6*, 227.
282. Shi, G.-Q.; Takagishi, S.; Schlosser, M. *Tetrahedron* **1994,** *50*, 1129.
283. Porwisiak, J.; Dmowski, W. *Tetrahedron* **1994,** *50*, 12259.
284. Rocca, P.; Marsais, F.; Godard, A.; Quéguiner, G. *Tetrahedron* **1993,** *49*, 3325.
285. Remuzon, P.; Bouzard, D.; Jacquet, J.-P. *Heterocycles* **1993,** *36*, 431.
286. Arzel, E.; Rocca, P.; Marsais, F.; Godard, A.; Quéguiner, G. *J. Heterocyclic Chem.* **1997,** *34*, 1205.
287. Arzel, E.; Rocca, P.; Marsais, F.; Godard, A.; Quéguiner, G. *Heterocycles* **1999,** *50*, 215.
288. Rocca, P.; Marsais, F.; Godard, A.; Quéguiner, G. *Tetrahedron* **1994,** *35*, 2003.
289. Rocca, P.; Marsais, F.; Godard, A.; Quéguiner, G.; Adams, L.; Alo, B. *J. Heterocyclic Chem.* **1995,** *32*, 1171.
290. Chambers, R. D.; Hall, C. W.; Hutchinson, J.; Millar, R. W. *J. Chem. Soc., Perkin Trans. 1* **1998**, 1705.
291. Ito, Y.; Kunimoto, K.; Miyachi, S.; Kako, T. *Tetrahedron* **1991,** *32*, 4007.
292. Bertini, V.; Lucchesini, F.; Pocci, M.; De Munno, A. *Heterocycles* **Heterocycles,** *41*, 675.
293. Radinov, R.; Chanev, C.; Haimova, H. *J. Org. Chem.* **1991,** *56*, 4793.
294. Marzi, E.; Bigi, A.; Schlosser, M. *Eur. J. Org. Chem.* **2001**, 1371.
295. Murtiashaw, C. W.; Breitenbach, R.; Goldstein, S. W.; Pezzullo, S. L.; Quallich, G. J.; Sarges, R. *J. Org. Chem.* **1992,** *57*, 1930.
296. Karig, G.; Spencer, J. A.; Gallagher, T. *Org. Lett.* **2001,** *3*, 835.
297. Effenberger, F.; Daub, W. *Chem. Ber.* **1991,** *124*, 2119.
298. Numata, A.; Kondo, Y.; Sakamoto, T. *Synthesis* **1999**, 306.
299. Gu, Y. G.; Bayburt, E. K. *Tetrahedron Lett.* **1996,** *37*, 2565.
300. Gribble, G. W.; Saulnier, M. G. *Heterocycles* **1993,** *35*, 151.
301. Mallet, M.; Quéguiner, G. *Tetrahedron* **1982,** *38*, 3035.
302. Rocca, P.; Cochennec, C.; Marsais, F.; Thomas-dit-Dumont, L.; Godard, A.; Quéguiner, G. *J. Org. Chem.* **1993,** *58*, 7832.
303. Comins, D. L.; Saha, J. K. *Org. Lett.* **1996,** *61*, 9623.
304. Cochennec, C.; Rocca, P.; Marsais, F.; Godard, A.; Quéguiner, G. *Synthesis* **1995**, 321.
305. Sammakia, T.; Hurley, T. B. *J. Org. Chem.* **1999,** *64*, 4652.

306. Trécourt, F.; Gervais, B.; Mallet, M.; Quéguiner, G. *J. Org. Chem.* **1996,** *61*, 1673.
307. Trécourt, F.; Mallet, M.; Mongin, O.; Quéguiner, G. *J. Org. Chem.* **1994,** *59*, 6173.
308. Fang, F. G.; Xie, S.; Lowery, M. W. *J. Org. Chem.* **1994,** *59*, 6142.
309. Muratake, H.; Tonegawa, M.; Natsume, M. *Chem. Pharm. Bull.* **1998,** *46*, 400.
310. Comins, D. L.; Joseph, S. P.; Goehring, R. R. *J. Am. Chem. Soc.* **1994,** *116*, 4719.
311. Shimano, M.; Shibata, T.; Kamei, N. *Tetrahedron Lett.* **1998,** *39*, 4363.
312. Comins, D. L.; Baevsky, M. F.; Hong, H. *J. Am. Chem. Soc.* **1992,** *114*, 10971.
313. Henegar, K. E.; Ashford, S. W.; Baughman, T. A.; Sih, J. C.; Gu, R.-L. *J. Org. Chem.* **1997,** *62*, 6588.
314. Martin, O.; de la Cuesta, E.; Avendaño, C. *Tetrahedron* **1995,** *51*, 7547.
315. Patel, P.; Joule, J. A. *J. Chem. Soc., Chem. Commun.* **1985**, 1021.
316. Mongin, O.; Rocca, P.; Thomas-dit-Dumont, L.; Trécourt, F.; Marsais, F.; Godard, A.; Quéguiner, G. *J. Chem. Soc., Perkin Trans. 1* **1995**, 2503.
317. Trécourt, F.; Gervais, B.; Mongin, O.; Le Gal, C.; Mongin, F.; Quéguiner, G. *J. Org. Chem.* **1998,** *63*, 2892.
318. Kessar, S. V.; Singh, P. *Chem. Rev.* **1997,** *97*, 721.
319. Vedejs, E.; Chen, X. *J. Am. Chem. Soc.* **1996,** *118*, 1809.
320. Taylor, S. L.; Lee, D. Y.; Martin, J. C. *J. Org. Chem.* **1983,** *43*, 4158.
321. Davies, S. G.; Shipton, M. R. *J. Chem. Soc., Perkin Trans. 1* **1991**, 501.
322. Kelly, T. R.; Xu, W.; Sundaresan, J. *Tetrahedron Lett.* **1993,** *34*, 6173.
323. Arzel, E.; Rocca, P.; Marsais, F.; Godard, A.; Quéguiner, G. *Tetrahedron Lett.* **1998,** *39*, 6465.
324. Arzel, E.; Rocca, P.; Marsais, F.; Godard, A.; Quéguiner, G. *Tetrahedron* **1999,** *55*, 12149.
325. Turck, A.; Plé, N.; Quéguiner, G. *Heterocycles* **1994,** *37*, 2149.
326. Plé, N.; Turck, A.; Heynderickx, A.; Quéguiner, G. *J. Heterocyclic Chem.* **1994,** *31*, 1311.
327. Plé, N.; Turck, A.; Heynderickx, A.; Quéguiner, G. *Tetrahedron* **1998,** *54*, 4899.
328. Turck, A.; Trohay, D.; Mojovic, L.; Plé, N.; Quéguiner, G. *J. Organomet. Chem.* **1991,** *412*, 301.
329. Ward, J. S.; Merritt, L. *J. Heterocyclic Chem.* **1991,** *28*, 765.
330. Turck, A.; Plé, N.; Dognon, D.; Harmoy, C.; Quéguiner, G. *J. Heterocyclic Chem.* **1994,** *31*, 1449.
331. Plé, N.; Turck, A.; Couture, K.; Quéguiner, G. *J. Org. Chem.* **1995,** *60*, 3781.
332. Plé, N.; Turck, A.; Heynderickx, A.; Quéguiner, G. *Tetrahedron* **1998,** *54*, 9701.
333. Plé, N.; Turck, A.; Bardin, F.; Quéguiner, G. *J. Heterocyclic Chem.* **1992,** *29*, 467.
334. Lee, J. J.; Cho, S. H. *Bull. Korean Chem. Soc.* **1996,** *17*, 868.
335. Plé, N.; Turck, A.; Couture, K.; Quéguiner, G. *Synthesis* **1996**, 838.
336. Liu, W.; Walker, J. A.; Chen, J. J.; Wise, D. S.; Townend, B. L. *Tetrahedron Lett.* **1996,** *37*, 5325.
337. Turck, A.; Plé, N.; Leprêtre-Gauquère, A.; Quéguiner, G. *Heterocycles* **1998,** *49*, 205.
338. Turck, A.; Plé, N.; Pollet, P.; Mojovic, L.; Duflos, J.; Quéguiner, G. *J. Heterocyclic Chem.* **1997,** *34*, 621.

339. Turck, A.; Plé, N.; Mojovic, L.; Ndzi, B.; Quéguiner, G.; Haider, N.; Schuller, H.; Heinisch, G. *J. Heterocyclic Chem.* **1995,** *32*, 841.
340. Turck, A.; Plé, N.; Trohay, D.; Ndzi, B.; Quéguiner, G. *J. Heterocyclic Chem.* **1992,** *29*, 699.
341. Grimaldi, T.; Romero, M.; Pujol, M. D. *Synlett* **2000**, 1788.
342. Rewcastle, G. W.; Katritzky, A. R. *Adv. Heterocycl. Chem.* **1993,** *56*, 172.
343. Katritzky, A. R.; Akutagawa, K. *Tetrahedron Lett.* **1985,** *26*, 5935.
344. Matsuzono, M.; Fukuda, T.; Iwao, M. *Tetrahedron Lett.* **2001,** *42*, 7621.
345. Davies, S. G. *Organotransition metal chemistry: Application to Organic Synthesis*; Pergamon: Oxford, 1982.
346. Rausch, M. D.; Gloth, R. E. *J. Organomet. Chem.* **1978,** *153*, 59.
347. Gibson (née Thomas), S. E.; Stead, J. W.; Sur, S. *J. Chem. Soc., Perkin Trans. 1* **2001**, 636.
348. Semmelhack, M. F.; Bisaha, J.; Czarny, J. *J. Am. Chem. Soc.* **1979,** *101*, 768.
349. Card, R. S.; Trajanovsky, W. S. *J. Org. Chem.* **1980,** *45*, 2560.
350. Kondo, Y.; Green, J. R.; Ho, J. *J. Org. Chem.* **1993,** *58*, 5182.
351. Aubé, J.; Heppert, J. A.; Milligan, M. C.; Smith, M. J.; Zenk, P. *J. Org. Chem.* **1992,** *57*, 3563.
352. Han, J. W.; Sou, S. V.; Cheung, Y. K. *J. Org. Chem.* **1997,** *62*, 8264.
353. Davies, S. G.; Hume, W. E. *J. Chem. Soc., Chem. Commun.* **1995**, 251.
354. Kündig, E. P.; Quattropani, A. *Tetrahedron Lett.* **1994,** *35*, 3497.
355. Uemura, M.; Hayashi, Y.; Hayashi, Y. *Tetrahedron Asymmetry* **1994,** *5*, 1427.
356. Price, D. A.; Simpkins, N. S.; McLeod, A. M.; Watt, A. P. *Tetrahedron Lett.* **1994,** *35*, 6159.
357. Nelson, S. G.; Hilfiker, M. A. *Org. Lett.* **1999,** *1*, 1379.
358. O'Brien, P. *J. Chem. Soc., Perkin Trans. 1* **1998**, 1439.
359. Tan, Y.-L.; Widdowson, D. A.; Wilhelm, R. *Synlett* **2001**, 1632.
360. Sander, R.; Mueller-Westerhoff, U. T. *J. Organomet. Chem.* **1996,** *512*, 219.
361. Rebière, F.; Riant, O.; Ricard, L.; Kagan, H. B. *Angew. Chem., Int. Ed. Engl.* **1996,** *35*, 568.
362. Tzukazaki, M.; Roglans, A.; Chapell, B. J.; Taylor, N. J.; Snieckus, V. *J. Am. Chem. Soc.* **1996,** *118*, 685.
363. Laufer, R. S.; Veith, U.; Taylor, N. J.; Snieckus, V. *Org. Lett.* **2000,** *2*, 629.
364. Price, D. A.; Simpkins, N. S. *Tetrahedron Lett.* **1995,** *36*, 6139.
365. Clark, R. D.; Jahangir, A. *Org. Reac.* **1995,** *47*, 1.
366. Broaddus, C. D. *J. Am. Chem. Soc.* **1966,** *88*, 4174.
367. Broaddus, C. D. *J. Org. Chem.* **1970,** *35*, 10.
368. Gilman, H.; Gaj, B. J. *J. Org. Chem.* **1963,** *28*, 1725.
369. Lochmann, L.; Petranek, J. *Tetrahedron Lett.* **1991,** *32*, 1482.
370. Jones, F. N.; Zinn, M. F.; Hauser, C. R. *J. Org. Chem.* **1963,** *28*, 663.
371. Vaulx, R. L.; Jones, F. N.; Hauser, C. R. *J. Org. Chem.* **1964,** *29*, 1387.
372. Clayden, J.; Darbyshire, M.; Pink, J. H.; Westlund, N.; Wilson, F. X. *Tetrahedron Lett.* **1998,** *38*, 8587.

373. Clayden, J.; Youssef, L. H. *unpublished work.*
374. Mao, C.-L.; Hauser, C. R. *J. Org. Chem.* **1970,** *35*, 3704.
375. Clark, R. D.; Miller, A. B.; Berger, J.; Repke, D. B.; Weinhardt, K. K.; Kowalczyk, B. A.; Eglen, R. M.; Bonhaus, D. W.; Lee, C. H.; Michel, A. D.; Smith, W. L.; Wong, E. H. F. *J. Med. Chem.* **1993,** *36*, 2645.
376. Fisher, L. E.; Muchowski, J. M.; Clark, R. D. *J. Org. Chem.* **1992,** *57*, 2700.
377. Clark, R. D.; Jahangir; Langston, J. A. *Can. J. Chem.* **1994,** *72*, 23.
378. Fisher, L. E.; Caroon, J. M.; Jahangir; Stavler, S. R.; Lundberg, S.; Muchowski, J. M. *J. Org. Chem.* **1993,** *58*, 3643.
379. Watanabe, M.; Sahara, M.; Kubo, M.; Furukawa, S.; Billedeau, R. J.; Snieckus, V. *J. Org. Chem.* **1984,** *49*, 742.
380. Clark, R. D.; Jahangir *J. Org. Chem.* **1987,** *52*, 5378.
381. Clayden, J.; Westlund, N.; Wilson, F. X. *Tetrahedron Lett.* **1999,** *40*, 3331.
382. Comins, D. L.; Brown, J. D. *J. Org. Chem.* **1983,** *51*, 3566.
383. Fraser, R. R.; Savard, S. *Can. J. Chem.* **1991,** *64*, 621.
384. Gschwend, H. W.; Hamdan, A. *J. Org. Chem.* **1975,** *40*, 2008.
385. Houlihan, W. J.; Parriano, V. A. *J. Org. Chem.* **1982,** *47*, 5177.
386. Flippin, L. A. *Tetrahedron Lett.* **1991,** *32*, 6857.
387. Creger, P. L. *J. Am. Chem. Soc.* **1970,** *29*, 1396.
388. Thompson, R. C.; Kallmerten, J. *J. Org. Chem.* **1990,** *55*, 6076.
389. Belletire, J. L.; Spletzer, E. G. *Synth. Commun.* **1986,** *16*, 575.
390. Schultz, A. G.; Green, N. J. *J. Am. Chem. Soc.* **1991,** *113*, 4931.
391. Pfeffer, P. E.; Silbert, L. S.; Chirinkoi, J. M. *J. Org. Chem.* **1970,** *37*, 451.
392. Hauser, F. M.; Rhee, R. P.; Prasanna, S.; Weinreb, S. M.; Dodd, J. H. *Synthesis* **1980,** 72.
393. Kraus, G. A. *J. Org. Chem.* **1981,** *46*, 201.
394. Regan, A. C.; Staunton, J. *J. Chem. Soc., Chem. Commun.* **1983**, 764.
395. Regan, A. C.; Staunton, J. *J. Chem. Soc., Chem. Commun.* **1987**, 520.
396. Klein, J.; Medlik-Balan, A. *J. Org. Chem.* **1976,** *41*, 3307.
397. Kobayashi, K.; Konishi, A.; Kanno, Y.; Suginome, H. *J. Chem. Soc., Perkin Trans. 1* **1993**, 111.
398. Comins, D. L.; Brown, J. D. *J. Org. Chem.* **1983,** *49*, 1078.
399. Harris, T. D.; Roth, G. P. *J. Org. Chem.* **1979,** *44*, 2004.
400. Flippin, L. A.; Muchowski, J. M.; Carter, D. S. *J. Org. Chem.* **1993,** *58*, 2463.
401. Braun, M.; Ringer, E. *Tetrahedron Lett.* **1972,** *24*, 1233.
402. Bates, R. B.; Siahaan, T. J. *J. Org. Chem.* **1982,** *51*, 1432.
403. Harmon, T. E.; Shirley, D. A. *J. Org. Chem.* **1974,** *21*, 3164.
404. Winkle, M. R.; Ronald, R. C. *J. Org. Chem.* **1982,** *47*, 2101.
405. Ronald, R. C.; Winkle, M. R. *Tetrahedron* **1983,** *39*, 2031.
406. Watanabe, M.; Date, M.; Kawanishi, K.; Hori, T.; Furukawa, S. *Chem. Pharm. Bull.* **1991,** *39*, 41.
407. Cabiddu, S.; Melis, S.; Piras, P. P.; Sotgiu, F. *J. Organomet. Chem.* **1979,** *178*, 291.
408. Cabiddu, S.; Floris, C.; Gelli, G.; Melis, S. *J. Organomet. Chem.* **1989,** *366*, 1.

409. García Ruano, J. L.; Carreño, M. C.; Toledo, M. A.; Aguirre, J. M.; Aranda, M. T.; Fischer, J. *Angew. Chem., Int. Ed. Engl.* **2000,** *39*, 2736.
410. Alo, B. I.; Familoni, O. B.; Marsais, F.; Quéguiner, G. *J. Chem. Soc., Perkin Trans. 1* **1990**, 1611.
411. Watanabe, H.; Hauser, C. R. *J. Org. Chem.* **1968,** *33*, 4278.
412. Watanabe, H.; Mao, C.-L.; Barnish, I. T.; Hauser, C. R. *J. Org. Chem.* **1969,** *34*, 919.
413. Watanabe, H.; Gay, R. L.; Hauser, C. R. *J. Org. Chem.* **1968,** *33*, 900.
414. Houlihan, W. J.; Parrino, V. A.; Uike, Y. *J. Org. Chem.* **1981,** *46*, 4511.
415. Spadoni, G.; Stankov, B.; Duranti, A.; Biella, G.; Lucini, V.; Salvatori, A.; Fraschini, F. *J. Med. Chem.* **1993,** *36*, 4069.
416. Ito, Y.; Kobayashi, K.; Saegusa, T. *J. Am. Chem. Soc.* **1977,** *99*, 3532.
417. Ito, Y.; Kobayashi, K.; Seko, N.; Saegusa, T. *Bull. Chem. Soc. Jap.* **1984,** *57*, 73.
418. Clark, R. D.; Muchowski, J. M.; Fisher, L. E.; Flippin, L. A.; Repke, D. B.; Souchet, M. *Synthesis* **1991**, 871.
419. Katritzky, A. R.; Black, M.; Fan, W.-Q. *J. Org. Chem.* **1991,** *56*, 5045.
420. Katritzky, A. R.; Fan, W.-Q.; Akutagawa, K.; Wang, J. *Heterocycles* **1990,** *30*, 407.
421. Smith, A. B.; Haseltine, J. N.; Visnick, M. *Tetrahedron* **1989,** *45*, 2431.
422. Kanazawa, A. M.; Correa, A.; Denis, J.-N.; Luche, J.-M.; Greene, A. E. *J. Org. Chem.* **1993,** *58*, 255.
423. Clark, R. D.; Jahangir *Tetrahedron* **1993,** *49*, 1351.
424. Fraser, R. R.; Mansour, T. S.; Savard, S. *J. Org. Chem.* **1985,** *50*, 3232.
425. Hamana, H.; Sugasawa, T. *Chem. Lett.* **1984**, 1591.
426. Mathieu, J.; Gros, P.; Fort, Y. *J. Chem. Soc., Chem. Commun.* **1999**, 951.
427. Katritzky, A. R.; Akutagawa, K. *J. Am. Chem. Soc.* **1986,** *108*, 6808.
428. Mali, R. S.; Jagtap, P. G. *Tetrahedron Lett.* **1992,** *33*, 1655.
429. Comins, D. L.; Killpack, M. O. *Tetrahedron Lett.* **1989,** *30*, 4337.
430. Schmidt, R. R.; Talbiersky, J. *Angew. Chem., Int. Ed. Engl.* **1976,** *15*, 171.
431. Schmidt, R. R.; Berger, G. *Chem. Ber.* **1976,** *109*, 2936.
432. Shiner, C. S.; Berks, A. H.; Fisher, A. M. *J. Am. Chem. Soc.* **1988,** *110*, 957.
433. James, C. A.; Snieckus, V. *Tetrahedron Lett.* **1997,** *38*, 8149.
434. Wang, X.; Snieckus, V. *Tetrahedron Lett.* **1991,** *32*, 4829.
435. Schlosser, M. *Pure Appl. Chem.* **1988,** *60*, 1627.
436. Lochmann, L.; Pospísil, J.; Lím, D. *Tetrahedron Lett.* **1966**, 257.
437. Schlosser, M. *J. Organomet. Chem.* **1967,** *8*, 9.
438. Lochmann, L.; Trekoval, J. *J. Organomet. Chem.* **1975,** *99*, 329.
439. Lochmann, L.; Lím, D. *J. Organomet. Chem.* **1971,** *28*, 153.
440. Hartmann, J.; Schlosser, M. *Helv. Chim. Acta* **1976,** *59*, 453.
441. Schlosser, M.; Choi, J. H.; Takagishi, S. *Tetrahedron* **1990,** *46*, 5633.
442. Schlosser, M.; Katsoulos, G.; Takagishi, S. *Synlett* **1990**, 747.
443. Katsoulos, G.; Takagishi, S.; Schlosser, M. *Synlett* **1991**, 731.
444. Faigl, F.; Schlosser, M. *Tetrahedron Lett.* **1991,** *32*, 3369.
445. Clayden, J.; Julia, M. *Synlett* **1995**, 103.
446. Gros, P.; Fort, Y.; Caubère, P. *J. Chem. Soc., Perkin Trans. 1* **1997**, 3071.

447. Bhida, B. H.; Narasimhan, N. S. *Chem. and Ind. (London)* **1974**, 75.
448. Wamada, K.; Yazawa, Y.; Uemura, D.; Toda, M.; Hirata, Y. *Tetrahedron* **1969,** *25*, 1969.
449. Liang, C. D. *Tetrahedron Lett.* **1986,** *27*, 1971.
450. Haenel, M. W.; Fieseler, A.; Jakubik, D.; Gabor, B.; Goddard, R.; Krüger, C. *Tetrahedron Lett.* **1993,** *34*, 2107.
451. MacNeil, S. L.; Snieckus, V. *J. Org. Chem.* **2001,** *66*, 3663.
452. Shimano, M.; Meyers, A. I. *J. Am. Chem. Soc.* **1994,** *116*, 10815.
453. Shimano, M.; Meyers, A. I. *Tetrahedron Lett.* **1997,** *38*, 5415.
454. Ahmed, A.; Clayden, J. *unpublished results*.
455. Massi, R.; Schlosser, M. *J. Org. Chem.* **1996,** *61*, 5430.
456. Simig, G.; Schlosser, M. *Tetrahedron Lett.* **1991,** *32*, 1963.
457. Moro-oka, Y.; Iwakiri, S.; Fukuda, T.; Iwao, M. *Tetrahedron Lett.* **2000,** *41*, 5225.
458. Davies, G. M.; Davies, P. S. *Tetrahedron Lett.* **1972,** *33*, 3507.
459. Court, J. J.; Hlasta, D. J. *Tetrahedron Lett.* **1996,** *37*, 1335.
460. Mills, R. J.; Taylor, N. J.; Snieckus, V. *J. Org. Chem.* **1989,** *54*, 4372.
461. Clayden, J.; Lai, L. W. *Tetrahedron Lett.* **2001,** *42*, 3163.
462. Hodges, J. C.; Patt, W. C.; Conolly, C. J. *J. Org. Chem.* **1991,** *56*, 449.
463. Dondoni, A.; Fantin, G.; Fagagnolo, M.; Medici, M.; Pedrini, P. *J. Org. Chem.* **1988,** *53*, 1748.
464. Iddon, B. *Heterocycles* **1985,** *23*, 417.
465. Ohta, S.; Matsukawa, M.; Ohashi, N.; Nagayama, K. *Synthesis* **1990**, 78.
466. Shapiro, G.; Gomez-Lor, B. *Heterocycles* **1995,** *41*, 215.

CHAPTER 3

Regioselective Synthesis of Organolithiums by X–Li Exchange

3.1 Halogen-lithium exchange

3.1.1 Reactivity

Wittig described, in 1938, the formation of organolithium **2** from **1** by bromine-lithium exchange with PhLi in Et_2O.[1] Hydrolysis of the product gave the debrominated product in 95% yield. Previous reports[2] of similar reactions had not identified the organolithium as an intermediate, and it was Gilman who started to use the reaction to synthesise, in a regiospecific manner, organolithiums which could be functionalised with other electrophiles – for example **5** from **4**.[3]

Most of the early work[4,5] on halogen-lithium exchange was carried out on aryl halides, and a few important features became evident. Firstly, the halogen-lithium exchange (as with other transmetallations) is an equilibrium process favouring (in most cases) formation of the more stable, less basic, organolithium.[6] Applequist[7] determined the equilibrium constants for transmetallation of PhI with various organolithiums, as shown below (Table 3.1.1), along with an estimate of the pK_a of the carbanion equivalents of the organolithiums.[8]

The equilibria mean that *n*-BuLi can be used to form organolithiums from aryl halides at low temperature (the subsequent reaction of ArLi with the BuX formed in the exchange is slow[6]); *t*-BuLi will form organolithiums from primary alkyl halides.[7] The formation of secondary organolithiums by halogen-metal exchange is difficult,[9] and reductive lithiation is usually preferable.

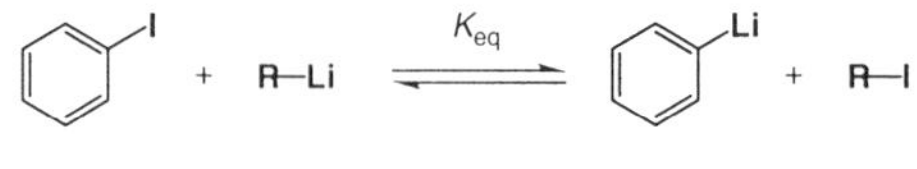

Organolithium	K_{eq}	Approx. pK_a of "carbanion"
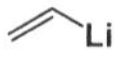 vinyllithium	0.004	36.5
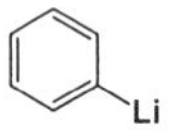 phenyllithium	1	37
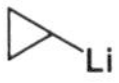 cyclopropyllithium	10	39
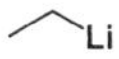 ethyllithium	3000	42
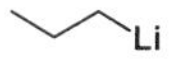 propyllithium	7500	42
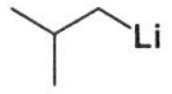 isobutyllithium	4 x 10^4	42
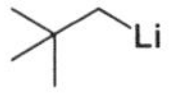 neopentyllithium	3 x 10^5	42
 cyclobutyllithium	10^6	43
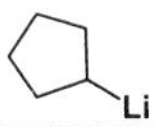 cyclopentyllithium	10^7	44

Table 3.1.1 Halogen-metal exchange and pK_a

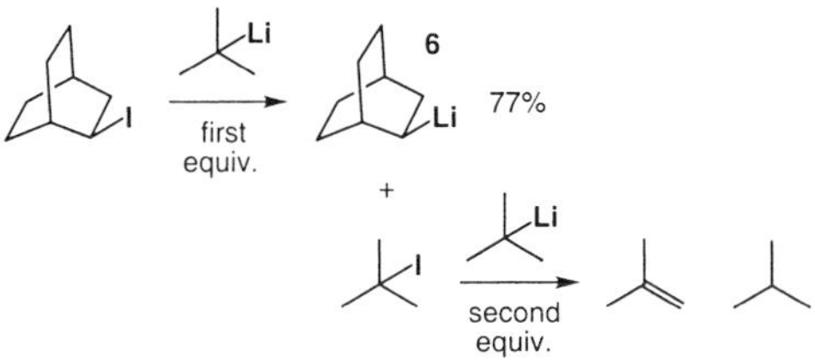

An advantage of using *t*-BuLi is that even exchanges whose equilibrium is only moderately favourable can be made irreversible by the use of two equivalents of *t*-BuLi (advisable in any case to avoid the danger of protonation of the new organolithium by *t*-BuX). The second equivalent drives the reaction forward by eliminating X^- from *t*-BuX, forming isobutane and

CHAPTER 3

Regioselective Synthesis of Organolithiums by X–Li Exchange

3.1 Halogen-lithium exchange

3.1.1 Reactivity

Wittig described, in 1938, the formation of organolithium **2** from **1** by bromine-lithium exchange with PhLi in Et_2O.[1] Hydrolysis of the product gave the debrominated product in 95% yield. Previous reports[2] of similar reactions had not identified the organolithium as an intermediate, and it was Gilman who started to use the reaction to synthesise, in a regiospecific manner, organolithiums which could be functionalised with other electrophiles – for example **5** from **4**.[3]

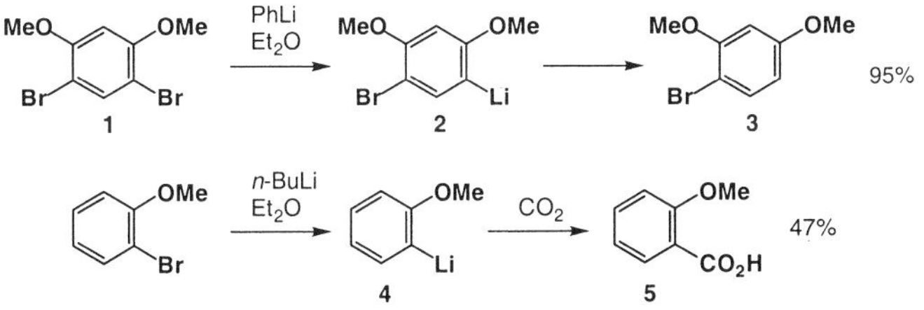

MeO OMe Br Br 1 PhLi Et_2O MeO OMe Br Li 2 → MeO OMe Br 3 95%

OMe Br n-BuLi Et_2O OMe Li 4 CO_2 OMe CO_2H 5 47%

Most of the early work[4,5] on halogen-lithium exchange was carried out on aryl halides, and a few important features became evident. Firstly, the halogen-lithium exchange (as with other transmetallations) is an equilibrium process favouring (in most cases) formation of the more stable, less basic, organolithium.[6] Applequist[7] determined the equilibrium constants for transmetallation of PhI with various organolithiums, as shown below (Table 3.1.1), along with an estimate of the pK_a of the carbanion equivalents of the organolithiums.[8]

The equilibria mean that n-BuLi can be used to form organolithiums from aryl halides at low temperature (the subsequent reaction of ArLi with the BuX formed in the exchange is slow[6]); t-BuLi will form organolithiums from primary alkyl halides.[7] The formation of secondary organolithiums by halogen-metal exchange is difficult,[9] and reductive lithiation is usually preferable.

Organolithium	K_{eq}	Approx. pK_a of "carbanion"
vinyl-Li	0.004	36.5
phenyl-Li	1	37
cyclopropyl-Li	10	39
ethyl-Li	3000	42
n-butyl-Li	7500	42
isobutyl-Li	4 x 10^4	42
neopentyl-Li	3 x 10^5	42
cyclobutyl-Li	10^6	43
cyclopentyl-Li	10^7	44

Table 3.1.1 Halogen-metal exchange and pK_a

An advantage of using *t*-BuLi is that even exchanges whose equilibrium is only moderately favourable can be made irreversible by the use of two equivalents of *t*-BuLi (advisable in any case to avoid the danger of protonation of the new organolithium by *t*-BuX). The second equivalent drives the reaction forward by eliminating X^- from *t*-BuX, forming isobutane and isobutylene,[10,11] and the product organolithium may be treated with relatively unreactive electrophiles without fear of alkylation or protonation by the alkyl halide by-product of the exchange. By this method even a relatively hindered secondary alkyllithium (2-adamantyllithium, **6**) may be formed using *t*-BuLi.[12]

Even PhLi can be used to transmetallate a primary alkyl iodide if the product organolithium is consumed by a subsequent reaction. For example, the norbornane derivative **8** forms in good yield on treatment of **7** with PhLi, despite the thermodynamic unfavourability of the transmetallation step.[13]

7 + PhLi ⇌ [+ PhI] ⟶ 8 + LiI

Decomposition of an organolithium after a halogen-lithium exchange step is more commonly used when the new alkyl or aryl halide is the useful product. For example, the iodination and bromination of aryllithiums is best carried out by halogen metal exchange with diiodomethane **9** or derivatives **10-12** of 1,2-dibromoethane. 1,2-Dibromoethane itself is prone to competing substitution reactions, yielding compounds which themselves eliminate to give vinylated products **13**. Even methyl iodide and other alkyliodides will occasionally act as an I^+ source with alkyllithiums.[11,14]

ArLi + 9 ⇌ ArI + ⟶ decomposition products

ArLi + {10 or 11 or 12} ⇌ ArBr + ⟶ decomposition products; ⇢ Ar–CH₂CH₂Br ⇢ 13

Secondly, the rate of the exchange decreases

$$\text{ArI} > \text{ArBr} > \text{ArCl} > \text{ArF}$$

with ArCl and ArF being not synthetically useful for exchange reactions.[15,16] Aryl chlorides and aryl fluorides tend to undergo deprotonation, leading to benzynes, rather than halogen-metal exchange (see section 2.3.2.5). Vinyl bromides and iodides are useful precursors to vinyllithiums (see section 3.1.5), but again alkyllithiums will deprotonate vinyl chlorides rather than engage in halogen-metal exchange.[17] Chlorine-lithium exchange will occur only if deprotonation is impossible and if there are other halogens to stabilise the vinyllithium. The schemes below demonstrate this behaviour: **14**[18] and **15**[19] are deprotonated to form chloro-stabilised vinyllithiums, while **16** cannot undergo deprotonation and a rare chlorine-lithium exchange generates a fluoro-stabilised vinyllithium.[20]

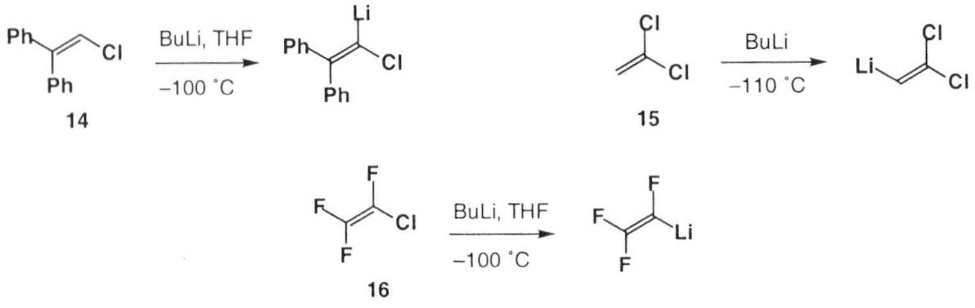

Alkyl iodides and alkyl bromides are similarly the only alkyl halides whose halogen-metal exchange is of general synthetic application. The conversion of alkyl chlorides to alkyllithiums is best carried out using reductive methods (section 4.1). Nonetheless, alkyl chlorides will undergo halogen-metal exchange if deprotonation α to Cl is impossible and if the organolithium is stabilised by additional substituents, as in **17**, **18** and **19**.[17,21,22]

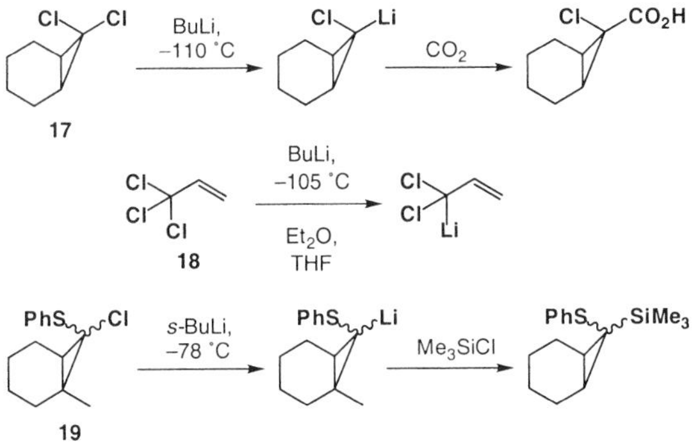

The third example is valuable particularly because of the ready availability of chlorocyclopropylsulfides such as **19** by reaction of a sulfide-stabilised carbene with an alkene.

Thirdly, the exchange reaction is accelerated by the presence of ether solvents, and may be an extremely fast reaction, even at temperatures close to the freezing point of THF and Et_2O (–100 °C) or at even lower temperatures in the extremely low-freezing Trapp solvent mixture (THF:Et_2O:pentane in a 4:1:1 ratio by volume).[23] The exchanges shown below are successful at the low temperatures required to avoid formation of an acetylene[24] or a benzyne:[25] in general, low temperatures can be used to favour halogen-metal exchange over deprotonation.

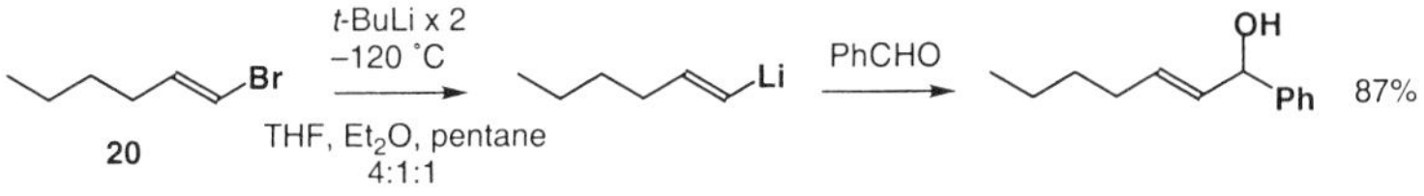

The best solvent systems for use with *alkyl* halides are Et_2O–pentane mixtures (usually in a ratio of about 3:2 by volume) since these lead to relatively few by-products.[7] TMEDA must usually be avoided, since although it accelerates halogen-lithium exchange, it usually accelerates deprotonation even more.[26]

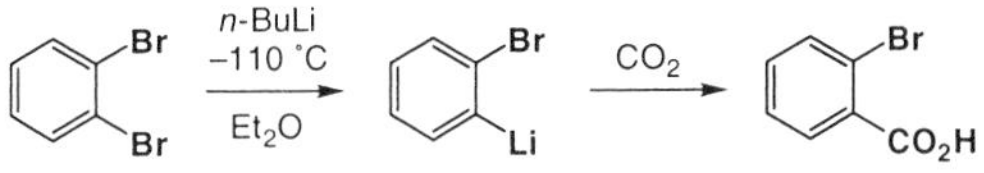

The rapidity of the reaction means that halogen-lithium exchange is the best way of making organolithiums containing other reactive functional groups – for example nitro groups,[27] esters (**22**),[28] epoxides (**23**),[29] or even ketones (**24**).[30] Many of the "Parham cyclisations"[31] described in section 7.2.1.1 rely on this high degree of chemoselectivity. The chloroalkyl group of **25** is sufficiently unreactive that it withstands bromine-lithium exchange,[32] being substituted on warming either directly or after addition of the aryllithium to benzonitrile.[33]

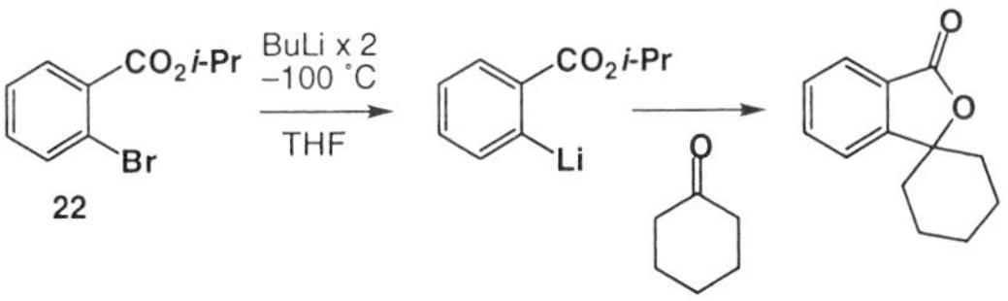

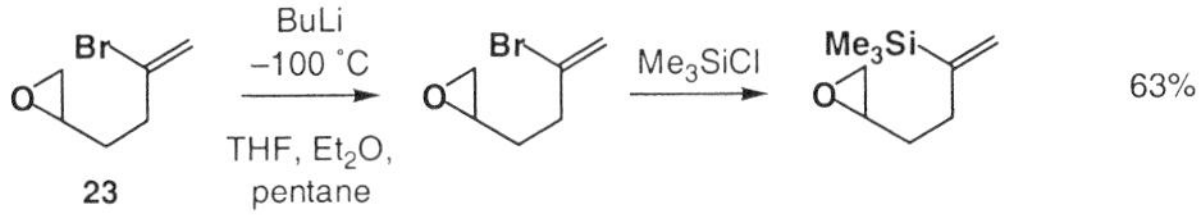

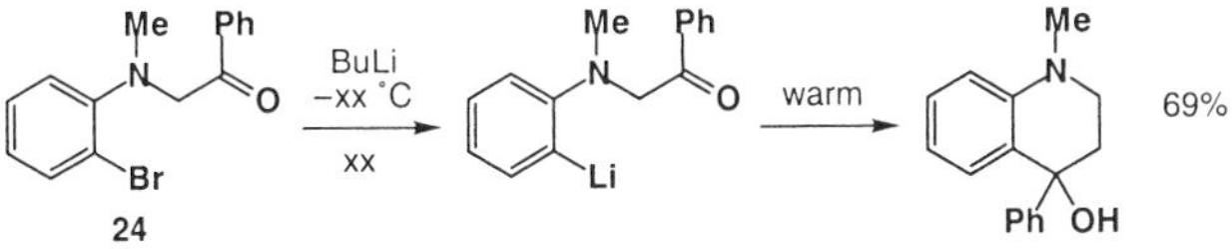

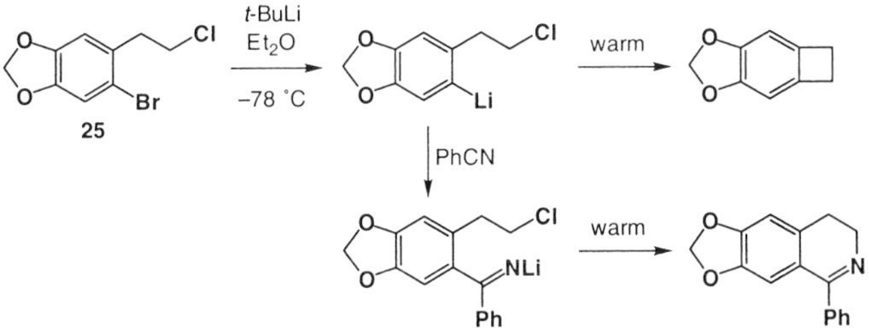

Impressive chemoselectivities are obtained by using mesityllithium **26**, which has been used to promote halogen-lithium exchange in the presence of a ketone in the synthesis of the precursor **27** of camptothecin.[34]

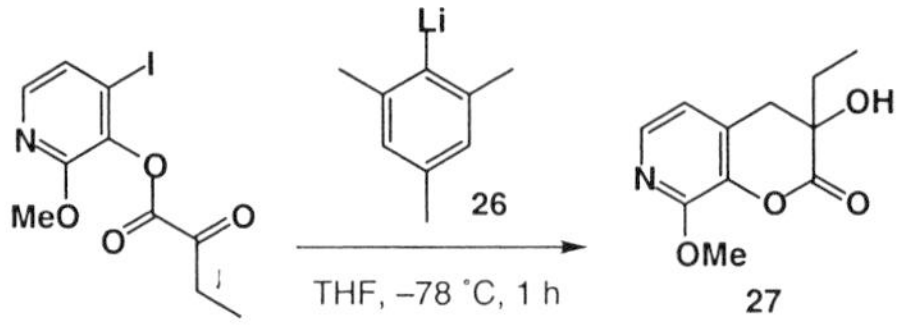

In some cases, it appears that halogen-metal exchange may even outstrip deprotonation of hydroxyl groups. It is certainly possible to carry out a halogen metal exchange in the presence of an alcohol or even water. For example, *t*-BuLi attacks the iodide **28** in the presence of MeOH, though the protonation by methanol competes with the subsequent elimination of HI from *t*-BuI.[35]

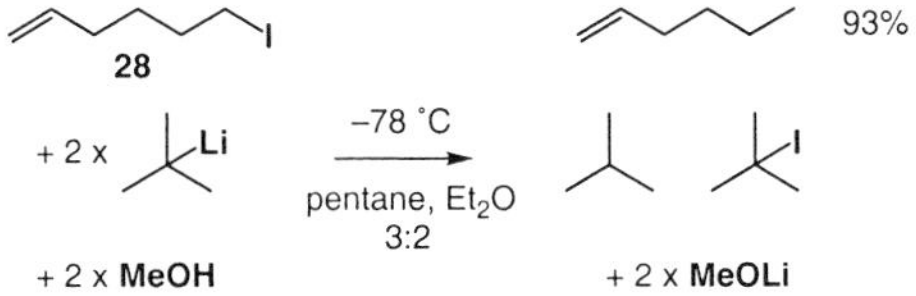

Similarly, when, attempted tritiation of the anthryllithium **30** with T_2O as an external quench failed because of competing protonation, the solution was to use T_2O as an internal quench: halogen-metal exchange of **29** proceeds in the presence of the T_2O at –70 °C.[36]

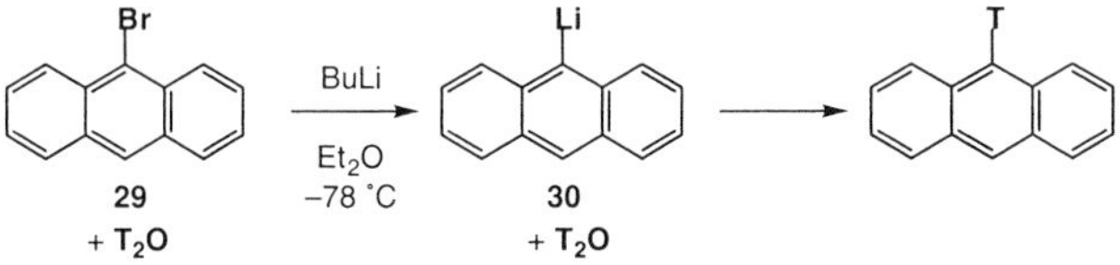

But can these results really be taken to mean that halogen-metal exchange is faster than deprotonation of OH? An alternative explanation is based on the local high concentration of organolithium which occurs as each drop of reagent is added to the substrate. This effect means that, for example, the extent of dibromo, monobromo, and bromine-free products arising from halogen-metal exchange of **31** is dependent on whether the solution is stirred or not.[37] Without stirring, local areas of high *t*-BuLi concentration allow double halogen-metal exchange to occur to a greater extent than is seen with stirring.

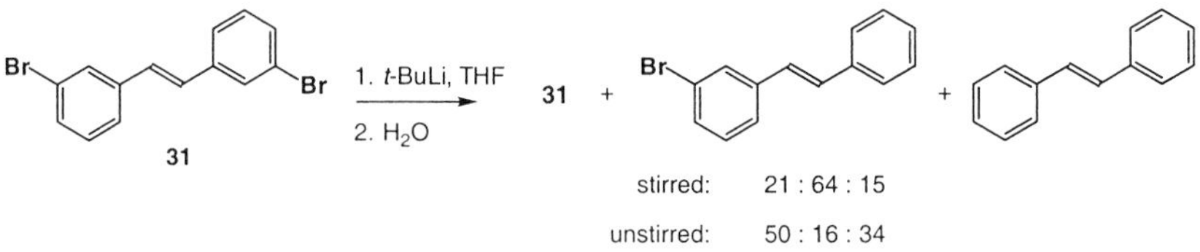

stirred: 21 : 64 : 15

unstirred: 50 : 16 : 34

Beak showed that halogen-metal exchange with an excess of alkyllithium in the presence of acidic protons can similarly be interpreted by considering the events within the local high concentration of organolithium reagent which occurs as each drop is added to the solution of alkyl halide (Scheme 3.1.1).[38,39] Halogen-metal exchange can be slower than deprotonation, but provided it is faster than mixing it will still take place in the locally high concentration of organolithium preference to further deprotonations. As the local high concentration of organolithium reagent disperses, the newly-formed organolithium becomes protonated intermolecularly by other molecules of alcohol. Evidence that this explanation accounts for halogen-metal exchange in the presence of alcohols *in specific cases*[40] is disputed,[41] but local concentration effects mean that results such as those above do not prove that halogen-metal exchange can outstrip deprotonation.[42]

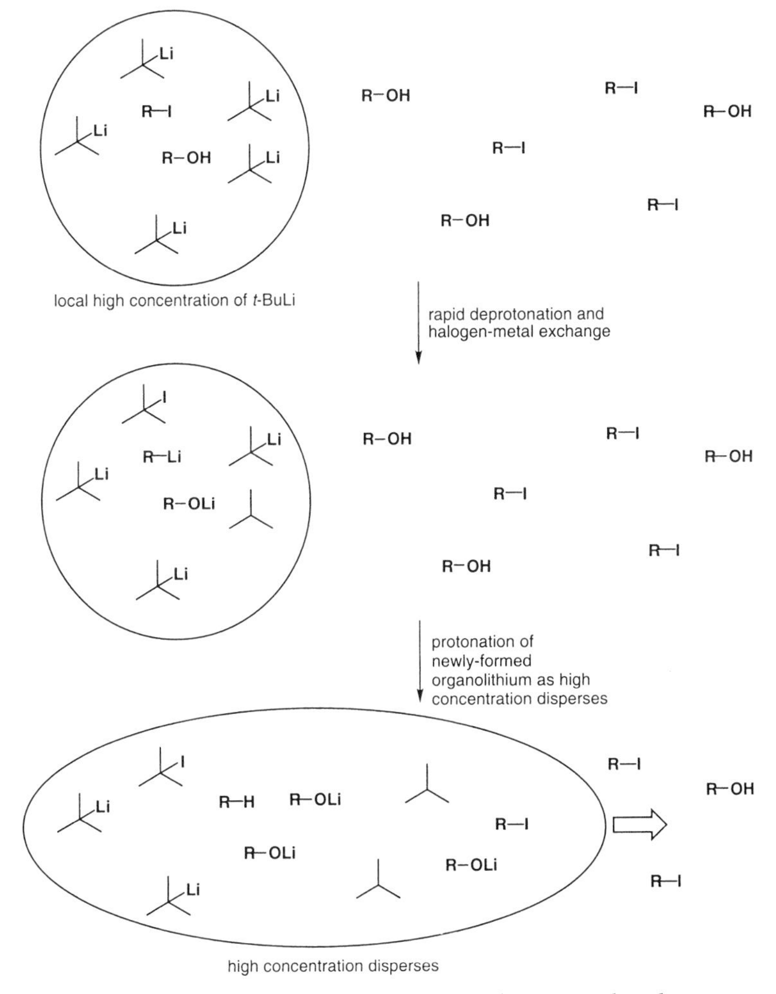

Scheme 3.1.1: Competition between deprotonation and halogen-metal exchange

3.1.2 Mechanism

Halogen-metal exchange is likely to proceed through one of two mechanisms, one involving radical intermediates – something like this:

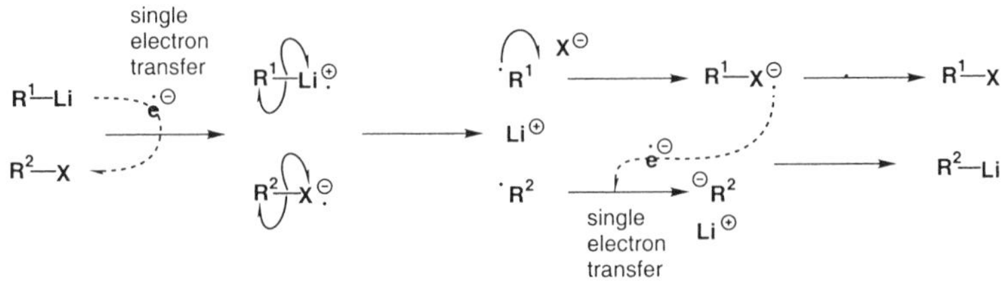

and one involving nucleophilic substitution at the halogen,[43] perhaps with an ate complex as an intermediate:[44]

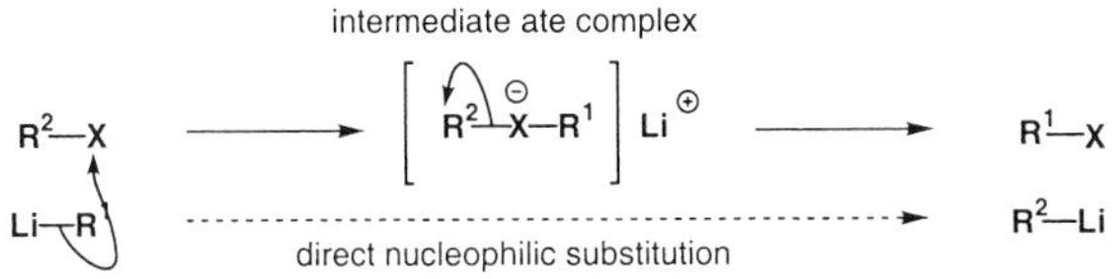

Proof that radicals are formed by treatment of an alkyl halide with an alkyllithium was first presented in 1956.[45] The degenerate exchange between BuBr and BuLi in cumene over an extended period was found to be accompanied not only by the formation of butane, butene and (in 43% yield) octane but also considerable quantities of the cumene dimer **32**.

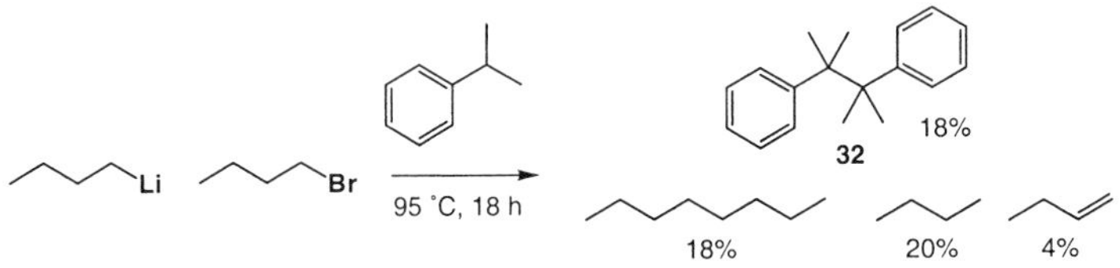

Dimer **32** presumably originates by abstraction of hydrogen from the solvent by radicals formed in the reaction, giving **33**, which undergoes radical dimerisation. Nonetheless, evidence that radicals are present in the reaction mixture is not evidence that they are an intermediate in the reaction pathway.

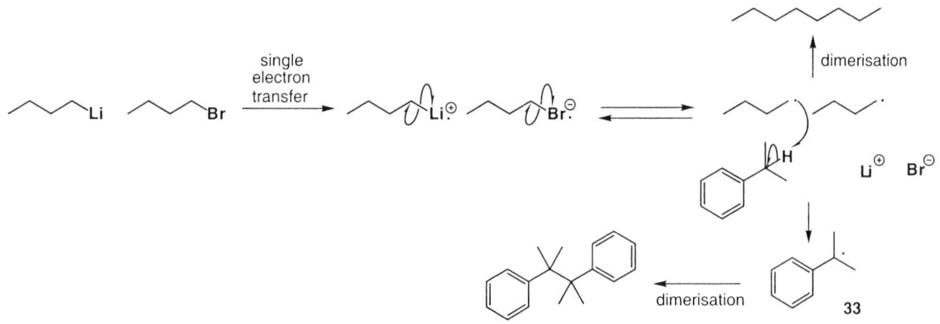

Evidence from ESR[46,47] and CIDNP[44] studies points to the same conclusion – that radicals are produced during the reactions of some alkyllithiums with some alkyl halides – but give no evidence that they are intermediates in the halogen-metal exchange reaction. Indeed, with ESR it was impossible to detect radicals during the halogen-lithium exchanges using primary alkyllithiums or phenyllithium, but that they were present during the reactions of secondary and tertiary alkyllithiums with alkyl halides.[46] Other studies demonstrated that radicals were formed from the reactions of primary and secondary alkyllithiums with alkyl iodides and bromides, but only in the presence of significant quantities (at least 1 equiv.) of Et_2O or TMEDA.[47]

A further group of experiments suggesting that radicals are produced during some, but not all, halogen-metal exchange reactions is based on the use of cyclisable radical probes. For example, the treatment of **34** with BuLi generates, in addition to coupling products, a 60% yield of cyclised material **35**.[48] This compound could be formed by radical or organolithium cyclisation (see section 7.2.4.1), but the product of an organolithium cyclisation would require a proton source to give **35**. Addition of D_2O at the end of the reaction gave a 25% yield of the deuterated product *d*-**35**, indicating a probable mixture of mechanisms.

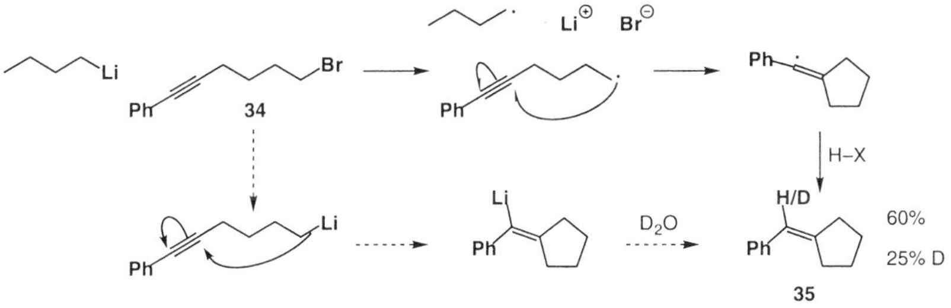

Since both hexenyllithiums and hexenylradicals cyclise in a similar manner, the use of compounds such as **36** and **38** as probes for radical mechanisms poses difficulties. Bailey showed that the cyclisation of the organolithium **36** to **37** is very slow at –78 °C,[49] but occurs at an appreciable rate close to room temperatures. A reinvestigation[49] of earlier work[50] carried out in the light of this information led to the conclusion that, with *t*-BuLi in 3:2 pentane:Et_2O the halogen-metal exchange of simple alkyl *bromides* generates radicals, and the cyclisation of bromohexene during its halogen-metal exchange probably proceeds with significant contribution from a radical mechanism.[51] Alkyl iodides on the other hand produce no cyclised material at low temperature and under these conditions must undergo the halogen-metal exchange without the intervention of radicals, despite being more readily reduced than alkyl bromides.[35] The formation of small amounts of reduced 1-hexene **39** in the reactions was proposed,[52] in contrast to previous suggestions,[50] to be due to the reaction of the newly formed alkyllithium and *t*-BuI. The use of d_9-*t*-BuLi gave deuterated 1-hexene *d*-**39**.

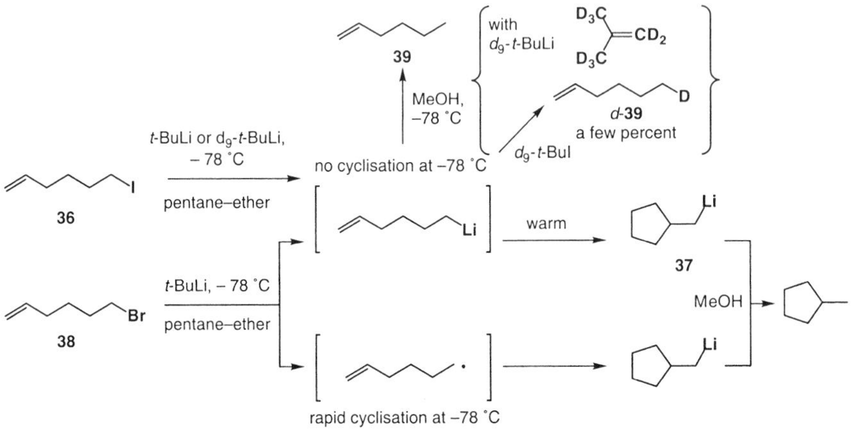

The mechanistic difference between alkyl iodides and alkyl bromides is also illustrated by the clean cyclisation of α,ω-diidodides such as **40** to cycloalkanes and the very messy reactions of α,ω-dibromides such as **41** under similar conditions.[53]

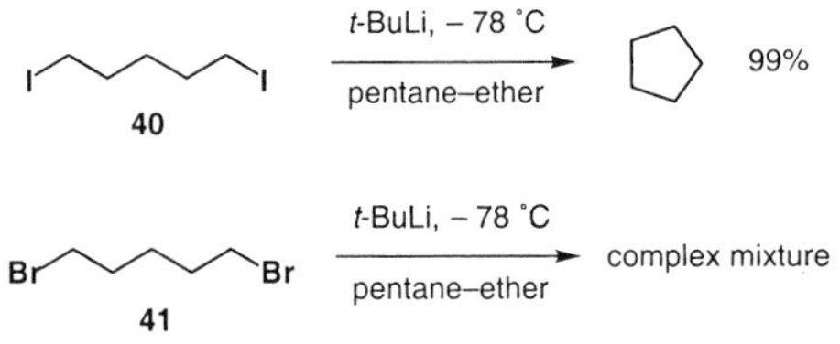

The formation of radicals during halogen-metal exchange reactions of alkyl *bromides* was confirmed by the use of a probe which can rearrange only by radical pathways, and not via organolithiums. Halogen-metal exchange of **42** with *t*-BuLi in 3:2 pentane:Et_2O at –23 °C produced, among other unidentified products, 13% of the alkene **43**.[54] Only a radical pathway can produce this compound.

t-BuLi, – 23 °C
pentane–ether
42
proposed product of polar mechanism
43
13%
product of radical mechanism

Parallel work on aryllithium and vinyllithium cyclisation onto alkenes highlighted further differences between radical and organolithium cyclisations which could be exploited as mechanistic probes. Halogen-metal exchange on aryl bromide **44** gave an organolithium **45**which cyclised very slowly indeed in THF at –78 °C (no cyclisation was detectable after 120 min), but with a useful rate at 23 °C (92% yield of the indane **48** after an hour).[55] Cyclisation in ether was slower (but gave less protonated material as a by-product), but rates in Et_2O–TMEDA were comparable to those in THF.

2 x n-BuLi
–78 °C, THF
23 °C
44
45
46
1. LiNp, –78 °C, 1 h
2. D_2O
D_2O
D_2O
46% **47** + 52% **48**
47
only product after 2 h at –78 °C
48
92% after 1 h at 23 °C

Formation of the organolithium **45** with lithium and naphthalene in THF, on the other hand, gave 52% cyclisation product after 1 h at –78 °C, conditions under which the cyclisation of organolithium **45** is only very slow. This reductive lithiation presumably does proceed to some degree via a single electron transfer, and the cyclisation is a radical one; lack of cyclisation at –78 °C using *n*-BuLi suggests that radicals do not intervene in the halogen-metal exchange version of the reaction and the cyclisation at 23 °C is a true anionic cyclisation. Similar experiments have been carried out with organosodium cyclisations – sodium-naphthalene bromine-metal exchanges involve radical intermediates while sodium metal mediated ones do not.[56]

For reactions which do not proceed via radical intermediates, the question of detail remains: does the polar substitution take place via an S_N2-like substitution at the halogen, or is there an intermediate – an ate complex? Gilman was the first to suggest a mechanism for halogen-metal exchange proceeding via nucleophilic attack by the alkyllithium on the halogen,[57] and Wittig later suggested an intermediate in the reaction, the ate complex **49**[58] (the first time this term had been used[59]). Since then, evidence has been accumulating that ate complexes are formed during the reactions of at least some alkyllithiums with alkyl halides and, in particular, with aryl halides. Linear free energy relationships show, for example, that charge accumulates in the transition state of the reaction of PhLi with substituted bromobenzenes, a reaction which is a clean equilibrium process between the two aryllithiums.[60,61] Similar results led to the proposal that ate complexes are involved in the attack of *n*-BuLi or aryl bromides.[62]

R = Bu or Ph

49

intermediate ate complex stabilised by electron-withdrawing X

Compelling evidence for the formation of ate complexes as true intermediates comes from the work of Reich,[63] who used the formation of butylbenzene by reaction of PhLi with BuI as a measure of reactive PhLi in solution. He found that addition of iodobenzene to the reaction slowed down the formation of butylbenzene. Since any exchange reaction involving iodobenzene would simply regenerate a molecule of PhLi, the only way PhI can be removing PhLi from the reactive sequence is by forming an ate complex **50**. HMPA reduced the yield of butylbenzene even further by stabilising the ate complex: coordination of the free Li^+ cation of the ate complex to HMPA is stronger than coordination of PhLi to HMPA, so HMPA increases the concentration of the ate complex **50**.

and/or HMPA

50

higher concentration of ate complex decreases rate of substitution

An ate complex was finally isolated and characterised in 1986.[64] Pentafluorophenyllithium and pentafluorophenyl iodide react together to form an isolable complex **51** which is nonetheless unstable above –78 °C. However, the addition of two equivalents of TMEDA stabilises the ate complex to the extent that it may be crystallised, and an X-ray crystal structure showed **52** to contain two nearly collinear and extremely long C–I bonds.

51
unstable at −78 °C

52
stable and crystalline at 20 °C

Analogous lithium ate complexes **53** are detected by NMR when solutions of PhLi are treated with HMPA,[65] and the formation of an intensely yellow solution in the halogen-metal exchange of the diiodide **54** with BuLi suggests the intermediacy of the ate complex **55**.[66]

53

54

55
intense yellow

Beak used the "endocyclic restriction test" (the study of the length of tether between two reactive sites necessary to allow a reaction to take place intramolecularly) to show that attack on the bromide atom during halogen-lithium exchange of an aryl bromide requires a more or less linear transition state. This result is compatible with either a direct S_N2 mechanism or one with an ate intermediate, but not with a process involving radical intermediates.[67] Using double × double isotope labelling, he found that halogen-metal exchange in **58** was intramolecular, but in **56** and **57** (which cannot attain a linear, intermolecular reaction geometry) the exchange is intermolecular. With radical intermediates, intermolecular exchange processes would be possible, and indeed favoured, in small rings.

56 57 58

The balance of evidence now[68] lies in favour of

- Aryl halides (bromides and iodides) reacting via ate complexes
- Primary alkyl iodides reacting via a polar mechanism, at least with *t*-BuLi in Et_2O–pentane mixtures
- Secondary alkyl iodides reacting sometimes via polar mechanisms, sometimes via radical mechanisms
- Alkyl bromides reacting via radical mechanisms.

3.1.3 Synthesis of aryllithiums

Several examples of halogen-metal exchange generating aryllithiums have already been given in this section: the reaction is one of the most reliable and commonly-used ways of generating organolithiums. The best conditions use the aryl bromide or iodide with two equivalents of *t*-BuLi (to avoid side-reactions possible with *n*-BuLi and *n*-BuX) in Et_2O or THF or mixtures of the two at temperatures of –78 °C or below. Occasionally, *n*-BuLi is preferable – for example, when selective lithiation by a precise number of equivalents of alkyllithium is required. Particularly useful are cases where bromine-lithium exchange gives regioisomers unavailable by ortholithiation.

In cases where stabilisation of the newly formed organolithium is available, multiply lithiated species may be formed from polybrominated starting materials. For example, treatment of **59** or **61** with four equivalents of *t*-BuLi generates dilithio species **60** and **62**.[69] Lithiation of the non-brominated amides forms only the mono-lithiated product.

CONEt2 Br Br 4 x t-BuLi THF, –78 °C → Li Li CONEt2 → MeI → CONEt2 84%
59 60

Et2NOC Et2NOC Br Br 4 x t-BuLi THF, –78 °C → Et2NOC Et2NOC Li Li → MeI → Et2NOC Et2NOC 69%
61 62

Sequential iodine lithium exchange from the 1,5-diiodonaphthalene **63** generates either the monoiodo-monolithio species **64** or the dilithio species **65**.[70] Clean monolithiation is obtained by allowing the initially formed mixture of mono- and dilithiated products to equilibrate with unreacted starting material over a period of 6 h at 5 °C. No equilibration is necessary on the second lithiation, which is complete in 1 h at –78 °C. 1,8-Diiodonaphthalene **66** can similarly be converted to the dilithio compound **67**.[71]

63 → t-BuLi x 2, Et2O, 5 °C, 6.5 h → 64 → t-BuLi x 2, Et2O, –78 °C, 1 h → 65
64 → DMF → CHO, I 55%
65 → ethylene oxide, CuI → OH, HO 56%

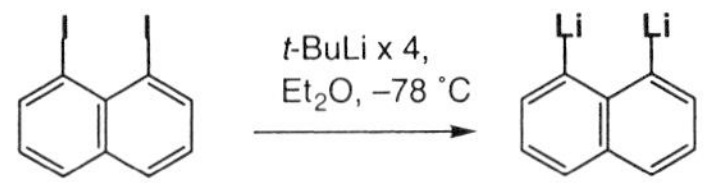

Clean monolithiation under equilibrating conditions means that the bromine atoms of 1,3,5-tribromobenzene may be replaced sequentially, one at a time.[72] More spectacularly, it is possible to use the selective formation of the most stable lithiated species to form almost any regioisomer of a substituted calixarene.[73] Tetrabrominated **68** may be monolithiated or tetralithiated, giving **69** and **71**. With two equivalents of BuLi, dilithiation takes place selectively on the alternate rings of the calixarene (**73**), from which the diacid **74** may be formed, or after protonation the monoacids **76** or **77**. The use of *n*-BuLi is advantageous where precise numbers of equivalents are required, since neither elimination nor substitution of BuBr happens at low temperature; butylation can be a problem on warming, however.

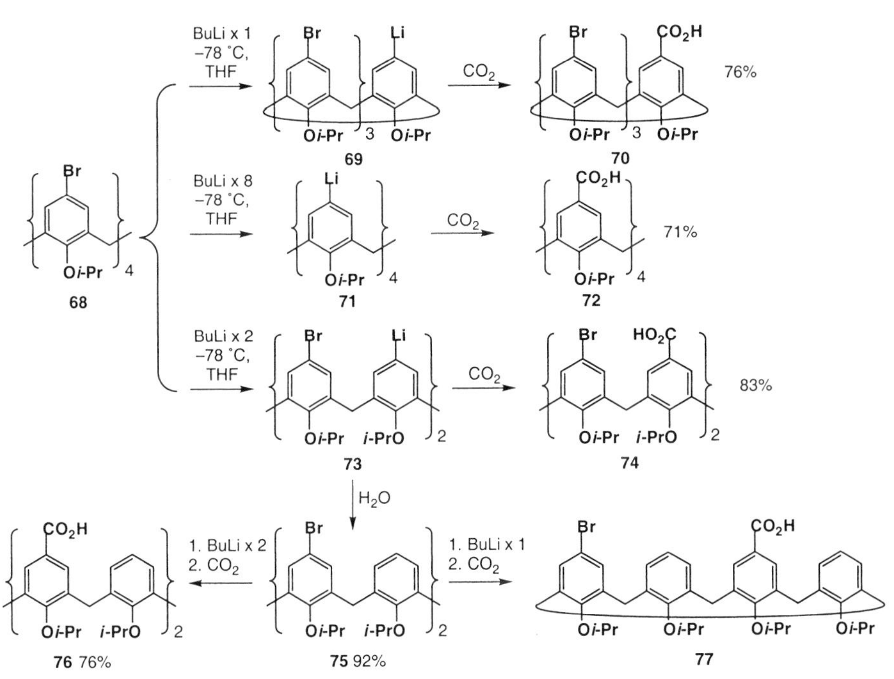

The speed of equilibration by halogen-metal exchange between different components of a reaction mixture means that the outcome of monolithiation of a dibromide is under thermodynamic control: when the bromo-substituents are non-identical, the result is always the more stable of the two organolithiums. For example, halogen-metal exchange between **78** or **79** and PhLi generates in each case the 2-lithio species in which the nitro group is more able to stabilise the lithium atom.[74] The stabilising effect of an adjacent amino or amido group leads to similar selectivity with **80** and **81**.[75]

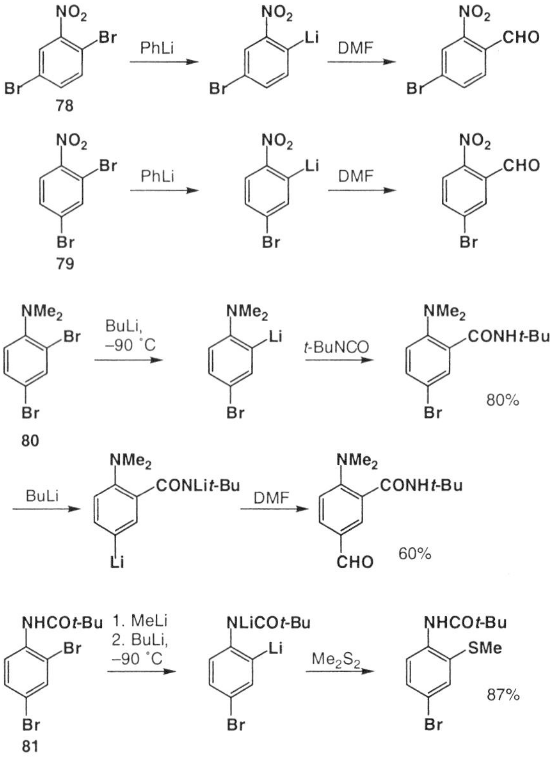

Selectivity in the double bromine-lithium exchange of **82** allows the straightforward synthesis of 1,2,3-trisubstituted benzene rings.[76] Metallation is apparently assisted by the OMe group, but presumably (unlike during ortholithiation) this is a purely thermodynamic effect, given the rate of exchange between different aryl bromides and aryllithiums. Although phenoxide $-O^-$ substituents can have a weak kinetic directing effect on deprotonation reactions (section 2.3.2.4.1), the lithiation of dibromooxine **84** indicates that they have a thermodynamic destabilising effect on an adjacent lithio substituent, since the 4-substituted product **85** is obtained. Halogen lithium exchange on **83** is successful, though its methyl ether is reported to undergo nucleophilic addition adjacent to N with alkyllithiums.[77]

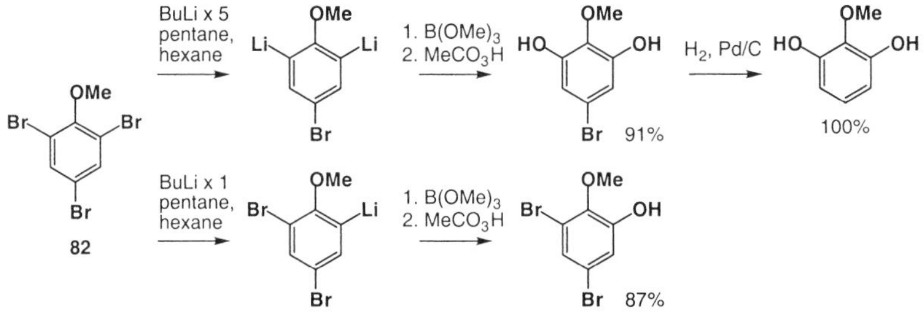

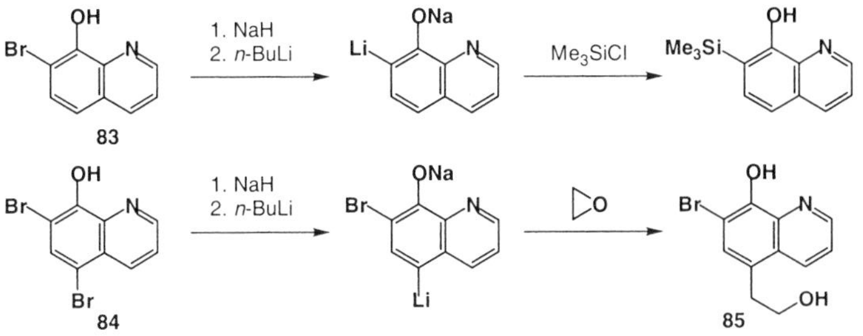

A recent synthesis of (±)-cordrastine used as a key step the formation of aryllithium **86** in the presence of the hydroxyl and amino functions. Carboxylation of **86** and lactonisation gave the natural product.[78]

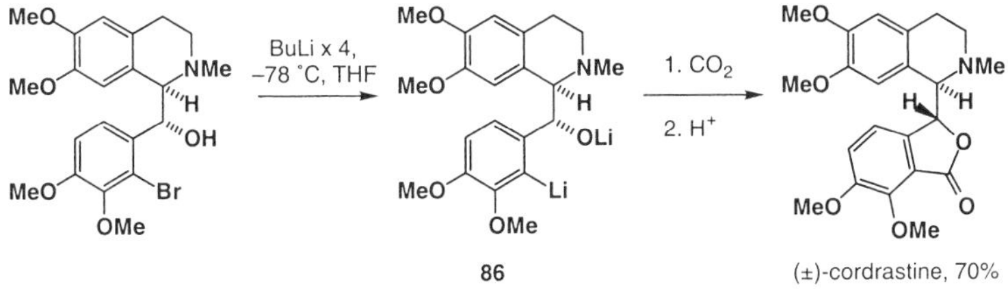

The anticancer compound combretastatin **87** has also been made via a halogen-metal exchange step.[79]

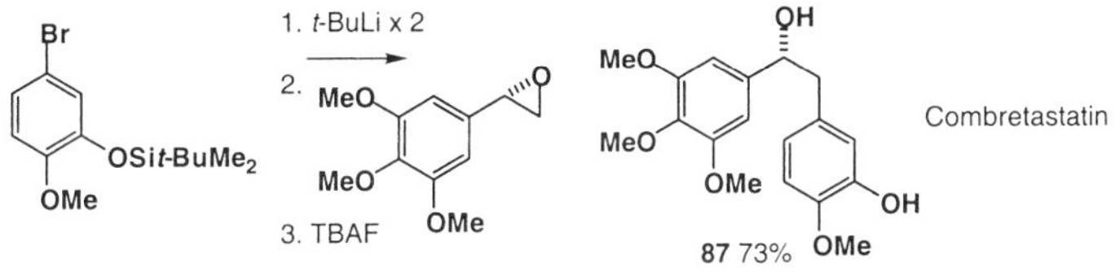

3.1.4 Synthesis of heteroaryllithiums

The ease with which brominated heterocycles may be prepared regioselectively makes the use of these compounds as starting materials for the synthesis of regioselectively lithiated heterocycles extremely attractive. Organolithium derivatives of all the simple heterocycles at all possible positions of substitution have been made by this method.[80] The scheme below illustrates some classical methods for forming 2-lithio-,[81] 3-lithio-,[81] and 4-lithiopyridines,[82] along with 4-lithioquinoline.[83] *n*-BuLi works well in these reactions, and indeed may be better than *t*-BuLi in reactions with electron deficient heterocycles, with which it tends to undergo addition reactions.

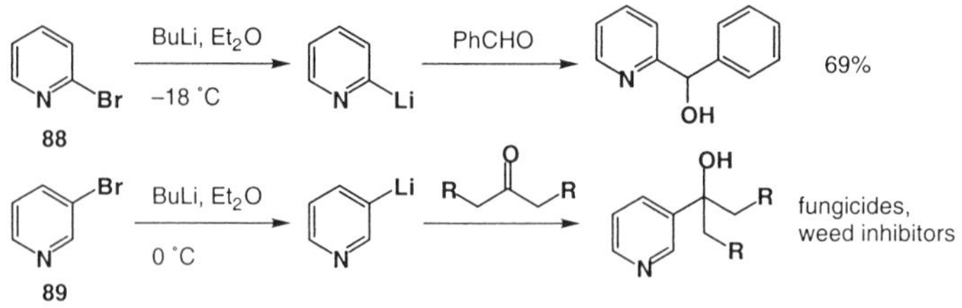

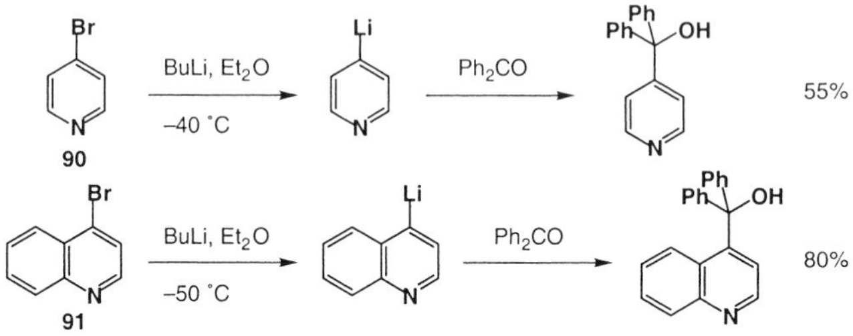

3-Bromopyridine **89** is particularly useful for the synthesis of agrochemicals based upon the 3-pyridinemethanol structure,[84] but its lithio-derivative has a tendency to isomerise by intermolecular halogen-metal exchange with unreacted 3-bromopyridine to give the more stable 3-bromo-4-lithio-derivative,[85] a process which can be turned to advantage (see section 2.3.3.1). 5-Lithiopyrimidine[86,87] **92** may be used for the synthesis of the fungicide fenarimol[84] – again, the 5-lithioderivative is less stable than the 4-lithio-derivative formed by deprotonation when the same starting material is treated with LDA at –10 °C.[88]

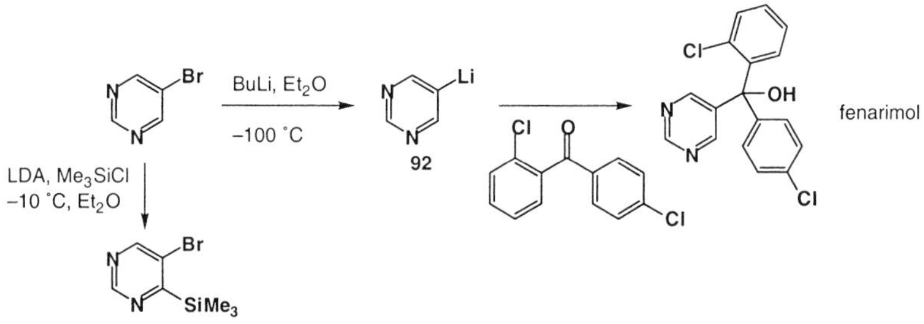

Halogen-metal exchange is useful particularly for the synthesis of heterocycles bearing lithium at less acidic sites – for example, the 3-position of both pyridine and of the 5-membered heterocycles. The schemes below show the synthesis of some β-lithiated 5-membered heterocycles (α-lithiation can be achieved using direct deprotonation methods - see section 2.2.3). 3-Lithiothiophene **93** in an important intermediate in the industrial synthesis of the vasodilator cetiedil **94**.[84]

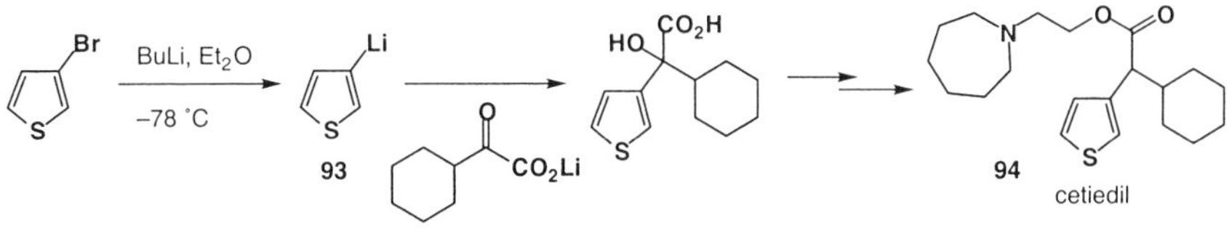

93 94

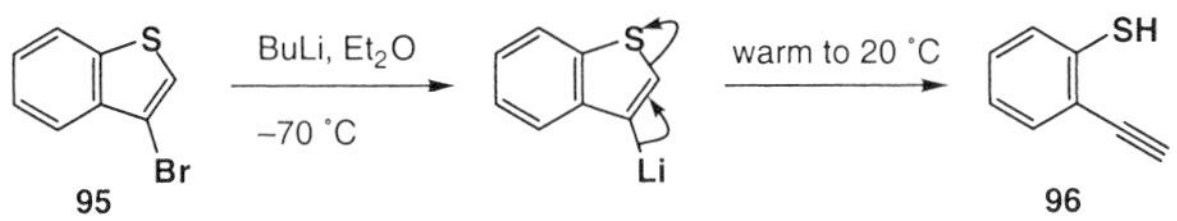

95 96

3-Lithiothiophenes and -furans in particular are unstable as the temperature rises above –78 °C. 3-Lithiothiophenes undergo ring-opening, so for example halogen metal exchange on 3-bromobenzothiophene **95** at –70 °C, followed by warming to room temperature, gives 2-ethynylthiophenol **96**.[89] 3-Lithiofurans are more prone simply to isomerisation to the more thermodynamically stable 2-lithiofurans: the transformation of **97** to **98** occurs at temperatures above –40 °C.[90] A similar process slowly transforms 3-lithiothiophene **99** to 2-lithiothiophene **100**, even at low temperature.[91]

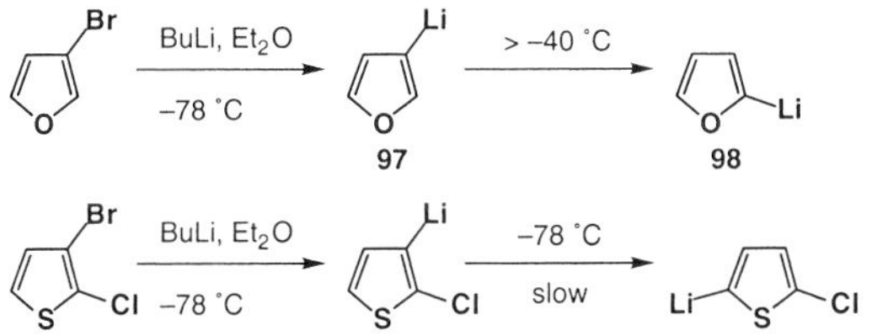

97 98 99 100

3-Lithiated pyrroles **101** and **102** readily form from the N-TIPS bromopyrroles, and have been used in the synthesis of analogues of the natural product lyngbyatoxin A.[92] Lithiated chloropurine **103** was used in the synthesis of a range of antiviral compounds.[93]

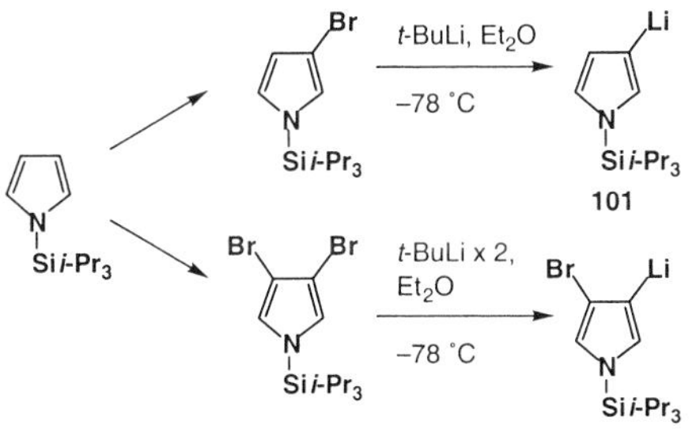

101 102

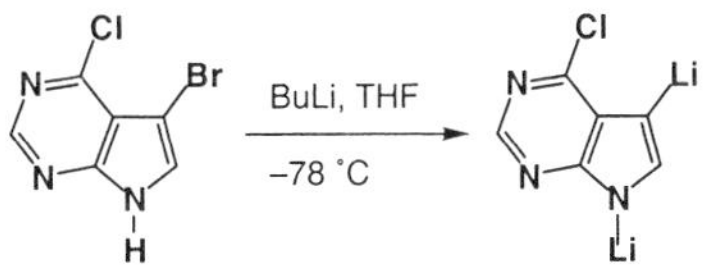

103

As with carbocyclic rings, the usual rules apply for avoiding deprotonation (and rearrangement) and promoting metallation: low temperatures and *t*- or *n*-BuLi as base.[94] LDA is used when deprotonation is required (the example below contrasts the deprotonation and halogen-metal exchange of **104**) and the halogen dance reactions described in section

2.3.3.1 depend for their selectivity on the relative rates of deprotonation and intermolecular halogen-metal exchange reactions.

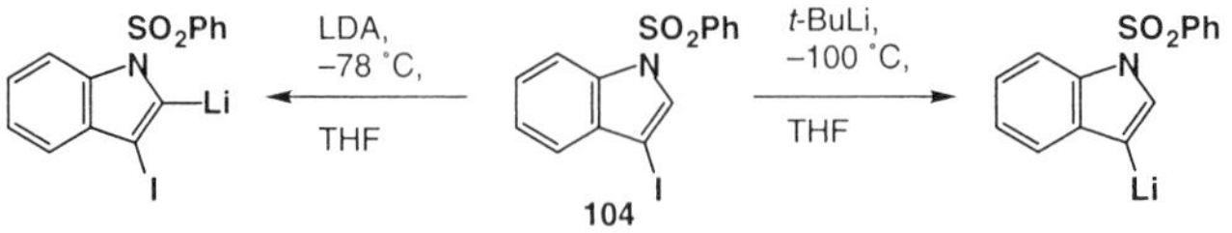

Silylation can alternatively be used to protect readily lithiated positions. For example, deprotonation and silylation of the pyrazole **105** with LDA/Me_3SiCl gives an intermediate **106** which is readily transmetallated, functionalised and desilylated.[95]

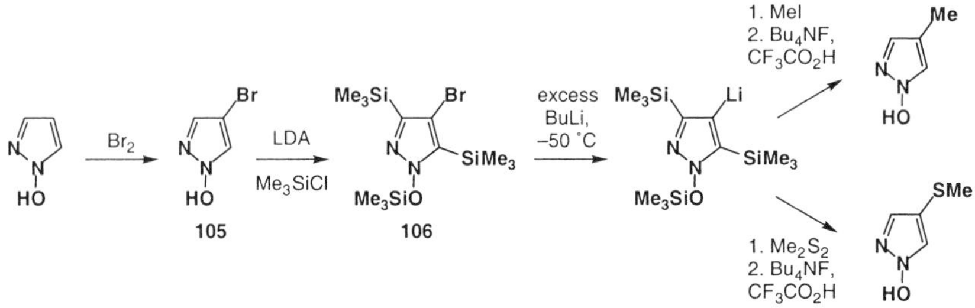

With polyhalogenated heterocycles, the most stabilised (usually the least basic) organolithium is formed first, allowing in many cases useful regioselectivity to be obtained. The least basic organolithiums are first of all those bearing lithium α to a heteroatom (or more than one), and secondly, in the pyridine series, those with lithium in the 4-position (though selectivities in the pyridine series are highly solvent-dependent[96,97]). Since sulfur is more anion-stabilising than nitrogen, the typical order of exchange in thiazoles is as shown below.[98] Imidazoles exchange the bromine α to the alkylated (pyrrole-like) nitrogen before the one α to the pyridine-like nitrogen, whose lone pair destabilises the adjacent anion.[99]

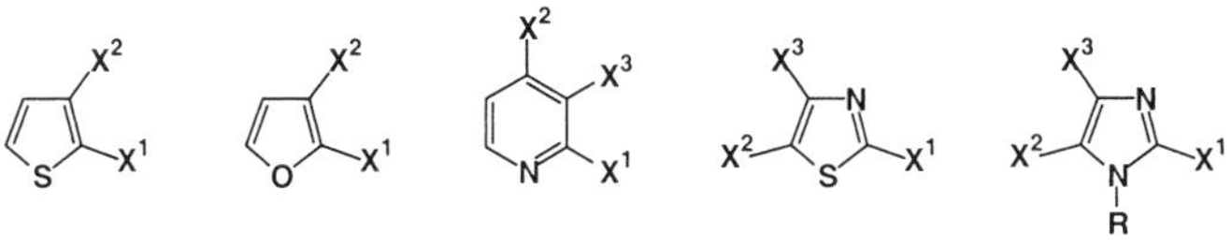

Typical order of exchange: $X^1/Li > X^2/Li > X^3/Li$

Some simple examples illustrate these points: on lithiation of dibromothiophene **107**, first the bromine next to sulfur is exchanged and then the bromine at the 3-position.[100]

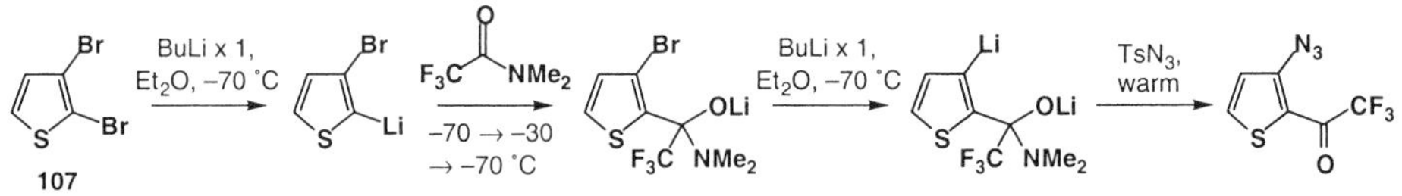

The dibromothiazole **108** undergoes exchange first at the 2-position and then at the 4-position.[101] Two different electrophiles can be introduced successively – in this case the silyl group can be removed to leave **109** regioselectively stannylated in the 4-position. Such use of silicon as a "protecting group" permits the synthesis of either 2- or 4-substituted thiazoles. Attempts to exchange Br for Li in **110** lead to problems with competing deprotonation at the 2-position, while **111** transmetallates cleanly.[98]

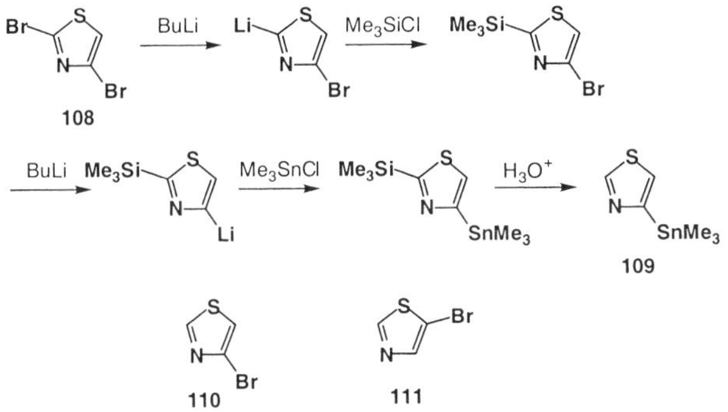

The three bromine atoms of the tribromoimidazole **112** can be replaced sequentially in one pot, again in order of increasing basicity of the resulting organolithium.[99] This method was applied to the synthesis of the antitumour agent carmethizole **113**.

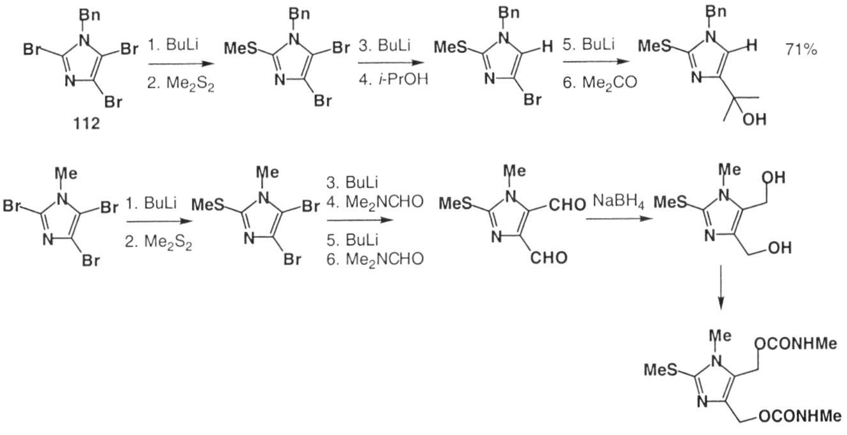

carmethizole **113**

Good selectivity for transmetallation at the 2-position of 2,5-dibromopyridine **114** is obtained by using toluene as a solvent.[96]

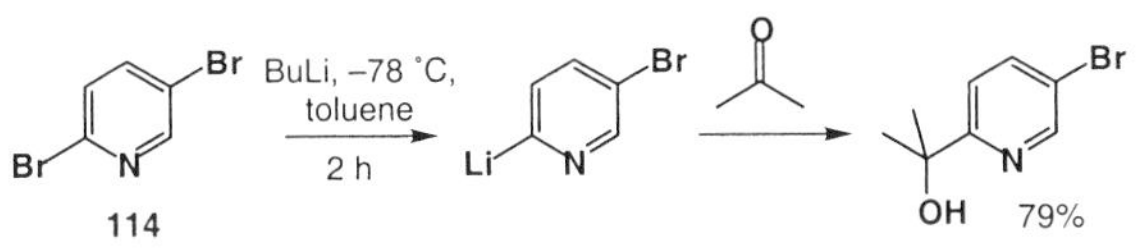

3.1.5 Synthesis of vinyllithiums

The synthesis of vinyllithiums **116** and **118** from vinyl bromides **115** and **117** with retention of double bond geometry is generally best done using the method of Seebach:[102] the vinyl bromide is treated with two equivalents of *t*-BuLi in Et_2O or THF at low temperature (–78 °C, for non-terminal bromides; <–110 °C for terminal bromides, which otherwise lithiate α to Br and generate alkynes). Choice of solvent is essential only with alkyl iodides as electrophiles, for which the use of THF leads to attack at iodine instead of at carbon. Clean alkylation in Et_2O is obtained on warming slowly to room temperature.[24]

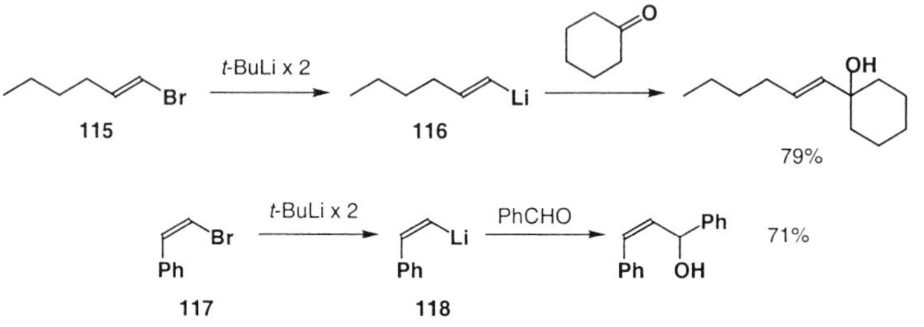

If the vinyl iodide is available, a simplified method may be used in which the iodide (**119** or **120**) is treated with 1.5 equivalents of *t*-BuLi at room temperature in hexane (the bromides work less well under these conditions). After 15 minutes, the electrophile is added; in hexane, even at this temperature, double bond geometry is maintained.[103,104] Successful electrophiles in this method include aldehydes, ketones and silyl halides; alkyl halides react with organolithiums very slowly in hexane, so alkylation by the BuI by-product is not a significant problem.

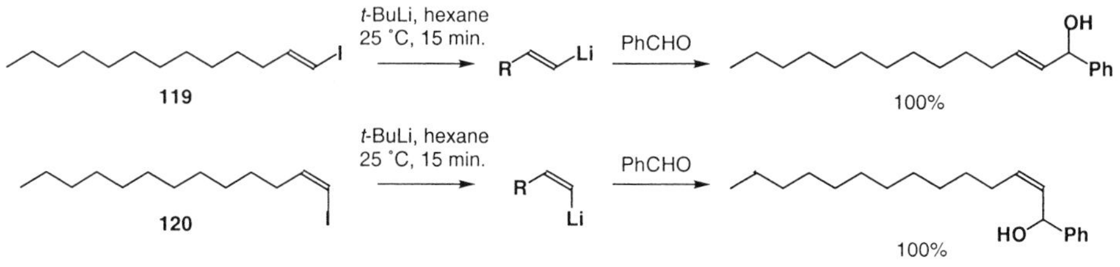

Certain functionalised vinyl bromides can similarly be lithiated: the *Z*-enol ether **121** gives **122,** which is stable up to –30 °C without elimination.[105] **121**[105] and **124**[106] are readily transmetallated, while the *E*-enol ether **123**[105] is preferentially lithiated α to oxygen and will not undergo transmetallation.

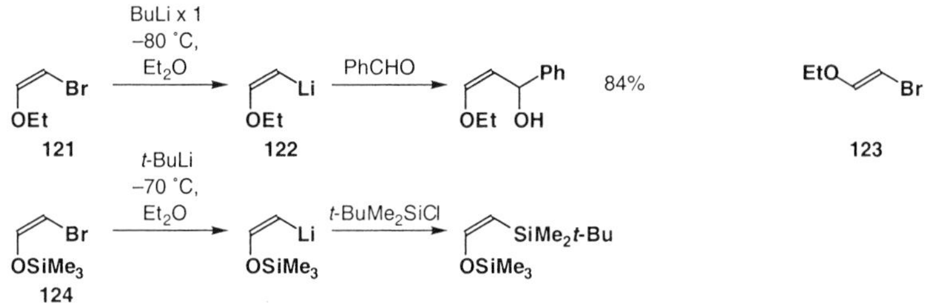

Competing deprotonation α to the halogen is a typical problem when heterosubstituents are in a position to direct lithiation to this position. So, for example, while both **125** and **126** transmetallate cleanly,[107] the *E* and *Z* isomers of **127** and **128** behave quite differently, with deprotonation of **127** leading to carbene formation and capture of *t*-BuLi while **128** undergoes transmetallation.[108] Clean transmetallation to an *E* vinyllithium can be achieved in this case by the use of the iodide (**129**) or by replacing one of the alkoxy groups of the acetal (**130**).

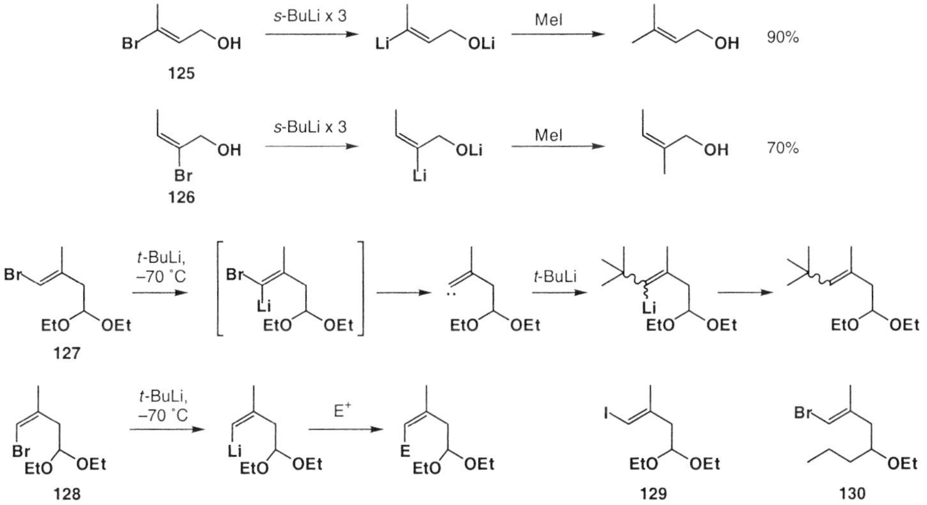

Lithiated silyl enol ethers related to **124** have been used in the synthesis of polyunsaturated aldehydes by chain extension, as shown below.[109,110] The stereospecificity (or otherwise) of the reaction is irrelevant to the stereochemistry of the products **131** and **132**, which is under thermodynamic control.

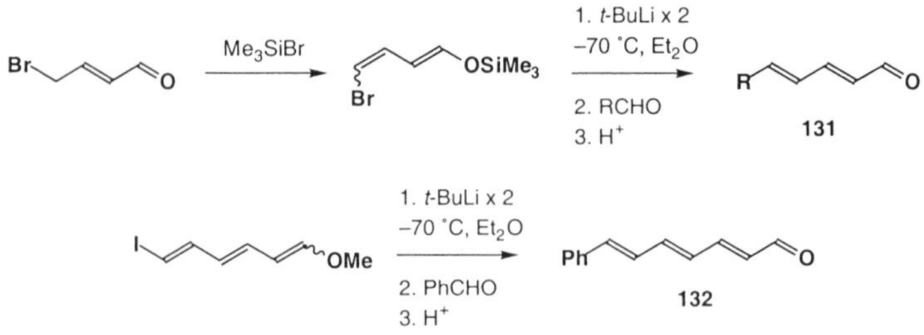

These methods have been applied to the synthesis of all *trans* retinal **133** by the route shown below:[111]

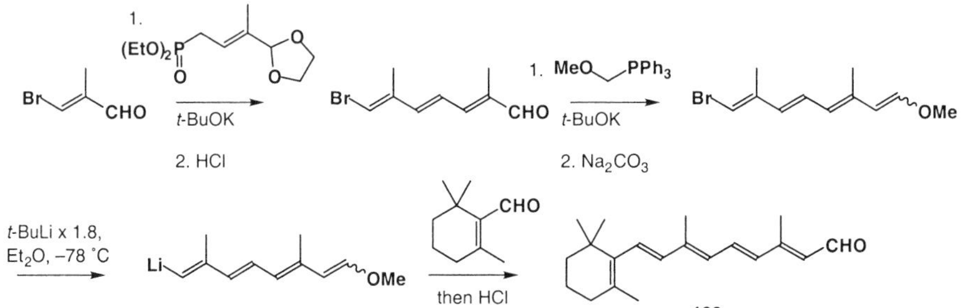

3.1.6 Synthesis of alkyllithiums

The standard way of making primary alkyllithiums from alkyl iodides is to use 2.2 equiv. of *t*-BuLi in ether at –78 °C, and then to warm to room temperature, before re-cooling ready for the subsequent reaction of the organolithium.[9] Negishi developed this method to make a range of organolithium reagents for transmetallation for other organometallics. The advantage of ether is that the –78 → +20 → –78 °C temperature régime selectively destroys any excess *t*-BuLi while leaving the primary organolithium **134** intact.

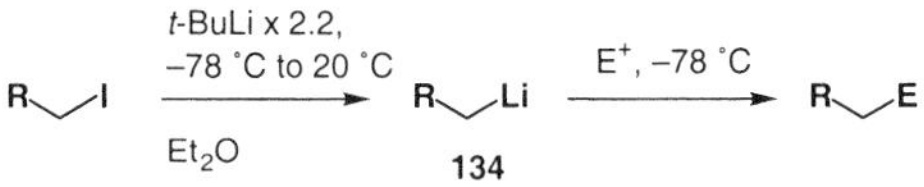

The method is less successful with benzylic and allylic halides, and problems of selectivity with these substrates have been overcome in the past via an initial substitution with lithium butyltelluride; the organolithium is then formed cleanly from the telluride **135** or **136** (which need not be isolated) by tellurium-lithium exchange.[112]

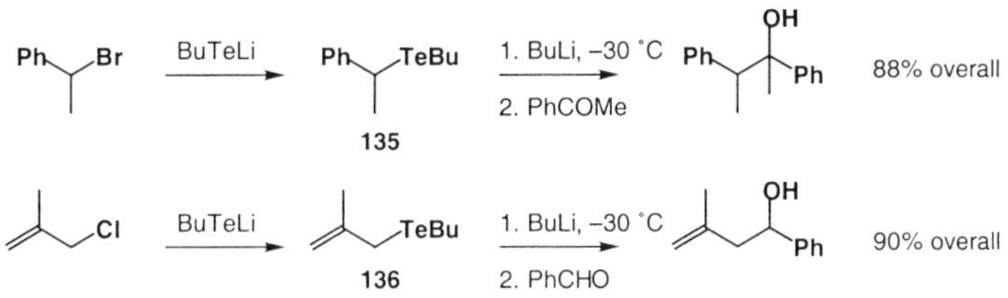

3.1.7 Diastereoselective halogen-lithium exchange

The halogen-metal exchange of one of a geminal pair of bromine atoms is readily achieved – a carbenoid is formed which may have sufficient stability to react as an organolithium before decomposing to a carbene by α-elimination.[17,113] In cyclic systems, diastereoselectivity in this process has long been known,[114-116] though the outcome of reactions such as that of **137** is under a mixture of kinetic and thermodynamic control.[117]

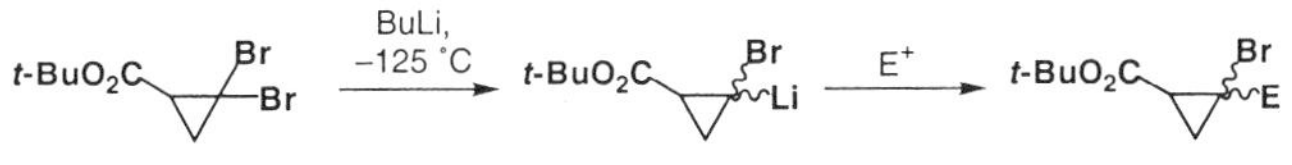

Hoffmann has shown that diastereoselective bromine-lithium exchange may be achieved even in acyclic systems.[118] Treatment of **138** with BuLi in the Trapp solvent mixture at –120 °C in the presence of acetone generates the epoxides **139** and **140** in a 94:6 ratio of diastereoisomers.[119,120] The selectivity was found to depend on the organolithium used for the bromine-lithium exchange, and it must therefore be under kinetic control.

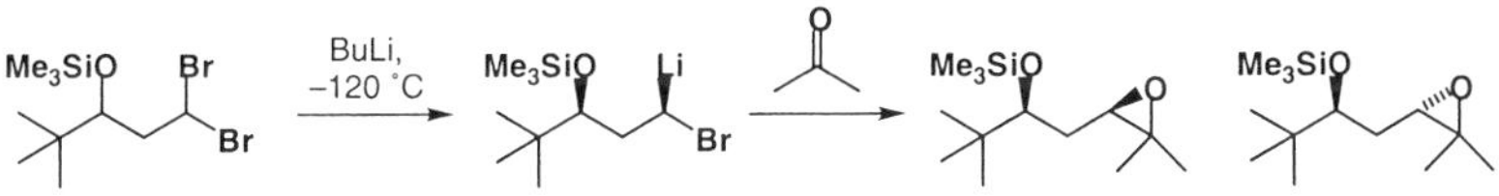

Changing the protecting group at O altered the ratio of products, particularly when a powerful Li-coordinating group (such as a carbamate) was introduced.[121] Reasonable levels of diastereoselectivity could also be obtained from **141**, with the alkoxy substituent in the 2-position.[122] Diastereoselective bromine-lithium exchange of this sort has been employed in the synthesis of a segment of the bryostatins (section 9.7).[123]

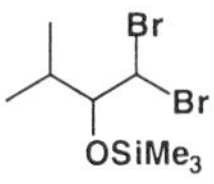

Stereoselective bromine-lithium exchange of one of a geminal pair of vinylic bromine atoms is also possible. Compounds **142-145** all lead to predominant exchange at the bromine *trans* to Ph when they are added to a solution of BuLi, with greater selectivity the more hindered that bromine becomes.[124] Selectivity does not change with time, but is inverted when the

BuLi is added to the substrate. These results are best rationalised by assuming the exchange is under kinetic control unless an excess of dibromide is present, in which case equilibration occurs, and that kinetic control favours exchange of the more hindered bromide atom, perhaps because of increased strain in the intermediate ate complex.

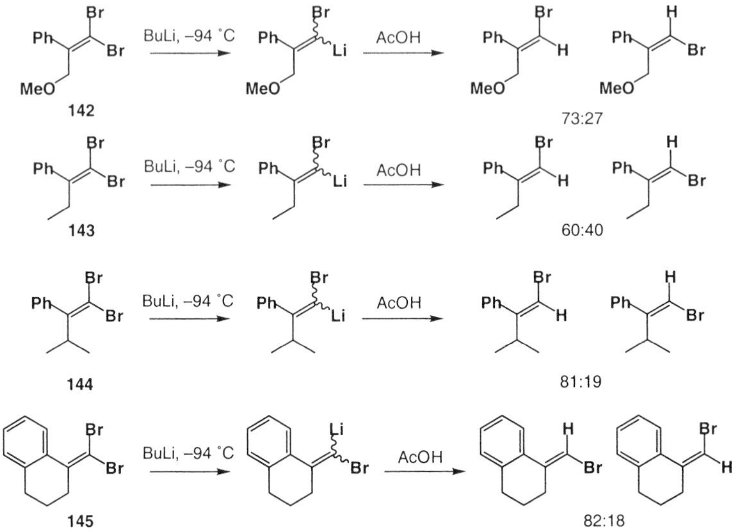

142 → (BuLi, –94 °C) → (AcOH) 73:27

143 → (BuLi, –94 °C) → (AcOH) 60:40

144 → (BuLi, –94 °C) → (AcOH) 81:19

145 → (BuLi, –94 °C) → (AcOH) 82:18

3.2 Tin-lithium exchange

Organolithiums react rapidly and reversibly with stannanes, exchanging the alkyl group of the organolithium for an alkyl group of the stannane. The reaction is under thermodynamic control and produces the most stable organolithium. For this reason, it is of particular value for the synthesis of aryllithiums, vinyllithiums, α-heterosubstituted organolithiums, and organolithiums stabilised in some way by intramolecular coordination, as these are all more stable than BuLi or MeLi.

There is considerable evidence that tin-lithium exchange proceeds via an ate complex, as proposed by Still[125] and Seebach,[126] among others. In certain cases, the ate complex is more stable than any of the possible organolithium products, and it can be characterised spectroscopically,[127] though NMR studies of most tin-lithium exchange reactions fail to show evidence of appreciable concentrations of the ate complex.[128,129] When methyllithium and tetramethylstannane are mixed in THF at –80 °C, ^{119}Sn NMR shows only a signal at δ 0, corresponding to Me_4Sn **146**. However, addition of HMPA causes a new peak at δ –277 to appear, in which the ^{119}Sn nuclei are equally coupled to fifteen H nuclei. Such upfield shifts for tin are characteristic of stannate complexes. Three equivalents of HMPA are sufficient to convert all the stannane to lithium pentamethylstannate **147**.[127]

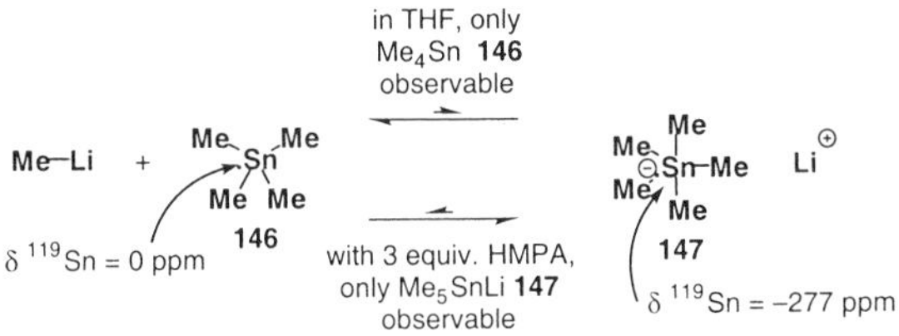

On warming the solution of stannate to –20 °C, coalescence occurs between the peaks in the ^{1}H NMR spectrum assigned to MeLi and to Me_5SnLi – the exchange processes between them is clearly fast at these temperatures even on the NMR timescale.

The ate complex **147** is only detectable in HMPA, and this is not proof that it is in fact on the reaction pathway for tin-lithium exchange in THF. However, in cases where the ate complex is more stable, for example where it is formed by cyclisation of an organolithium onto a stannane, it may be observed even in THF. The bis-stannane **148** reacts with methyllithium in THF at –80 °C to give Me_4Sn plus a symmetrical species **149** which has a high-field ^{119}Sn NMR signal characteristic of a stannate complex. Hydrolysis of **149** returns the stannane **150**, while warming to –40 °C generates a second equivalent of Me_4Sn and a further ate complex **151**. Hydrolysis of **151** generates principally the stannane **152**.[130]

Only on warming to +25 °C does the stannate **151** finally decompose to generate two equivalents of the doubly lithiated species **153**, plus a third equivalent of Me_4Sn, in an equilibrium concentration. The same mixture of ate complex **151**, organolithium **153** and Me_4Sn can be reproduced by reacting stannane **154** with MeLi or the independently synthesised **153** with Me_3SnCl.

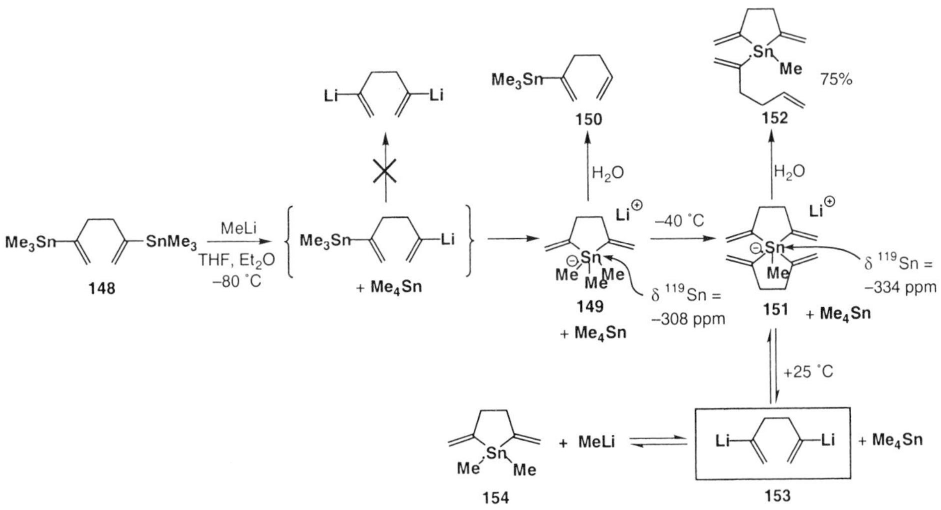

The conjugated dienyl bis-stannane **155** behaved very similarly, but in this case the final doubly lithiated diene **157** was much more stable than its precursor spirostannate **156**, and was formed in quantitative yield.[130]

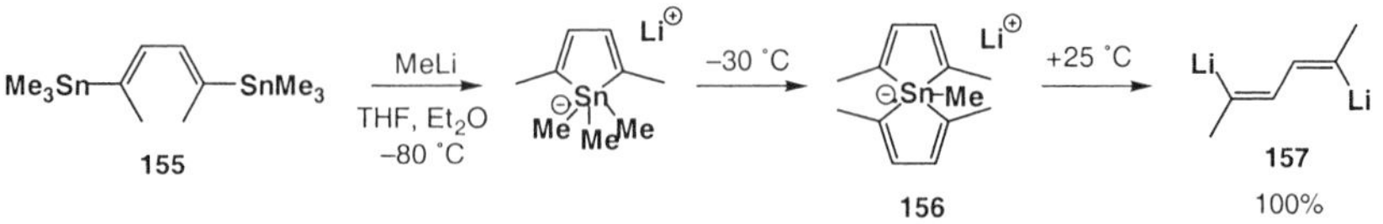

A similar stable "stannolate" **160** was produced on attempted transmetallation of benzostannole **159**, itself produced by tellurium-lithium exchange from **158**.[131]

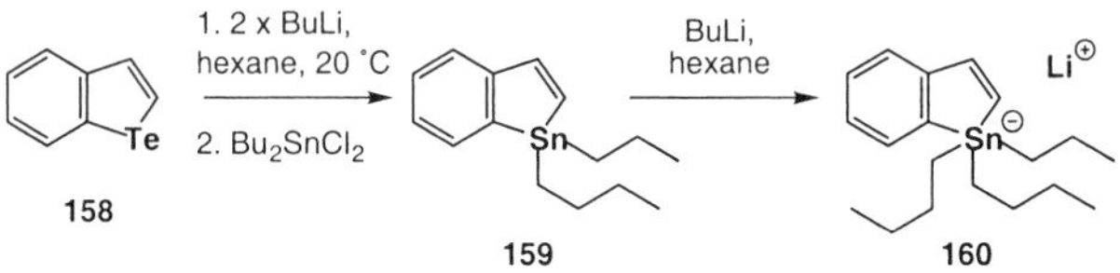

The possibility that the ate complex is not only on the reaction pathway but that it, rather than a product organolithium, is in fact the reactive species in subsequent electrophilic substitution reactions was addressed by Still.[125] He and others[127] have concluded that known ate complexes have much lower reactivity towards electrophiles than that observed for organolithiums, and it is generally considered that this possibility is unlikely.

Tin-lithium exchange has been widely used in synthesis, particularly to form stereodefined organolithiums (section 5.2.1). For example, Sinaÿ used a radical reaction to displace the sulfone from the glycoside **161**, substituting for a Bu_3Sn– group. Tin-lithium exchange of **162** gave a vinyllithium **163** which reacted with electrophiles to yield products such as **164**.[132]

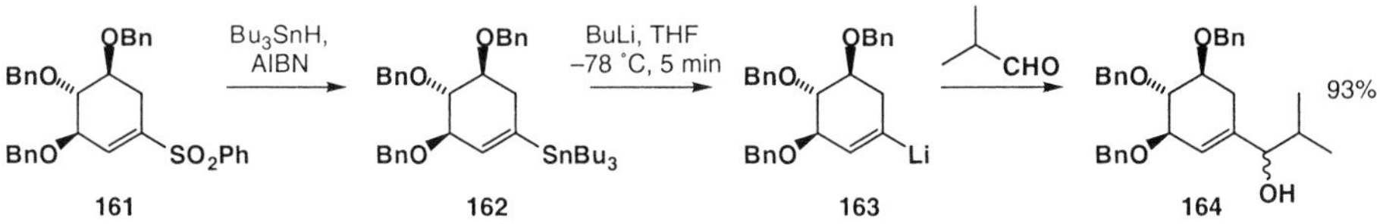

Unlike organolithiums, lithium amides do not attack tin, so functionalisation of tin-containing molecules by deprotonation with LDA is still feasible:[133]

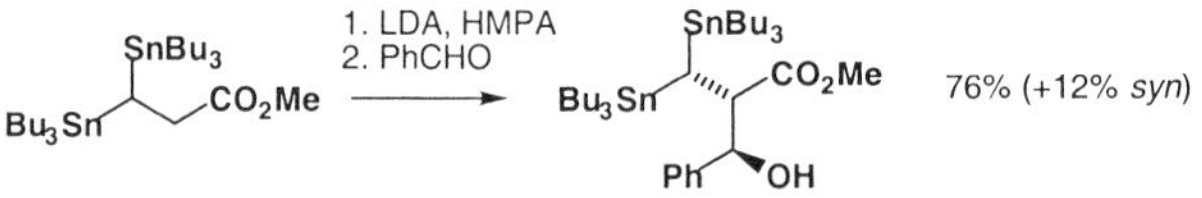

3.3 Chalcogen-lithium exchange

3.3.1 Selenium-lithium exchange

Gilman noted in 1949 that butyllithium slowly displaces phenyllithium from phenylselenides in ether.[134] But it was the work of Krief, who reported in 1974[135] that in THF this reaction becomes much faster (complete in 30 min at 0 °C) that gave selenium-lithium exchange synthetic possibilities. The reaction of a dialkylselenide – or a trialkylselenonium salt – with an alkyllithium, displacing the most stable organolithium, is effective, but of rather limited use, and there is insufficient difference in stability between Me and Bu for clean displacement of one by the other:

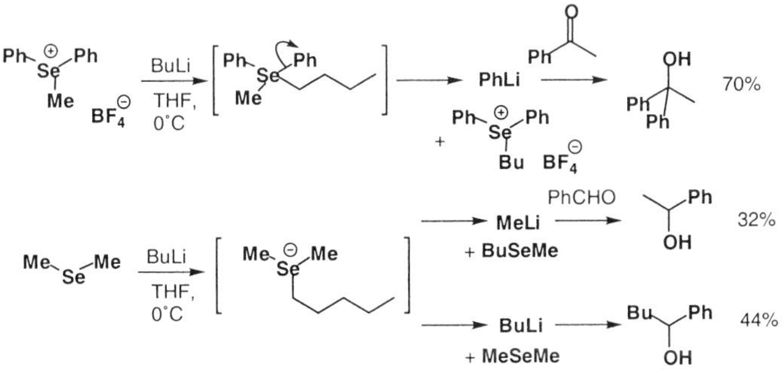

When the starting selenide carries a benzyl or allyl group, this method does however become a useful way of generating benzyllithiums[136] or allyllithiums,[137] which cannot be made by halogen-lithium exchange because of coupling between the organolithium formed and the unreacted halide (selenium is therefore an alternative to tin). Allylic and benzylic phenylselenides or methylselenides such as **165-167** are equally successful: the choice of starting material is best determined by the desirability or otherwise of a volatile selenide by-product. Deprotonation of the benzylic centre competes with substitution if the base is added to the selenide rather than vice-versa, but deprotonation can be deliberately achieved with LDA, LDA–*t*-BuOK or LiTMP.[137]

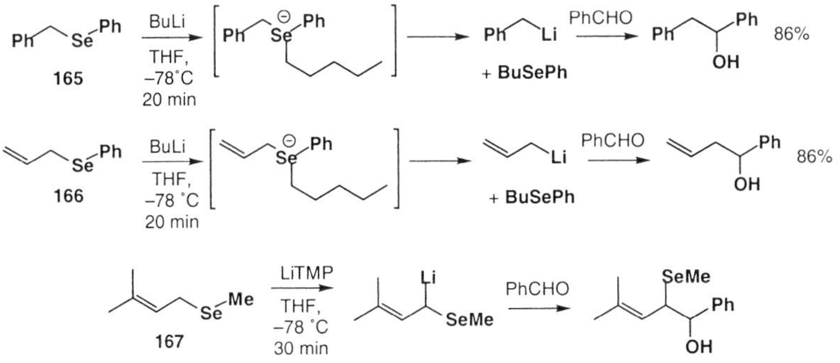

A benzylic group is displaced even if it is fully substituted,[138] and the resulting tertiary benzylic organolithiums such as **168** react reliably with electrophiles, making this a general way of constructing the quaternary carbon centre of molecules such as **169** or **170** – provided one of the substituents is aryl.

1. PhLi
2. MeSH, $ZnCl_2$
BuLi or *s*-BuLi
−78 °C, THF, 30 min
RI, 20 °C, 30 min
97%
168
169
170

Even easier to make than simple selenides are the selenoacetals **171**and **175**, which, as first reported by Seebach,[139,140] undergo very clean selenium-lithium exchange at −78 °C.[141] In this respect they are quite different from thioacetals, which are deprotonated by strong bases (and of course acetals themselves, which are resistant to any attack by base). Both $(PhSe)_2$ and $(MeSe)_2$ selenoacetals, even those formed from ketones (**175**), undergo exchange with *t*-BuLi, *s*-BuLi or (slightly more slowly) *n*-BuLi in THF. The selenium-stabilised organolithiums are stable for hours at −78 °C, and neither undergo rearrangement to the more stable lithiomethyl or lithiophenyl isomers **173** or **174** nor lose selenolate to give carbenes. They react well with aldehydes and ketones, and the products can be transformed to a number of functional groups, including epoxides and allylic alcohols. As with benzyl and allylselenides, use of a less nucleophilic reagent such as LDA–*t*-BuOK leads to deprotonation.[142]

R^2 = Me or Ph
R^2SeH
BuLi, −78 °C
E^+
171
172
stable for hours
173
174
1. LDA, *t*-BuOK
2. R^3X
175
176
$(PhS)_2$
177
178

With a sulfur electrophile, mixed thioselenoacetals **177** form which may then undergo a second selenium-lithium exchange, leading to fully substituted α-thio organolithiums **178**.[143]

It is impossible to generate these species directly by lithiation of sulfides because of competing ring-lithiation of the sulfides.

3.3.2 Tellurium-lithium exchange

Tellurium-lithium exchange proceeds in much the same way as selenium-lithium exchange but is rarely used.[144] Vinyltellurides such as **179** can be converted stereospecifically into vinyllithiums **180**,[145] and alkyltelluro-heterocycles have been used as precursors to lithio-species.[146]

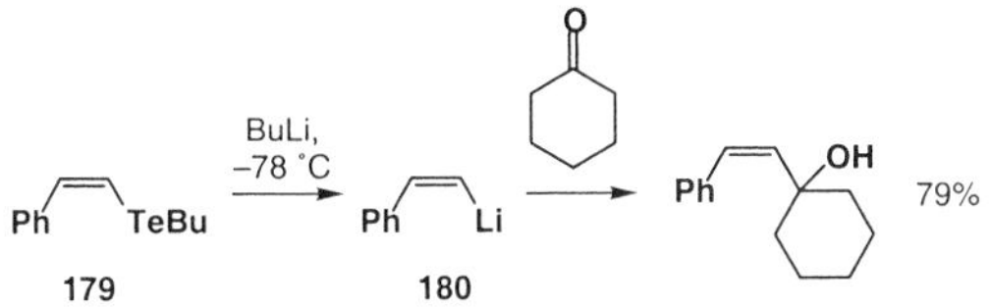

There is one important class of exchange which is unique to tellurium: telluroesters such as **181** undergo tellurium-lithium exchange to generate the highly unstable acyllithiums such as **183**.[147] These must be formed in the presence of an electrophile – typically pinacolone (*t*-butyl methyl ketone **182**) or they decompose.

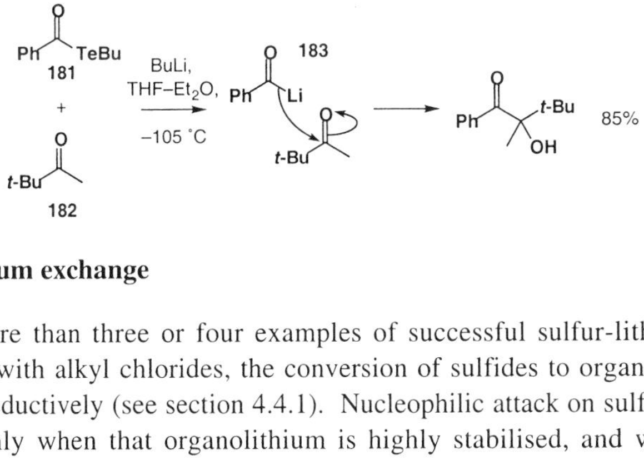

3.3.3 Sulfur-lithium exchange

There are no more than three or four examples of successful sulfur-lithium exchange in sulfides, and, as with alkyl chlorides, the conversion of sulfides to organolithiums is much better achieved reductively (see section 4.4.1). Nucleophilic attack on sulfide S displaces an organolithium only when that organolithium is highly stabilised, and was first noted by Brandsma,[148] who found that PhLi displaced the alkynyllithium **185** from **184**.

The tetrathioorthocarbonate **186** undergoes a similar reaction to produce a useful one carbon nucleophile at the carboxylic acid oxidation level,[149] and substitution on the cyclopropananone thioacetal **188** provides an alternative way of making Trost's[150] lithiated sulfide **189**.[151]

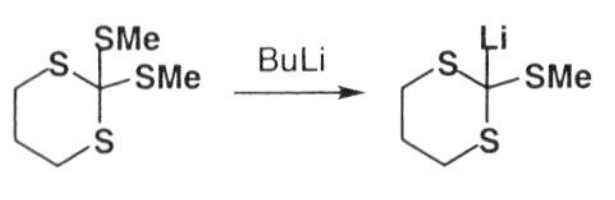

Sulfur-lithium exchange is easier and has much greater potential (much of it still unrealised) when the sulfur is at the sulfoxide oxidation level. It has long been known that organolithiums, like Grignard reagents, will attack a sulfoxide, displacing with inversion at sulfur the substituent best able to support an anion. The reaction has been commonly used to form sulfoxides with defined stereochemistry:[152-157]

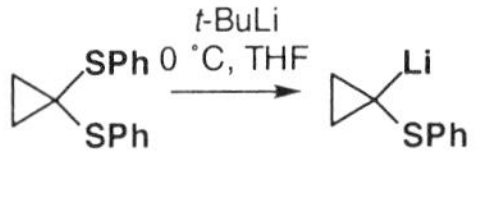

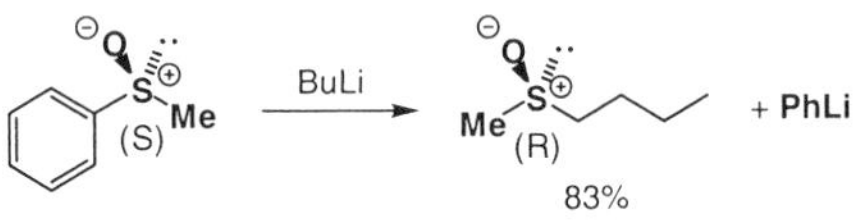

However, only recently has the reaction been used as a way to generate a useful aryllithium, with the sulfoxide being merely a by-product of the S–Li exchange.[158-160] Kagan used the substitution at ferrocenylsulfoxide **190**, for example (previously noted as a side reaction in the attempted ortholithiation of ferrocenylsulfoxides[161]), to generate the enantiomerically pure organolithium **191** and hence derivatives such as **192**.[158,159] The sulfoxide by-product is almost racemic when *t*-BuLi is used to carry out the substitution. Note the selectivity for attack at S rather than Sn.

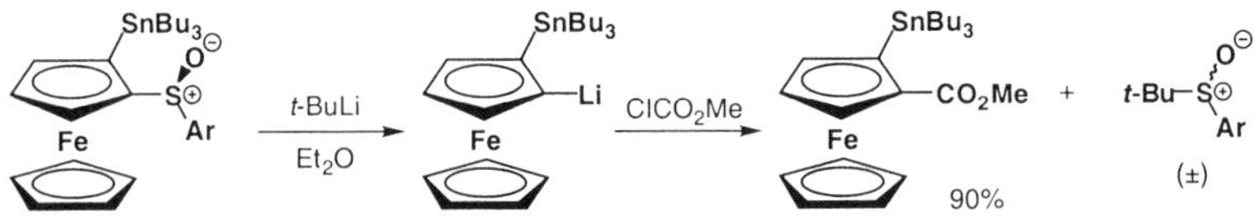

3.4 Phosphorus-lithium exchange

Hayashi showed that binaphthyl phosphine oxides such as **193** undergo phosphorus-lithium exchange with BuLi to generate binaphthylllithiums such as **194**.[162]

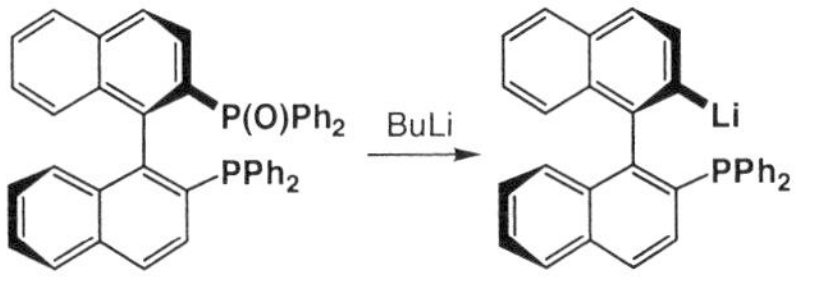

References

1. Wittig, G.; Pockels, U.; Dröge, H. *Chem. Ber.* **1938,** *71*, 1903.
2. Marvel, C. S.; Hager, F. D.; Coffman, D. D. *J. Am. Chem. Soc.* **1927,** *49*, 2323.
3. Gilman, H.; Langham, W.; Jacoby, A. L. *J. Am. Chem. Soc.* **1939,** *61*, 106.

4. Gilman, H.; Jones, R. G. *Org. Reac.* **1951,** *6*, 339.
5. Jones, R. G.; Gilman, H. *Chem. Rev.* **1954,** *54*, 835.
6. Gilman, H.; Jones, R. G. *J. Am. Chem. Soc.* **1941,** *63*, 1441.
7. Applequist, D. E.; O'Brien, D. F. *J. Am. Chem. Soc.* **1963,** *85*, 743.
8. Cram, D. J. *Fundamentals of Carbanion Chemistry*; Academic Press: New York, 1965.
9. Negishi, E.-i.; Swanson, D. R.; Rousset, C. J. *J. Org. Chem.* **1990,** *55*, 5406.
10. Corey, E. J.; Beames, D. J. *J. Am. Chem. Soc.* **1972,** *94*, 7210.
11. Seebach, D.; Neumann, H. *Chem. Ber.* **1974,** *107*, 847.
12. Wieringa, J. H.; Wynberg, H.; Strating, J. *Synth. Commun.* **1971,** *1*, 7.
13. Sauers, R. R.; Schlosberg, S. B.; Pfeffer, P. E. *J. Org. Chem.* **1968,** *33*, 2175.
14. Harrowven, D. C.; Nunn, M. I. T. *Tetrahedron Lett.* **2001,** *42*, 7501.
15. Gilman, H.; More, F. W. *J. Am. Chem. Soc.* **1940,** *62*, 1843.
16. Langham, W.; Brewster, R. Q.; Gilman, H. *J. Am. Chem. Soc.* **1941,** *63*, 545.
17. Köbrich, G. *Angew. Chem., Int. Ed. Engl.* **1967,** *6*, 41.
18. Köbrich, G.; Trapp, H.; Hornke, I. *Tetrahedron Lett.* **1964**, 1131.
19. Köbrich, G.; Flory, K. *Chem. Ber.* **1966,** *99*, 1773.
20. Tellier, F.; Sauvêtre, R.; Normant, J. F.; Deromsee, Y.; Jeannin, Y. *J. Organomet. Chem.* **1987,** *331*, 281.
21. Seyferth, D.; Murphy, G. J.; Woodruff, R. A. *J. Organomet. Chem.* **1974,** *66*, c29.
22. Shook, C. A.; Romberger, M. L.; Jung, S.-H.; Xiao, M.; Sherbine, J. P.; Zhang, B.; Lin, M.-T.; Cohen, T. *J. Am. Chem. Soc.* **1993,** *115*, 10754.
23. Köbrich, G.; Trapp, H. *Chem. Ber.* **1966,** *99*, 670.
24. Seebach, D.; Neumann, H. *Tetrahedron Lett.* **1976**, 4839.
25. Gilman, H.; Gorsich, R. D. *J. Am. Chem. Soc.* **1956,** *78*, 2217.
26. Gjös, N.; Gronowitz, S. *Acta Chem. Scand.* **1971,** *25*, 2596.
27. Köbrich, G.; Buck, P. *Chem. Ber.* **1970,** *103*, 1412.
28. Parham, W. E.; Jones, L. D. *J. Org. Chem.* **1976,** *41*, 2704.
29. Peterson, P. E.; Nelson, D. J.; Risener, R. *J. Org. Chem.* **1986,** *51*, 2381.
30. Kihara, M.; Kashimoto, M.; Kobayashi, Y.; Kobayashi, S. *Tetrahedron Lett.* **1990,** *31*, 5347.
31. Parham, W. E.; Bradsher, C. K. *Acc. Chem. Res.* **1982,** *15*, 300.
32. Parham, W. E.; Bradsher, C. K.; Reames, D. C.; Gross, P. M. *J. Org. Chem.* **1981,** *46*, 4804.
33. Jakiela, D. J.; Helquist, P.; Jones, L. D. *Org. Synth.* **1984,** *62*, 74.
34. Kondo, Y.; Asai, M.; Miura, T.; Uchiyama, M.; Sakamoto, T. *Org. Lett.* **2001,** *3*, 13.
35. Bailey, W. F.; Patricia, J. J.; Nurmi, T. T.; Wang, W. *Tetrahedron Lett.* **1986,** *27*, 1861.
36. Taylor, R. *Tetrahedron Lett.* **1975**, 435.
37. Beak, P.; Liu, C. *Tetrahedron* **1994,** *50*, 5999.
38. Gallagher, D. J.; Beak, P. *J. Am. Chem. Soc.* **1991,** *113*, 7984.
39. Beak, P.; Musick, T. J.; Chen, C.-W. *J. Am. Chem. Soc.* **1988,** *110*, 3538.
40. Narasimhan, N. S.; Ammanamanchi, R. *J. Chem. Soc., Chem. Commun.* **1985**, 1368.

41. Narasimhan, N. S.; Sunder, N. M.; Ammanamanchi, R.; Bonde, B. D. *J. Am. Chem. Soc.* **1990,** *112*, 4431.
42. Frye, S. V.; Johnson, M. C.; Valvano, N. L. *J. Org. Chem.* **1911,** *56*, 3750.
43. Zefirov, N. S.; Makhanolov, D. I. *Chem. Rev.* **1982,** *82*, 615.
44. Bailey, W. F.; Patricia, J. J. *J. Organomet. Chem.* **1988,** *352*, 1.
45. Bryce-Smith, D. *J. Chem. Soc.* **1956**, 1603.
46. Fischer, H. *J. Org. Chem.* **1969,** *73*, 3834.
47. Russell, G. A.; Lamson, D. W. *J. Am. Chem. Soc.* **1969,** *91*, 3967.
48. Ward, H. R. *J. Am. Chem. Soc.* **1967,** *89*, 5517.
49. Bailey, W. F.; Patricia, J. J.; DelGobbo, V. C.; Jarret, R. M.; Okarma, P. J. *J. Org. Chem.* **1985,** *50*, 1999.
50. Ashby, E. C.; Pham, T. N.; Park, B. *Tetrahedron Lett.* **1985,** *26*, 4691.
51. Ashby, E. C.; Pham, T. N. *J. Org. Chem.* **1987,** *52*, 1291.
52. Bailey, W. F.; Nurmi, T. T.; Patricia, J. J.; Wang, W. *J. Am. Chem. Soc.* **1987,** *109*, 2442.
53. Bailey, W. F.; Gagnier, X.; Patricia, J. J. *J. Org. Chem.* **1984,** *49*, 2092.
54. Newcomb, M.; Williams, W. G.; Crumpacker, E. L. *Tetrahedron Lett.* **1989,** *26*, 1183.
55. Ross, G. A.; Koppang, M. D.; Bartak, D. E.; Woolsey, N. F. *J. Am. Chem. Soc.* **1985,** *107*, 6742.
56. Garst, J. F.; Hines, J. B.; Bruhnke, J. D. *Tetrahedron Lett.* **1986,** *27*, 1963.
57. Sunthankar, S. V.; Gilman, H. *J. Org. Chem.* **1951,** *16*, 8.
58. Wittig, G.; Schöllkopf, U. *Tetrahedron* **1958,** *3*, 91.
59. Wittig, G. *Angew. Chem.* **1958,** *70*, 65.
60. Winkler, H. J. S.; Winkler, H. *J. Am. Chem. Soc.* **1966,** *88*, 964.
61. Winkler, H. J. S.; Winkler, H. *J. Am. Chem. Soc.* **1966,** *88*, 969.
62. Rogers, H. R.; Houk, J. *J. Am. Chem. Soc.* **1982,** *104*, 522.
63. Reich, H. J.; Phillips, N. H.; Reich, I. L. *J. Am. Chem. Soc.* **1985,** *107*, 4101.
64. Farnham, W. B.; Calabrese, J. C. *J. Am. Chem. Soc.* **1986,** *108*, 2449.
65. Reich, H. J.; Green, D. P.; Medina, M. A.; Goldenberg, W. J.; Gudmundsson, B. O.; Dykstra, R. R.; Phillips, N. H. *J. Am. Chem. Soc.* **1998,** *120*, 7201.
66. Müller, M.; Stiasny, H.-C.; Brönstrup, M.; Burton, A.; Hoffmann, R. W. *J. Chem. Soc., Perkin Trans. 2* **1999**, 731.
67. Beak, P.; Allen, D. J. *J. Am. Chem. Soc.* **1992,** *114*, 3420.
68. Adcock, W.; Clark, C. I.; Trout, N. A. *J. Org. Chem.* **2001,** *66*, 3362.
69. Mills, R. J.; Horvath, R. F.; Sibi, M. P.; Snieckus, V. *Tetrahedron Lett.* **1985,** *26*, 1145.
70. Wong, W.; D'Andrea, S. V.; Freeman, J. P.; Szmuszkowicz, J. *J. Org. Chem.* **1991,** *56*, 2914.
71. Katz, H. E. *J. Org. Chem.* **1985,** *50*, 5027.
72. Tamborski, C.; Chen, L. S. *J. Fluorine Chem.* **1995,** *75*, 117.
73. Larsen, M.; Jørgensen, M. *J. Org. Chem.* **1996,** *61*, 6651.
74. Voss, G.; Gerlach, H. *Chem. Ber.* **1989,** *122*, 1199.
75. Stanetty, P.; Krumpak, B.; Rodler, I. K. *J. Chem. Res. (S)* **1995,** *9*, 342.

76. Green, K. *J. Org. Chem.* **1991,** *56*, 4325.
77. Mongin, F.; Fouquet, J.-M.; Rault, S.; Levacher, V.; Godard, A.; Trécourt, F.; Quéguiner, G. *Tetrahedron Lett.* **1995,** *36*, 8415.
78. Orito, K.; Miyazawa, M.; Suginome, H. *Tetrahedron* **1995,** *51*, 2489.
79. Ramacciotti, A.; Fiaschi, R.; Napolitano, E. *Tetrahedron Asymmetry* **1996,** *7*, 1101.
80. Schlosser, M. In *Organometallics in Synthesis;* Schlosser, M. Ed.; Wiley: New York, 1994; pp. 1-166.
81. Gilman, H.; Spatz, S. M. *J. Org. Chem.* **1951,** *16*, 1485.
82. Wibaut, J.; Heeringa, L. G. *Rec. Trav. Chim. Pays-Bas* **1955,** *74*, 1003.
83. Gilman, H.; Soddy, T. S. *J. Org. Chem.* **1958,** *23*, 1584.
84. Totter, F.; Rittmeyer, P. In *Organometallics in Synthesis;* Schlosser, M. Ed.; Wiley: New York, 1994; pp. 167-194.
85. Mallet, M.; Quéguiner, G. *Tetrahedron* **1985,** *41*, 3433.
86. Kress, T. J. *J. Org. Chem.* **1979,** *44*, 2081.
87. Frissen, A. E.; Marcelis, A. T. M.; Buurman, D. G.; Pollmann, C. A. M.; Vanderplas, H. C. *Tetrahedron* **1989,** *45*, 5611.
88. Turck, A.; Plé, N.; Majovic, L.; Quéguiner, G. *J. Heterocyclic Chem.* **1990,** *27*, 1377.
89. Dickinson, R. P.; Iddon, B. *J. Chem. Soc. (C)* **1971**, 3447.
90. Haarmann, H.; Eberbach, W. *Tetrahedron Lett.* **1991,** *32*, 903.
91. Gronowitz, S.; Frejd, T. *Acta. Chem. Scand. Ser. B* **1976**, 485.
92. Shum, P. W.; Kozikowski, A. P. *Tetrahedron Lett.* **1990,** *31*, 6785.
93. Pudlo, J. S.; Nassiri, M. R.; Kern, E. R.; Wotring, L. L.; Drach, J. C.; Townend, L. B. *J. Med. Chem.* **1990,** *33*, 1984.
94. Saulnier, M. G.; Gribble, G. W. *J. Org. Chem.* **1982,** *47*, 757.
95. Balle, T.; Vedsø, P.; Bergstrup, M. *J. Org. Chem.* **1999,** *64*, 5366.
96. Wang, X.; Rabbat, P.; O'Shea, P.; Tillyer, R.; Grabowski, E. J. J.; Reider, P. J. *Tetrahedron Lett.* **2000,** *2000*, 4335.
97. Parham, W. E.; Piccirilli, R. M. *J. Org. Chem.* **1977,** *42*, 257.
98. Athmani, S.; Bruce, A.; Iddon, B. *J. Chem. Soc., Perkin Trans. 1* **1992**, 215.
99. Lipshutz, B. H.; Hagen, W. *Tetrahedron Lett.* **1992,** *33*, 5865.
100. Spagnolo, P.; Zanirato, P. *J. Chem. Soc., Perkin Trans. 1* **1996**, 963.
101. Dondoni, A.; Mastellari, A. R.; Medici, A.; Negrini, E.; Pedrini, P. *Synthesis* **1986**, 757.
102. Neumann, H.; Seebach, D. *Chem. Ber.* **1978,** *111*, 2785.
103. Shinokubo, H.; Miki, H.; Yokoo, T.; Oshima, K.; Utimoto, K. *Tetrahedron* **1995,** *51*, 11681.
104. Yokoo, T.; Shinokubo, H.; Oshima, K.; Utimoto, K. *Synlett* **1994**, 645.
105. Lau, K. S. Y.; Schlosser, M. *J. Org. Chem.* **1978,** *43*, 1595.
106. Duhamel, L.; Gralek, J.; Ngono, B. *J. Organomet. Chem.* **1989,** *363*, c4.
107. Schlosser, M.; Hannmer, E. *Helv. Chim. Acta* **1974,** *57*, 2547.
108. Bonnet, B.; Plé, G.; Duhamel, L. *Synlett* **1996**, 221.
109. Contreras, B.; Duhamel, L.; Plé, G. *Synth. Commun.* **1990,** *20*, 2983.
110. Soullez, D.; Plé, G.; Duhamel, L. *J. Chem. Soc., Perkin Trans. 1* **1997**, 1639.

111. Duhamel, L.; Duhamel, P.; Le Gallic, Y. *Tetrahedron Lett.* **1993,** *34*, 319.
112. Kanda, T.; Sugino, T.; Kambe, N.; Sonoda, N. *J. Organomet. Chem.* **1994,** *473*, 71.
113. Taylor, K. G. *Tetrahedron* **1982,** *38*, 2751.
114. Taylor, K. G.; Hobbs, W. E.; Clark, M. S.; Chaney, J. *J. Org. Chem.* **1972,** *37*, 2436.
115. Baird, M. S.; Baxter, A. G. W. *J. Chem. Soc., Perkin Trans. 1* **1979**, 2317.
116. Harada, T.; Yamaura, Y. *Bull. Chem. Soc. Jap.* **1987,** *60*, 1715.
117. Schmidt, A.; Köbrich, G.; Hoffmann, R. W. *Chem. Ber.* **1991,** *124*, 1253.
118. Hoffmann, R. W.; Stiasny, H.-C.; Krüger, J. *Tetrahedron* **1996,** *52*, 7421.
119. Hoffmann, R. W.; Bewersdorf, M.; Ditrich, K.; Krüger, K.; Stürmer, R. *Angew. Chem., Int. Ed. Engl.* **1988,** *27*, 1176.
120. Hoffmann, R. W.; Bewersdorf, M.; Krüger, M.; Mikolaiski, W.; Stürmer, R. *Chem. Ber.* **1991,** *124*, 1243.
121. Hoffmann, R. W.; Brumm, K.; Bewersdorf, M.; Mikolaiski, W.; Kusche, A. *Chem. Ber.* **1992,** *125*, 2741.
122. Hoffmann, R. W.; Julius, M. *Liebigs Ann. Chem.* **1991**, 811.
123. Hoffmann, R. W.; Stiasny, H.-C. *Tetrahedron Lett.* **1995,** *36*, 4595.
124. Harada, T.; Katsuhira, T.; Oku, A. *J. Org. Chem.* **1992,** *57*, 5805.
125. Still, W. C. *J. Am. Chem. Soc.* **1978,** *100*, 1481.
126. Meyer, N.; Seebach, D. *Chem. Ber.* **1980,** *113*, 1290.
127. Reich, H. J.; Phillips, N. H. *J. Am. Chem. Soc.* **1986,** *108*, 2102.
128. Sawyer, J. S.; Macdonald, T. L.; McGarvey, G. J. *J. Am. Chem. Soc.* **1984,** *106*, 3376.
129. Sawyer, J. S.; Kucerovy, A.; Macdonald, T. L.; McGarvey, G. J. *J. Am. Chem. Soc.* **1988,** *110*, 842.
130. Ashe, A. J.; Lohr, L. L.; Al-Taweel, S. M. *Organometallics* **1991,** *10*, 2424.
131. Maercker, A.; Bodenstedt, H.; Brandsma, L. *Angew. Chem., Int. Ed. Engl.* **1992,** *31*, 1339.
132. Lesimple, P.; Beau, J.-M.; Jaurand, G.; Sinaÿ, P. *Tetrahedron Lett.* **1986,** *27*, 6201.
133. Isono, N.; Mori, M. *J. Org. Chem.* **1996,** *61*, 7867.
134. Gilman, H.; Webb, F. J. *J. Am. Chem. Soc.* **1949,** *71*, 4062.
135. Dumont, W.; Bayet, P.; Krief, A. *Angew. Chem., Int. Ed. Engl.* **1974,** *13*, 804.
136. Clarembeau, M.; Krief, A. *Tetrahedron Lett.* **1985,** *26*, 1093.
137. Clarembeau, M.; Krief, A. *Tetrahedron Lett.* **1984,** *25*, 3629.
138. Krief, A.; Clarembeau, M.; Barbeaux, P. *J. Chem. Soc., Chem. Commun.* **1986**, 457.
139. Seebach, D.; Peleties, N. *Angew. Chem., Int. Ed. Engl.* **1969,** *8*, 450.
140. Seebach, D.; Peleties, N. *Chem. Ber.* **1972,** *105*, 511.
141. Krief, A.; Dumont, W.; Clarembeau, M.; Bernard, G.; Badaoui, E. *Tetrahedron* **1989,** *45*, 2005.
142. Clarembeau, M.; Krief, A. *Tetrahedron Lett.* **1986,** *27*, 1723.
143. Seebach, D.; Beck, A. K. *Angew. Chem., Int. Ed. Engl.* **1974,** *13*, 806.
144. Barros, S. M.; Comasseto, J. V.; Berriel, J. *Tetrahedron Lett.* **1989,** *30*, 7353.
145. Petragnani, N.; Comasseto, S. V. *Synthesis* **1991**, 909.
146. Sugimoto, O.; Sudo, M.; Tanji, K.-i. *Tetrahedron* **2001,** *57*, 2133.

147. Hiiro, T.; Morita, Y.; Inoue, T.; Kambe, N.; Ogawa, A.; Ryu, I.; Sonoda, N. *J. Am. Chem. Soc.* **1990,** *112*, 455.
148. Brandsma, L. *Rec. Trav. Chim. Pays-Bas* **1964,** *83*, 307.
149. Seebach, D. *Angew. Chem., Int. Ed. Engl.* **1967,** *6*, 442.
150. Trost, B. M.; Keeley, D. E.; Arndt, H.; Bogdanowicz, M. S. *J. Am. Chem. Soc.* **1977,** *99*, 3088.
151. Cohen, T.; Daniewski, W. M.; Weisenfeld, R. B. *Tetrahedron Lett.* **1978**, 4665.
152. Zhao, S. H.; Samuel, O.; Kagan, H. B. *Tetrahedron* **1987,** *43*, 5135.
153. Annunziata, M.; Caponi, M.; Cardellicchio, C.; Naso, F.; Tortorella, P. *J. Org. Chem.* **2000,** *65*, 2843.
154. Farnum, D. G.; Veysoglu, T.; Carde, A. M.; Duhn-Emswiler, B.; Pancoast, T. A.; Reitz, T. J.; Carde, R. T. *Tetrahedron Lett.* **1977**, 4009.
155. Durst, T.; LeBelle, M. J.; van den Elzen, R.; Tin, K.-C. *Can. J. Chem.* **1974,** *52*, 761.
156. Lockard, J. P.; Schroeck, C. W.; Johnson, C. R. *Synthesis* **1973**, 485.
157. Jacobus, J.; Mislow, K. *J. Am. Chem. Soc.* **1967,** *89*, 5228.
158. Riant, O.; Argouarch, G.; Guillaneux, D.; Samuel, O.; Kagan, H. B. *J. Org. Chem.* **1998,** *63*, 3511.
159. Argouarch, G.; Samuel, O.; Riant, O.; Daran, J.-C.; Kagan, H. B. *Eur. J. Org. Chem.* **2000**, 2893.
160. Pedersen, H. L.; Johanssen, M. *J. Chem. Soc., Chem. Commun.* **1999**, 2517.
161. Hua, D. H.; Lagneau, N. M.; Chen, Y.; Robben, P. M.; Clapham, G.; Robinson, P. D. *J. Org. Chem.* **1996,** *61*, 4508.
162. Shimada, T.; Kurushima, H.; Cho, Y.-H.; Hayashi, T. *J. Org. Chem.* **2001,** *66*, 8854.

CHAPTER 4

Regioselective Synthesis of Organolithiums by C–X Reduction

4.1 Reductive lithiation of alkyl and aryl halides

4.1.1 Reductive lithiation with lithium metal

Simple, unfunctionalised organolithiums (particularly those which are commercially available) are usually prepared by reductive lithiation of alkyl halides with lithium metal at ambient temperature or above.[1] However, the use of lithium metal can pose problems, primarily because of the temperatures required. The newly formed organolithiums may attack unreacted starting material, leading to coupled products. Coupling is fastest with alkyl iodides and bromides, even though their rate of reduction is also fast, so the best substrates for reductive lithiation with lithium metal are alkyl chlorides (though allyl chlorides and benzyl chlorides also undergo coupling too fast to be useful – sections 4.2 and 4.4 describe solutions to this problem). The lower solubility of lithium chloride in ether solvents is an additional bonus of using the chlorides. The nature of the lithium metal is also critical: best results in the synthesis of *t*-BuLi are obtained with lithium containing 1-2% sodium.[2] Many reductive lithiations of compounds no more complex than the butyl halides are subject to widely varying yields and require a certain amount of experimental "magic",[3-5] and for new preparations the arene-promoted lithiations are generally preferable.

t-BuCl —(2 x Li; 1% Na, pentane, 30-40 °C)→ t-BuLi + LiCl

The mechanisms of the reduction of alkyl halides and aryl halides differ slightly. In both, a single electron is transferred to the substrate. In alkyl halides, the electron enters the C–X σ^* orbital, and bond-breakage to give halide and an alkyl radical is concerted with electron transfer.[6] With aryl halides, reversible electron transfer to an antibonding orbital of the aromatic ring precedes rate-determining breakage of the C–X bond to X^- and $Ar^\bullet$.[7] The radical in each case is trapped by a second atom of lithium to give the organolithium.

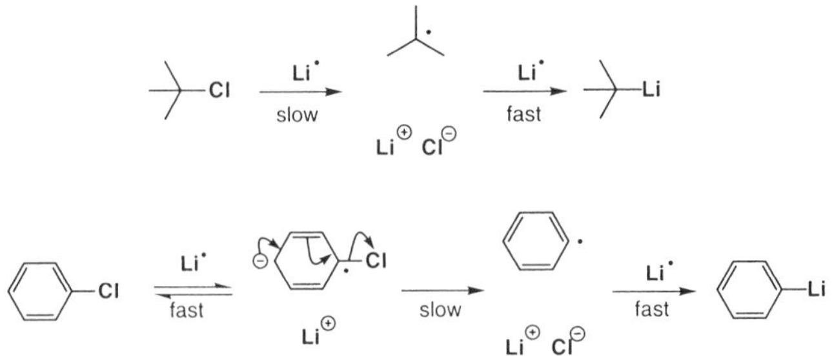

Reductive lithiation is neatly complementary to deprotonation as a method for the synthesis of organolithiums: in contrast with deprotonation, reductive formation is fastest for alkyllithiums (the more substituted the better) and slowest for aryllithiums. The order of reactivity follows logically from the relative stabilities of the intermediate radicals, whose formation is the rate-determining step of the sequence.[8]

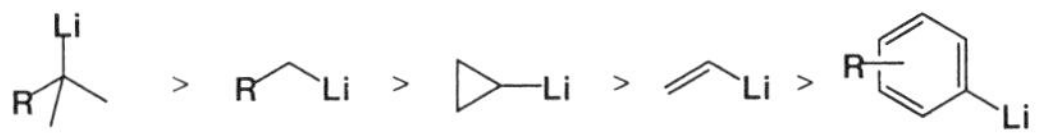

Relative rate of formation by reductive lithiation:

4.1.2 Reductive lithiation promoted by arenes[9]

In 1972, Screttas[10] reported that the reductive lithiation of aryl halides (including fluorobenzene) and (to a lesser extent) alkyl halides was successful even at –50 °C when carried out in THF in the presence of a stoichiometric quantity of naphthalene. Carbonation of the products gave carboxylic acids in high yields.[11]

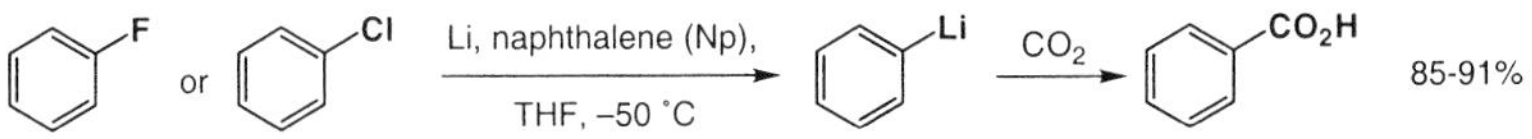

Although this was the first reported use of *lithium* in conjunction with naphthalene,[12] it was already well-known that naphthalene and other simple aromatic compounds promoted the reduction of alkyl halides by sodium.[13] Naphthalene acts as an acceptor of sodium's valence electron, and the resulting radical anion **1** very rapidly reduces the alkyl halide to generate a radical which accepts a second electron to form an alkylmetal. However, alkylsodiums RNa are far too reactive to wait to be trapped with electrophiles, and are immediately protonated by solvent or starting material to give RH in 20-70% yield, plus a considerable amount of naphthalene addition product **2**.[13,14] By adding magnesium chloride to the mixture (E = MgCl), it had been shown that Grignard reagents could be formed directly.[15] More simply though, using lithium as the metal reductant gives the more stable alkyllithium RLi which can then react further with electrophiles.

M⊕ M⊕ Cl⊖ E⊕ R–E
M• 1 R–Cl R• M• R–M Stable if M = Li
Too reactive if M = Na
RH + 2

There were still problems though, mainly arising from the susceptibility of naphthalene to attack by the intermediate radical or the newly formed organolithium, and the separation of naphthalene and naphthalene-derived by-products from the desired reaction product. Freeman found a solution to this problem by using, in the place of naphthalene, 4,4'-di-*t*-butylbiphenyl (DBB) **3**.[16] Since electron transfers may take place across 7-9 Å, while bond formation required distances of <2 Å, the *t*-butyl groups of this simple-to-make compound prevent the participation of DBB in the formation of by-products while allowing it to take part in single electron reductions. Li and stoichiometric DBB in THF at –78 °C promote the formation of >90% yields (by GC) of the carboxylic acids derived from the three chlorides **4**, **5** and **6**. The reaction is instantaneous at –78 °C, and fast even at –100 °C, lessening even further potential problems from by-products. DBB can typically be recovered in 97% yield from these reactions, while naphthalene is typically carbonated to give dicarboxylic acids which may themselves react with the product organolithiums, lowering the yield of organolithium.[14]

DBB 3 Li• –78 °C, THF M⊕ R–Cl M⊕ Cl⊖ R• DBB 3 Li• R–Li CO_2 R–CO_2H >90% (GC) from

Cl 4 Cl 5 Cl 6

An alternative solution to the problem of purification is to use 1-dimethylaminonaphthalene 7 (DMAN),[17] which can be removed from the product mixture by an acid wash. Reactions in the presence of this reagent must take place below –45 °C: above this temperature the DMAN radical anion decomposes to give 1-lithionaphthalene.[18] Polymer-supported arene catalysts can also be used.[19,20]

DMAN, **7**

Later work by Meyers[21] not only confirmed that organolithiums form very rapidly in the presence of DBB (some could be formed only with DBB: Li metal alone or with naphthalene failed) but also that they have higher reactivity, perhaps due to a difference in aggregation state, than those formed using other methods. For example, butyl chloride and bromide gave good yields of organolithium with DBB even at 0 °C in THF. The tertiary alkyl halides **8** gave organolithiums only with DBB, and none with naphthalene (Np). Cyclopropyl bromide, which fails to undergo halogen-metal exchange with *t*-BuLi, generates good yields of cyclopropyllithium with DBB.

X = Cl or Br
8 → Li, DBB, –78 °C, THF
9 → Li, DBB, –78 °C, THF

Barluenga and Yus showed that reductive lithation with naphthalene was nonetheless an effective way of making functionalised organolithiums. The β-oxygenated species such as **11** are stable below –78 °C provided the lithium is at a primary centre (above this temperature they decompose with elimination of Li_2O) and can be formed by reductive lithiation of the lithium alkoxide **10**.[22-25] The amide **12** behaves similarly,[26] and protected aldehyde **14** yields homoenolate equivalent **15**.[27]

BuLi → **10** → Li, Np, –78 °C, THF → **11** → Me_2S_2 → 85%; > –78 °C

12 → 1. BuLi; 2. Li, Np, –78 °C, THF → **13** →

14 → Li, Np, –78 °C, THF → **15** → Ph=NPh → 70%

An important development followed in 1991,[28] when Yus and Ramón showed that naphthalene, biphenyl and DBB could be used *catalytically* in the reduction of functionalised alkyl chlorides.[29,9] Only 1% of the arene catalyst is required, and organolithiums are produced at –78 °C which may react with a range of electrophiles (as in the synthesis of **16** and **17**). The advantage of using catalytic quantities are both obvious (removal of the arene at the end of the reaction is no longer a problem) and less obvious (the reactions are self-titrating: the initial colour – green for naphthalene, brown for biphenyl and blue/violet for DBB – dissipates on addition of the substrate to the mixture but reappears when the reduction is complete and the electrophile may be added).

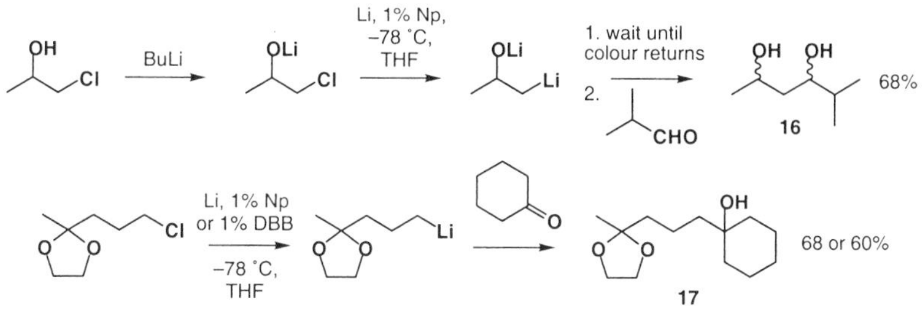

Functionalised vinyl lithiums prepared from vinyl chlorides are a little more stable: **18** can be made and functionalised at –78 °C for example (though the corresponding alkoxy-substituted vinyllithiums are unstable),[30] and the low-temperature lithiation of **19** provides a d^3 reagent **20**.[31]

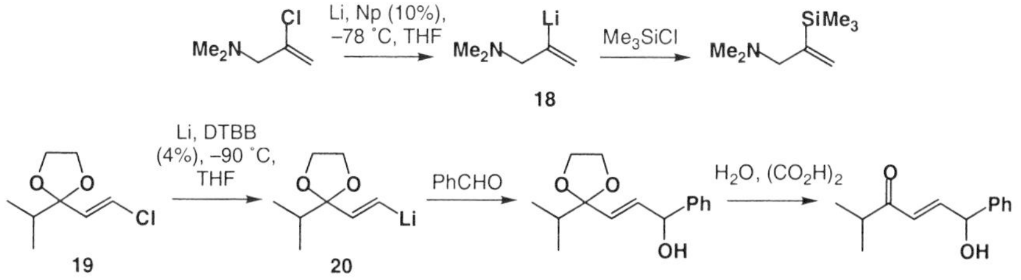

With allyl chlorides, trapping of the product with an electrophile *in situ* (which can even be a reactive ketone such as cyclohexanone) prevents homocoupling reactions:[32]

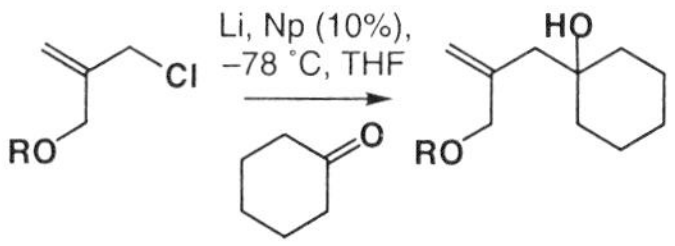

Although the method will rarely be preferable to bromine-lithium exchange, catalytic naphthalene also promotes the reductive lithiation of chlorophenols or chloroanilides such as **21**,[33] and *m*- and *p*-dichlorobenzene can be reductively lithiated twice, introducing two different electrophiles sequentially to give compounds like **22**.[33]

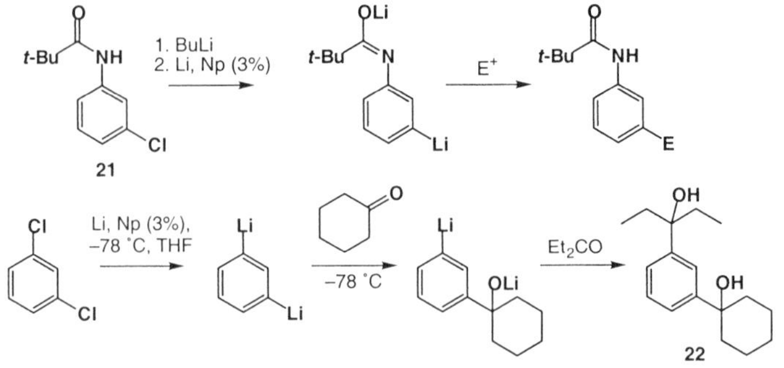

One-pot multiple reductive lithiations[34] are in fact surprisingly easy to make with lithium and catalytic DBB, providing the electrophile is available *in situ*. Dichloroalkanes will react twice with carbonyl electrophiles (**23**), and even CCl_4 will give a tetrasilylated product **24** in 80% yield.[35,36]

An α-functionalised carbenoid intermediate **26** can be made from **25**,[37,38] and acyllithiums **28** and **30** are available from the carbamoyl chloride **27** and thiocarbamoyl chloride **29**.[39] Lithiated sugars may also be made from chloropyranosides (see below).[40]

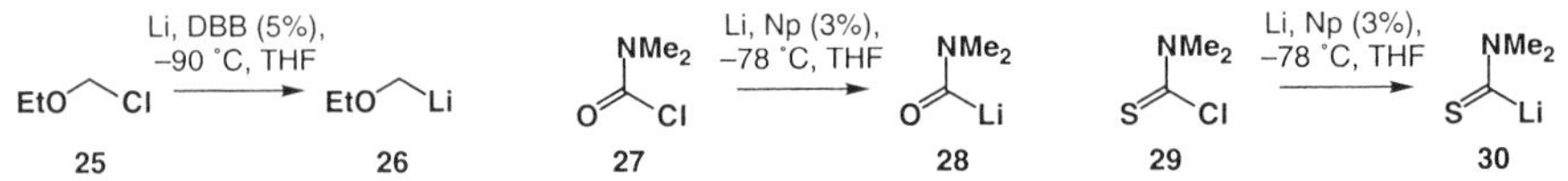

4.2 Reductive lithiation of C–O bonds[9]

The cleavage of anisole to lithium phenoxide by lithium in refluxing THF in the presence of biphenyl has been known since 1961,[41,42] and in 1963 Eisch published a practical method for forming allyllithium **32** from allyl phenyl ether **31**, again using lithium in the presence of biphenyl, but at 0 °C or below.[43] Twelve equivalents (a six-fold excess) of lithium metal was required for good yields in this reaction. A similar reaction takes place between benzyl phenyl ether and lithium.[44]

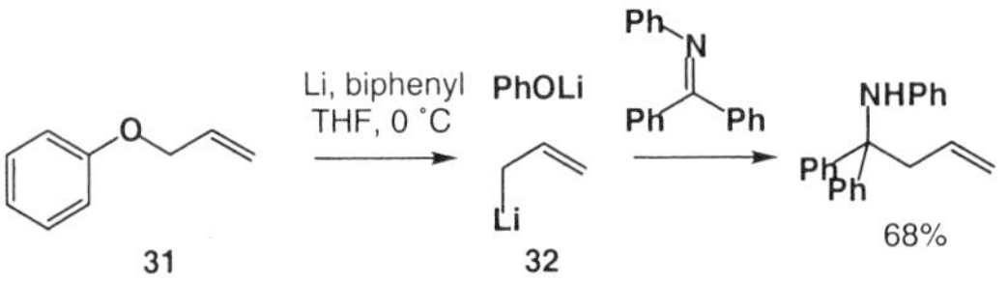

Several other examples of ether cleavage by lithium metal are known, but in many cases they are surprisingly unselective and exhibit powerful solvent effects.[45] For example, in heptane, anisole gives (very slowly) not lithium phenoxide but phenyllithium and lithium methoxide. The most useful cleavages are those of alkyl aryl ethers,[45] and in general, the alkyl oxygen bond is cleaved more readily in THF or with more substituted alkyl groups (*t*-butyl phenyl ether **33** gives phenoxide only[46]) and aryl-oxygen cleavage is promoted when the alkyl group lies out of the plane of the aromatic ring (as in **35**) or when the ether is an acetal (as in **34**).[46]

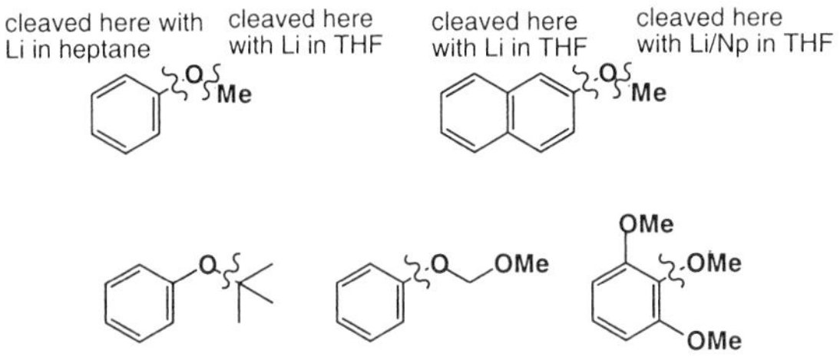

cleaved here with Li in heptane; cleaved here with Li in THF; cleaved here with Li in THF; cleaved here with Li/Np in THF

OMe

OMe

OMe

33 **34** **35**

In diaryl ethers (and dibenzofurans), the bond to the less substituted aromatic ring is cleaved.[47]

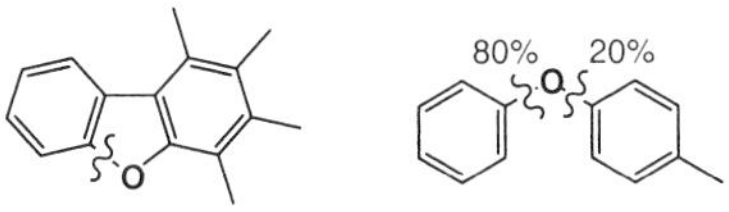

80% 20%

Catalytic DBB[29] avoids the need for such vigorous conditions, and allows benzyl[48] and allyl lithiums to be formed from silyl ethers,[49] mesylates[50] triflates,[51] carbamates and carbonates,[9] and even directly from lithium alkoxides.[49]

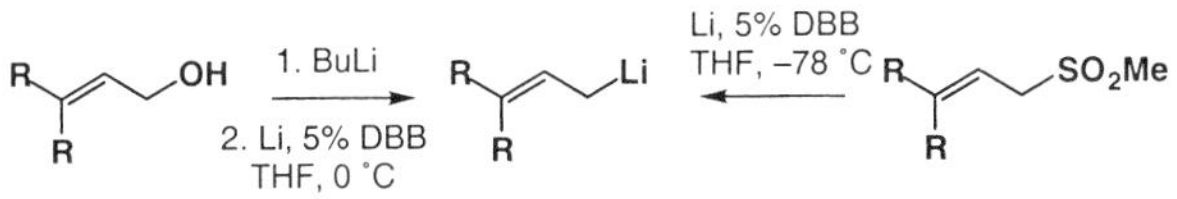

1. BuLi; 2. Li, 5% DBB, THF, 0 °C

Li, 5% DBB, THF, –78 °C

Catalytic DBB at much higher temperatures also allows the reductive lithiation of phenyl ethers to give alkyllithiums.[52,53] Functionalised benzylic organolithiums are produced from the heterocycles **36**[54] and **38**[55] and an alkyllithium is produced from **37**.[52]

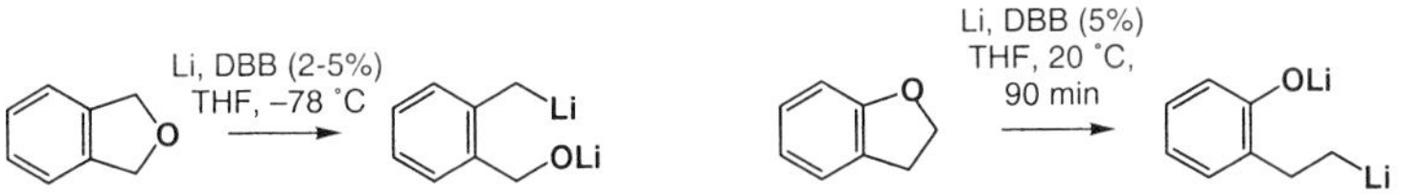

Li, DBB (2-5%), THF, –78 °C

36

Li, DBB (5%), THF, 20 °C, 90 min

37

Li, DBB (2-5%), THF, –78 °C

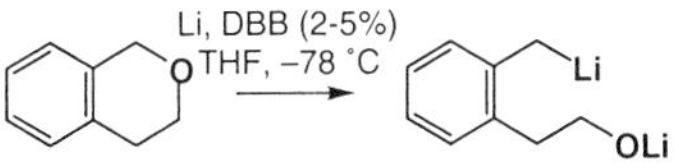

38

Extending the arene-catalysed cleavage to the production of alkyllithiums other than allyl- or benzyllithiums requires more susceptible leaving groups such as cyclic sulfates[56,57] or phosphates.[58] Reductive cleavage of cyclic sulfates from 1,3-diols generates cyclopropanes:[59]

1. $SOCl_2$
2. $NaIO_4$, $RuCl_3$

Li, 5% DBB
THF, 0 °C

85%

Early cleavages of epoxide C–O bonds used vigorous conditions which did not allow the organolithium product to be isolated: it was either protonated to give an alcohol[60,61] or Li_2O was eliminated to give an alkene.[62] However, arene-catalysed epoxide cleavage by lithium is fast even at –78 °C, and these conditions allow the product lithioalkoxide such as **39** and **40** to be trapped with electrophiles.[63,64]

Li, biphenyl THF, –78 °C or Li, DBB THF, –78 °C → **39**; Me_2S_2 57%; $R_2C=O$ 65%

Li, DBB THF, –95 °C 5 min → **40**; ArCHO 71%

Alkyl substituted terminal epoxides **41** cleave to give primary organolithiums, while phenyl or vinyl substituted epoxides **42** and **43** yield the stabilised allyl or benzyllithium. It is probable that the cleavage of an alkyl-substituted epoxide is not fully regioselective in all cases, but that the more substituted regioisomer is not observed as it is less stable and undergoes elimination of Li_2O to give an alkene.[64]

41 **42** **43**

Reductive cleavage of epoxides is successful primarily for terminal epoxides: 1,2-disubstituted epoxides give alkenes by elimination of Li_2O, unless they are cyclic, in which case the secondary organolithium may be trapped with stereoselectivity at the former C–O bond:

Li, DBB THF → ; CHO → 62%; 58:42 diastereoisomers

The aspect of epoxide cleavage with the most potential is its ability to use the products of asymmetric epoxidation.[64,65] Organolithium **45** is formed from **44** with Li and 5% DBB at –78 °C, and reacts with aldehydes and ketones to give enantiomerically enriched 1,2,4-triol derivatives such as **46**.

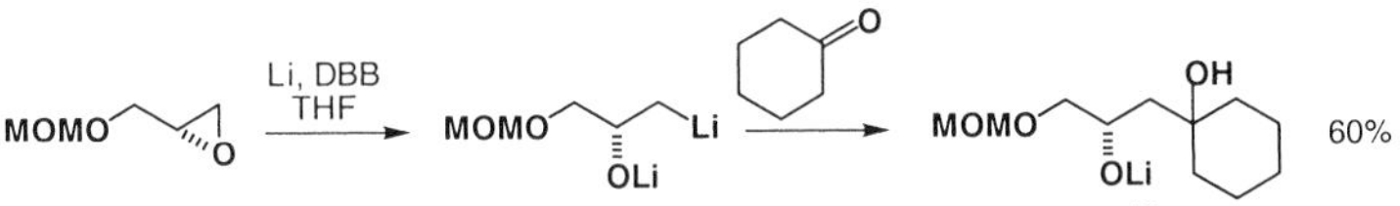

MOMO (44) —Li, DBB, THF→ MOMO, OLi, Li (45) —cyclohexanone→ MOMO, OLi, OH (46) 60%

Oxetanes are cleaved in a similar way, though a temperature of 0 °C is required.[66] The γ-lithioalkoxide products **47** are significantly more stable than the β-lithioalkoxides from the epoxide opening, and give moderate to good yields of a wide range of important, principally 1,4-difunctionalised products of quenching with electrophiles.

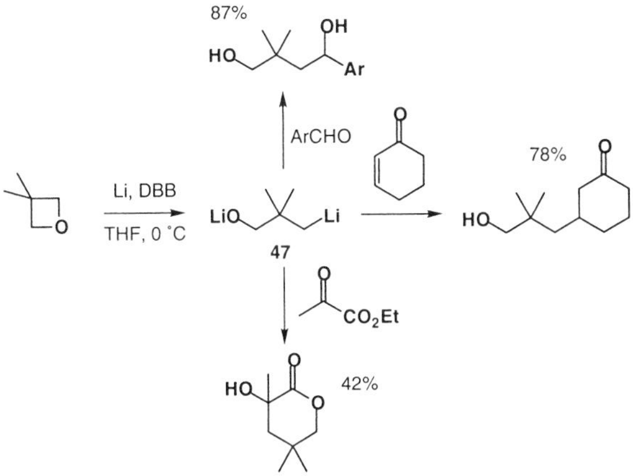

Li, DBB, THF, 0 °C → LiO, Li (47)

ArCHO → HO, OH, Ar 87%

cyclohexenone → HO 78%

CO_2Et → HO 42%

As with epoxides, the less substituted organolithium is formed, unless there is a benzylic alternative. Benzylic C–O bonds are the most reactive of all:

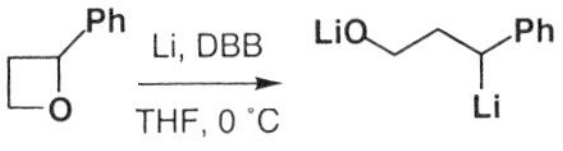

Ph —Li, DBB, THF, 0 °C→ LiO, Ph, Li

In the presence of BF_3, even THF can be cleaved to give δ-lithioalkoxide **48**,[67] and with catalytic naphthalene[68] the organolithium can be trapped with carbonyl compounds. Acid-catalysed ring closure generates tetrahydropyrans such as **49** – a convenient ring expansion of THF.

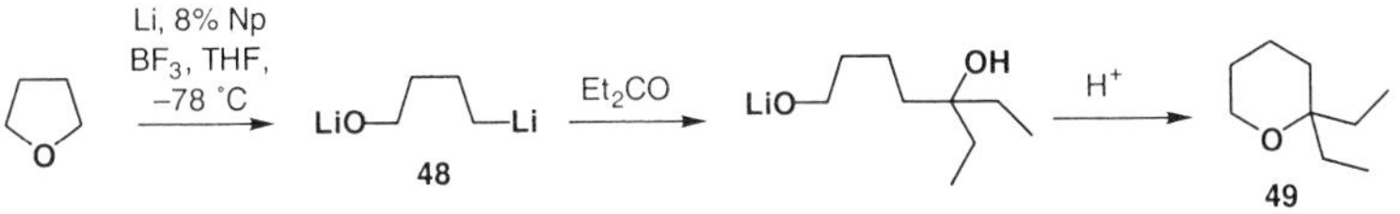

—Li, 8% Np, BF_3, THF, –78 °C→ LiO, Li (48) —Et_2CO→ LiO, OH —H^+→ O (49)

The reductive lithiation of dioxolanes requires the formation of an allylic or benzylic organolithium, and represents a valuable umpolung method for the functionalisation of carbonyl compounds.

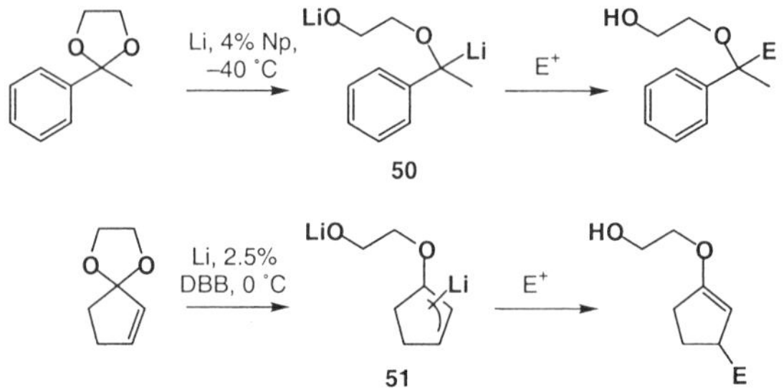

4.3 Reductive lithiation of C–N bonds

C–N bonds are much harder to lithiate reductively, and the reaction is most successful when accompanied by relief of strain. Aziridines (**52**) are opened to β-lithioamines (**53**) provided only a catalytic amount of naphthalene is used. Enantiomerically pure products are easily obtained using ephedrine-derived aziridines such as **54**.[69,70]

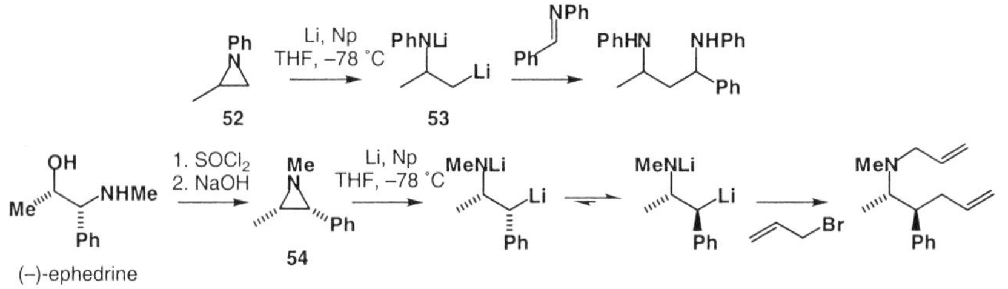

Azetidines are similar, and open only with an N-phenyl substituent (**55**). The use of aziridines or azetidines as nitrogenated d^3 or d^4 synthons has great potential for the synthesis of diamines.[71]

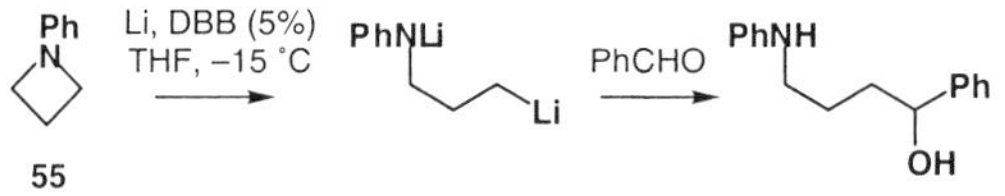

Acyclic C–N bonds of allyl or benzyl triflamides,[51] pivalamides and carbamate derivatives[9] can be reductively lithiated, as can N-benzyl benzotriazoles.[72]

4.4 Reductive lithiation of C–S bonds

4.4.1 Reduction of sulfides

The synthesis of organolithiums by reductive lithiation of phenylsulfides owes its importance to the ease with which the sulfides can be formed.[8] Unlike halides, sulfides are not electrophilic, and self-coupling side reactions do not pose a problem.

With sulfides as intermediates, alkenes can be used as precursors to organolithiums with regioselectivity in the formation of **56** determined by whether a radical[11] or polar[73] thiol addition is employed. Easy lithiation of phenyl benzyl sulfide **57** makes substituted benzyllithiums such as **58** readily available.[73] Reductive C–S cleavage is probably the best way of making benzylic organolithiums.

PhSH, AIBN; Li, Np; $Ph_2C=O$ — **56** — 68%

1. BuLi 2. *s*-BuI; Li, Np; E^+ — **57**, **58** — up to 97%

Although the majority of sulfides used as precursors to organolithiums have been thioacetals or sulfides bearing other α-substituents, the usefulness even of simple phenylsulfides related to **56** was demonstrated in the synthesis of dihydroerythronolide A by Stork,[74] which used a sulfide intermediate **59** in the conversion of an alcohol to a Grignard reagent.

$(PhS)_2$, Bu_3P; Li, DBB; 1. $MgBr_2$ 2. R^2COMe — **59**

Cyclic arylsulfides, such as the heterocycle **60**, undergo ring opening with Li and DBB.[75] The sequence to **61** illustrates a useful ring expansion. Simple phenylsulfides have also been used as starting materials for anionic cyclisations,[76] and functionalised sulfides have been converted to functionalised organolithiums.[77,78]

Li, DBB, THF, −10 °C; *i*-PrCHO; 1. TsCl 2. $NaBH_4$, py — **60**, 78%, **61**

However, most widely used of all are sulfides made by thioacetalisation procedures, and it was in developing a route to substituted versions of Trost's reagent **64** (only the unsubstituted cyclopropane can be lithiated by deprotonation[79]) that Cohen discovered the transformation of dithioacetals **62** to α-lithiosulfides **63** using lithium and naphthalene.[80] Dithioacetals of larger rings also cleanly give organolithiums.[81]

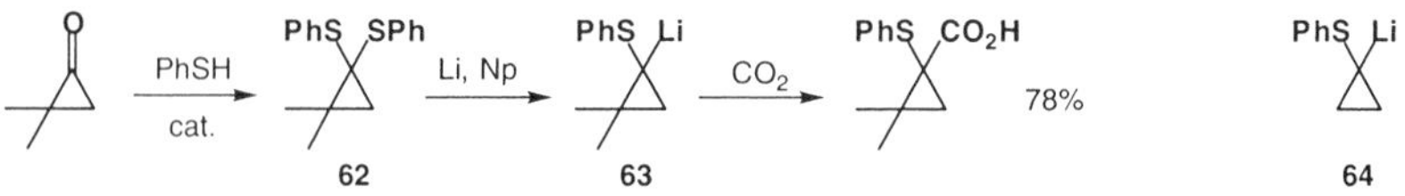

1,1-Bis(alkylthio)alkenes such as **65** are also reductively lithiated, allowing the synthesis of tri- and tetrasubstituted double bonds (**66**).[82,83]

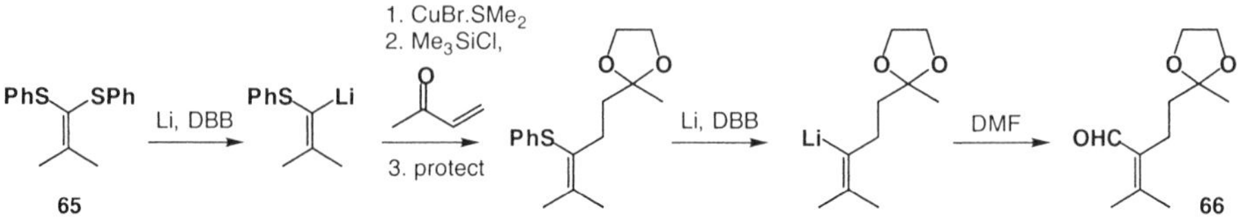

Reductive lithiation of the vinylsulfide **67** derived from a cyclic ketone provides a valuable alternative to the Shapiro reaction (section 8.1),[84] and the sequence of events leading to **68** or **69** determines the regioselectivity.

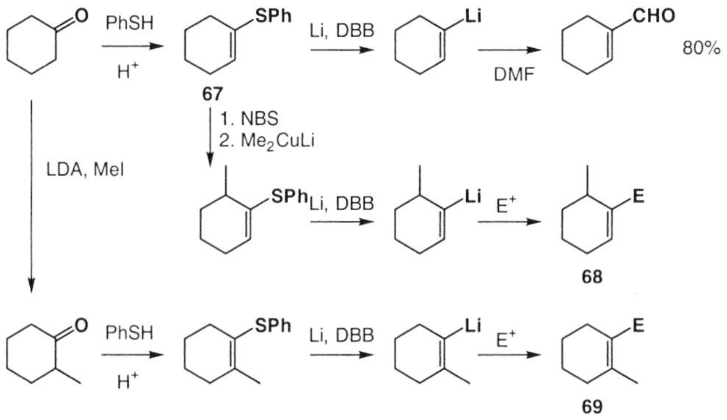

The simple reductive lithiation and quench of dithioacetals means that α-functionalised – and in particular, α-silylated sulfides are very easy to prepare. Cohen has exploited this in a number of ways: for example, a ketenedithioacetal **70** can be used to generate an α-lithiovinylsilane **71** if a silyl halide is used as the electrophile in the first step.[85]

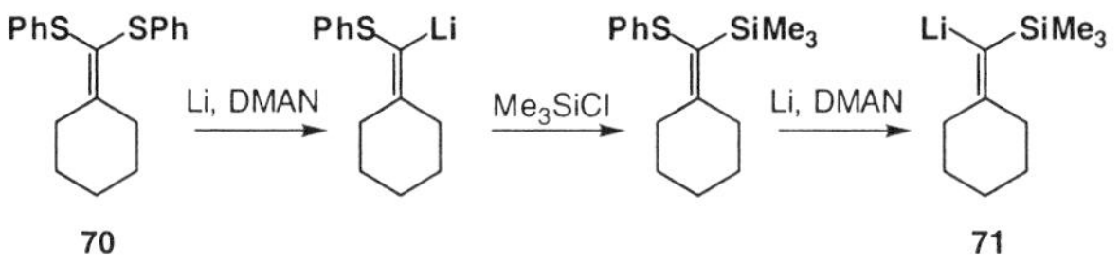

Lithiosilanes derived from cyclopropane dithiocetals add to aldehydes to give precursors for Peterson olefinations – one of the best ways of making alkylidene cyclopropanes. In the example below, the lithiated allyl sulfide **72** adds cleanly to a ketene dithioacetal to give cyclopropane **73**. Successive reductive lithiations give silane **74** and then a mixture of diastereoisomers of hydroxysilane **75**. Purification of the major diastereoisomer and elimination in base gives **76**.[85]

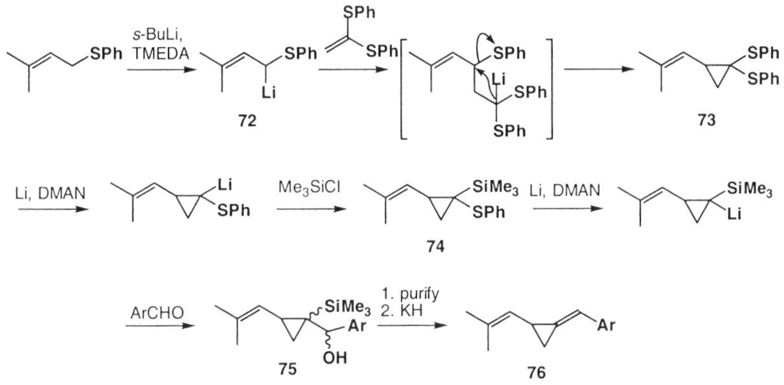

With unsaturated aldehydes as the carbonyl component in the Peterson step, dienes are produced which undergo a double ring expansion via the allylidenecyclopropane rearrangement to give 7,5-fused ring systems such as **77**.[86]

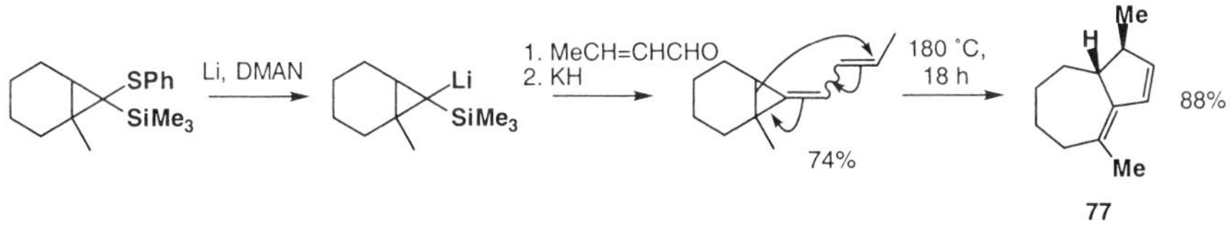

Other thioacetal-type functional groups are good starting materials for reductive lithiation: for example, α-lithioether **79** is readily made by oxidation of a sulfide to a thioacetal **78** followed by reductive lithiation.[87]

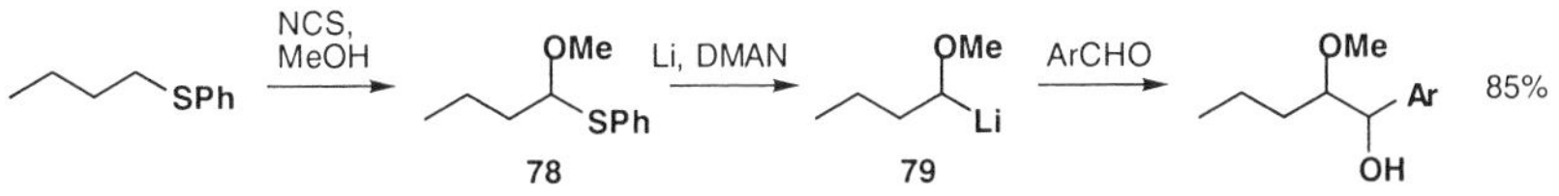

Reductive lithiation of a thioacetal provided the first method for the synthesis of 2-lithiotetrahydropyran. The scheme below shows a lithiated dihydropyran **80** being used in an efficient synthesis of two diastereoisomers of the insect pheromone brevicomin **81**.[87]

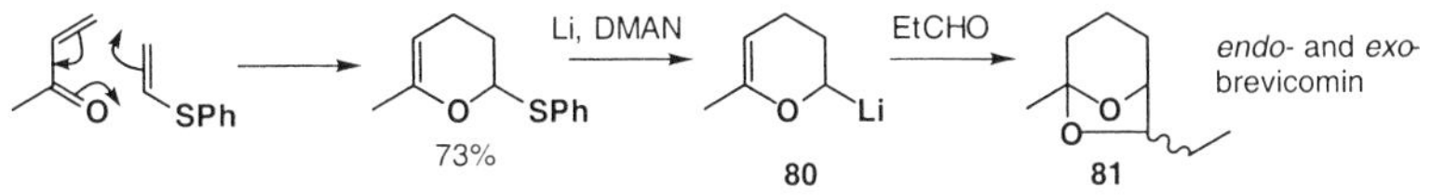

The reductive lithiation of substituted tetrahydropyrans such as **82** is stereoselective, producing principally the axial organolithium at –78 °C.[81] Since reductive lithiation proceeds by fast reduction of a more slowly formed radical, the stereochemical outcome of the reaction is determined by the preferred conformation of the radical intermediate. Radical **83** prefers to exist in this conformation because of anomeric stabilisation by interaction of the SOMO with the adjacent oxygen lone pair. However, in the organolithium **84** itself, this interaction is destabilising, and **84**'s thermodynamic preference is for the equatorially substituted stereoisomer. At temperatures above –30 °C, therefore, *ax*-**84** inverts its configuration to give *eq*-**84**. The stereochemistry of the product derives stereospecifically from that of **84** so either *ax*-**85** or *eq*-**85** may be made according to the temperature régime of the sequence.

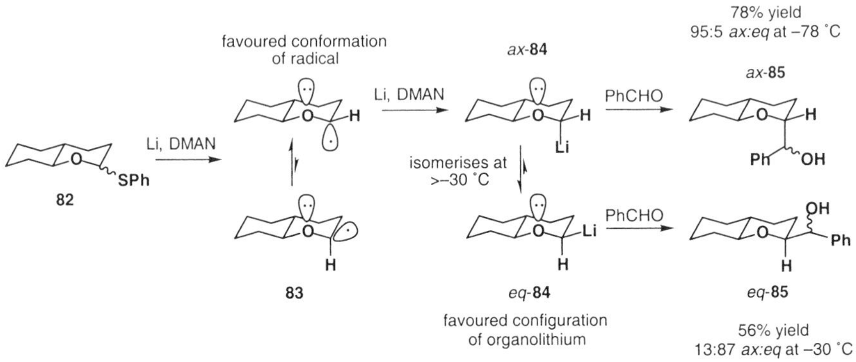

For tetrahydropyrans which are not conformationally locked, as the oxadecalin system **82** is, an equatorial lithium substituent is obtained by allowing the ring to flip (rather than by inversion at the C–Li centre), placing the methyl group axial. Kinetic control in the formation of the configurationally stable organolithium **86** allows the stereoselective synthesis of **87**, whose xanthate ester undergoes rearrangement to **88**. Radical desulfurisation gives the natural product *trans*-rosoxide **89**.[81]

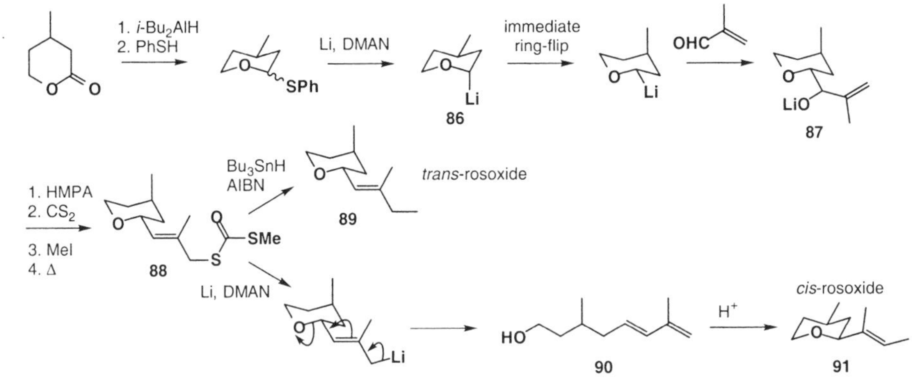

Cis-rosoxide **91** is available by reductive lithiation of **88**: the allyllithium produced is unstable, and the ring opens to give the hydroxydiene **90**. Acid catalysis re-closes the ring to its more stable diequatorial diastereoisomer **91**.[81]

Among the most important of the lithiated tetrahydropyrans are the lithiated deoxysugars, readily made by reduction of a precursor thioglycoside. Starting from the glucopyranoside **92**, reductive lithiation leads to elimination, but the product glucal **93** is an ideal starting material for the synthesis of the lithiated deoxysugars. HCl gives a chloropyranoside **94** whose reductive lithiation gives an axial organolithium **95** (for the reasons outlined above) which reacts with electrophiles to give axially substituted deoxy sugars **96**.[40] The same organolithium is produced by reductive lithiation of the thioglycoside **97**.

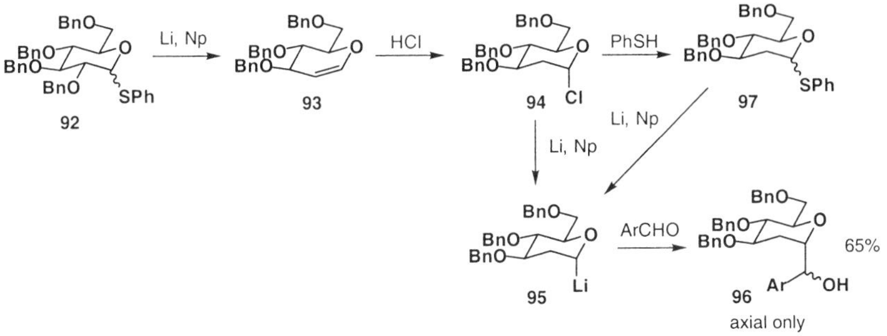

By oxidising the sulfide to a sulfone, the synthetic versatility of this class of compounds is further increased. Deprotonation of either or both diastereoisomers of **98** leads, under thermodynamic control, to the equatorial organolithium **101** in which a destabilising interaction between the oxygen lone pair and the lithio substituent is avoided. However, lithium-naphthalene reduction of **102** to the organolithium **103** is axially selective because of the stabilisation afforded to the intermediate radical by the axial lone pair. Protonation of the product gives **104**.[88]

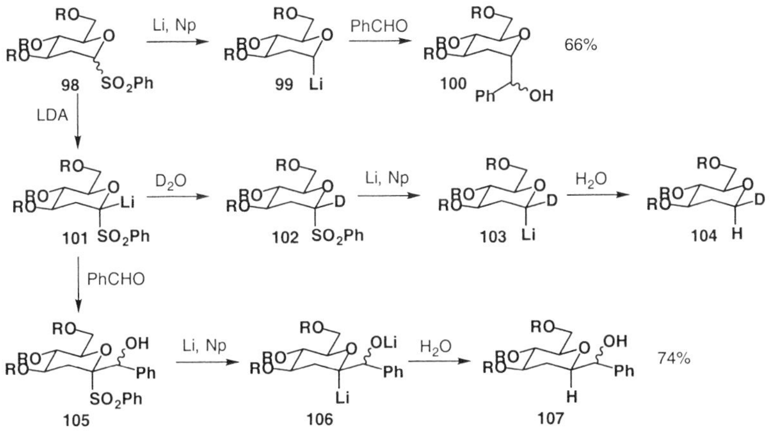

A sequence involving formation of two organolithiums by successive deprotonation and reduction steps allows the introduction of an electrophile either axially or equatorially at the anomeric position of a glycoside.[89] Direct reductive lithiation of **98** gives axial **99** and hence

100, while deprotonation to give **101** and therefore **105** can be followed by reductive lithiation and protonation in the axial position to give the equatorially substituted **107**.

Reductive lithiation of allyl phenyl sulfides is a general and very versatile way of making allyllithiums.[90] Allyllithiums themselves frequently react non-regioselectively with electrophiles, but transmetallation to an allyltitanium species improves regioselectivity in favour of electrophilic attack at the more substituted terminus;[91] transmetallation with cerium chloride favours electrophilic attack at the less substituted terminus, giving, for example, **109**.[90] Allyllithiums **108** and **110** exist predominantly in a *cis*-configuration, and low temperature transmetallation gives *cis*-allylceriums **111** and hence *cis*-alkenes by reaction with electrophiles. The scheme below shows the synthesis of a *cis,cis*-skipped diene *Z,Z*-**112** using this method.[90]

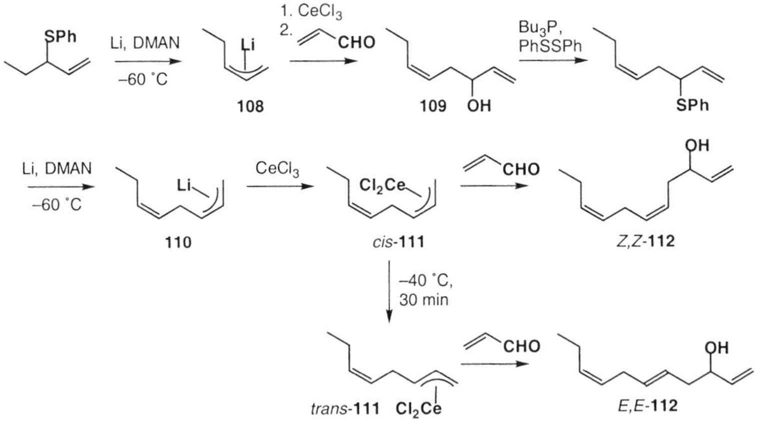

Unlike allyllithiums, the allylceriums prefer the *trans* configuration, so warming the intermediate **111** after transmetallation but before electrophilic quench leads to *trans* alkenes such as *E,E*-**112**.

4.4.2 Reduction of sulfones

1,1-Bissulfones have been reductively converted to organolithiums (lithiated sulfones) such as **113** and **114**, using lithium – naphthalene.[92] Two successive lithiations lead to the pheromone **115** of the lesser tea tartrix.

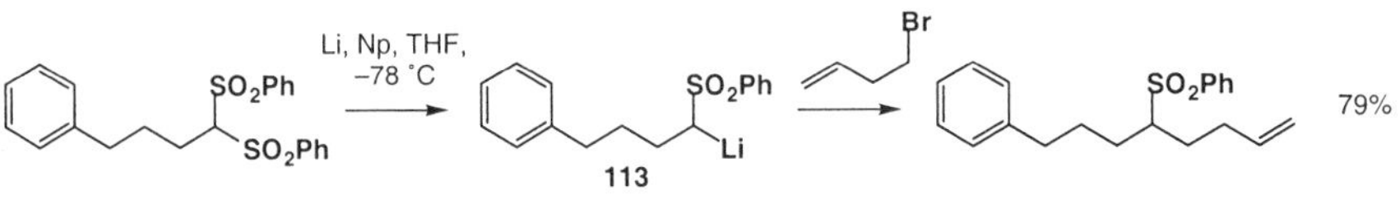

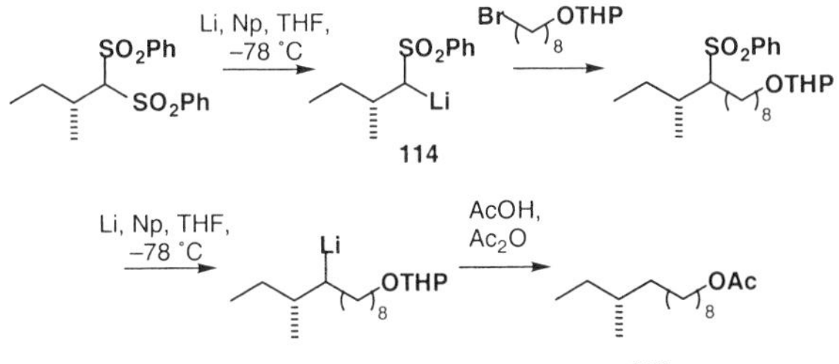

α-Heterosubstituted monosulfones[93,94] and even simple arylsulfones[95,96] will just about undergo reductive lithiation, but the reactions must be carried out under "Barbier" conditions, with the electrophile present as the reduction takes place. Vinyl sulfones cannot be reduced. The synthesis of lithiosugars from sulfones was discussed above.

4.5 Reductive lithiation of C–C bonds and π-bonds

α-Functionalised organolithiums can be made by trapping the ketyl formed on reduction of a phenylketone[97] or imine[98] with lithium and naphthalene – effectively nucleophilic addition of "Li_2" across a π bond:

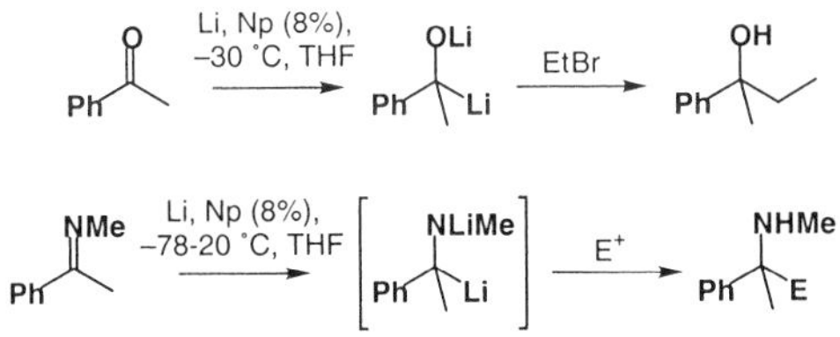

A similar reaction presumably underlies one of the most remarkable transformations yet achieved by reductive lithiation with catalytic DBB: the transformation of a nitrile into an organolithium with cleavage of a C–C bond.[99] The electrophile needs to be present in the reaction mixture (Barbier conditions), but silylation or reaction with a carbonyl compound competes successfully with deprotonation of the nitrile starting material:

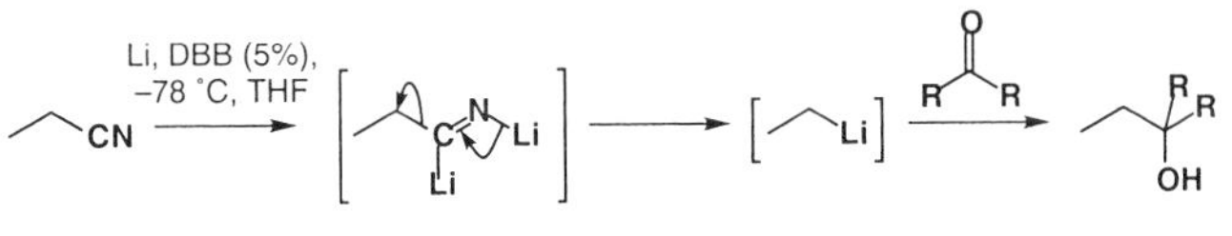

C=C bonds conjugated with carbonyl groups have been reductively lithiated with lithium and naphthalene in a reaction closely corresponding to the Birch reduction.[9]

References

1. Wakefield, B. J. *Organolithium Methods*; Academic Press: London, 1988.
2. Smith, W. N. *J. Organomet. Chem.* **1974,** *82*, 1.
3. Applequist, D. E.; Chmurny, G. N. *J. Am. Chem. Soc.* **1967,** *89*, 875.

4. Alexandrou, N. E. *J. Organomet. Chem.* **1966,** *5*, 301.
5. Glaze, W. H.; Selman, C. M. *J. Organomet. Chem.* **1968,** *11*, P3.
6. Garst, J. F. *Acc. Chem. Res.* **1974,** *4*, 480.
7. Andrieux, C. P.; Gallardo, I.; Savéant, J.-M. *J. Am. Chem. Soc.* **1986,** *108*, 638.
8. Cohen, T.; Bhupathy, M. *Acc. Chem. Res.* **1989,** *22*, 152.
9. Ramón, D. J.; Yus, M. *Eur. J. Org. Chem.* **2000**, 225.
10. Screttas, C. G. *J. Chem. Soc., Chem. Commun.* **1972**, 752.
11. Screttas, C. G.; Micha-Screttas, M. *J. Org. Chem.* **1978,** *43*, 1064.
12. Zieger, H. E.; Angres, I.; Maresca, L. *J. Am. Chem. Soc.* **1973,** *95*, 8201.
13. Holy, N. L. *Chem. Rev.* **1974,** *74*, 243.
14. Freeman, P. K.; Hutchinson, L. L. *J. Org. Chem.* **1980,** *45*, 1924.
15. Ford, W. T.; Burke, G. *J. Am. Chem. Soc.* **1974,** *96*, 621.
16. Freeman, P. K.; Hutchinson, L. L. *Tetrahedron Lett.* **1976**, 1849.
17. Cohen, T.; Matz, J. R. *Synth. Commun.* **1980,** *10*, 371.
18. Cohen, T.; Sherbine, J. P.; Hutchins, R. R.; Lin, M.-T. *Organometallic Syntheses* **1986,** *3*, 361.
19. Gómez, C.; Ruiz, S.; Yus, M. *Tetrahedron Lett.* **1998,** *39*, 1397.
20. Gómez, C.; Ruiz, S.; Yus, M. *Tetrahedron* **1999,** *55*, 7017.
21. Rawson, D. J.; Meyers, A. I. *Tetrahedron Lett.* **1991,** *32*, 2095.
22. Barluenga, J.; Flórez, J.; Yus, M. *J. Chem. Soc., Chem. Commun.* **1982**, 1153.
23. Barluenga, J.; Flórez, J.; Yus, M. *J. Chem. Soc., Perkin Trans. 1* **1983**, 3019.
24. Fernández-Simón, J. L.; Concellón, J. M.; Yus, M. *J. Chem. Soc., Perkin Trans. 1* **1988**, 3339.
25. Nájera, C.; Yus, M.; Seebach, D. *Helv. Chim. Acta* **1984,** *67*, 289.
26. Barluenga, J.; Foubelo, F.; Fañanas, F. J.; Yus, M. *Tetrahedron Lett.* **1988,** *29*, 2859.
27. Barluenga, J.; Rubiera, C.; Fernández, J. R.; Yus, M. *J. Chem. Soc., Chem. Commun.* **1987**, 425.
28. Yus, M.; Ramón, D. J. *J. Chem. Soc., Chem. Commun.* **1991**, 398.
29. Yus, M. *Chem. Soc. Rev.* **1996**, 155.
30. Huerta, F. F.; Gómez, C.; Yus, M. *Tetrahedron* **1995,** *51*, 3375.
31. Bachki, A.; Foubelo, F.; Yus, M. *Tetrahedron Lett.* **1994,** *35*, 7643.
32. Ramón, D. J.; Yus, M. *Tetrahedron* **1993,** *49*, 10103.
33. Guijarro, A.; Ramón, D. J.; Yus, M. *Tetrahedron* **1993,** *49*, 469.
34. Köbrich, G. *Angew. Chem., Int. Ed. Engl.* **1967,** *6*, 41.
35. Guijarro, A.; Yus, M. *Tetrahedron Lett.* **1994,** *35*, 253.
36. Guijarro, A.; Yus, M. *Tetrahedron* **1996,** *52*, 1797.
37. Guijarro, D.; Yus, M. *Tetrahedron Lett.* **1993,** *34*, 3487.
38. Guijarro, A.; Mancheño, B.; Ortiz, J.; Yus, M. *Tetrahedron* **1996,** *52*, 1643.
39. Ramón, D. J.; Yus, M. *Tetrahedron Lett.* **1993,** *34*, 7115.
40. Lancelin, J.-M.; Morin-Allory, L.; Sinaÿ, P. *J. Chem. Soc., Chem. Commun.* **1984**, 355.
41. Eisch, J. J.; Kaska, W. *Chem. and Ind. (London)* **1961**, 470.
42. Eisch, J. J. *J. Org. Chem.* **1963,** *28*, 707.

43. Eisch, J. J.; Jacobs, A. M. *J. Org. Chem.* **1963,** *28*, 2145.
44. Gilman, H.; Schwebke, G. L. *J. Org. Chem.* **1962,** *27*, 4259.
45. Maercker, A. *Angew. Chem., Int. Ed. Engl.* **1987,** *26*, 972.
46. Cherkasov, A. N.; Pivnitskii, K. K. *Zh. Org. Khim.* **1972,** *8*, 211.
47. Keumi, T.; Murata, C.; Sasaki, Y.; Kitajima, H. *Synthesis* **1980**, 634.
48. Azzena, U.; Demartis, S.; Fiori, M. G.; Melloni, G.; Pisano, L. *Tetrahedron Lett.* **1995,** *36*, 5641.
49. Alonso, E.; Guijarro, D.; Yus, M. *Tetrahedron* **1995,** *51*, 11457.
50. Guijarro, D.; Mancheño, B.; Yus, M. *Tetrahedron* **1995,** *48*, 4593.
51. Alonso, E.; Ramón, D. J.; Yus, M. *Tetrahedron* **1996,** *52*, 14341.
52. Bachki, A.; Foubelo, F.; Yus, M. *Tetrahedron Lett.* **1998,** *39*, 7759.
53. Foubelo, F.; Saleh, S. A.; Yus, M. *J. Org. Chem.* **2000,** *65*, 3478.
54. Almena, J.; Foubelo, F.; Yus, M. *Tetrahedron* **1995,** *51*, 3351.
55. Almena, J.; Foubelo, F.; Yus, M. *Tetrahedron* **1995,** *51*, 3365.
56. Guijarro, D.; Mancheño, M.; Yus, M. *Tetrahedron Lett.* **1992,** *33*, 5597.
57. Guijarro, D.; Guillena, G.; Mancheño, B.; Yus, M. *Tetrahedron* **1994,** *50*, 3427.
58. Guijarro, D.; Mancheño, B.; Yus, M. *Tetrahedron* **1994,** *50*, 8551.
59. Guijarro, D.; Yus, M. *Tetrahedron* **1995,** *51*, 11445.
60. Brown, H. C.; Ikegami, S.; Kawakami, J. H. *J. Org. Chem.* **1970,** *35*, 3243.
61. Benkeser, R. A.; Rappa, A.; Wolsieffer, L. A. *J. Org. Chem.* **1986,** *51*, 3391.
62. Gurudutt, K. N.; Ravindranath, B. *Tetrahedron Lett.* **1980,** *21*, 1173.
63. Bartmann, E. *Angew. Chem., Int. Ed. Engl.* **1986,** *25*, 653.
64. Cohen, T.; Jeung, I. H.; Mudryk, B.; Bhupathy, M.; Awad, M. M. A. *J. Org. Chem.* **1990,** *55*, 1528.
65. Yus, M. *Tetrahedron Asymmetry* **1995,** *6*, 1907.
66. Mudryk, B.; Cohen, T. *J. Org. Chem.* **1989,** *54*, 5657.
67. Mudryk, B.; Cohen, T. *J. Am. Chem. Soc.* **1991,** *113*, 866.
68. Ramón, D. J.; Yus, M. *Tetrahedron* **1992,** *48*, 3585.
69. Almena, J.; Foubelo, F.; Yus, M. *Tetrahedron Lett.* **1993,** *34*, 1649.
70. Almena, J.; Foubelo, F.; Yus, M. *J. Org. Chem.* **1994,** *59*, 3210.
71. Almena, J.; Foubelo, F.; Yus, M. *Tetrahedron* **1995,** *50*, 5775.
72. Katritzky, A. R.; Qi, M. *J. Org. Chem.* **1997,** *62*, 4116.
73. Screttas, C. G.; Micha-Screttas, M. *J. Org. Chem.* **1979,** *44*, 713.
74. Stork, G.; Rychnovsky, S. D. *J. Am. Chem. Soc.* **1987,** *109*, 1565.
75. Cohen, T.; Chen, F.; Kulinski, T.; Florio, S.; Capriati, V. *Tetrahedron Lett.* **1995,** *36*, 4459.
76. Chen, F.; Mudryk, B.; Cohen, T. *Tetrahedron* **1994,** *50*, 12793.
77. Foubelo, F.; Gutierrez, A.; Yus, M. *Tetrahedron Lett.* **1997,** *38*, 4837.
78. Foubelo, F.; Gutierrez, A.; Yus, M. *Synthesis* **1999**, 503.
79. Trost, B. M.; Keeley, D. E.; Arndt, H.; Bogdanowicz, M. S. *J. Am. Chem. Soc.* **1977,** *99*, 3088.
80. Cohen, T.; Daniewski, W. M.; Weisenfeld, R. B. *Tetrahedron Lett.* **1978**, 4665.
81. Cohen, T.; Lin, M.-T. *J. Am. Chem. Soc.* **1984,** *106*, 1130.

82. Cohen, T.; Weisenfeld, R. B. *J. Org. Chem.* **1979,** *44*, 3601.
83. Cohen, T.; Zhang, B.; Cheskauskas, J. P. *Tetrahedron* **1994,** *50*, 11569.
84. Cohen, T.; Doubleday, M. D. *J. Org. Chem.* **1990,** *55*, 4784.
85. Cohen, T.; Sherbine, J. P.; Matz, J. R.; Hutchins, R. R.; McHenry, B. M.; Willey, P. R. *J. Am. Chem. Soc.* **1984,** *106*, 3245.
86. Shook, C. A.; Romberger, M. L.; Jung, S.-H.; Xiao, M.; Sherbine, J. P.; Zhang, B.; Lin, M.-T.; Cohen, T. *J. Am. Chem. Soc.* **1993,** *115*, 10754.
87. Cohen, T.; Matz, J. R. *J. Am. Chem. Soc.* **1980,** *102*, 6900.
88. Beau, J.-M. *Tetrahedron Lett.* **1985,** *26*, 6185.
89. Beau, J.-M.; Sinaÿ, P. *Tetrahedron Lett.* **1985,** *26*, 6189.
90. Guo, B.-S.; Doubleday, W.; Cohen, T. *J. Am. Chem. Soc.* **1987,** *109*, 4710.
91. Cohen, T.; Guo, B.-S. *Tetrahedron* **1986,** *42*, 2803.
92. Yu, J.; Cho, H.-S.; Chandrasekhar, S.; Falck, J. R.; Mioskowski, C. *Tetrahedron Lett.* **1994,** *35*, 5437.
93. Alonso, D. A.; Alonso, E.; Nájera, C.; Yus, M. *Synlett* **1997**, 491.
94. Alonso, D. A.; Alonso, E.; Nájera, C.; Ramón, D. J.; Yus, M. *Tetrahedron* **1997,** *53*, 4835.
95. Guijarro, D.; Yus, M. *Tetrahedron Lett.* **1994,** *35*, 2965.
96. Alonso, E.; Guijarro, D.; Yus, M. *Tetrahedron* **1995,** *51*, 2697.
97. Guijarro, D.; Mancheño, B.; Yus, M. *Tetrahedron* **1993,** *49*, 1327.
98. Guijarro, D.; Yus, M. *Tetrahedron* **1993,** *49*, 7761.
99. Guijarro, D.; Yus, M. *Tetrahedron* **1994,** *50*, 3447.

CHAPTER 5

Stereoselective and Stereospecific Synthesis of Organolithiums

5.1 Configurational Stability of Organolithiums

5.1.1 Determining configurational stability

The possibility of making organolithiums with stereochemistry residing at a lithium-bearing centre, and of exploiting their stereochemistry in synthesis, is intimately associated with the configurational stability of the lithium-bearing stereogenic centre. There are three main ways of determining configurational stability:

1. *Synthesis of single enantiomers.* For chiral compounds which racemise rather slowly – over a period of minutes or hours – stereochemical stability must be determined by synthesis of one enantiomer and frequent sampling to assess the ee of the remaining material. Electrophilic quenches over a period of time provide the most practical method for assessing the ee of an organolithium, providing stereospecificity is not ee-dependent. Polarimetry is not a realistic option with air- and water-sensitive organometallics, and aggregation makes linear relationship between organolithium ee and $[\alpha]_D$ unlikely.

2. *Variable Temperature NMR.* Rates of interconversion of diastereotopic signals, determined at different temperatures, allow the rate of inversion of a stereogenic centre to be determined at those temperatures, and hence the configurational stability of the centre to be evaluated. This method is useful for relatively rapid racemisations, with half-lives of seconds or less.[1-3]

3. *The Hoffmann test.* The Hoffmann test,[4] while not quantitative, gives a qualitative guide to the configurational stability of an organolithium on the timescale of its addition to an electrophile.

The essence of the Test is this: the organometallic under investigation is made to react with a chiral, and enantiomerically pure, electrophile (commonly an aldehyde or amide). A characteristic mixture of diastereoisomers is obtained, each arising from one enantiomer of the starting organolithium. If the organolithium is configurationally stable on the timescale of the addition reaction, 50% of the product will arise from the *R* enantiomer and 50% from the *S* enantiomer (scheme 5.1.1).

However, because the transition states for reactions of the two enantiomeric organolithiums with the chiral electrophile are diastereoisomeric, one product must be formed faster than the other. If the organolithium is configurationally stable on the timescale of addition to the aldehyde, this rate difference will be of no consequence: eventually both enantiomers will react to completion. If, on the other hand, the organometallic is configurationally *unstable* on this timescale, this difference in rate will be more consequential, because the ratio of diastereoisomers will be weighted according to the rate at which each enantiomer reacts with the electrophile. In the limiting case, one enantiomer reacts much faster than the other, and since configurational instability continually restores a concentration of that enantiomer, only the product arising from reaction of that enantiomer will be observed (Scheme 5.1.2).

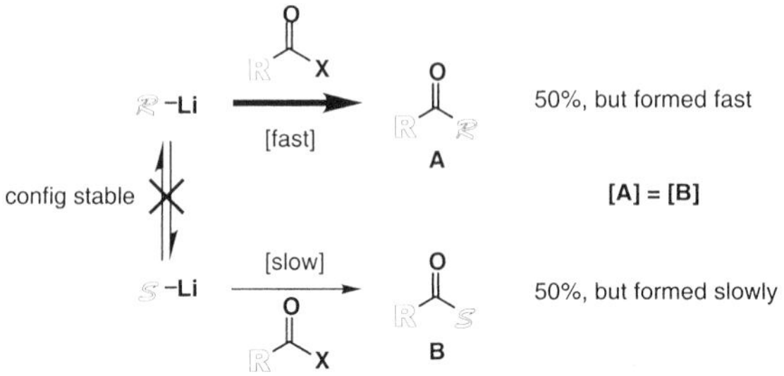

Scheme 5.1.1: Configurationally ***stable*** *racemic organolithium reacting with an* ***enantiomerically pure*** *electrophile*

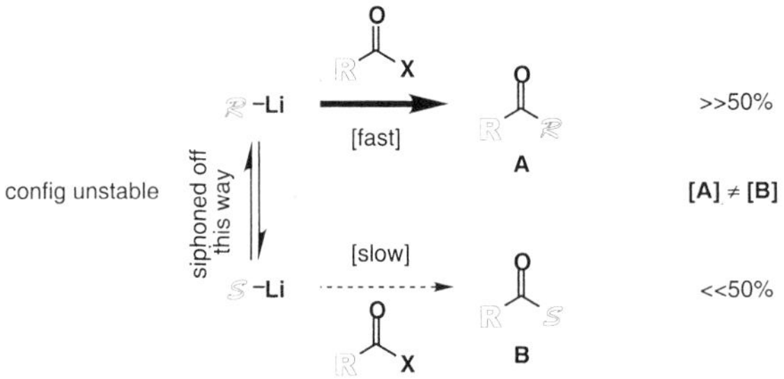

Scheme 5.1.2: Configurationally ***unstable*** *racemic organolithium reacting with an* ***enantiomerically pure*** *electrophile*

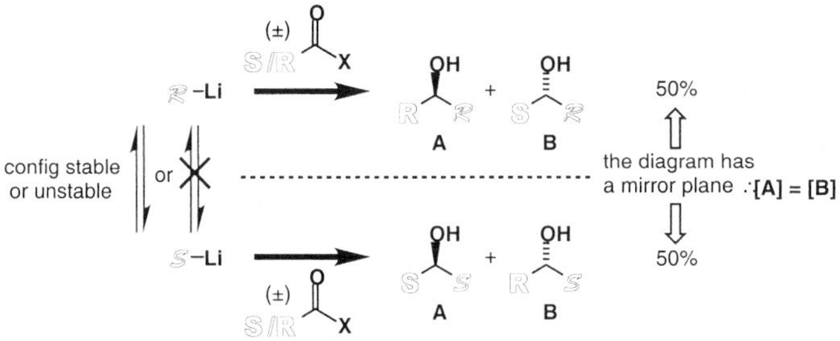

Scheme 5.1.3: Any racemic organolithium reacting with a ***racemic*** *electrophile*

The ratio obtained from this one experiment gives some information, therefore, which may be used to deduce that a reagent is configurationally unstable (if only one diastereoisomer is obtained for example). However, the power of the Hoffmann test is that neither interpretation of the ratio of diastereoisomers, nor even identification of the diastereoisomers, is required. It merely suffices to repeat the reaction with same aldehyde in *racemic* form. Since both enantiomers of the electrophile are now available for reaction, both enantiomers of the chiral organometallic must react identically, whether the organometallic is configurationally stable or not. Scheme 5.1.3 shows this: effectively Scheme 5.1.3 has a plane of symmetry across the middle, so necessarily the reactions of both enantiomers of the organolithium must yield the same results. For a configurationally unstable organometallic, the situation is no different from before, because rapid inversion means that the organometallic can effectively ignore the configuration of the electrophile – the same ratio of diastereoisomers ensues from the racemic as from the enantiomerically pure electrophile. For a configurationally *stable* organometallic, both enantiomers now have the opportunity to react at the rate of the faster-reacting enantiomer from before. The ratio will differ from the ratio obtained with the enantiomerically pure electrophile. *Any difference between the two ratios* (the one from the enantiomerically pure electrophile vs. the one from the racemic electrophile*) indicates some degree of configurational stability in the organometallic under investigation.*

The application of the test is illustrated diagrammatically below:[5]

Experiment 1

substrate (±) + electrophile (±) ⟶ diastereoisomer **A** + diastereoisomer **B**

Ratio A:B = p

If $p = 1$, test is insensitive; choose a different electrophile

If $p \neq 1$ (best value = 1.5-3.0) go to Experiment 2

Experiment 2

substrate (±) + electrophile *enantiomerically pure* ⟶ diastereoisomer **A** + diastereoisomer **B**

Ratio A:B = q

If $q = p$, substrate is configurationally *unstable*

If $q \neq p$ (q will usually be close to 1), substrate is configurationally *stable*

Some examples help to clarify the reasoning. Firstly, an organolithium which lacks configurational stability. The benzyllithium **1** reacts with the racemic Weinreb amide **2** to give a 63:37 ratio of diastereoisomeric ketones **3** and **4**, a ratio in the ideal range of 1.5-3.0. Repeating the reaction with enantiomerically pure amide gives exactly the same ratio: the benzyllithium **1** must therefore be configurationally unstable.[5]

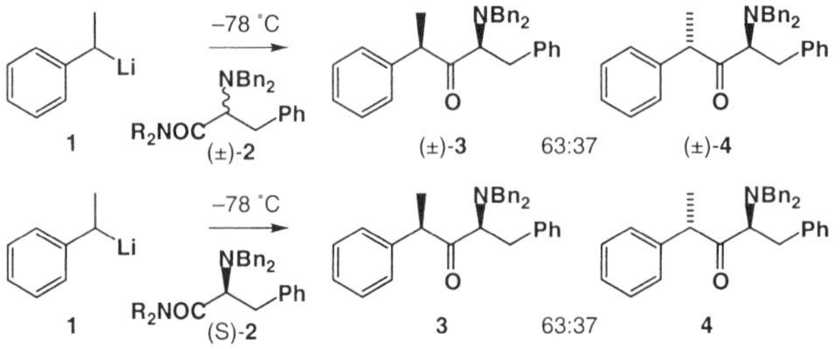

The lithiated carbamate **5** gives a similar result with the racemic aldehyde **6**: a 68:32 ratio of two diastereoisomers **7** and **8** (the innate diastereoselectivity of the aldehyde ensures only one diastereoisomer at the new OH bearing centre). But with the enantiomerically pure aldehyde, a 54:46 ratio arises, significantly different from the ratio with racemic aldehyde, and indicative of configurational stability on the microscopic timescale.[6]

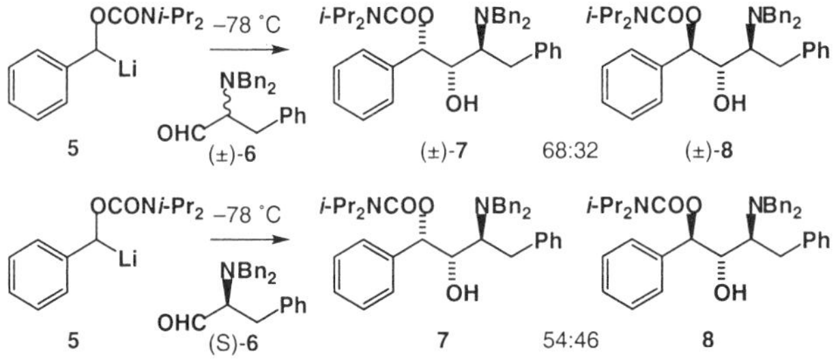

In some cases the Hoffmann test fails to be useful. For example, the lithiated amine **9** gives a 50:50 mixture of diastereoisomers **10** and **11** on reaction with (±)-**6**. No useful information can be drawn in such an instance.[6]

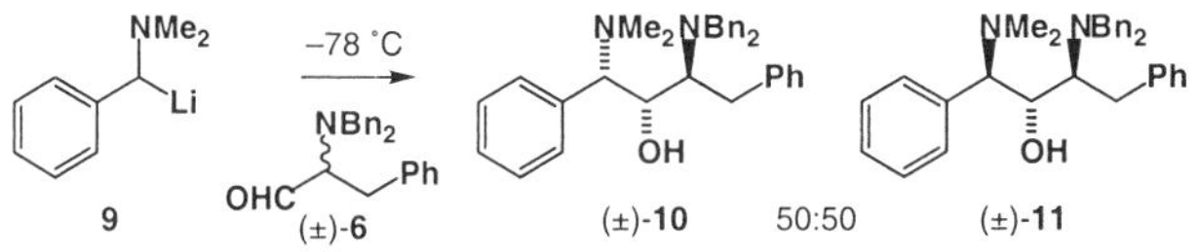

Beak has described what he calls the "poor man's Hoffmann test" because is does not require use of an enantiomerically pure electrophile.[7] This version of the Test also relies on a rate difference between diastereoisomeric transition states, but it quantifies the rate difference by the outcome of a reaction in which enantiomeric organolithiums compete for a deficit of achiral electrophile in the presence of an enantiomerically pure chiral ligand, usually (–)-sparteine **12**.

H
N
N
H
=
N
N
(–)-sparteine **12** [= "(–)-sp."]

(–)-sparteine
(–)-sp.
R -Li
R -Li
E+ (excess)
[fast]
R -E
A
50%, but formed fast
config stable
A = B: racemic product
S -Li
(–)-sparteine
S -Li
(–)-sp.
E+ (excess)
[slow]
S -E
B
50%, but formed slowly

Scheme 5.1.4: Configurationally ***stable*** *organolithium reacting with an* ***excess*** *of electrophile.*

(–)-sparteine
(–)-sp.
R -Li
R -Li
E+ (deficit)
[fast]
R -E
A
>50%
config stable
A ≠ B: non-racemic product
S -Li
(–)-sparteine
S -Li
(–)-sp.
E+ (deficit)
[slow]
S -E
B
<50%

Scheme 5.1.5: Configurationally ***stable*** *organolithium reacting with* ***deficit*** *of electrophile.*

(–)-sparteine
(–)-sp.
R -Li
R -Li
E+ (excess or deficit)
[fast]
R -E
A
>>50%
config unstable
siphoned off this way
A ≠ B: non-racemic product
S -Li
(–)-sparteine
S -Li
(–)-sp.
E+ (excess or deficit)
[slow]
S -E
B
<<50%

Scheme 5.1.6: Configurationally ***unstable*** *organolithium reacting with* ***any quantity*** *of electrophile*

When the organolithium is configurationally stable, an excess of a chiral electrophile will allow both diastereoisomeric organolithium-(–)-sparteine complexes to react to completion, if at different rates. The ee of the product varies through the course of the reaction, but ultimately racemic product results (Scheme 5.1.4). With a deficit of electrophile, however, the reaction stops before completion, and the faster-reacting organolithium-(–)-sparteine complex forms more product that than slower-reacting complex. Some level of enantiomeric excess in the product results (Scheme 5.1.5), but the key diagnostic point is that the results with an excess and with a deficit of electrophile differ.

With a configurationally unstable organolithium, the equilibrium between the two diastereoisomeric organolithium-(–)-sparteine complexes is continually maintained, so the enantioselectivity is independent of the extent of reaction: the same result is obtained with a deficit of electrophile as with an excess (Scheme 5.1.6).

"Configurationally stable" is a term which has meaning only when it is associated with a temperature and a timescale. The three methods allow us to define three timescales on which configurational inversion may take place. Firstly, the NMR timescale, which is dependent on peak separation, but typically is sensitive to rates of the order of 100-10 s^{-1}.[8] The temperature limits on dynamic NMR experiments mean that studies of configurational stability on the NMR timescale are applicable only to compounds which racemise with within seconds at room temperature.

This is about the bottom end of the next timescale - the *microscopic* timescale - which is associated with the rate at which organolithium nucleophiles will add to electrophiles (usually a matter of seconds at the most). Configurational stability on the microscopic timescale is studied by the Hoffmann test, or by formation of organolithiums in the presence of electrophiles (such as Me_3SiCl) with which they react immediately – *in situ* quench conditions.

The third timescale is the laboratory, or *macroscopic*, timescale of minutes or more: if an organolithium can be formed as a single enantiomer and subsequently reacted with an electrophile to produce a product with at least some enantiomeric excess, then it has configurational stability on the macroscopic timescale.

Finally, it is important to distinguish between, on the one hand, the kinetic stability of either a chiral organolithium which retains its stereochemistry in the absence of an external chiral influence or a pair of diastereoisomeric organolithiums which do not interconvert, and on the other hand the thermodynamic configurational stability of an organolithium whose configuration is maintained by some external stereogenic influence – complexation to a chiral ligand for example. The first type of configurational stability is the more interesting and the more important, and is the one we discuss in this section. We shall use the term "configurationally stable" only of organolithiums that have *kinetic* configurational stability.

In general, the possibility of handling stereochemically stable organolithiums with lithium-bearing stereogenic centres is now well established for a range of α-oxy and α-amino organolithiums, a few α-thioorganolithiums and organolithiums on small rings.

5.1.2 Unfunctionalised organolithiums

5.1.2.1 Secondary organolithiums

It was an organolithium compound – 2-lithiooctane **14** – which first demonstrated the ability of a carbon-metal bond to retain its configuration, if only partially. Letsinger,[9] in 1950, reported the synthesis of **14** from (–)-**13** in 94:6 petrol:ether by transmetallation with *s*-BuLi. Carbonation after 2 minutes at –70 °C gave a product with an ee of about 20% (by optical rotation), with overall partial retention of configuration. Allowing the organolithium to warm to 0 °C for 20 minutes returned entirely racemic material. Racemic material is also formed if the organolithium **14** is made by reductive lithiation of enantiomerically pure chloride,[10] presumably because radicals intervene in the mechanism. Other reports of optically active organometallics at this stage turned out to be erroneous interpretations of the presence of optically active impurities.[11]

(–)-**13** I —(*s*-BuLi, petrol/ether 94:6, –70 °C, 2 min or 0 °C, 20 min)→ **14** Li —(CO_2)→ **15** CO_2H 20% ee or 0% ee

The problem with Letsinger's organolithium – the reason for its half-life of racemisation measured in seconds at –70 °C – was the presence of ether, which is essential for the transmetallation step. In 1962, Curtin[12] managed to synthesise optically active *s*-BuLi **16** in the absence of ether by transmetallation of an organomercury compound in pure pentane. Even after 4 hours at –40 °C, **16** had retained 83% of its optical activity; after 30 min at –8 °C, it had retained 55% of its optical activity, and it was only on adding 6% ether at this temperature that it very rapidly racemised. Secondary, unfunctionalised organolithiums are configurationally stable in non-polar solvents (half-life for racemisation measured in hours at –40 °C and minutes at 0 °C), but configurationally unstable in the presence of ether (half-life of seconds at –70 °C).

5.1.2.2 Primary organolithiums

The configurational stability of primary organolithiums, while of no synthetic consequence, has been determined by NMR methods. Analysis of the line shape of the AA'BB' system of **17**[13] or of the ABX system of **18**[14] gave half-lives for inversion of about 0.01 s for **17** or **18** in ether at 30 °C, increasing (for **17**) to about 0.1 s at 0 °C and 1 s at –18 °C. The rate of inversion was decreased by a factor of 8 in pentane or 10 in toluene, but the addition of 1 equiv. TMEDA to pentane restored a rate of inversion more or less equal to that in pentane.

Increasing the degree of substitution at the lithium-bearing centre, though it makes the organolithium more reactive, also increases its configurational stability.

AA'BB' ABX

16 17 18 19

5.1.2.3 Solvent effects

Ether, by increasing the polarity of the solution, promotes the formation of ion pairs, and hence racemisation by migration of the Li^+ cation from one face of the anion to the other.[11] Its effect on primary organolithiums, which presumably require much less charge separation to invert, is much smaller than its effect on secondary organolithiums. The role of charge separation in organometallic racemisation is supported by the fact that the greater the ionic character of the C–Metal bond, the faster the racemisation (Li>Mg>Al>Hg).[13] For comparison, **18** inverts in ether three times as fast as its –MgBr analogue **19**.[15] In general (but not always) polar solvents lead to faster racemisation (see section 5.1.4.4), and individual examples are mentioned in the text. The main exceptions are cyclic alkyllithiums lacking intramolecular coordiantion, in which TMEDA tends to *increase* configurational stability (see section 5.1.5.2).

5.1.3 Cyclopropyllithiums

Cyclopropyllithiums have complete configurational stability at 0 °C, even in ether. For example, the two diastereoisomeric cyclopropyl bromides *cis*- and *trans*-**20**, on treatment with isopropyllithium at 0 °C in 94:6 pentane:ether for a few minutes, and then CO_2, reacted with >95% stereospecificity, and with overall retention of configuration.[16] Similar results were obtained with ethylene oxide as the electrophile, transmetallating with BuLi.[17] An enantiomerically pure tertiary cyclopropyl bromide, on treatment with BuLi, gives an organolithium **23** which is similarly configurationally stable at –8 °C even in ether or THF.[18]

94:6 pentane:ether, 0 °C, a few minutes; CO_2

cis-**20** → *cis*-**21** → *cis*-**22**: 10% yield, <5% *trans*

trans-**20** → *trans*-**21** → *trans*-**22**: 15% yield, <5% *cis*

23

In a small ring, the pyramidal anion formed by charge separation is much slower to invert because ring strain inhibits formation of a trigonal, planar carbon, so even if charge separation takes place in a polar solvent, the Li^+ is likely to return to the same face from which it departed.

It is essential, for successful retention of stereochemistry, to form the organolithiums by halogen-metal exchange, and not by reductive lithiation. While transmetallation of *cis*- and *trans*-**24** with BuLi, quenching the organolithium **25** with ethylene oxide, gives complete retention of stereochemistry in **26**. Owing to the intervention of radicals, reductive lithiation leads to considerable (but not total) epimerisation:

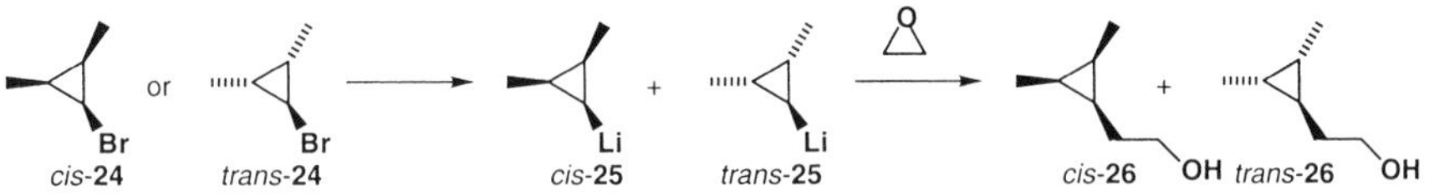

cis-**24**	BuLi	100:0
cis-**24**	Li metal	54:46
trans-**24**	BuLi	0:100
trans-**4**	Li metal	

Corey exploited the remarkable configurational stability of cyclopropyllithiums in his synthesis of hybridalactone. The stannane **28** was made by Simmons-Smith cyclopropanation of the allylic alcohol **27** and resolved by formation of an O-methyl mandelate ester. Transmetallation of **29** with 2 equiv. BuLi gave an organolithium which retained its stereochemistry even in THF over a period of 3 h at 0 °C, finally adding to **31** to give **32**.

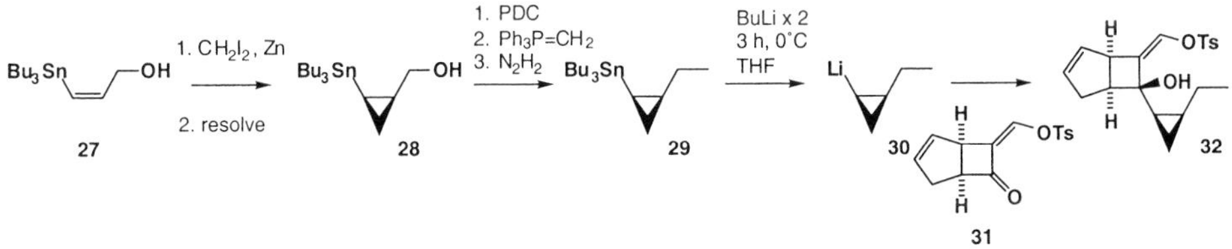

The amide-substituted cyclopropyllithium *trans*-**33** is configurationally stable at –78 °C, but epimerises to the more stable, chelated *cis* diastereoisomer on warming to 0 °C.[19]

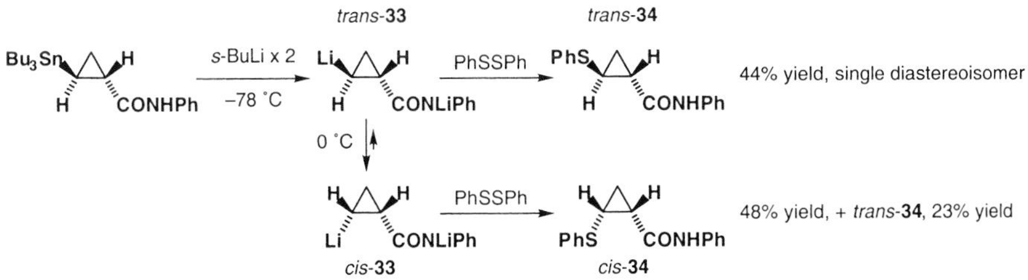

Cyclopropyllithiums stabilised additionally by conjugating groups (alkynes[20] or nitriles,[21,22] – but not isonitriles, which maintain configurational stability[23] – for example) do not retain their stereochemistry, even though they retain some pyramidal character.

One example of a cyclobutyllithium (**36**) appears to indicate configurational stability. However, the stereospecific transformation of **35** to **37** could equally be explained by thermodynamic stabilisation of one epimer of an otherwise configurationally unstable organolithium.[24]

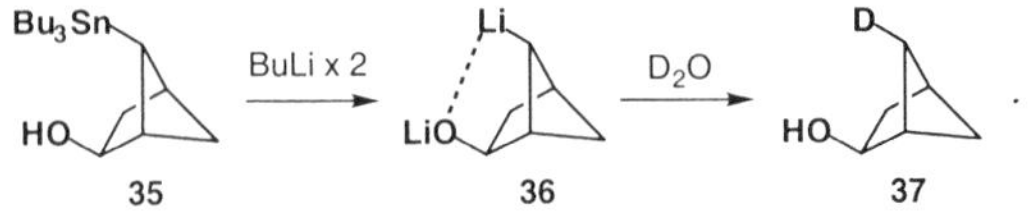

5.1.4 Organolithiums α to oxygen

5.1.4.1 Simple acyclic α-alkoxy organolithiums

In 1980, Still discovered that transmetallation of chiral α-alkoxystannanes is stereospecific,[25] and this result set the scene for the exploitatoin of chiral α-alkoxyorganolithiums. Not only are the organolithiums formed stereospecifically, but they are also configurationally stable over periods of *at least minutes at –30 °C*. Transmetallation of *syn*- and *anti*-**38**, quenching with acetone, gave clearly different products **40** each with no trace of the other.

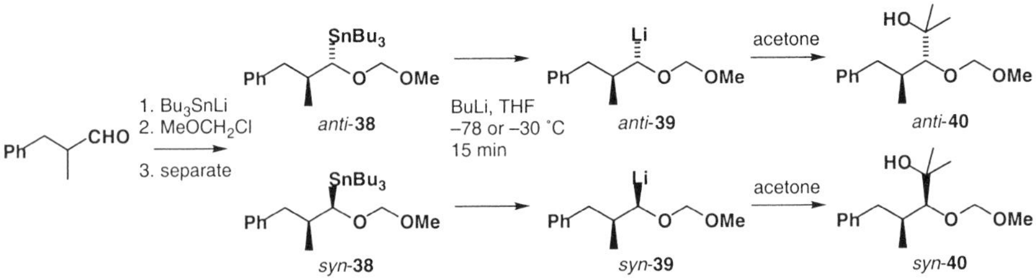

Other α-alkoxyorganolithiums which are similarly configurationally stable over a period of up to 30 minutes at –78 °C include **41**[26], **42**[27], **43**[28], **44** and **45**.[29] For example, transmetallation of enantiomerically pure **46** with BuLi in THF at –78 °C for 10 min gave the organolithium **43** which underwent conjugate addition to **47** to give the lactone **48** in 92% ee.[28]

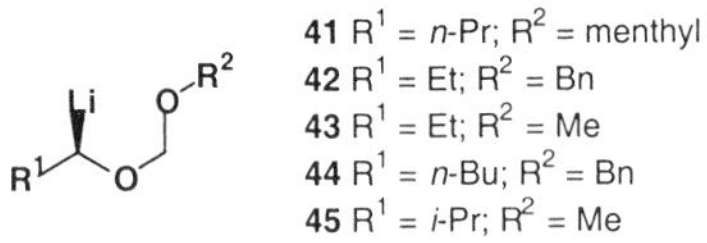

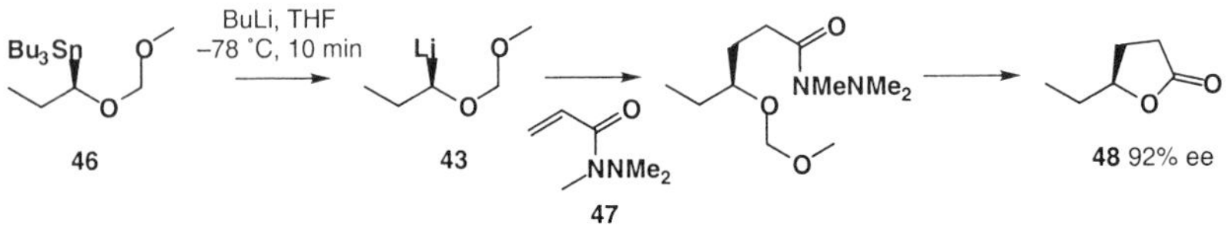

Organolithiums α to *O*-carbamoyl groups have greater configurational stability. For example, organolithium **50** formed by treating the enantiomerically pure stannane **49** with BuLi at –78 °C gives products with almost complete retention of enantiomeric purity (**51** is formed with 96% ee).[30] Similarly organolithium **52** is configurationally stable over a period of 2.5 h in TMEDA/ether at –78 °C.[31]

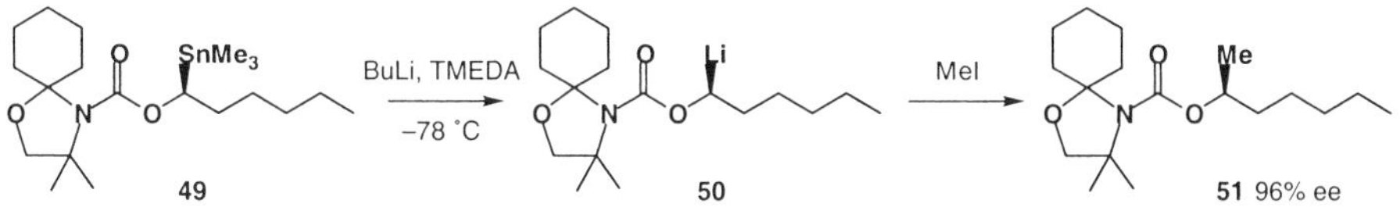

Kinetic-isotope controlled deprotonation of **53** can be used to form a single enantiomer of the organolithium **54**, which is configurationally stable over 4 h at –78 °C in TMEDA/ether.

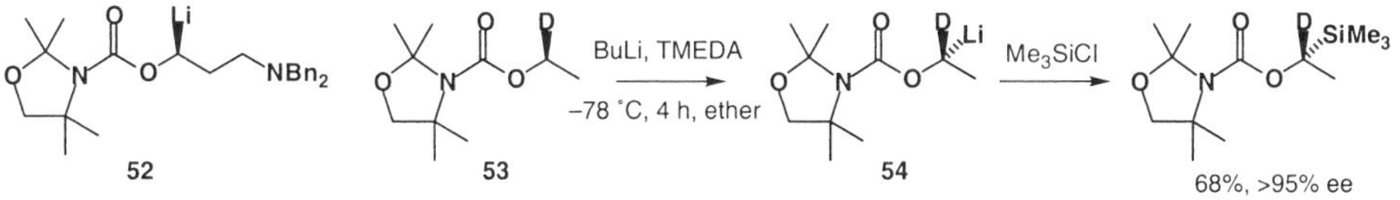

In contrast, the benzylic organolithium **55** is configurationally unstable: formed by kinetic isotope effect-governed deprotonation at –78 °C, it is completely racemic after 85 s, but it is estimated to have a half-life for racemisation measured in milliseconds.[32] The key difference is the benzylic nature of the anion, which enormously increases the rate of racemisation; the change from OCH_2OR to OMe also contributes since intramolecular coordination favours configurational stability (see later).

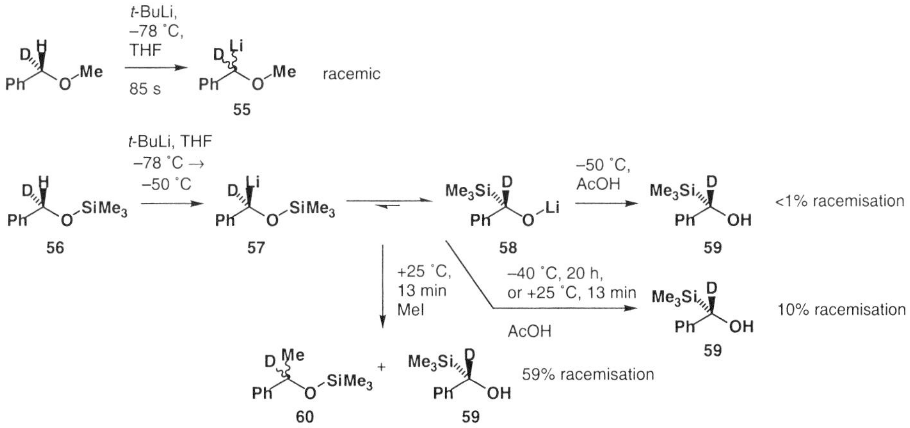

Remarkably, **57**, the trimethylsilyl analogue of **55**, appears to racemise much more slowly. At –50 °C, it undergoes a Brook rearrangement (section 8.2) to give **59** with 99% inversion of stereochemistry. If the anion **58** is held at –40 °C for 20 h, or 25 °C for 13 min, the product is formed with about 10% racemisation, indicating the rearrangement is reversible and that racemisation of the organolithium is occurring. The rate of racemisation is increased by the presence of methyl iodide (or LiI) because a methyl iodide quench after 13 min at 25 °C gave 40% racemic **60** plus 60% **59** which was now 59% racemised. The fact that racemisation is so much slower in **57** than **55** is presumably because the anion is very short-lived, and rearrangement back to the alkoxide **58** is much faster than inversion.

Organolithium **62**, the intermediate in the Brook rearrangement of **61**, must also be configurationally stable on the timescale of the rearrangement – though here the reaction is stereospecific with *retention* of stereochemistry (see section 8.2).[33]

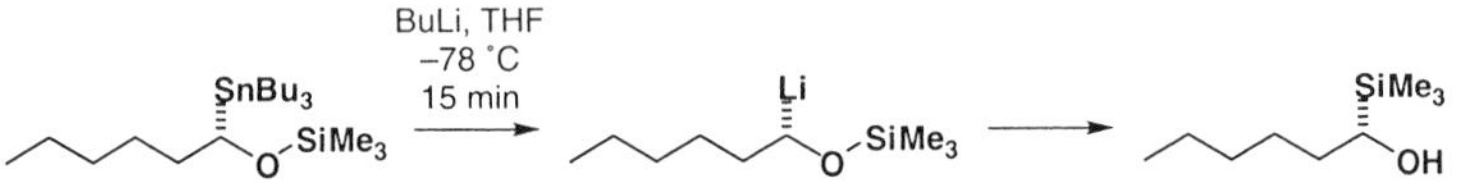

BuLi, THF
–78 °C
15 min

$SnBu_3$ / $SiMe_3$ **61** → Li / $SiMe_3$ **62** → $SiMe_3$ / OH **63**

The stereochemistry of another rearrangement – the [2,3]-Wittig (discussed in section 8.4) – indicates that the organolithiums **64**,[34] **65**,[35] **66** and **67**[36] have complete configurational stability over a period of minutes at –78 °C despite the lack of a second coordinating atom within the alkoxy group.

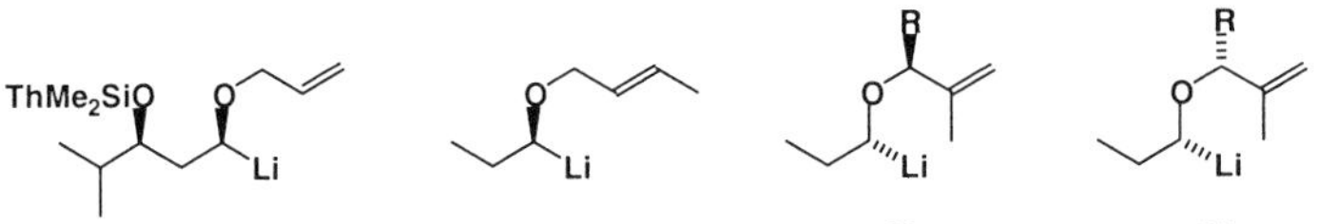

$ThMe_2SiO$ Li **64** Li **65** R Li **66** R Li **67**

5.1.4.2 *Cyclic α-alkoxy organolithiums*

The axial and equatorial organolithiums *ax*-**69** and *eq*-**69** are both configurationally stable over a period of minutes at –60 °C.[37] They react stereospecifically and with retention with electrophiles, as shown below.

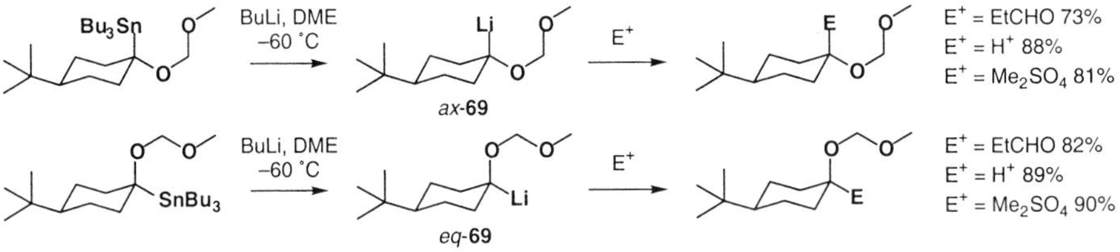

Bu_3Sn —BuLi, DME, –60 °C→ Li *ax*-**69** —E^+→ E; E^+ = EtCHO 73%, E^+ = H^+ 88%, E^+ = Me_2SO_4 81%

$SnBu_3$ —BuLi, DME, –60 °C→ Li *eq*-**69** —E^+→ E; E^+ = EtCHO 82%, E^+ = H^+ 89%, E^+ = Me_2SO_4 90%

Organolithiums α to oxygen atoms within the ring are also configurationally stable at –78 °C in THF. The axial and equatorial stannanes *ax*-**70** and *eq*-**70**, for example, give organolithiums **71** which retain their stereochemistry over periods of up to 50 min in THF at –78 °C, and can be quenched stereospecifically with aldehydes.[38]

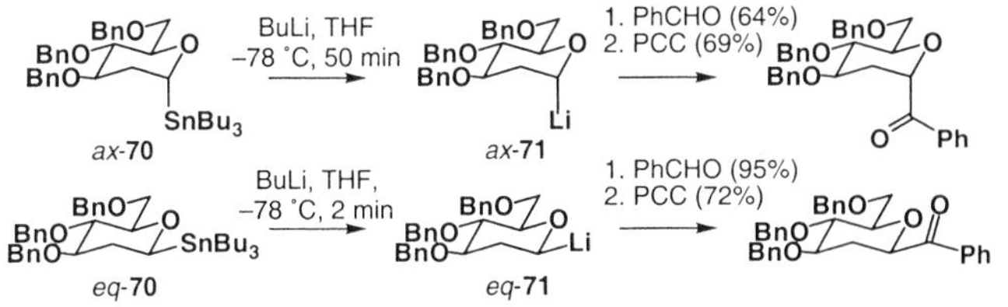

Very usefully, these cyclic organolithiums can be *epimerised at temperatures about –30 °C*. For example, Cohen[39] showed that both the axial and equatorial diastereoisomers of **72** can be reduced with lithium and dimethylaminonaphthalene (DMAN) to give the axial organolithium *ax*-**73**, which is quenched stereospecifically by benzaldehyde to give *ax*-**74**. Adding TMEDA and warming to –30 °C allows the organolithium to equilibrate to the more thermodynamically stable equatorial isomer *eq*-**73**, leading to *eq*-**74** (see section 4.4.1).

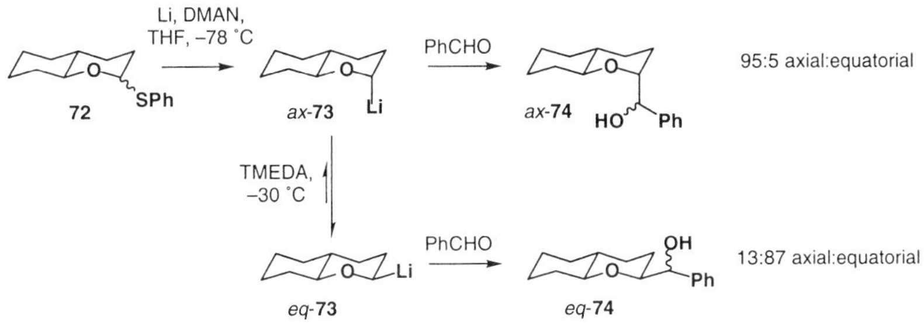

Rychnovsky[40] has exploited this controlled configurational instability in the following way. Reductive lithiation (lithium and di-*tert*-butylbiphenyl, DBB) of either diastereoisomer of **75** at –78 °C gives the axial organolithium *ax*-**76**, which can be quenched with acetone to yield the axial alcohol *ax*-**77** with a 98:2 axial:equatorial ratio.[41] Allowing the solution of *ax*-**76** to warm to –30 °C starts its epimerisation (38:62 axial:equatorial after 30 min at –30 °C), and when it reaches –20 °C the same acetone quench returns a 95:5 equatorial:axial ratio of alcohols **77**.

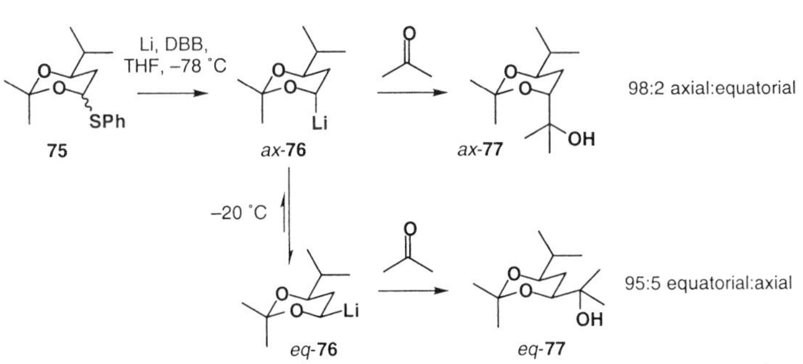

Similarly, the axial organolithium *ax*-**79** produced by reduction of **78** under kinetic control can be quenched to give 98:2 axial:equatorial products in 81% yield, or epimerised (–20 °C ,

45 min) to the more stable equatorial organolithium *eq*-**79**, giving 99:1 equatorial:axial products **80** in 59% yield.[41] The axial organolithium *ax*-**82** produced by reduction of **81** under kinetic control can be quenched to give 97:3 axial:equatorial products **83** in 53% yield, or epimerised (+20 °C, 20 min) to give 95:5 equatorial:axial products **83** in 30% yield. Yields of the equatorial products *eq*-**83** produced thermodynamic control are lowered by the slow deprotonation of THF, and in the last example, a mixture of 25% THF in hexane was used to lessen competition from this pathway.

Li, DBB, THF, −78 °C
78
ax-79
ax-80
81%, 98:2 axial:equatorial
−20 °C, 45 min
eq-79
eq-80
59%, 99:1 equatorial:axial

Li, DBB, THF, hexane (1:3), −78 °C
81
ax-82
ax-83
53%, 97:3 axial:equatorial
+20 °C, 20 min
eq-82
eq-83
30%, 95:5 equatorial:axial

The ability to select the stereochemical outcome of these reactions allows the synthesis of polyacetate-type structures with complete control of stereochemistry. The scheme below shows an example.[42] The malic acid-derived **84** is lithiated to give *ax*-**85**. This axial organolithium is quenched with ethylene oxide, deprotected and converted to the epoxide **86**. Meanwhile, another sample of *ax*-**85** is epimerised to the more stable equatorial organolithium *eq*-**85** and quenched with Bu_3SnCl. The stannane **87** acts as a convenient source of the organolithium which avoids the production of LiSPh, which competes in reactions with electrophiles (a point of concern when the electrophile is more valuable than acetone or ethylene oxide). Regeneration of the equatorial organolithium *eq*-**85** from **87** and attack on the epoxide **86** gives the polyacetate fragment **88** – clearly other stereoisomers are equally possible by choice of conditions and starting materials.

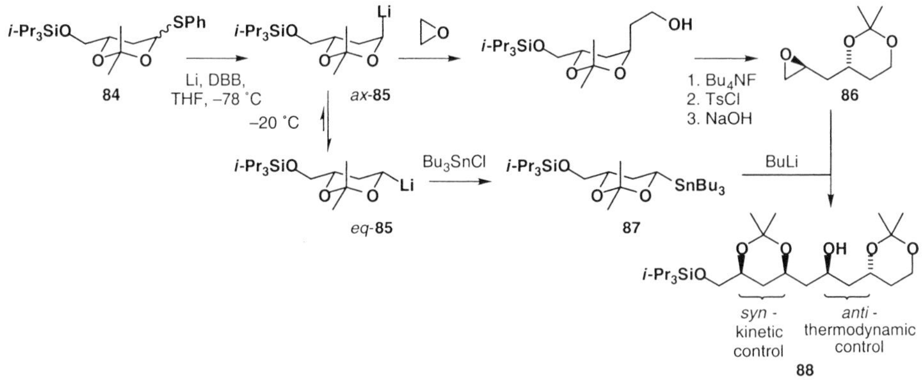

The stannanes **89** and **91** both generate equatorial organolithiums **90** and **92**, which can interconvert only by a ring-flipping process. The interconversion does not take place at –78 °C, and the organolithium with the equatorial lithium is clearly the more stable as it quenches with no trace of other stereoisomers even at 0 °C.[43] The axially-substituted **92** starts to invert at –40 °C (96% retention); but at –20 °C 35% has inverted, and at 0 °C inversion is complete.

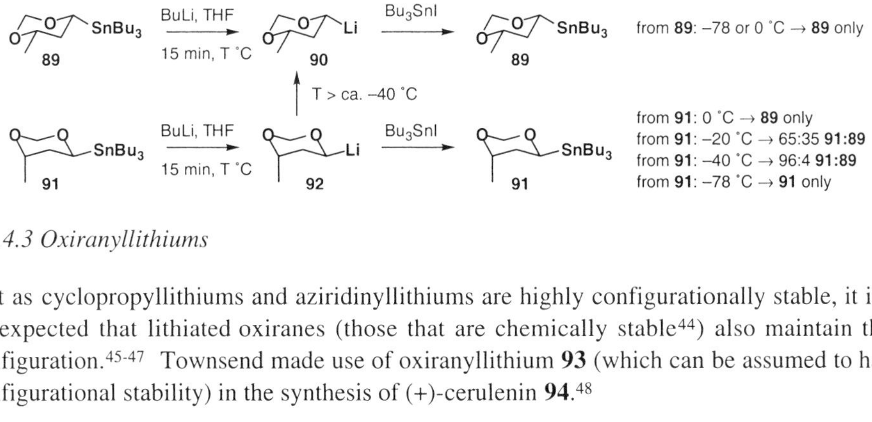

5.1.4.3 Oxiranyllithiums

Just as cyclopropyllithiums and aziridinyllithiums are highly configurationally stable, it is to be expected that lithiated oxiranes (those that are chemically stable[44]) also maintain their configuration.[45-47] Townsend made use of oxiranyllithium **93** (which can be assumed to have configurational stability) in the synthesis of (+)-cerulenin **94**.[48]

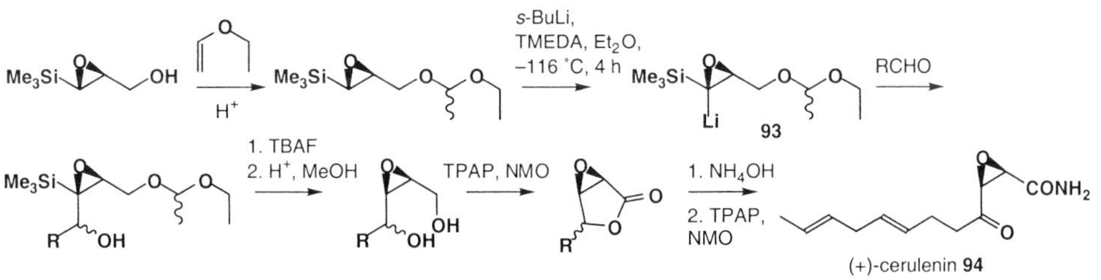

5.1.4.4 Allylic and benzylic α-alkoxy organolithiums

An unfortunate feature, pointed out by Hoppe,[49] of asymmetric organolithium chemistry is that those features required for appropriate reactivity tend to work against configurational stability. The polar solvents (ether, THF or TMEDA) required for successful halogen-metal exchange reactions also make the products of these reactions, in the absence of a stabilising α heteroatom or small ring, configurationally unstable. A similar problem met early attempts to form configurationally stable organolithiums by deprotonation: the features of the starting material which tend to acidify the protons and favour deprotonation turn out to be the same features that favour racemisation of the organolithium once formed. Hoppe found that this problem could be overcome by using a carbamate group: carbamate-stabilised *tertiary* allylic α-alkoxyorganolithiums may be formed by deprotonation and are configurationally stable over a period of a few hours at –78 °C.[49] Enantiomerically enriched **95** is slowly deprotonated by BuLi in a mixture of ether, hexane and TMEDA. After 7 h, the product **97** of electrophilic quench has 52% ee, indicating a half-life for racemisation of **96** at –78 °C of about 3 h (assuming the electrophilic quench is fully stereospecific). Hoppe later found that the organolithium could be formed with improved configurational stability by leaving ether out of the solvent mixture.[50]

BuLi (0.95 equiv.) ether, hexane, TMEDA
7 h, –78 °C
N*i*-Pr$_2$ **95** → N*i*-Pr$_2$ **96** → (CHO) **97** N*i*-Pr$_2$ + starting material
24% yield 52% ee; 25% yield 68% ee

The same is true of tertiary *benzylic* α-alkoxyorganolithiums. Enantiomerically enriched (97% ee) **98** is deprotonated by *s*-BuLi in TMEDA/hexane in 30 min at –70 °C, and a Me_3SiCl quench gave almost no loss of enantiomeric purity (96% ee) in the silylated product **100**.[51,52] In THF, however, **99** racemises rapidly. The analogous compounds **101** and **102** are similarly configurationally stable in the presence of TMEDA and Et_2O over 5 min at –78 °C.[53] The naphthalene-substituted **103**, though, racemises under these conditions (though it is configurationally stable in the absence of TMEDA). The stereospecificity of this reaction is discussed in section 6.1.4.

s-BuLi, TMEDA, hexane, 30 min, –70 °C; Me_3SiCl
98 97% ee → **99** → **100** 96% ee (configuration unknown)

101 102 103

Small variations in the nature of the intramolecular chelating group can affect the configurational stability of the organolithium, as shown by the analogous ester-stabilised **105**. The chiral organolithium **105** could be made by treating optically active hindered ester **104** with *s*-BuLi in toluene or hexane in the presence of TMEDA or Et_2O (but not at all in their absence) or in THF. Quenches with various electrophiles gave some enantiomeric excess in all cases except in the presence of THF, which caused racemisation of the organolithium within 30 min.[54] After 5 min in 4:1 toluene/ether, the ee of the product **106** formed by deuteration of the anion (AcOD) was >98%; this dropped to 95% after 10 min and 91% after 60 min. Racemisation was somewhat faster in hexane in the presence of 1 equiv. TMEDA: deuteration of the organolithium (MeOD) gave 65% ee after 2 min, 44% ee after 5 min and 32% ee after 30 min. The carbamate-stabilised organolithium **99** is configurationally stable under these conditions.[51,52]

s-BuLi solvent, t min, –78 °C; MeOD or AcOD

104 105 106

4:1 toluene:ether; 5 min	>98% ee
4:1 toluene:ether; 10 min	95% ee
4:1 toluene:ether; 60 min	91% ee [half-life of hours]
TMEDA, hexane; 2 min	65% ee
TMEDA, hexane; 5 min	44% ee
TMEDA, hexane; 30 min	32% ee [half-life of minutes]
THF, 30 min	0% ee [half-life of seconds]

The similar tertiary benzylic α-alkoxyorganolithium **108** has some configurational stability at –78 °C in THF, but it is chemically unstable and after 2 h has formed the ketone **109**.[54] This ketone is formed in 20% ee, but without knowledge of the timescale of the dimerisation it is not possible to assess the half-life for racemisation of **108** in THF

LiTMP THF, 2 h, –78 °C

107 108 109 20% ee

On the other hand, α-alkoxyorganolithiums are not configurationally stable on a macroscopic timescale when they are *secondary* and allylic or benzylic. For example, despite the known (see section 5.2.1) stereospecificity of the tin-lithium and lithium-tin exchanges of similar compounds, tin-lithium exchange of **110** with *n*-BuLi/TMEDA at –78 °C gives an organolithium **111** which has completely racemised after 10 min: stannylation returns racemic stannane **112**.[55] Similarly, **111** racemises rapidly at –70 °C in pentane/cyclohexane in the presence of (–)-sparteine.[56] Section 6.2.5 discusses how this allows it to be crystallised as a single enantiomer in the form of a diastereoisomeric complex with (–)-sparteine **12**. The configurationally unstable secondary benzylic organolithium **55** has already been discussed.

Me_3Sn ... **110** 79% ee —(s-BuLi, TMEDA)→ **111** —(Bu_3SnCl)→ (±)-**112**

5

Nonetheless, secondary benzylic α-alkoxyorganolithiums may have configurational stability on the *microscopic* timescale, as indicated by a Hoffmann test.[6] As discussed above (section 5.1.1), the lithiated benzylcarbamate **5** is configurationally stable by the test. A Hoffmann test on **99**, the tertiary benzylic analogue of **5**, confirmed the configurational stability previously demonstrated by Hoppe.[51,52]

5.1.5 Organolithiums α to nitrogen[57]

5.1.5.1 Cyclic α-aminoorganolithiums: 3 membered rings

It has already been noted that the presence of a small ring favours configurational stability in organolithiums. The first organolithium α to nitrogen of proven configurational stability was the isonitrile-substituted cyclopropyllithium **114**. Treatment of the enantiomerically pure isonitrile **113** with LDA for 30 min at –72 °C followed by methylation returned the product cyclopropane **115** in 98% ee.[23] **114** maintained its configuration up to –52 °C (the ee was eroded to 93% after 30 min at this temperature) but racemised within 30 min at –5 °C.

113 —(LDA, T °C, 30 min)→ **114** —(MeI)→ **115**

T = –72 °C: 98% ee
T = –52 °C: 93% ee
T = –25 °C: 58% ee
T = –5 °C: 0.3% ee

The same results were obtained in THF, ether, DME or in the presence of TMEDA, triglyme, HMPA or crown ethers.

Aziridinyllithiums are configurationally stable even when adjacent to a carbonyl group.[58] For example, treatment of the ester **116** with LDA and then BuLi at –75 °C, maintaining the organolithium **117** at this temperature for 20 min, and then quenching with MeI, MeOD or MeOH, returned the aziridine **118** with no loss of enantiomeric purity.[59] However, the thioester-substituted aziridinyllithiums **120** and **123** showed some degree of interconversion as the temperature was raised. **120** is clearly more stable than **123**, since **123** demonstrated configurational instability even within 15 min at –95 °C. **120** showed only small amounts of inversion even at –60 °C.

Tin-lithium exchange of **125** gives an intermediate **126** that retains its stereochemistry, presumably (though this was not proven) under kinetic and not simply thermodynamic control. A variety of electrophiles to give various *cis*-substituted aziridines **127**.[60]

The borane-complexed aziridines **128** can be lithiated diastereoselectively and quenched stereospecifically with electrophiles.[61] This one result does not prove the kinetic configurational stability of the intermediates, but a related reaction sequence does show that they have at least some configurational stability. The same pair of organolithiums **131a** and **131b** were made by two routes: one by methylation of the parent aziridine **129**; the other by lithiation of the methylaziridine **132**. On quenching with Bu_3SnCl, the ratio of **130a**:**130b** produced in each of the reactions was different – certain proof that the organolithiums **131** do not undergo rapid epimerisation.

5.1.5.2 Pyrrolidinyllithiums and piperidinyllithiums: 5- and 6-membered rings

Lithiated acylpiperidines were one of the first classes of α-aminoorganolithiums to be studied in detail, and it has long been known, for example, that organolithiums such as **135** are configurationally stable in the sense that the thermodynamically favoured configuration of the C–Li bond is equatorial, perpendicular to the π-orbitals of the amide system.[62-64] However, the early work on these compounds was unable to give any evidence about the *kinetic* stability of the organolithiums to configurational inversion. Gawley found that chiral oxazolines could be used to direct the stereoselective deprotonation of non-benzo fused piperidines (an advantage of the oxazolines over the formamidines), and that lithiation and alkylation of **133** gave a single stereoisomer.[65] At that time it was proposed that **135** was kinetically configurationally stable, because MNDO calculations suggested that both diastereoisomers of **135** were of equal energy. Later work[57] showed that the reactions of **135** with Bu_3SnCl are non-stereoselective, allowing both diastereoisomers of **134** to be formed. Each can be transmetallated with BuLi at –78 °C for 1 h and quenched with Bu_3SnCl or MeI (**136**). Since the outcome of the reaction of each organolithium is different, this provided final proof that **135a** and **135b** have some configurational stability over 1 h at –78 °C.

1. *s*-BuLi THF/ether 1 h, –23 °C 2. Bu_3SnCl 3. separate

133 → 134a (Bu_3Sn) —BuLi, THF, –78 °C→ 135a (Li) —MeI→ 136a (Me) from **134a**: 6:1

133 → 134b (Bu_3Sn) —BuLi, THF, –78 °C→ 135b (Li) —MeI→ 136b (Me) from **134b**: 1:1

In 1991, Beak showed that *N*-Boc pyrrolidines may be deprotonated enantioselectively with *s*-BuLi complexed with (–)-sparteine **12** (see section 5.4).[66] The product organolithiums are of proven configurational stablility, as shown by the following sequence.[67] The stannane **137** (96% ee) undergoes transmetallation with *s*-BuLi/TMEDA in Et_2O at –78 °C, and the organolithium **138** thus formed, after 30 min, reacts with Me_3SiCl to give the silylated pyrrolidine **139** in 38% yield and 74% ee. In the absence of TMEDA, the yield of the product is only 15%, but the ee is still 93%. TMEDA appears to decrease the configurational stability of the intermediate organolithium. (The preliminary report of this reaction[66] contained an error, perpetuated elsewhere,[68] suggesting that TMEDA increases the configurational stability of **138**.)

137 96% ee → s-BuLi, (TMEDA), Et_2O, –78 °C, 30 min → 138 → Me_3SiCl → 139

with TMEDA: 36% yield, 74% ee
without TMEDA: 15% yield, 93% ee

The related indolines **142** and **145** are similarly configurationally stable at –78 °C, and **142** is stable even over a period of 1.25 h at –40 °C.[69] **140** required the diamine **141** (di-*n*-butyl bispidine, an achiral analogue of (–)-sparteine) for successful transmetallation, but **142** is configurationally stable at –78 °C in the presence or the absence of TMEDA.

140 98% ee → BuLi, MTBE, –78 °C or –40 °C, 1.25 h, 141 → 142 → Me_2SO_4 → 143 94-96% ee

144 74% ee → BuLi, –78 °C, MTBE, TMEDA or Et_2O → 145 → $Ph_2C{=}O$ → 146 76% ee

The organolithium **148**, formed diastereoselectively from **147** on deprotonation with *s*-BuLi-TMEDA, is configurationally stable at –78 °C, but over a period of hours at –45 or –30 °C slowly epimerises.[70]

147 → s-BuLi, TMEDA, T °C → 148 → MeI, T °C → 149

>15:1 diastereoselectivity (T = –90 °C)
2:1 (T = –45 °C)
1:1 (T = –30 °C)

Meyers found that the formamide-stabilised organolithium **150**, while chemically less stable than the Boc-protected **138**, is also configurationally stable.[68] At –78 °C, in the absence of TMEDA, **150** was completely configurationally stable over 30 min, and showed good configurational stability even at –55 °C. Racemisation was slightly faster in DME, probably because of increased solvation of the Li cation.

88% ee → s-BuLi, Et_2O, –78 °C, 30 min → 150 → Me_2SO_4 →

after 0.5-30 min: 88% ee (30-50% yield)
after 10 min at –55 °C: 82% ee (36% yield)

In 1993, Gawley published work on the configurational stability of pyrrolidinyllithiums and piperidinyllithiums in the absence of a dipolar or chelating stabilising group (an amide or carbamate for example).[71,57] He made the stannane **134b** from **133** (see above), purified the major diastereoisomer and reduced it to the *N*-methylated piperidine **151**. Similarly, in the pyrrolidine series, he reduced Beak's Boc-protected stannane **137** to **154**. Both stannanes **151** and **154** were transmetallated with BuLi and organolithiums **152** and **155** were maintained for a period of time at a temperature indicated in the scheme below. They were carboxylated and their ees determined by reduction to the alcohol and formation of the MTPA esters **153** and **156**.

1. *s*-BuLi 2. Bu_3SnCl 3. purify; 1. AcCHO 2. $LiAlH_4$; BuLi, THF, t min, T °C; 1. $(MeO)_2CO$ 2. $LiAlH_4$ 3. R*COCl

133 → **134b** ($SnBu_3$) → **151** ($SnBu_3$, Me) → **152** (Li, Me) → **153** (OCOR*, Me)

THF	–80 °C, 45 or 75 min	99% de
THF	–60 °C, 45 min	95% de
THF	–60 °C, 75 min	93% de (10% yield)
THF/TMEDA	–80, –60 or –40°C, 45 min	99% de
THF/TMEDA	–20 °C, 15 min	28% de (38% yield)

DIBAL; BuLi, TMEDA, THF, t min, T °C; 1. CO_2 2. $LiAlH_4$ 3. R*COCl

137 94% ee ($SnBu_3$, Boc) → **154** ($SnBu_3$, Me) → **155** (Li, Me) → **156** (OCOR*, Me)

–80 , –60 or –40 °C, 45 min	94% de
–20 °C, 15 min	80% de (34% yield)

The pyrrolidinyllithium **155** requires the presence of TMEDA for acceptable yields, and under these conditions (THF/TMEDA) is fully configurationally stable up to almost –20 °C, where decomposition is severe. The piperidinyllithium **152** is similarly stable in THF/TMEDA at least as warm as –40 °C, but racemises and decomposes at –20 °C. Without TMEDA, it is both chemically and configurationally less stable: racemisation and decomposition occur over a period of hours at –60 °C. The two organolithiums are remarkable in being unstabilised by further chelation within the molecule. Their configurational stability is *increased* by the presence of TMEDA (contrast the effect of TMEDA on acyclic organolithiums and on the lithiated Boc-pyrrolidine **138**). It is possible that the higher energy lone-pair of the amine nitrogen disfavours racemisation to a greater extent than the carbamate-type nitrogen lone-pairs of **138** and **150**, and that bridging of the Li between C and N is greater in the absence of intramolecular stabilisation (see section 5.1.5.6).

To assess the effect of intramolecular chelation in this class of organolithium, Gawley also made **157** and treated it under similar conditions.[57] In THF alone, the MEM-protected **158** has greater chemical stability than **155**, and is configurationally stable up to about –60 °C. Like the lithiated Boc-pyrrolidine **138** (but unlike the lithiated *N*-methyl pyrrolidines **155**) TMEDA tends to *decrease* its configurational stability and a direct comparison between MEM protected **158** and **155** in the presence of TMEDA shows that the MEM group also lessens (slightly) the configurational stability of the α-amino organolithium. Nonetheless, it is considerably more configurationally stable than a directly analogous acyclic organolithium.

1. catecholBBr 2. $MeOCH_2COCl$ 3. DIBAL; BuLi, THF, t min, T °C; 1. CO_2 2. $LiAlH_4$ 3. R*COCl

137 94% ee → **157** → **158** → **159**

Solvent	Conditions	Result
THF	–78°C, 10 or 75 min	92% de
THF	–60 °C, 45 min	86% de
THF	–40 °C, 45 min	69% de
THF/TMEDA	–78°C, 75 min or –60 °C, 45 min	92% de
THF/TMEDA	–40 °C, 34 min	84% de (34% yield)
DME	–78 °C, 10 min	92% de
DME	–60 °C, 45 min	0% de

The ultimate in configurational stability in any organolithium thus far discovered was found by Coldham during the cyclisation reaction of the organolithium **161**.[72] Despite the high temperature (20 °C) required for the tin-lithium exchange step of **160** in hexane–ether, the cyclisation to pseudoheliotridane **162** is entirely stereospecific, indicating that **161** is configurationally stable at 20 °C on the timescale of the reaction.

BuLi x 2, 10:1 hexane:Et_2O –78 - 20 °C, 1 h; then MeOH

160 94% ee → **161** 94% ee → pseudoheliotridane **162** 87% 94% ee

5.1.5.3 Lithiated formamidines

1. BuLi, –80 °C 2. MeI; 1. N_2H_4 2. Me_2N–CH=N–R; *s*-BuLi, –78 °C, THF; fast at –78 °C; E^+, –100 °C

163 → **164** → **165** → **166** ⇌ *ent*-**166** → product, <6% ee

R = H or OMe
E = D or Bn

The potential for the use of the lithiated derivatives of formamidines **163** in synthesis meant they received a considerable amount of study during the 1980's.[73-75] However, only in 1991 was the question of their configurational stability finally resolved. It was already known that **166** and similar compounds had pyramidalised lithium-bearing carbon atoms[75] and that they were formed by a highly diastereoselective deprotonation.[73,74] However, the presence of the chiral formamidine (or oxazoline) group had always clouded the issue of configurational stability. Meyers therefore made the chiral amidine **164** using the formamidine technique. He then re-protected it with an achiral formamidine group to give **165**. Treatment with *s*-BuLi at –100 °C gave a red organolithium **166** which was quenched at that temperature with benzyl chloride or $(CD_3)_2SO$. In both cases the products has ees of less that 6%, indicating complete lack of configurational stability in the organolithium **166** in THF at –78 °C.[76]

5.1.5.4 Acyclic α-amino organolithiums

Chong made the enantiomerically enriched α-amino organolithium **170** (94% ee) by enantioselective reduction of the acylstannane **167** followed by Mitsunobu inversion of the stannylcarbinol **168**. Transmetallation of **169** with BuLi in THF at –95 °C gives an organolithium **170** of almost complete configurational stability.[77] For example, after 10 min, quenching with CO_2 gave the acid **171** in 94% ee; after 180 min the ee was still 92%. At –78 °C configurational stability was just beginning to fade – after 10 min **171** was recovered in 92% ee; after 180 min the ee was 80%. Replacing THF with DME increased the rate of racemisation, but adding HMPA to the THF causes complete racemisation in less than 2 h. After 120 min at –55 °C, only 24% ee was obtained. The effect of HMPA could again be polar, or alternatively it may be disrupting the intramolecular coordination which assists configurational stability. Very similar results were obtained with isopropyl-substituted **172**.

167 →[(*R*)-BINAL-H] **168** →[1. phthalimide, PPh_3, DEAD; 2. N_2H_4; 3. Boc_2O; 4. NaH, MeI] **169** 94% ee →[BuLi, THF, T °C, t min] **170** →[1. CO_2; 2. H^+] **171**

172

Conditions	**171**
–95 °C, 10 min	97% yield; 94% ee
–95 °C, 180 min	75% yield; 92% ee
–78 °C, 10 min	95% yield; 92% ee
–78 °C, 180 min	76% yield; 80% ee
–78 °C, 180 min (DME)	54% yield; 68% ee
–78 °C, 180 min + HMPA	50% yield; 0% ee
–55 °C, 120 min	76% yield; 24% ee

Chong also studied α-amino organolithiums more directly comparable to Still's acyclic α-alkoxy organolithiums. Enantioselective reduction of an acylstannane led to stannane **173** in 94% ee. Transmetallation with BuLi gave an organolithium **174** which was maintained at –95 °C or –78 °C for a period of time, before being quenched with CO_2 and reduced to the alcohol. As with the Boc-protected organolithiums **170** and **172**, configurational stability was almost complete at –95 °C, but racemisation took place with a half-life of somewhat more than 1 h at –78 °C. Racemisation was much faster in DME than in THF.

173 94% ee →[BuLi, THF, T °C, t min] **174** →[1. CO_2; 2. $LiAlH_4$] **175**

Conditions	**175**
–95 °C, 10 min	94% ee
–95 °C, 180 min	90% ee
–78 °C, 20 min	83% ee

Similar conclusions with regard to configurational stability of organolithiums α to carbamate-type nitrogen atoms were arrived at by Pearson,[78,79] who made and separated the diastereoisomeric stannanes **176a** and **176b**. Both transmetallate to distinguishable organolithiums **177a** and **177b** with BuLi in THF at –78 °C, and can be recaptured stereospecifically with Bu_3SnCl. However, the configurational stability of **177b** at –78 °C is marginal, at best. After 5 min, a 2.4:1 ratio of diastereoisomers is recovered; this becomes 1.2:1 after 15 min and has inverted to 1:2.4 after 30 min. Clearly, the other diastereoisomer **177a** is the more stable. Indeed, this diastereoisomer showed no signs of epimerisation to **177b**, and gave essentially diastereoisomerically pure products after 15 or 45 min at –78 °C. The effect of TMEDA on the configurational stability of **177b** was also studied: in the presence of TMEDA, **177b** has epimerised completely (it gave only **176a** on quench) to **177a**. As with other acyclic α-amino organolithiums, TMEDA decreases the conformational stability.

SnBu3 · BuLi, THF, –78 °C · Li · Bu_3SnCl → **176a** from **176a**: sole product
176a · **177a**
(minutes - hours at –78 °C)
SnBu3 · BuLi, THF, –78 °C · Li · Bu_3SnCl → **176b** from **176b**: 5 min: 2.4:1 **176b:176a**
15 min: 1.2:1 **176b:176a**
30 min: 1:2.4 **176b:176a**
45 min: 1:6.9 **176b:176a**
176b · **177b**
with TMEDA, 5 min: **176a** only

5.1.5.5 Benzylic and allylic α-aminoorganolithiums

Despite the often capricious stereospecificity of their reactions with electrophiles (see section 6.1.4), benzylic α-amino organolithiums can display configurational stability on a macroscopic timescale at –78 °C. In one of a series of experiments, Beak showed that transmetallation of the stannane **179** with BuLi in the presence of either TMEDA or (–)-sparteine **12** generated the organolithium **180**.[80] After 30 min – 10 h in the presence of sparteine – ee was conserved. With TMEDA, configurational stability was lower, since almost racemic material was recovered after 10 h at –78 °C. The effect of sparteine is independent of its chirality (ie this is a kinetic and not a thermodynamic effect) because the enantiomeric organolithium does not racemise in the presence of sparteine. A related compound formed by deprotonation using a chiral lithium amide was much less configurationally stable.[81]

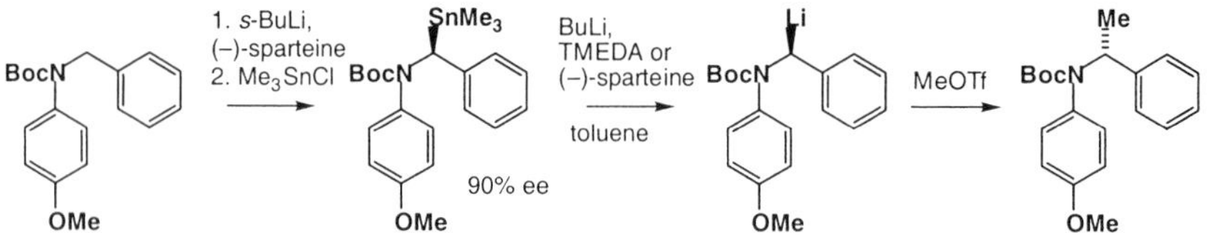

1. *s*-BuLi, (–)-sparteine 2. Me_3SnCl → 179 (90% ee); BuLi, TMEDA or (–)-sparteine, toluene → 180; MeOTf → 181

178 179 180 181

with TMEDA: 0.5 h: 40% ee
10 h: 8% ee
with (–)-sparteine: 0.5 h: 90% ee
10 h: 90% ee

Tertiary analogues **183** of **180** are also configurationally stable at –78 °C.[80] **182** can be deprotonated and kept for 8 h at –78 °C with little loss of enantiomeric purity. It is consequently possible to make either enantiomer of the protected amine **184** from the appropriate amine **182**, of which both enantiomers are available by virtue of the invertive substitution reactions characteristic of tin electrophiles (see section 6.1.4).[82,83] Tertiary *O*-substituted benzyllithiums also have high configurational stability (see above).

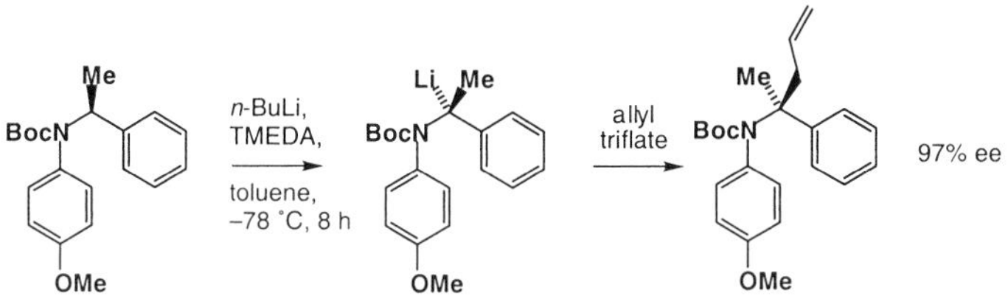

n-BuLi, TMEDA, toluene, –78 °C, 8 h → 183; allyl triflate → 184 (97% ee)

182 183 184

Hamada showed, in 1998, that benzylic α-aminoorganolithium **186** could maintain its configuration for long enough to undergo an acyl transfer rearrangement giving **187** in 94% ee from enantiomerically pure **185**.[84] There is no data on the rate of this reaction, however, so it is not clear what the implications are for the configurational stability of benzylic organolithiums **186**.

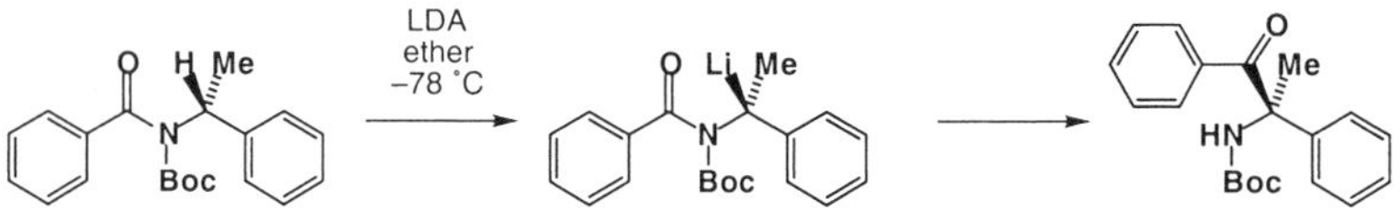

LDA, ether, –78 °C

185 186 **187** 94% ee

The configurational stability of allylic α-amino organolithiums depends strongly on the nature of the inter- and intramolecular coordination available to the lithium. The organolithium **189**, for example, may also be configurationally stable, but only in the presence of sparteine **12**.[85] This is evident by comparing the two reactions below, both of which go through an allyl organolithium intermediate complexed with sparteine. If the intermediate were not configurationally stable, the same enantiomer of the product **190** would be produced in each case. Irrespective of the details of the reaction, the fact that enantiomeric products **190** are produced proves configurational stability in both diastereoisomeric **189**-(–)-sparteine complexes. The organolithium **189** is not stable, however, in the presence of TMEDA: transmetallation / quench of **191** using *n*-BuLi/TMEDA then allyl bromide returns racemic material **190**.

By NMR, Hoffmann determined the configurational stability of a simple lithiated benzylamine **192** to be 9.0 kcal mol^{-1}.[2] However, an attempted Hoffmann test on **193** was inconclusive due to the low level of kinetic resolution observed in its reaction with chiral aldehydes.[6]

5.1.5.6 Crystallographic and theoretical data

Boche has published some crystallographic and theoretical data which shed light on the particularly high configurational stability of oxygen- and nitrogen-substituted organolithiums, which are worth discussing at this stage.[86] Considering that the inversion of an organolithium will require the formation of a pyramidal carbanion–Li$^+$ ion pair, followed by inversion of the carbanion through a planar transition state, the relative energies of the most stable pyramidal and planar conformations of a series of heterosubstituted carbanions were calculated. The energy differences are presented below.

X =	ΔE/kJ mol^{-1}
H	12.6
SiH_3	3.3
SH	4.6
OH	17.6
NH_2	21.7
F	56.9

Silicon and sulfur substituents tended to lower the energy difference relative to the unsubstituted anion, while oxygen and, even more so, nitrogen and fluoro substituents raised the energy difference. These calculated figures follow closely the known pattern of poor configurational stability in *S*-substituted organolithiums and the high configurational stability in the *O*- and *N*-substituted anions, and suggest that the stability of the O- and N-substituted anions results from lone-pair repulsion in the planar form.

Crystallography also shows up some features which are of value in interpreting the configurational stability of α-alkoxy organolithium[86] compounds. α-Alkoxy organolithiums whose crystal structures have been determined fall into three classes: (1) those in which the lithium atom bridges between the anion and the oxygen; (2) those in which the lithium atom shows no bonding to the α-oxygen, but instead is stabilised by chelation; (3) those in which no C–Li can be seen: the lithium is bonded solely to the oxygen and the anion is stabilised by conjugation. Schematic illustrations of three of these structures **194-196** are shown below. The three structural classes also find corresponding structures in calculated local minima for the structure of $LiCH_2OH$, shown below with calculated bond-lengths.

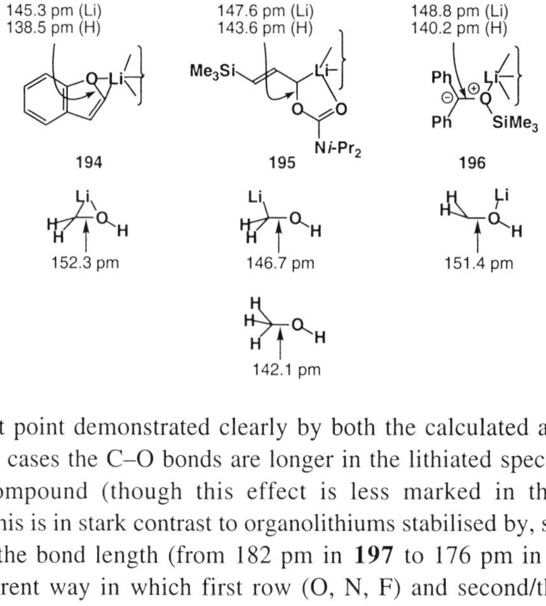

145.3 pm (Li)
138.5 pm (H)
194
152.3 pm

147.6 pm (Li)
143.6 pm (H)
Me_3Si ... $Ni\text{-}Pr_2$
195
146.7 pm

148.8 pm (Li)
140.2 pm (H)
Ph, Ph, $SiMe_3$
196
151.4 pm

142.1 pm

A further important point demonstrated clearly by both the calculated and crystallographic results is that in all cases the C–O bonds are longer in the lithiated species than in the non-lithiated parent compound (though this effect is less marked in the amino-stabilised organolithiums). This is in stark contrast to organolithiums stabilised by, say, sulfur, in which lithiation shortens the bond length (from 182 pm in **197** to 176 pm in **198**). This feature illustrates the different way in which first row (O, N, F) and second/third row (S, P, Se) elements stabilise the organolithiums: the earlier elements do so largely by inductive electron-withdrawal, while the later ones do so by σ-conjugation with the X–C σ* orbitals.

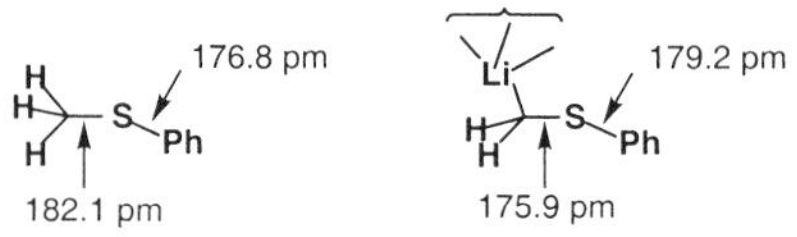

176.8 pm
182.1 pm
197

179.2 pm
175.9 pm
198

α-Amino organolithiums[87] show some similar features: they similarly divide into structures with bridging Li atoms and structures in which there is no Li–N bond. Two examples are shown in **199** and **200**. Both of these have strongly pyramidalised lithium-bearing benzylic carbon atoms. NMR studies show a further interesting aspect to the structure of **201** in THF: at temperatures above 0 °C the organolithium has an η^3 structure **202** with a C–Li bond to the *ipso* and *ortho* carbons of the benzene ring.[88] At lower temperatures the organolithium reverts to an η^1 structure **201**.

199 200 201 202

5.1.6 Organolithiums α to halogens

Köbrich[89] outlined, in 1969, the fact that the *exo* α-chlorocyclopropyllithium **203** was configurationally stable at –110 °C in THF, but that isomerisation takes place at –80 °C.[90]

BuLi –110 °C; CO_2; 203; slow isomerisation at –80 °C

However, it was Seyferth[91,92] who proved unequivocally that α-bromocyclopropyllithiums are configurationally stable below –78 °C in THF or THF/ether by making both stereoisomers of the compound **206**. Halogen-lithium exchange of **204** with an excess of *n*-BuLi followed by an electrophilic quench with any of a wide range of electrophiles consistently gives a single, *endo* isomer of the products, presumably via the *endo* cyclopropyllithium **205a**. However, in the presence of excess dibromide **204** the *endo* organolithium **205a** equilibrates to give some *exo* isomer **205b**, presumably via a series of successive bromine-lithium exchanges.

204; BuLi or MeLi –107––93 °C, THF / Et_2O; 205a; Me_3SiCl; 206a; 76% with BuLi in excess, 61% with **204** in excess

excess 204; 205b; Me_3SiCl; 206b; 16% with **204** in excess

This method allowed the synthesis of both diastereoisomeric stannanes **207a** and **207b**, which underwent tin-lithium exchange and electrophilic quench with almost complete stereospecificity at –103 °C, indicating complete configurational stability at this temperature.[92]

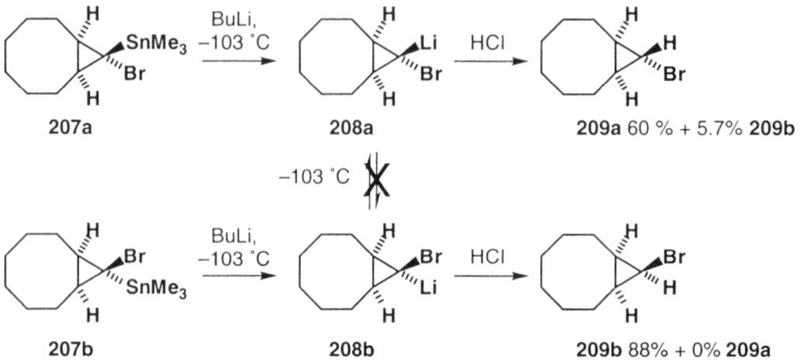

Similar results were obtained in the cyclohexene-fused cyclopropane series, where Warner showed the epimerisation of **210b** was slow, but detectable above –78 °C.[93].

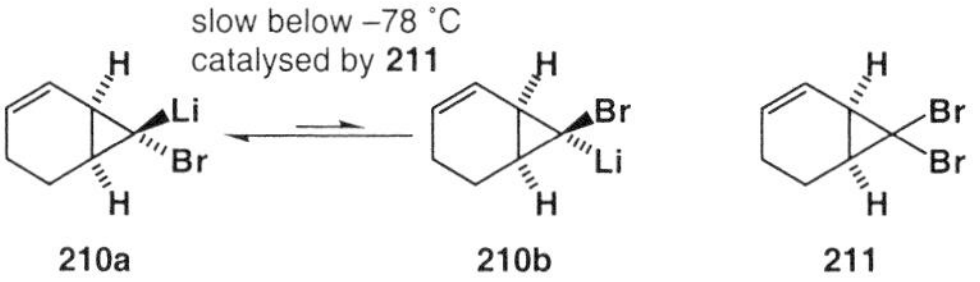

In all of these cases, the α-bromocyclopropyllithium with the lithium atom *endo* is more stable than that with the lithium *exo*, and epimerisation in the presence of excess dibromide was exploited in the synthesis of a number of cyclopropane-containing targets.[94-96] In other cases, the endo isomer is configurationally stable for thermodynamic reasons, being further stabilised by intramolecular coordination.[97,98]

Given the established configurational stability of cyclopropyllithiums (see section 5.1.3) the first report, by Hoffmann in 1991, that acyclic α-bromoorganolithiums had configurational stability, was highly significant. Hoffmann made both diastereoisomers of the organolithium **213** at –110 °C in the Trapp solvent mixture[99] and found that they had measurable configurational stability over 10 min at this temperature. Pure **212a** led to a 98.8:1.2 mixture of diastereoisomeric products after quenching with acetone and formation of the epoxide, while a 15:85 mixture of **212a** and **212b** gave a 25:75 mixture of the products **214a** and **214b**.[100] Similar results were obtained with a silyl ether analogue of **212**.[101] The previous observation that α-bromoorganolithiums react with similar selectivity irrespective of the electrophile also pointed towards this result (rapidly equilibrating organolithiums would be expected to react with electrophile-dependent stereoselectivity).[102] These reactions show that tin-lithium exchange can outpace bromine-lithium exchange.

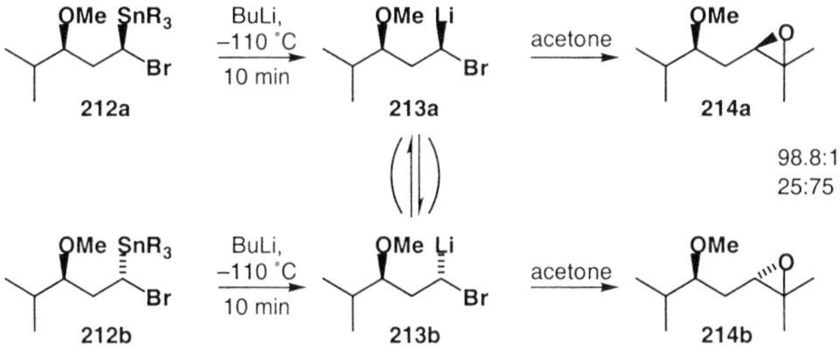

98.8:1.2 **214a**:**214b** from **212a**
25:75 **214a**:**214b** from 15:85 **212a**:**212b**

The Hoffmann test, as expected, similarly showed **215** to be configurationally stable on the timescale of its addition to an aldehyde at –110 °C.[103]

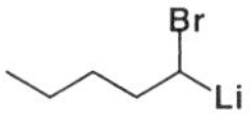

215

5.1.7 Organolithiums α to sulfur

5.1.7.1 Lithiated sulfides

Simple organolithiums α to sulfide functional groups are typically configurationally stable on the timescale of nucleophilic additions, so if the addition takes place intramolecularly or if the electrophile is present *in situ* during the formation of the organolithium then stereospecific reactions may occur. On the macroscopic timescale, however, organolithiums α to sulfur usually racemise or epimerise rapidly. Beak showed that the tertiary α-thioorganolithiums **217** racemise to give the same ratio of products after 1 h at –78 °C.[104]

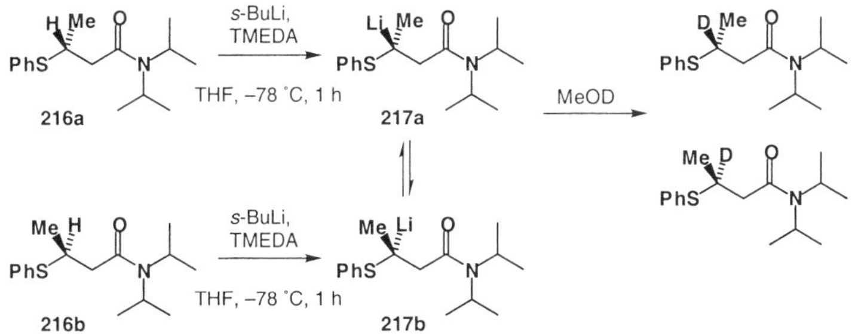

87:13 from **216a**; 89% D
84:16 from **216b**; 98% D

McDougal showed that **219** epimerises rapidly on a macroscopic timescale, and that both diastereoisomers of **218** give the same 98:2 ratio of diastereoisomers after transmetallation and silylation.[105]

BuLi, THF −78 °C
218a
219a
Me$_3$SiCl
98:2 ratio from either **218a** or **218b**
218b
219b

A 95:5 diastereoisomeric ratio of **222a**:**222b** is similarly obtained on reduction and silylation of the dithioacetal **220**. However, this ratio changes to 61:39 if the quench is *in situ*, indicating that under these conditions thermodynamic equilibrium between the two diastereoisomeric organolithiums **221** has not been reached on the timescale of the reactions.

Li/Np, THF −78 °C, or Li/Np, Me$_3$SiCl −78 °C
220
221a
221b
Me$_3$SiCl
222a
222b
95:5 (2 stage)
61:39 (*in situ*)

Reich defined limits to the configurational stability of the α-methylthiocyclohexyllithium **223** by finding that it could be trapped stereospecifically by Me_3SiCl (i.e. it epimerised slower than it reacted with Me_3SiCl) but that epimerisation was complete within 1 minute at −78 °C.[106] Similarly, both diastereoisomers of **224** cyclise (by attack on the ketone) faster than they epimerise.[107] The *p*-tolylthio analogue epimerised even faster than **224**, and could not be trapped stereospecifically.[106]

R = Me
R = *p*-tol
223
224

Brückner[108] made both diastereoisomers of the stannane **225** and found that the diastereoisomeric organolithiums **226** derived from them underwent stereospecific [2,3]-sigmatropic rearrangement to different thiols **227a** and **227b** over a period of 30 min at −78 °C in THF. Full stereospecificity was not maintained, however, in the slower benzylic [2,3]-sigmatropic rearrangement of **228**, indicating that the α-alkylthio organolithium is only just configurationally stable over periods of a few minutes at −78 °C.

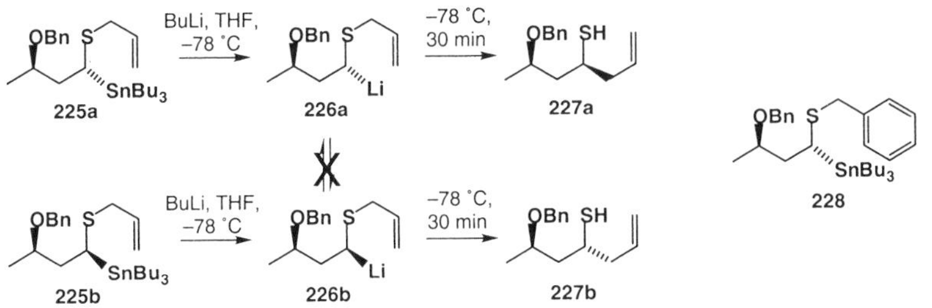

Hoffmann[103] confirmed the configurational stability of α-alkylthioorganolithiums on the reaction timescale by the Hoffmann test, showing that the simple lithiated sulfide **229** is configurationally stable on the timescale of the addition to the aldehyde **6**. The reaction of **229** at –120 °C in methyl-THF with racemic **6** gave a 70:30 ratio of diastereoisomers **230a** and **230b**; with (*S*)-**6** a 52:48 ratio of diastereoisomers was formed.

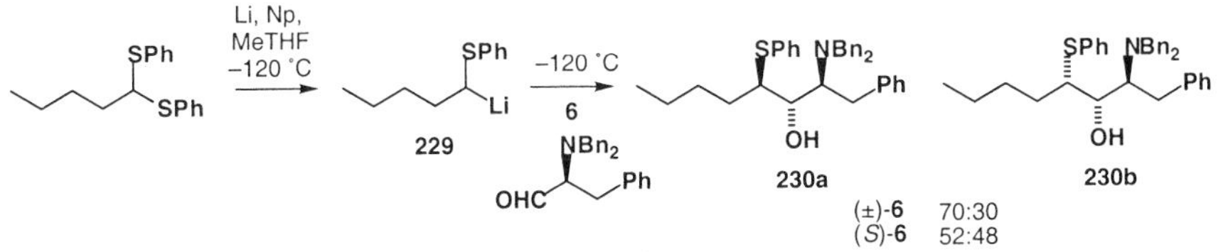

Hoffmann found that the bulky duryl (2,3,5,6-tetramethylphenyl) group, by slowing the rate of C–S rotation (see below), also slowed considerably the rate of racemisation of the lithiated sulfides. For example, while the lithiated phenylsulfide **231** has a barrier to racemisation of 11.3 kcal mol^{-1} at 263K, the durylsulfide **232** shows a barrier greater than 13.9 kcal mol-1.[109-111]

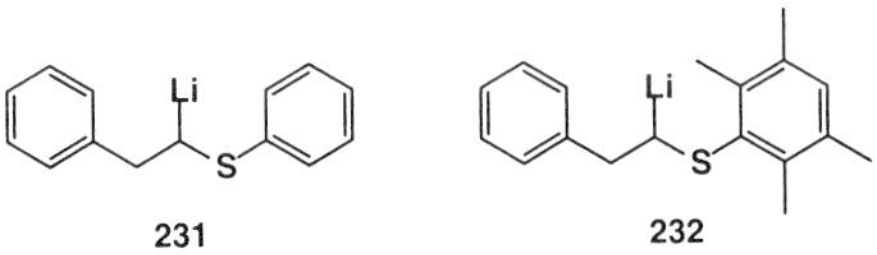

The limited configurational stability of α-alkylthio organolithiums does not extend to those with benzylic lithium-bearing carbon atoms.[112] A Hoffmann test reaction of **233** with **6**, for example, gives a 40:60 ratio of stereoisomers **234** whether enantiomerically pure or racemic **6** is used.[6] Dynamic NMR experiments[2] quantified the barrier to racemisation in **233** as 9.95 kcal mol^{-1} at 213 K, the temperature at which the diastereotopic CH_2 group coalesces.

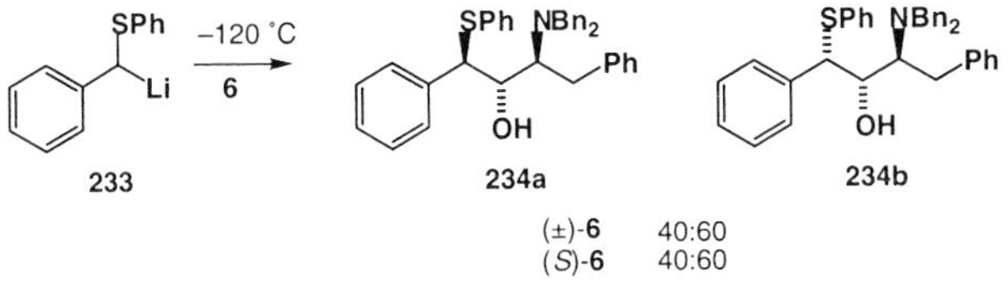

Nonetheless, a "poor man's Hoffmann test" (see section 5.1.1) on **235** indicates that it does possess some degree of configurational stability.[112]

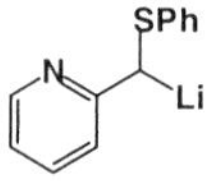

SPh
N
Li
235

5.1.7.2 Lithiated thiocarbamates

One class of α-sulfur-substituted organolithiums are in a league of their own with regard to configurational stability – thiocarbamates. In common with other organolithiums, the most configurationally stable are those which have tertiary lithium-bearing centres. For example, the compound **236** can be kept in Et_2O at –78 °C for several hours without racemisation, and the benzylic organolithium **237** maintains its configuration even over 10 min at 0 °C.[113] The secondary analogue **238** is configurationally unstable at –78 °C.

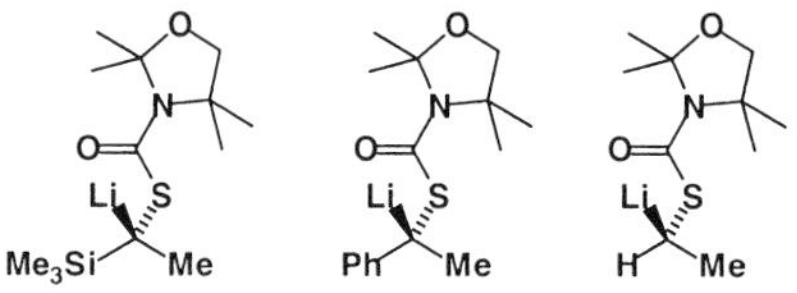

Me_3Si Me **236** Ph Me **237** H Me **238**

The dianion **239** is also configurationally stable over several hours at –78 °C in THF; it reacts non-regioselectively, and the ees of the two regioisomers **240** and **241**, though different, do not change on prologed storage of the organolithium intermediate, indicating that loss of ee is due to non-stereospecific quenching rather than slow racemisation.[114,115]

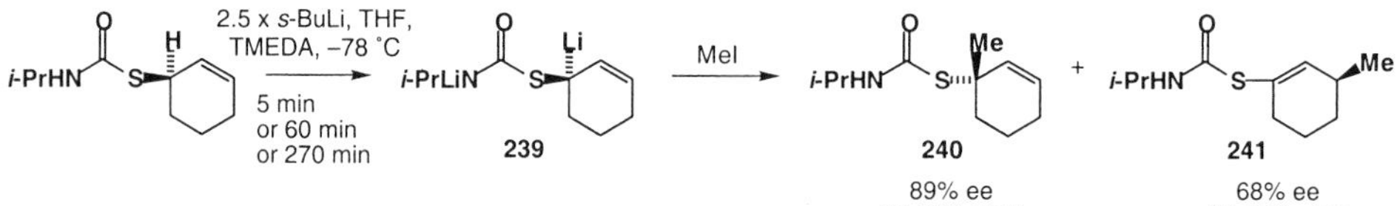

2.5 x *s*-BuLi, THF, TMEDA, –78 °C; 5 min or 60 min or 270 min → **239**; MeI → **240** (89% ee, 17-21% yield) + **241** (68% ee, 21-44% yield)

5.1.7.3 Lithiated sulfones

In contrast with lithiated sulfides, lithiated sulfones – even those with benzylic organolithium centres – are configurationally stable on at least a microscopic timescale. Lithiated benzyl phenylsulfone **242** reacted with (±)-**6** to give a 38:62 ratio of **243a**:**243b** but with (*S*)-**6** to give a 45:55 ratio of **243a**:**243b**.[6] For convenience, the lithiated sulfones have been shown with a Li–C bond, one which may not be present in the true solution structure of the compounds.[116]

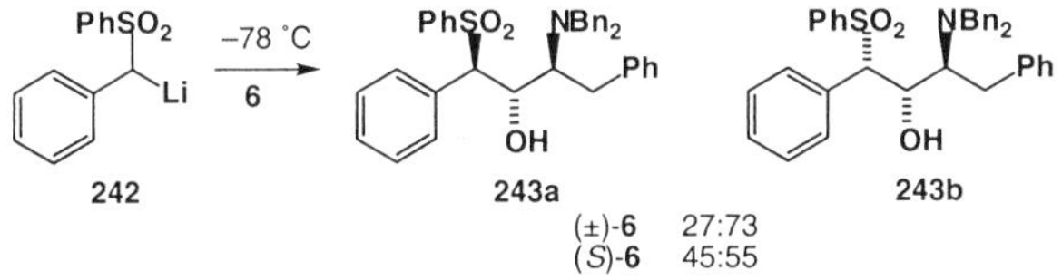

Gais studied the rate of inversion of the lithiated sulfones **244** and **245**, and found that they were remarkably similar: about 40 kJ mol^{-1} at 215 K.[117] Given the difference in structure, these results suggest strongly that, for sulfones, formation of an ion pair is not the rate-limiting step of the racemisation. The trifluoromethylsulfone **246** has an even higher barrier to racemisation: 66 kJ mol^{-1} at 333 K.[117]

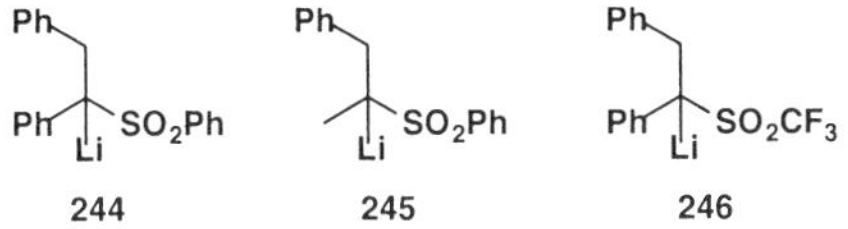

Lithiated sulfones are configurationally stable even on the macroscopic timescale. The enantiomerically pure sulfones **247** and **251** may be deprotonated and react with electrophiles after 3 min at –105 °C to return highly enantiomerically enriched products.[118] Polarimetry indicates a barrier to racemisation of 71 mol^{-1} at 239 K for **248** (a half-life of 30 days at –78 °C) and of 53.5 kJ mol^{-1} at 298 K for **252** (a half-life of 3 h at –105 °C).

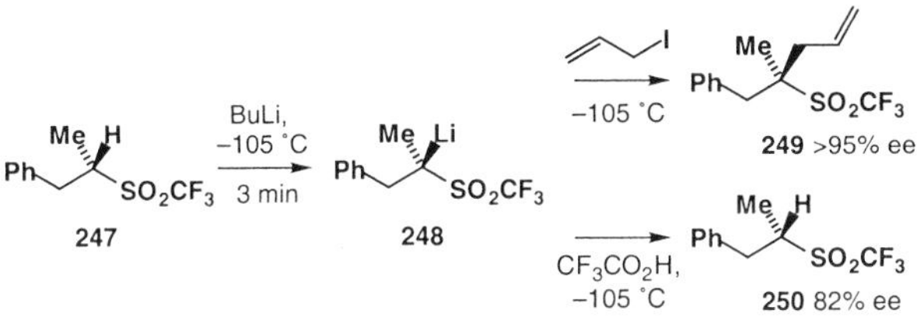

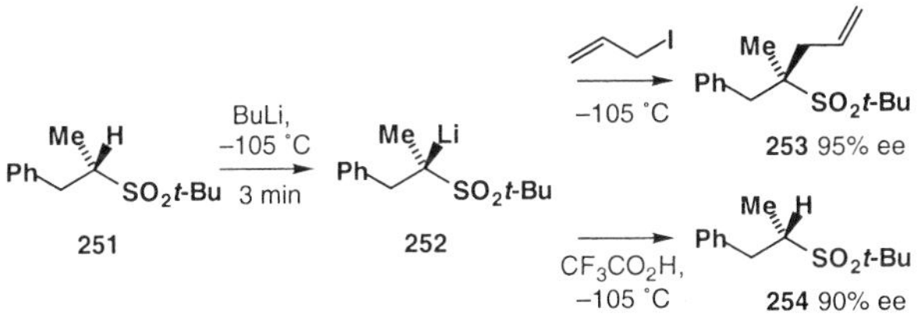

Lithiated sulfoxides also have exceptionally high configurational stability at their anionic centre. For example, the diastereoisomers of **255** are configurationally stable over a period of one hour even at room temperature.[119]

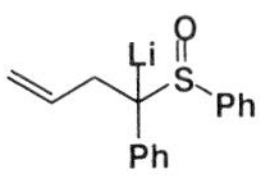

Lithiated sulfones[120,121] and sulfoxides are arguably not organolithiums since they probably exist as ion pairs with no covalent C–Li bond, benzylic sulfones being planar anions and non-benzylic anions being tetrahedral.[116] But the key lesson to be learnt from these results is that racemisation of a sulfur-substituted organolithium may be strongly dependent on the size and nature of the sulfur's other substituent because the rate of racemisation may be limited not by inversion of the anion but by a slow rotation about the C–S bond (below). Bulky S-substituents slow this rotation, as do electron-withdrawing S-substituents, since they stabilise the conformation placing the S–R σ^* orbital antiperiplanar to the C–Li bond or the carbanionic sp^3 orbital.[122]

5.1.7.4 Mechanism of racemisation

Reich obtained experimental data on the mechanism of racemisation of sulfur-substituted organolithiums by studying the coalescence of the diastereotopic methyl groups of **256** by dynamic NMR.[123] He calculated a barrier to racemisation $\Delta G^{\ddagger}$ of 33 kJ mol^{-1} for **256** in THF at –100 °C, which was lowered to 32 kcal mol^{-1} in the presence of 1 equiv. HMPA. Adding three equivalents of HMPA however slowed the racemisation to one twentieth of the rate in THF alone: three equivalents of HMPA is also sufficient to separate the organolithium fully into a solvent-separated ion pair. It cannot be, therefore, that the organolithium racemises via a solvent-separated ion pair, but it is quite possible that it racemises via a contact ion pair, to which the solvent-separated ion pair has to revert before it can invert. The slower rate of racemisation of the ion pair is fully in accordance with a mechanism in which the rate determining step is C-S bond rotation, since greater $n \rightarrow \sigma^*$ conjugation ("negative hyperconjugation") is expected in the ion pair than in the organolithium.

Li
PhS Si Ph

256

This mechanism for racemisation, which consists of four steps – formation of a contact ion pair, inversion of the anion, rotation of the C–S bond to restore $n \rightarrow \sigma^*$ conjugation, and re-formation of the C–Li bond – is supported by theoretical calculations,[3] and explains the higher barriers to racemisation of the lithiated duryl sulfides such as **232** (by steric hindrance) and the trifluoromethylsulfones such as **248** (by increased $n \rightarrow \sigma^*$ conjugation).

form ion pair → invert anion → (slow, expecially if Ar is bulky) rotate S–C → collapse ion pair

X = O or lone pairs (sulfones, sulfoxides, sulfides)

5.1.8 Organolithiums α to selenium

Trapping with electrophiles shows that **257** epimerises within a minute at –78 °C in THF.[124]

Li
SeR
257

Reich[106] found that the similar organolithium **259a** was configurationally stable on the timescale of its reaction with electrophiles, but not on a macroscopic timescale. Selenium-lithium exchange of **258** gave a pair of organolithiums **259a** and **259b** which gave the same 92:8 ratio of axial:equatorial products with a wide range of electrophiles, suggesting that it does not epimerise faster than it can react with the electrophiles (or the ratio would vary). The ratio is, however, not the kinetic ratio of organolithiums **259** formed by the reaction, because selenium-lithium exchange *in the presence* of Me_3SiCl gives a kinetic 98:2 ratio of axial:equatorial isomers.

SePh
Ph
SePh
Ph
258
n-BuLi, –78 °C
Li
Ph
SePh
Ph
259a
Me_3SiCl (in situ)
$SiMe_3$
Ph
SeR
Ph
261
98:2 axial:equatorial Si
E+
E
Ph
SeR
Ph
262
92:8 axial:equatorial E
SePh
Ph
Li
Ph
259b

The epimerisation of acyclic selenium-substituted organolithiums is even faster at –78 °C[125] or –100 °C, and **264a** only becomes configurationally stable over periods of an hour or so at –120 °C or less.[126] For example, the two tributylstannanes **263a** and **263b**, after transmetallation at–78 °C for 2 h and methyl iodide quench, give the same 3:1 ratio of diastereoisomeric products.

–78 °C, THF R = Bu
–120 °C, MeTHF R = Me
OMe SnR_3
SePh
263a
RLi
OMe Li
SePh
264a
MeI
OMe Me
SePh
265a
92:8 from **263a**
65:35 from either **263a** or **263b**
OMe SnR_3
SePh
263b
RLi
OMe Li
SePh
264b
MeI
OMe Me
SePh
265b
16:84 from **263b**

The trimethylstannanes, on the other hand (which can be transmetallated at –120 °C in methyl THF) each give a ratio of products reflecting the stereochemistry of the starting material. The stereospecificity of the reactions indicates considerable configurational stability in these α-seleno organolithiums at –120 °C. The same ratios are obtained after 3 h or 6 h, indicating that the epimerisation takes place during the exothermic addition of the methyllithium.

Hoffmann studied the configurational stability of the simple organolithium **267** using the Hoffmann test.[103] The organolithium **267** was made by Se-Li exchange of the selenoacetal **266**, and reacted at –80 °C in a Trapp solvent mixture[99] with the chiral but racemic aldehyde **6**. Only two of the four possible diastereoisomeric addition products were formed, **268a** and **268b**, which differ in configuration only at the Se-bearing centre, in a ratio of 67.4:32.6. The addition was repeated with an excess of enantiomerically pure **6**, and the product ratio changed to 56.1:43.9, indicating that **267** is configurationally stable on the timescale of its addition to the aldehyde. At lower temperatures, the ratio approached 50:50 (52:48 in ether at –105 °C) and at higher temperatures approached the 67:33 ratio obtained with the enantiomerically pure aldehyde, suggesting that the rate of racemisation begins to approach that of the addition as the temperature rises – not unreasonable given the less negative activation entropy expected for a unimolecular racemisation. In contrast, a Hoffmann test on the lithiated benzylselenide **270** indicates that it is configurationally unstable on the timescale of its addition to the aldehyde **6**.[6]

266 → (BuLi, Et_2O/THF, –105 °C) → 267 → (–80 °C, 6) → 268a + 268b

(±)-**6** 67.4:32.6
(*S*)-**6** 56.1:43.9

269 → (BuLi, Et_2O/THF, –78 °C) → 270 → (–78 °C, 6) → 271a + 271b

(±)-**6** 34:66
(*S*)-**6** 34:66

Dynamic NMR studies[127] quantified the barrier to racemisation of the lithiated selenides 272-277 on lithiation with BuLi in C_6D_6/d_8-THF. These barriers are slightly (**6 kJ** mol^{-1}) higher than those in the analogous sulfides. As with the sulfides, increased steric hindrance to C–Se bond rotation slowed the barrier to racemisation. A similar effect was evident in tellurium- but not silicon-substituted organolithiums.[109] Lithiated benzylselenides **278** were also studied by NMR,[2] and their (lower) barriers to racemisation are shown below. The sulfur- and selenium-substituted benzylic organolithiums had very similar barriers to racemisation.

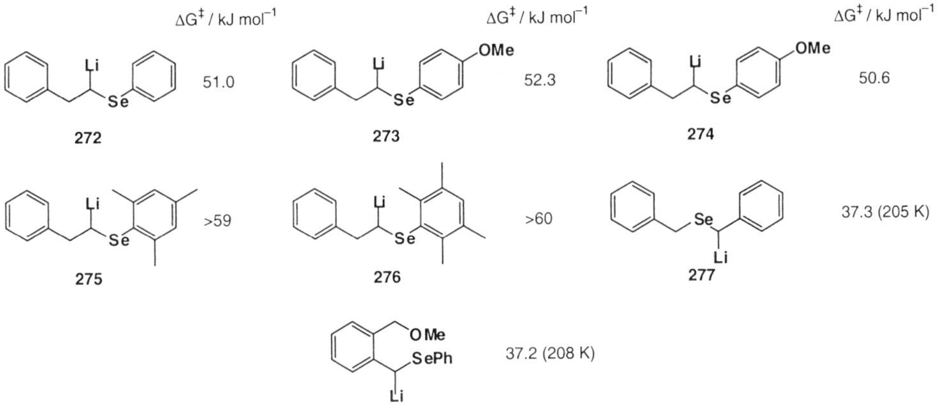

ΔG‡ / kJ mol⁻¹

272 51.0
273 52.3
274 50.6
275 >59
276 >60
277 37.3 (205 K)
278 37.2 (208 K)

5.1.9 Organolithiums α to phosphorus

No organolithium α to phosphorus has been shown unequivocally to be configurationally stable. The phosphonamide **279** is configurationally unstable on a macroscopic timescale,[128] the phosphine oxide **280** gives racemic products on lithiation even in the presence of an internal quench,[129] and in a Hoffmann test the phosphine oxide **281** gave the same ratio of diastereoisomers with either racemic or enantiomerically pure **6**.[129]

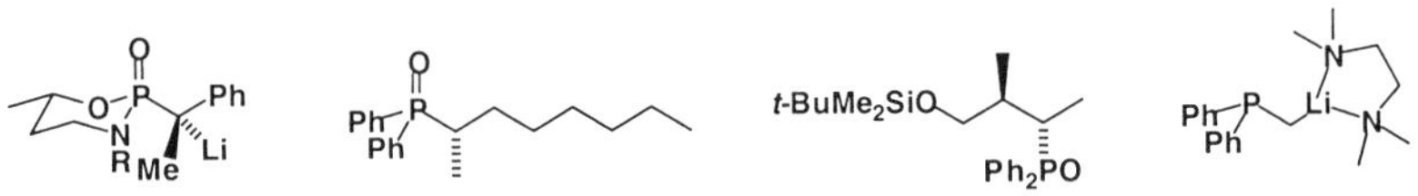

279 280 281 282

One report suggests that configurational stability on the NMR timescale may be evident in **282**, since at very low temperatures the CH_2 group gives rise to a pair of diastereotopic signals.[130]

5.1.10 Organolithiums α to silicon

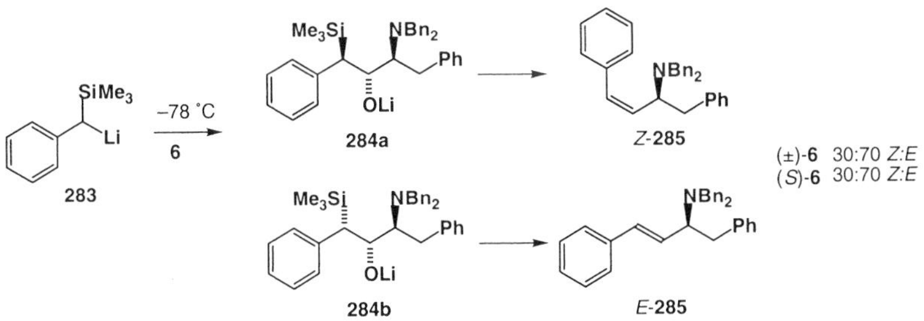

283 → (−78 °C, 6) 284a → Z-285; 284b → E-285

(±)-6 30:70 Z:E
(S)-6 30:70 Z:E

A Hoffmann test on the lithiated benzylsilane **283** indicated that it had no configurational stability on the timescale of its addition to the aldehyde **6**: the same 30:70 ratio was obtained whether the aldehyde was enantiomerically pure or racemic.[6] The diastereoisomeric ratio in this case was determined using the stereospecificity of the Peterson elimination to give *E*- or *Z*-**285**. Fraenkel[1] used NMR to show that an α-lithio silane had a barrier to inversion of 16-20 kJ mol^{-1}.

5.1.11 Benzyllithiums

5.1.11.1 Secondary benzyllithiums

Benzyllithiums generally have very low configurational stability,[52] with stabilisation of the anion contributing to increased planarity of the lithium-bearing centre,[131,75,76] and increased tendency to form solvent-separated ion pairs, and an increased rate of racemisation. For simple secondary benzyllithiums, Hoffmann tests have shown racemisation taking place faster than addition to electrophiles. So, for example, a Hoffmann test on **1** returns the same ratio of diastereoisomeric ketones on addition to the Weinreb amide **2**, whether the amide is racemic or enantiomerically enriched. Similar results were obtained using ketones or aldehydes as the electrophile: **1** racemises faster than it adds to amides, ketones or aldehydes.[5]

Li
1
Me N OMe NBn2
(*S*)- or (±)-**2**
NBn2
63:37 ratio from both (*S*)- and (±)-**2**

A number of heterosubstituted secondary benzylic organolithiums have already been discussed in section 5.1, and it should be enough here simply to summarise these results by saying that rapid racemisation is generally observed for secondary benzyllithiums substituted by sulfur,[5] selenium,[5] silicon,[5] or nitrogen.[76] Trials on the macroscopic timescale of organolithiums containing secondary lithium-bearing centres demonstrate that it is impossible to react these centres stereospecifically. For example, reductive cleavage of the aziridines **286** and reaction with electrophiles gives the same product irrespective of the stereochemistry of the starting aziridine.[132] Similarly, the organolithium **287** may be made in several ways, which should give varying initial ratios of diastereoisomeric organolithiums, but the organolithiums always give the same ratio of diastereoisomers of products on reaction with electrophiles.[133]

Me N Ph Me
286a
Li, Np
−78 °C, THF, 6 h
Li NLiMe Ph Me
Me N Ph Me
286b
Li, Np
−78 °C, THF, 6 h
Li NLiMe Ph Me
E⁺
E NMe Ph Me
Li Ph OLi R
287

Exceptions are the few secondary benzyllithiums which show configurational stability on the microscopic (but not macroscopic) timescale – in other words, they racemise slower than they react with electrophiles, though they still cannot be maintained in stereoisomerically pure form for periods of minutes or more. These are the carbamates **288** and the sulfones **289**, discussed in sections 5.1.4 and 5.1.7.[6] It is significant that both of these compound classes contain powerful lithium-coordinating oxygen atoms, which may hold the lithium counterion close to one face of the benzylic system.[134,120]

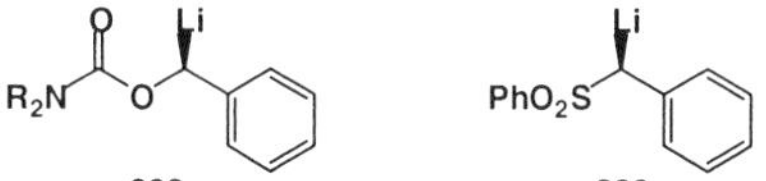

The non-heterosubstituted secondary organolithiums **292**, and their tertiary analogues **293**, both generated by selenium-lithium exchange from the two diastereoisomers **290** and **291**, cyclise stereospecifically, suggesting the intermediacy of a configurationally stable secondary benzylic organolithium.[135] However, the possibility that this reaction goes via initial attack by selenium on the electrophilic centre and not via an organolithium cannot be ruled out.[136,137]

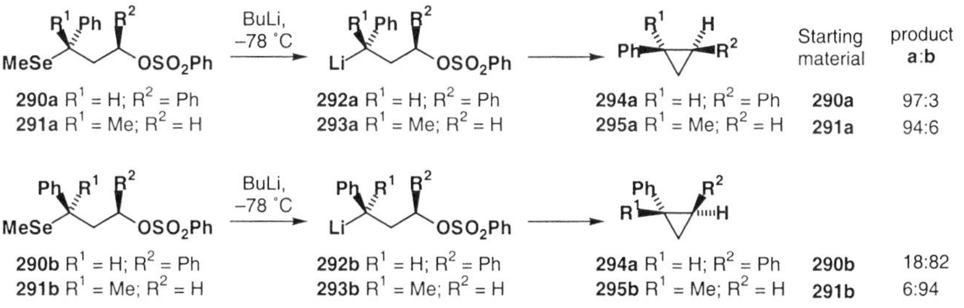

Three types of secondary benzyllithium are known which are configurationally stable on the macroscopic timescale, and they too each contain powerful lithium-coordinating oxygen atoms. One is the *N*-Boc amino substituted **180** discussed above.[80] Partial configurational stability is evident in a close analogue of **180**.[81]

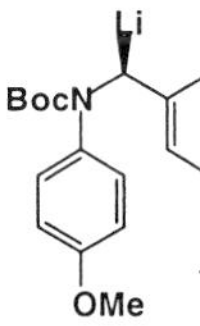

The second is the non-heterosubstituted lithiated anilide **297**. This compound presents a remarkable example because it is configurationally unstable in the presence of TMEDA, but configurationally stable in the presence of (–)-sparteine or its achiral analogue di-*n*-butylbispidine **299**.[138] Treatment of the enantiomerically enriched stannane **296** with *s*-BuLi returns, after 1 h, racemic material in the presence of no diamine additive, or in the presence of TMEDA, indicating a configurationally unstable intermediate on this timescale. However, in the presence of di-*n*-butylbispidine **299**, enantiomeric enrichment is retained – the **297–299** complex is configurationally stable over a period of one hour at –78 °C.

ee **296**	additive	ee **298**
64%	–	2%
70%	TMEDA	4%
68%	**299**	68%

The third is a pair of atropisomeric non-heterosubstituted lithiated naphthamides **301a** and **301b**.[139] Both diastereoisomers of the stannane **300** may be synthesised, and transmetallation leads to organolithiums **301** which react to give different products **302**. The ^{1}H NMR spectra of these organolithiums at –40 °C are clearly different (see section 5.2.1), and they show no sign of epimerisation over a period of 60 min at –40 °C.[140] The powerfully coordinating amide substituent in **301** must be holding the lithium tightly bound to one face of the benzyl system.

starting ratio **300a**:**300b**	**300a**:**300b** (by NMR)	**302a**:**302b** (by NMR)
>98:2	>98:2	>98:2
5:95	35:65	60:40

Remarkably, the closely related benzyllithium **304** is configurationally *unstable* even at –78 °C.[138] Transmetallation of **303** (88% ee) at –78 °C gave an organolithium which reacted to give racemic product **305** in the presence or absence of TMEDA. Furthermore, the reaction of the **304**-(–)-sparteine complex with each of racemic or enantiomerically pure **2** in a Hoffmann test gave the same 1:1.6 ratio of diastereoisomers. It is not yet clear whether this unexpected difference between **301** and **304** is due to an electronic difference between the naphthyl and phenyl systems, or whether it arises from the difference in steric hindrance, and therefore the dihedral angle between the ring and the amide, in the two compounds.

303 88% ee → BuLi, 1.5 h, –78 ˚C → 304 → (±)-305; 2

5.1.11.2 Tertiary benzyllithiums

Tertiary benzyllithiums racemise significantly more slowly than their secondary counterparts. NMR shows that the simple lithiated hydrocarbon **306** inverts with a barrier of 39.3 kJ mol^{-1} at –70 ˚C in THF[141] and the non-heterosubstituted tertiary benzyllithiums **308a** and **308b** – formed by Halle-Bauer cleavage of the ketones **307** – show some degree of configurational stability on the macroscopic timescale, being protonated at a rate comparable with their rate of epimerisation in refluxing benzene in a partially stereospecific, partially stereoselective reaction.[142]

306; 307a → LiNH$_2$, PhH, 120 h → 308a → 309a 85:15 axial: equatorial (24%)
307b → LiNH$_2$, PhH, 120 h → 308b → 309b 55:45 axial: equatorial (7.3%)

Incorporation of the benzylic centre into a cyclopropyl ring – unsurprisingly – increases the barrier to inversion: **320** racemises with a barrier as high as 55.6 kJ mol^{-1}.[143]

310

Heterosubstituted tertiary benzyllithiums show significant configurational stability on the microscopic, but not always the macroscopic timescale. Illustrating the latter point, the variously substituted alcohols **311** and **312** can each be formed using two equivalents of BuLi at –78 ˚C or 0 ˚C, but they all epimerise rapidly before reaction with electrophiles.[144,133] Even intramolecular reverse [1,4]-Brook rearrangements (section 8.2) cannot compete with the rate of epimerisation (though Brook rearrangements are typically slow relative to many other types of intramolecular reactions).

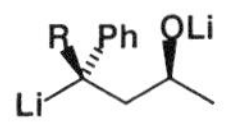

R = Me, MeO, MeS, MeSe

311 312

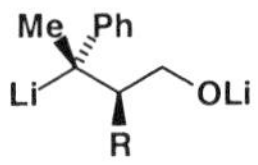

Heterosubstituted tertiary benzyllithiums containing lithium-coordinating substituents frequently (though not always[76]) show configurational stability on the macroscopic timescale. Many of these have been discussed above, and they include the carbamates **99** (which maintain their configuration for minutes at –78 °C),[52,53] the thiocarbamate **237** (which is configurationally stable even at 0 °C),[113] the *N*-Boc benzylamine **183** (which is configurationally stable at –78 °C),[80] the trifluoromethylsulfone **246**,[117] and the sulfoxide **255** (which is configurationally stable over a period of an hour even at room temperature).[119]

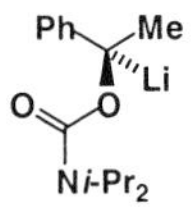

99 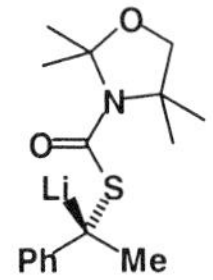 237 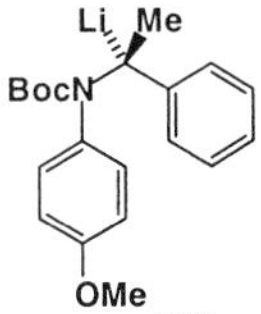 183 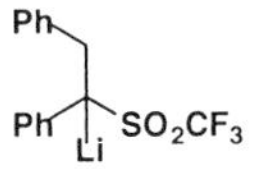 246 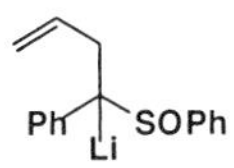 255

Benzylic sulfones and sulfoxides, whose tertiary anions are not truly organolithiums, also exhibit high configurational stability, as discussed above.

5.1.12 Vinyllithiums

It is well established that vinyllithiums of the general structure **313**, where R ≠ aryl or silyl, are configurationally stable in solution.[145,146] Aryl- and silyl-substituted vinyllithiums may however isomerise in solution in the presence of coordinating solvents such as THF. For example, **314a** and **314b** are configurationally stable at room temperature in hexane in the presence of less than 1 equiv. THF, but equilibrate at –30 °C in the presence of more than 5 equivalents of THF.[147,148] A silyl substituent at R^1 of **313** also induces configurational instability.[149]

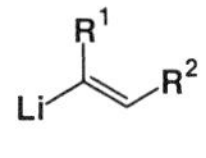

313a 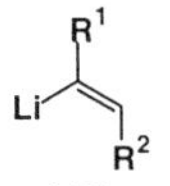 313b 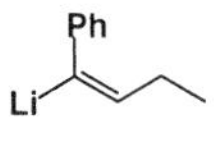 314a 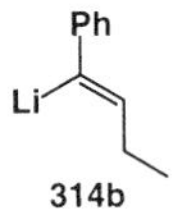 314b

5.1.13 Summary

The chart gives a qualitative guide to the configurational stability typical of organolithiums with various structural features. All values are approximate.

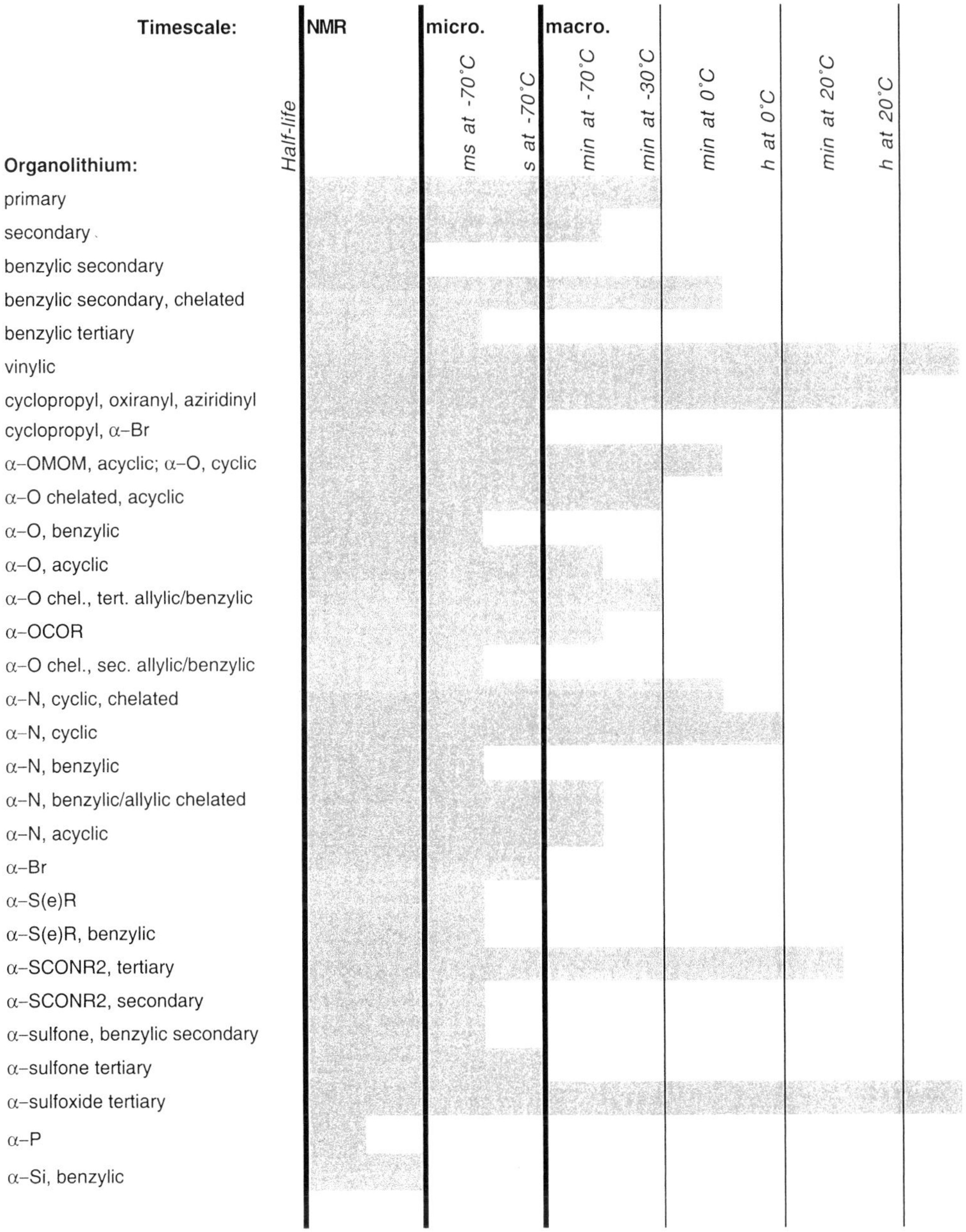

Timescale: NMR | micro. | macro.

Half-life: ms at -70°C | s at -70°C | min at -70°C | min at -30°C | min at 0°C | h at 0°C | min at 20°C | h at 20°C

Organolithium:

primary
secondary
benzylic secondary
benzylic secondary, chelated
benzylic tertiary
vinylic
cyclopropyl, oxiranyl, aziridinyl
cyclopropyl, α–Br
α–OMOM, acyclic; α–O, cyclic
α–O chelated, acyclic
α–O, benzylic
α–O, acyclic
α–O chel., tert. allylic/benzylic
α–OCOR
α–O chel., sec. allylic/benzylic
α–N, cyclic, chelated
α–N, cyclic
α–N, benzylic
α–N, benzylic/allylic chelated
α–N, acyclic
α–Br
α–S(e)R
α–S(e)R, benzylic
α–SCONR2, tertiary
α–SCONR2, secondary
α–sulfone, benzylic secondary
α–sulfone tertiary
α–sulfoxide tertiary
α–P
α–Si, benzylic

5.2 Stereospecific synthesis of organolithiums by X-Li exchange

5.2.1 Tin-lithium exchange

As described in section 3.2, tin-lithium exchange is a rapid, thermodynamically-controlled process by which ligands at tin and lithium may interchange. Apart from the enantioselective deprotonation methods discussed in section 5.4, tin-lithium exchange is the most important stereocontrolled way of making organolithiums because – with one exception – all stereochemical evidence is consistent with the process being fully stereospecific and proceeding with retention. However, as with studies of the stereospecificity of organolithium reactions, much of the "evidence" for fully retentive stereospecificity in the tin-lithium exchange is based on assumptions which have themselves never been rigorously proved. (Retention of stereochemistry at tin during tin-lithium exchange has been proved in some cases.[150])

5.2.1.1 Vinylstannanes

To start with what is absolutely certain: tin-lithium exchange of vinylstannanes for vinyllithiums proceeds with retention of double bond geometry. Previous evidence[146] had pointed towards this fact, but had always relied on assumptions of retention in subsequent electrophilic reactions of the vinyllithium.[151] Seyferth[145] studied both stannanes **318** and vinyllithiums **317** directly by NMR, assigning their geometry by coupling constant: the reactions unambiguously proceeds with retention, also confirming the retentive nature of all previously studied electrophilic substitutions of vinyllithiums.

E-315 —(Li, Et_2O)→ *E*-317 (J = 22.2 Hz) ←(BuLi)— *E*-318

Z-316 —(Li, Et_2O)→ *Z*-317 (J = 17.4 Hz) ←(BuLi)— *Z*-318

5.2.1.2 α-Heterosubstituted stannanes

The current picture with regard to the generation of three-dimensional stereochemistry by tin-lithium exchange is very much as it was in the two-dimensional world in 1960. There is plenty of compelling evidence that almost all known examples of tin-lithium exchange at tetrahedral carbon proceed with retention, but little firm proof. The first demonstration of the reaction's stereospecificity was provided by Still in 1980 in a seminal paper on the stereospecific synthesis of α-alkoxyorganostannanes.[25] His sequence of reactions started with the stannane **320**, formed by purification of the major product of nucleophilic addition of Bu_3SnLi to aldehyde **319**. The stannane transmetallates stereospecifically to an organolithium

321 which reacts stereospecifically with electrophiles, and the *overall* course of the reaction was proved to proceed with retention by re-stannylation with Bu_3SnI.

1. Bu_3SnLi 2. $MeOCH_2Cl$ 3. separate; **319** → **320**; BuLi, THF, −78 °C → **321**; Bu_3SnI, −50 °C ←

Still used a similar sequence, starting from the enantiomerically pure stannane **322**, to prepare enantiomerically pure organolithium **323** which methylates stereospecifically. The product **324** of this sequence was shown to have *R* configuration – but the configuration of the starting material, and of the intermediate organolithium, could only be inferred by assuming retention in each step.

1. Bu_3SnLi 2. resolve 3. $BnOCH_2Cl$ → **322**; BuLi → **323**; Me_2SO_4 → (*R*)-**324**

In a similar vein, transmetallation-methylation of the major diastereoisomer **326a** produced by nucleophilic attack on **325** generated the major diastereoisomer **327a** from attack by MeMgBr on the same aldehyde. Even *overall* retention from **326a** to **327a** can only be inferred from the assumption that MeMgBr and Bu_3SnLi attack chiral aldehydes from the same direction.

325: 1. Bu_3SnLi 2. $BnOCH_2Cl$ → **326a** major 3:1 **326b** minor; **326a**: 1. BuLi 2. Me_2SO_4 → **327a**

325: 1. MeMgBr 2. $BnOCH_2Cl$ → **327a** major 2.5:1 **327b** minor

That *overall* retention in similar systems is the norm later received firmer confirmation from studies of compounds where the stereochemistrry of both stannane and final product could be unambiguously assigned.[37,152] This has been widely assumed to mean that each individual step (transmetallation and quench) is retentive, but the same result could arise from a double inversion. This possibility may not withstand the vigorous slicings of Ockham's razor,[153] but it must be borne in mind as a conceivable mechanistic pathway in some instances.

Still was careful not to make any claims for the stereochemistry of the intermediate organolithium,[25] but subsequent workers have frequently cited this paper in support of assumptions of retention in the tin-lithium exchange step. Statements such as "the method of Still and Sreekumar leads to α-lithio ethers with retention";[29] "retention of stereochemistry is well known for the generation of α-heterosubstituted organometallics";[79] "enantiomerically defined α-alkoxyorganolithiums ... can be generated stereospecifically (retention of configuration) from easily obtainable enantio-enriched stannanes"[154] are common, they are probably correct in most cases, but they have bot been proved!

The accumulated evidence that non-benzylic α-heterosubstituted stannanes typically undergo overall retentive Sn–Li–E^+ substitutions is now overwhelming, and many examples have been cited in the sections on configurational stability and on the stereospecific reactions of organolithiums. The exceptions (mainly arising when the intermediate organolithium is allylic or benzylic) to the rule of overall retention are best explained by invertive substitutions rather than invertive tin-lithium exchanges (see section 6.1). An early and very clear example was provided by Macdonald,[37,152] who was able to make the two stannanes *ax*-**328** and *eq*-**328** by using kinetic control (equatorial attack) or thermodynamic control (axial Bu_3Sn: the A-value of R_3Sn is remarkably small,[155] and the solvated lithium alkoxide prefers to lie equatorial). Each be transmetallated and quenched with electrophiles to give axial or equatorial products with provable overall retention.

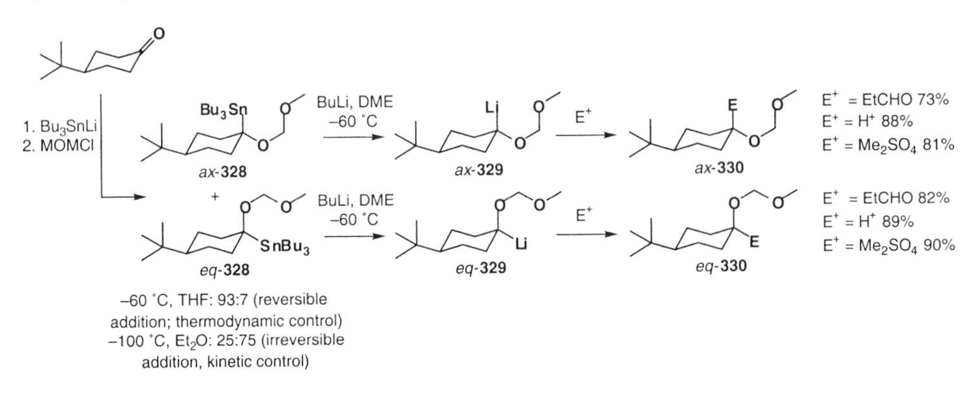

5.2.1.3 Non-heterosubstituted stannanes

The stereochemistry of the tin-lithium exchange of non-α-heterosubstitued stannanes has been studied only to a limited extent because the reaction itself typically lacks the driving force that favours the formation of stabilised α-heteroorganolithiums, aryllithiums or vinyllithiums. The examples that do exist lead to either (a) cyclopropyl or cyclobutyllithiums or (b) benzyllithiums.

Tin-lithium exchange giving cyclopropyllithiums is stereospecific, and appears to proceed with retention, though the examples that exist are complicated by the presence of potential lithium-coordinating heteroatoms attached to the ring. For example, the cylopropylstannane **331** is transmetallated stereospecifically to an organolithium **332a** which reacts with electrophiles to give **333a** with overall retention.[19] On warming the intermediate **332a**, overall stereospecificity is lost, presumably because of epimerisation to a more stable, chelated organolithium **332b**.

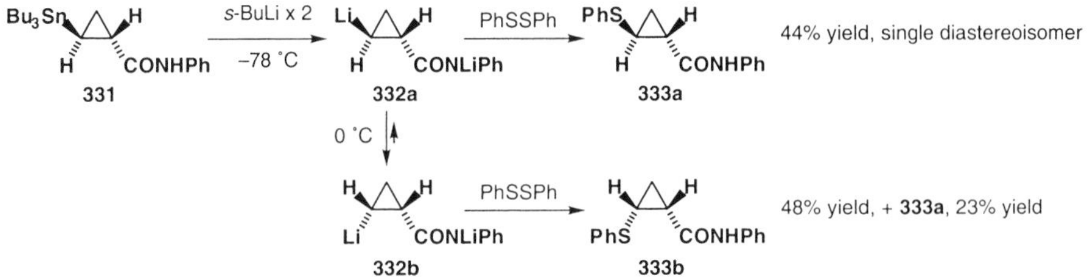

Similarly, the stannane **334** transmetallates with apparent stereospecificity and the product organolithium **335** reacts with overall retention.[156] The origin of stereochemistry in the product can only be assumed here because of the unknown role of chelation by the –OLi group.

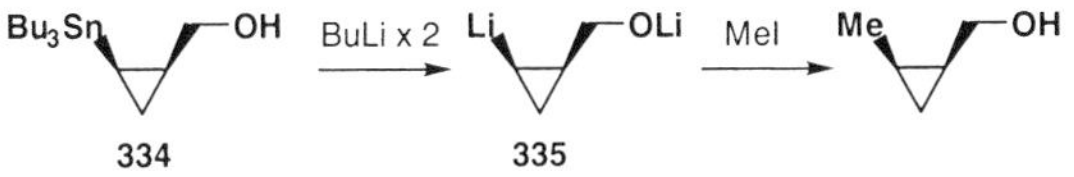

Similar points of uncertainty surround the stereospecific tin-lithium exchange of cyclobutylstannane **336**. The expected epimerisation of the intermediate organolithium **337** may be prevented by O–Li coordination, but the fact that the *exo* stannane **339** does not undergo transmetallation adds weight to the suggestion that the initial tin-lithium exchange is stereospecific and retentive.[24]

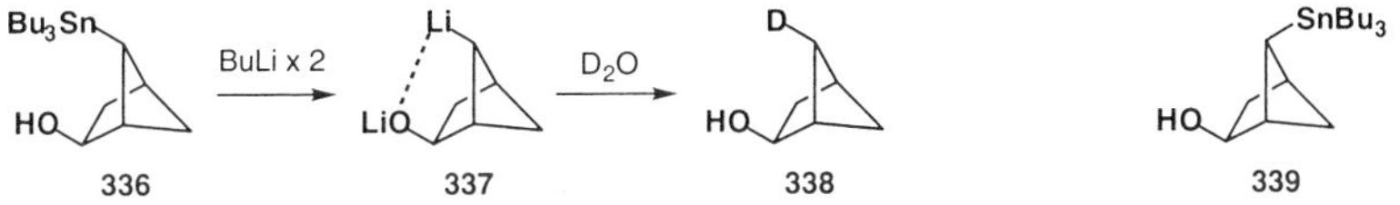

Non-α-heterosubstituted organolithiums in which the lithium finds itself in a benzylic position are equally readily synthesised by tin-lithium exchange. The known capriciousness of the stereospecificity of electrophilic substitution of benzylic organolithiums (see section 6.1.4.1) might lead us to expect that among these reactions, if anywhere, is where we should expect exceptions to the general rule of retentive tin-lithium exchange. Two types of molecule in this class have been investigated, and in one of them, tin-lithium exchange behaves perfectly normally. The stannane **340** may be formed enantioselectively using (–)-sparteine-directed lithiation and substitution (section 6.2).[157] Transmetallation and re-stannylation returns product **342** (E = $SnMe_3$) with overall clean inversion, almost certainly because of an invertive electrophilic substitution of the organolithium.

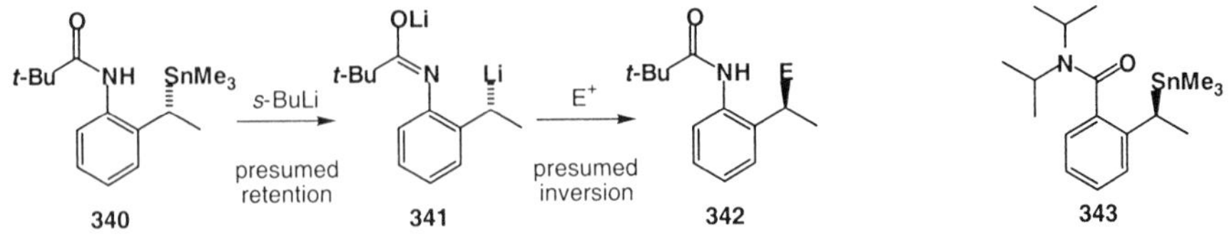

Transmetallation of stannane **343** has been carried out, but because of the configurational instability of the organolithium product, the stereospecificity of the reaction cannot be known.[158] However, the related stannanes **345**, while superficially similar to **343**, differ in two important ways: organolithiums **346** are configurationally stable at the lithium-bearing centre (section 5.1.11), and can exist as either of two diastereoisomeric atropisomers since rotation about the Ar–CO bond is slow.[139,159] These stannanes provided the first ever evidence that tin-lithium exchange may be non-stereospecific. Lithiation of **344** is diastereoselective and generates naphthyllithium **346b**, which reacts with Bu_3SnCl with inversion of stereochemistry to produce **345a** and **345b** in an 89:11 ratio. This kinetic ratio is by no means the equilibrium ratio between the two atropisomeric diastereoisomers, and heating the mixture for 48 h at 65 °C converts them to a 4:96 ratio of **345a** and **345b**. In this way, either diastereoisomer of the stannane may be produced.

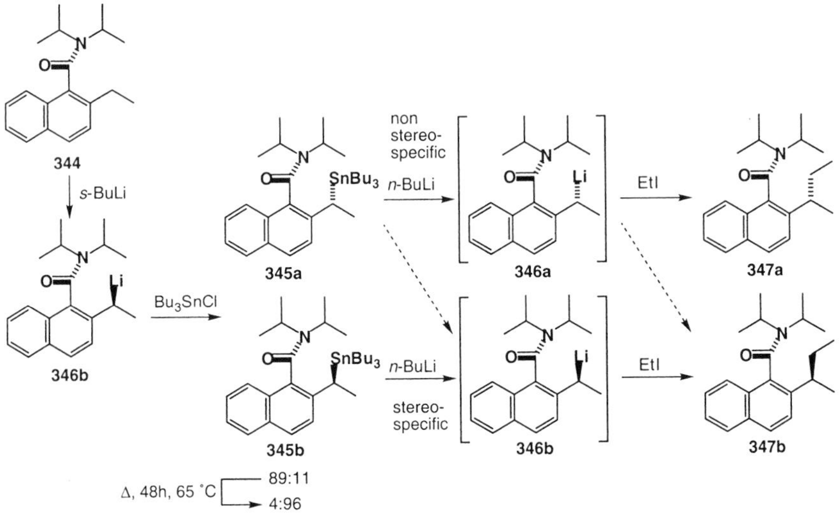

starting ratio **345a:b**	temperature	yield **347a+b**	product ratio **347a:b**
0:100	–78 °C	97	1:99
94:6	–78 °C	87	40:60
0:100	–40 °C	94	2:98
95:5	–40 °C	83	40:60

Transmetallation and alkylation of each diastereoisomer provided evidence for lack of stereospecificity in a tin-lithium exchange. The thermodynamically favoured stannane **345b** reacted entirely as expected, giving, after transmetallation and alkylation at –78 °C or at –40

°C, essentially a single diastereoisomer of the product **347b**. However, when a sample containing 95% of the stannane **345a** was transmetallated and alkylated under either of these conditions, the product contained a mixture of 40:60 **347a**:**347b**.

Given that the intermediate organolithiums are configurationally stable (see section 5.1.11), the loss of stereospecificity could be occurring in either of the two steps in the sequence. NMR experiments indicated that both steps were in fact non-stereospecific. By carrying out the deprotonation and transmetallations in d_8-THF in an NMR tube at –40 °C it was possible to acquire ^{1}H NMR spectra of both organolithiums **346a** and **346b**. Figure 5.2.1 shows the spectrum of **346b** produced by deprotonating **344** with *s*-BuLi; Figure 5.2.2 the spectrum of the same compound produced by transmetallation of **345b**. They each show a single organolithium, and prove both that transmetallation of **345b** is stereospecific and that deprotonation of **344** is stereoselective.

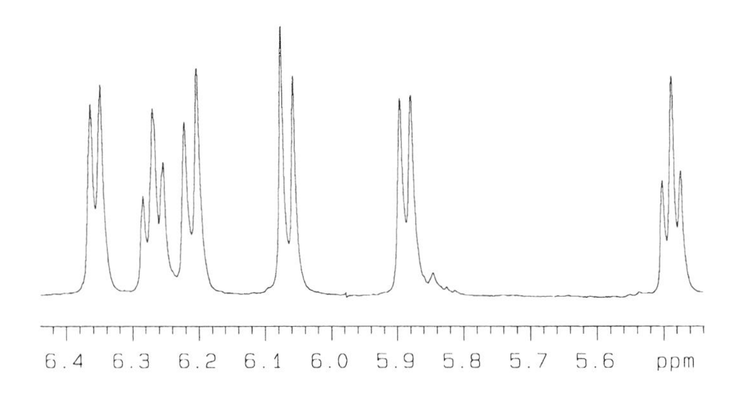

*Figure 5.2.1: Portion of ^{1}H NMR spectrum of **346b** produced by deprotonation of **344***

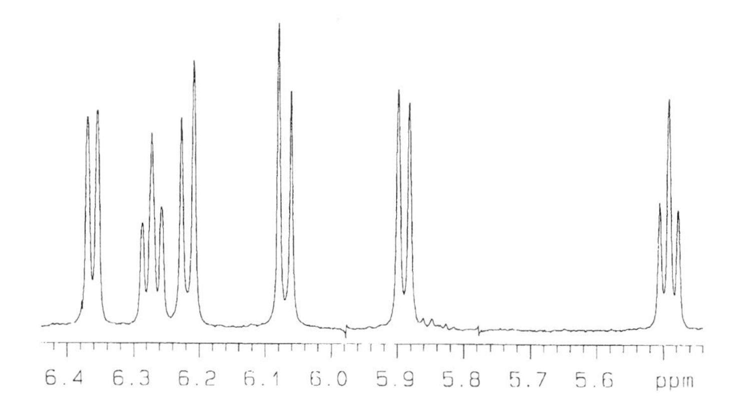

*Figure 5.2.2: Portion of ^{1}H NMR spectrum of **346b** produced by transmetallation of **345b***

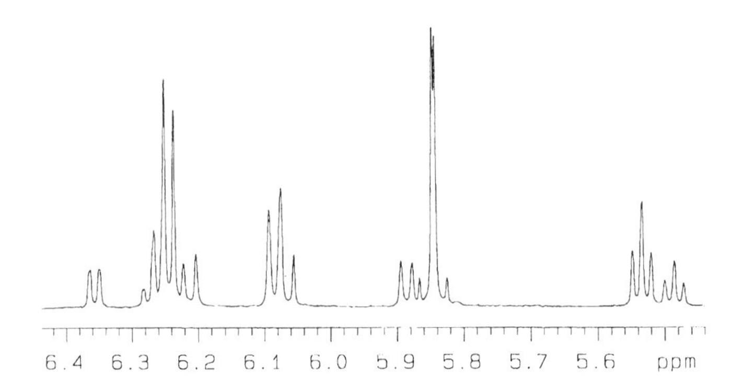

*Figure 5.2.3: Portion of ^1H NMR spectrum of **346a** and **346b** produced by transmetallation of **345a**.*

Figure 5.2.3 shows the spectrum of the organolithiums formed by transmetallation of **345a**: it clearly contains some of **346b**, but is largely another compound we assign as organolithium **346a**. The ratio of organolithiums – about 65:35 **346a:b** – is not the ratio of starting stannanes, so the transmetallation is not stereospecific – nor is it the ratio of the products, so neither is the electrophilic quench stereospecific.

Given the mechanism of the tin-lithium exchange, it is worth considering why the reaction is so generally stereospecific. Typically, we would expect **347** and RLi an intermediate ate complex **348**, which, for the exchange to be retentive, must collapse with retention to give **349**. Presumably the organolithium **349** must form from a very short lived ion pair, and the stereochemistry of the organolithium is dictated by the stereoselectivity of the attack of Li^+ on the carbanion. Even if an ion pair is not an intermediate, retentive and invertive electrophilic substitutions of organotin compounds are both known. So why is the reaction so generally retentive? Because the exchange is only thermodynamically favourable when the product organolithium is stabilised in some way, there is almost always a lithium-coordinating heteroatom X involved in the ion-pair collapse. The effect of the heteroatom will be to anchor the lithium cation during the collapse of the ion pair, leading to overall retention. Retention might be much rarer in the formation of non-heteroatom-stabilised organolithiums.

$Li^{\oplus}$

$SnBu_3$ BuLi $^{\ominus}SnBu_4$ Li

R X R X R X

347 **348** **349**

The transmetallation of **345b** probably proceeds via a stannate intermediate **350b** (see section 3.2) which may collapse to give either diastereoisomeric organolithium depending on the direction of attack on the C–Sn bond by Li^+. Likely conformations of the ate complexes **350a** and **350b** derived from **345a** and **345b** are shown below. In the absence of data on the relative rates of formation, conformational change, and collapse of the ate complexes, it is impossible to propose a detailed rationale for the variation in stereospecificity. However, we expect that **350b**, in common with **345b**, unambiguously prefers the *syn* conformation, in which delivery of either RLi or Li^+ can take place in such a way that retentive transmetallation occurs. The preferred conformation of **350a** (and **345a**), on the other hand, is presumably less clear-cut: either the tin or the methyl substituent must lie close to the bulky NR_2 group on the amide, and population of the *anti* conformer could lead to invertive transmetallation. It is interesting to note that a higher proportion of inversion was observed in the transmetallation of **345a** than in the transmetallation of its trimethylstannyl analogue.[140] More inversion is also observed with smaller organolithiums RLi, due perhaps to a shift of the conformational equilibrium of **350a** to the right with smaller tin substituents, or to preferential attack by larger organolithiums on the less crowded tin substituent in the *syn* conformation.

345b
RLi
syn conformation **350b** *anti* conformation
retention inversion
Li Me H R_4Sn SnR_4
retention inversion
346b **346a**

345a
RLi
syn conformation **350a** *anti* conformation
Li Me H R_4Sn larger R smaller R SnR_4
retention inversion
346a **346b**

In the case of **345a**, the diastereoisomer **350a** has a heteroatom capable of anchoring the lithium close to the rear face of the stannate as it collapse, presumably the origin of the inverted organolithium product. The prospect of designing organic molecules deliberately to deliver lithium to the rear face of a stannate in the hope of promoting an invertive tin-lithium exchange is an interesting one.

5.2.2 Halogen-lithium exchange

Letsinger's early work[9] on stereospecificity during iodine-lithium exchange established that the reaction proceeded with at least some degree of stereospecificity for simple secondary alkyl iodides. The acid **351** was obtained in 20% ee from the sequence below. Unfortunately, ether is necessary for an efficient halogen metal exchange, and the presence of ether catalyses rapid racemisation of the organolithium, hence the low ee.

s-BuLi, petrol/ether 94:6, −70 °C, 2 min; CO_2

351 20% ee

Both Walborsky[18] and Applequist[16] showed that bromine-lithium exchange of a cyclopropyl bromide **352** – giving a highly configurationally stable cyclopropyllithium **353** – proceeds with reliable stereospecificity, presumably retention.

BuLi, Et_2O, 28 °C

352 → **353** → 53%; 100% optical purity

It is important to contrast the stereospecific nature of halogen-metal exchange with the loss of stereochemistry typical in reductive lithiations.[10] For example, transmetallation of *cis* and *trans*-**354** with BuLi, quenching with ethylene oxide, gives complete retention of stereochemistry, while reductive lithiation leads to considerable (but not total) epimerisation.[17]

cis-**354** or *trans*-**354** → *cis*-**355** + *trans*-**355** → *cis*-**356** + *trans*-**356**

Substrate	Reagent	*cis*-356 : *trans*-356
cis-**354**	BuLi	100:0
cis-**354**	Li metal	54:46
trans-**354**	BuLi	0:100
trans-**354**	Li metal	31:69

Halogen-lithium exchange of vinylbromides retains double bond geometry.[160]

5.2.3 Selenium-lithium exchange

Selenium-lithium exchange to form α-thio or α-seleno organolithiums proceeds with retention of configuration. This was proved by Reich in an elegant series of experiments, of which an important one is described below.[106] The epimerisation of *eq*-**358** to *ax*-**358** is fast, but trapping of the organolithium with Me_3SiCl is competitive. With high concentrations of

Me_3SiCl, it is possible to record the initial stereospecificity of the Se–Li exchange – at least 96% retention. Proof that this is the result of stereospecificity and not stereoselectivity comes from related reactions which showed that, in diselenoacetals, attack of BuLi at the axial heteroatom is preferred. These experiments also provided evidence that, though a seleno-ate complex is probably an intermediate in the sequence, it is not the final reactive species itself, since the ratios of axial and equatorial products were not dependent on the nature of the starting selenide (SePh or SeMe) - i.e. the selenium is lost before the stereoselective step.

1. n-BuLi, –78 °C; 2. Me_3SiCl: 357 → [eq-358 ⇌ ax-358] → 360 96%

n-BuLi, –78 °C, Me_3SiCl (in situ quench): 357 → 359 96%

5.2.4 Sulfur-lithium exchange

Sulfur-lithium exchange in sulfides is much slower[11] than selenium-lithium exchange, and nothing is known of its stereospecificity. More common is the reductive lithiation of sulfides to give organolithiums (section 4.4), which is apparently non-stereospecific, presumably because of the intermediacy of free radicals. The cyclopropyllithium **361**, for example, expected to be configurationally stable, is formed stereoselectively but probably not stereospecifically (though this was not proved).[19] The apparent inversion to form **362** (assuming retentive quench) is likely to be due to the stereoselective formation of the more stable chelated diastereoisomer of the organolithium.

LDMAN → 361 → H_2O → 362, 96% yield, 95:5 diastereoselectivity

5.2.5 Other metal-lithium exchanges

Mercury-lithium exchange, like other soft metal-lithium exchanges, is stereospecific and presumably retentive.[11]

5.2.6 Stereospecific deprotonation

Since the thermodynamic acidity of C–H bonds and the configurational stability of the resulting organolithium tend to be opposed to one another, relatively few examples of provable stereospecific deprotonation are known. Those that are known, such as deprotonation of **363** to give configurationally stable tertiary benzylic organolithium **364**,[161] are stereospecific and are all assumed to proceed with retention of stereochemistry.

s-BuLi, TMEDA; Me_2R^*SnBr (reacts with inversion)

363 >97% ee **364** >97% de

The carbamate **366** can be made from enantiomerically pure alcohol **365** and deprotonated stereospecifically to give organolithium **367**, as shown by formation of esters **368** with retention of stereochemistry.[53]

NaH, THF; s-BuLi, TMEDA; $(MeO)_2C{=}O$

365 **366** **367** **368**

The kinetic isotope effect can be used to direct the steric course of the deprotonation at a chiral deuteromethylene group and form a secondary organolithium **370** with retention of stereochemistry.[162,163]

s-BuLi, TMEDA; Me_3SiCl

369 **370** **371** 95% ee

5.3 Diastereoselective deprotonation

5.3.1 Diastereoselective lateral lithiation

Most benzylic organolithiums, unless they bear an α heteroatom, are configurationally unstable over a period of seconds or more (see section 5.1.11), so any stereoselectivity in lateral lithiation is rarely detectable. However, as implied above, the lateral lithiation of tertiary 1-naphthamides **344** is stereoselective, and yields a single diastereoisomeric atropisomer of the organolithium **346b**.[159,139,140] Both diastereoisomers of **346** were characterised by NMR. These organolithiums react with most electrophiles to give **372** with retention of stereochemistry, but with trialkyltin halides with inversion (see section 6.1.4).

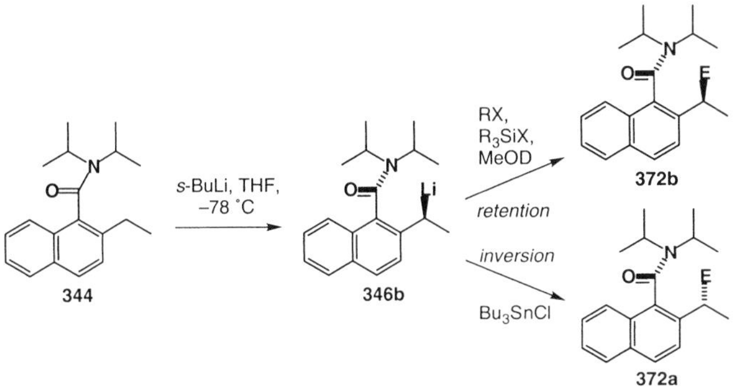

By carrying out a subsequent ortholithiation at low temperature, it was possible to show that tertiary benzamides also react atroposelectively in laterally lithiation-electrophilic quench sequences.[159] Either atropisomer **375** or **378** could be made starting from **373** or **376**.

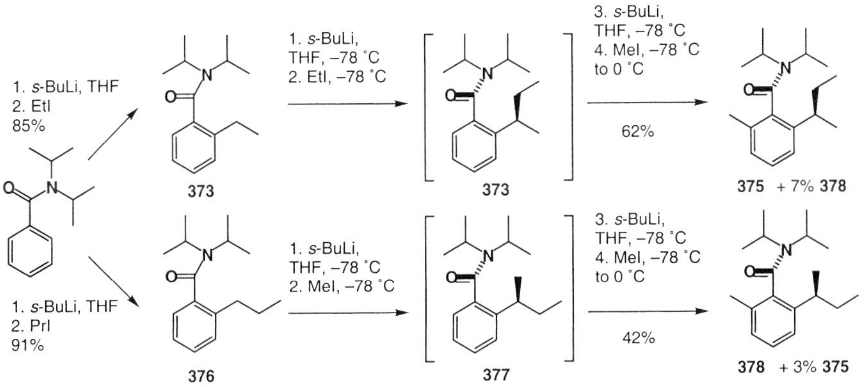

The double lateral lithiation–silylation of **379** allows the construction of remote stereogenic centres of **380** in a single step.[164]

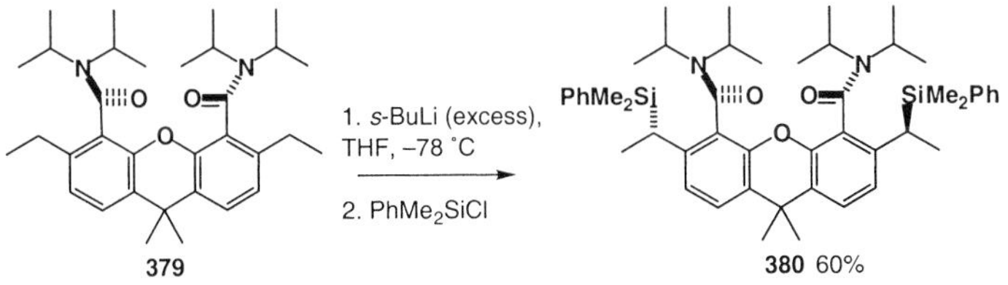

5.3.2 Diastereoselective ortholithiation

Stereochemistry is rarely an issue in ortholithiation reactions unless the directing group is chiral and a prochiral electrophile is used which gives rise to a new stereogenic centre. Most attempts to use metallation-directing groups containing stereogenic centres have failed to give any useful level of stereoselectivity – for example, lithiated sulfoxides[165] and lithiated chiral amides[166] and hydrazines[167] all add to benzaldehyde to give close to 1:1 mixtures of diastereoisomers. In the heterocyclic and ferrocenyl series, by contrast, additions of ortholithiated sulfoxides to aldehydes can lead to good levels of stereoselectivity (see section 2.3.4.2). Asymmetric lithiation of prochiral ferrocenes or arenechromium tricarbonyl complexes with chiral bases is also effective (see section 5.4).

In aromatic compounds bearing two rotationally restricted amide groups, diastereoisomers can arise because of the relative orientation of the amides. Ortholithiation can therefore lead to diastereoselectivity if the ortholithiation forms one of the two diastereoisomers. A simple case is **381**, whose double lithiation–ethylation leads only to the C_2-symmetric diamide **379**, indicating the probable preferred conformation of the starting material.[164]

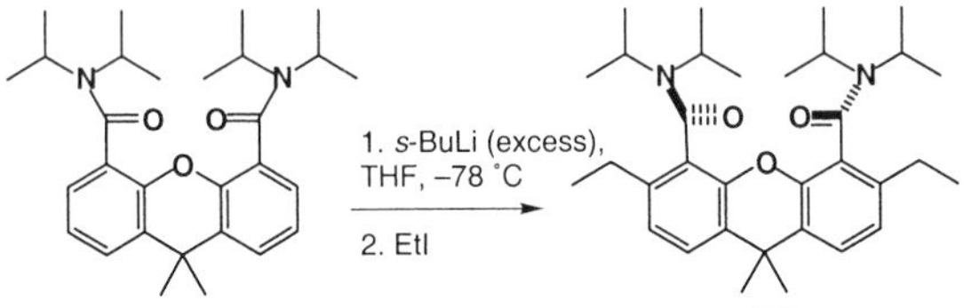

1. s-BuLi (excess), THF, –78 °C
2. EtI

381 379

The biphenyl diamide **382** also displays diastereoselectivity in its ortholithiation–electrophilic quench, giving the C_2-symmetric, chiral diastereoisomer of the diamide **384**.[168] Heating the product converts it mainly to the more stable achiral, centrosymmetric diamide **385**.

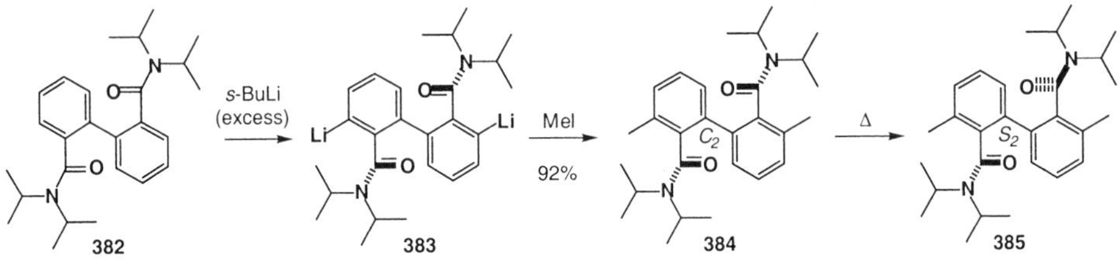

s-BuLi (excess) MeI 92% Δ

382 383 384 385

5.4 Enantioselective Deprotonation

In the next chapter we shall consider in detail the reactions of organolithiums directed by the presence of chiral ligands, such as (–)-sparteine, which govern their enantioselectivity. Detailed discussion of the structures and mechanisms involved will be reserved until then. However, conceptually the simplest sort of asymmetric organolithium reaction is an enantioselective deprotonation, and we shall consider such reactions first in this chapter devoted to the stereoselective synthesis of organolithiums. An example is shown below: the complex of *s*-BuLi with (–)-sparteine **12** selectively removes the pro-*S* proton from one of the two pairs of enantiotopic protons in *N*-Boc pyrrolidine **386**. The chiral, configurationally stable organolithium[169] **387** reacts with electrophiles to give products with 88-94% ee.[66,170]

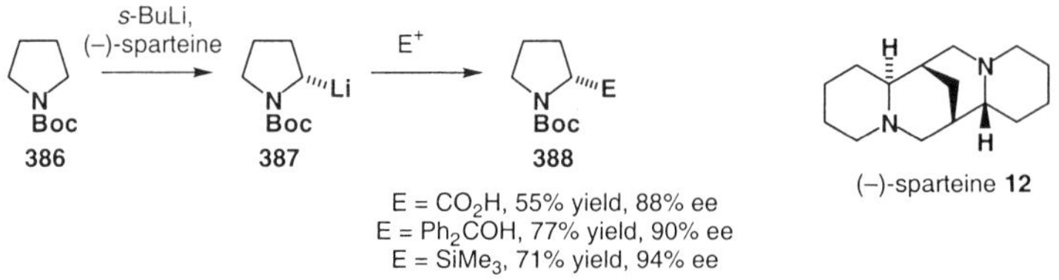

s-BuLi, (–)-sparteine; E^+

386 → 387 → 388

E = CO_2H, 55% yield, 88% ee
E = Ph_2COH, 77% yield, 90% ee
E = $SiMe_3$, 71% yield, 94% ee

(–)-sparteine **12**

Enantiomeric excess in the final product has to arise in the first (deprotonation) step and not the second because **387** is configurationally stable under the conditions of the reaction. This condition is general for all organolithiums: if an enantiomerically enriched product is formed in good yield by electrophilic quench of an organolithium of known configurational stability, the enantiodetermining step must necessarily be the formation of the organolithium. The stereochemistry of this organolithium is deduced in this case from the presumed retentive nature of the electrophilic quench at its tetrahedral lithium-bearing centre: this is discussed in detail in section 6.1.

Further confirmation that the deprotonation determines the ee of the product came from the result shown below. When the intermediate **387** is formed as a racemate, either by deprotonation with *s*-BuLi–TMEDA or by tin-lithium exchange from a racemic stannane, the products **390** of electrophilic quench are racemic too, even if the electrophile is added in the presence of (–)-sparteine.

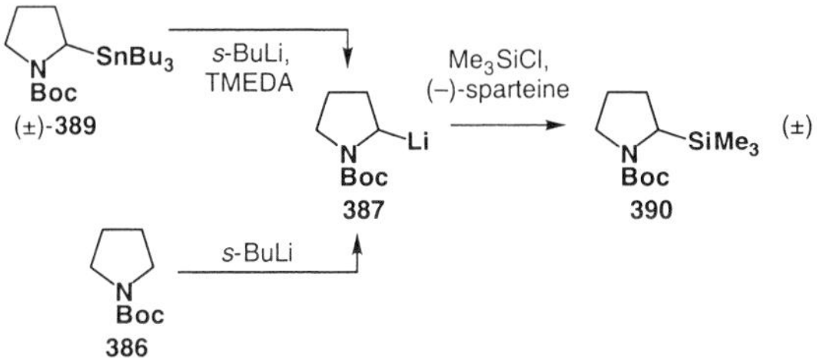

(±)-389 —*s*-BuLi, TMEDA→ 387; 386 —*s*-BuLi→ 387; 387 —Me_3SiCl, (–)-sparteine→ 390 (±)

A second *s*-BuLi-(–)-sparteine deprotonation amplifies the enantiomeric excess of the product **388**. The sequence below gives the 2,5-dimethylpyrrolidine derivative **391** in >99% ee.[67] A phosphorus analogue **392** is also deprotonated enantioselectively.[171]

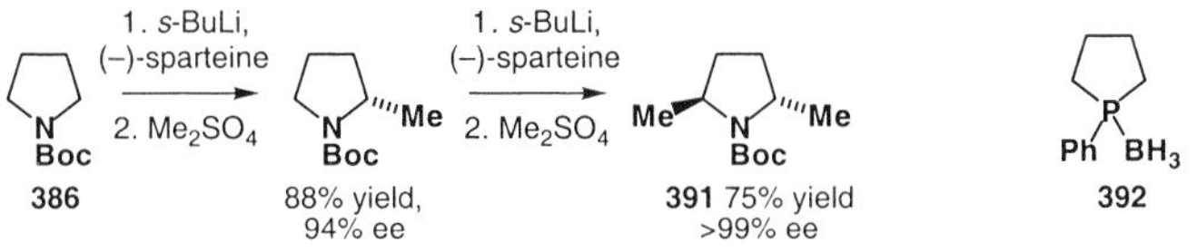

386 —1. *s*-BuLi, (–)-sparteine; 2. Me_2SO_4→ 88% yield, 94% ee —1. *s*-BuLi, (–)-sparteine; 2. Me_2SO_4→ **391** 75% yield, >99% ee

392 (Ph, BH_3)

The piperidines **393** are deprotonated enantioselectively, but the organolithiums have significantly lower levels of configurational stability, and high enantiomeric excesses result only if the electrophilic quench is a fast, intramolecular reaction, as in the formation of **394**.

OTs, N, Boc, 393 → s-BuLi, (–)-sparteine → N, Boc, 394 — 77% yield; 55% ee, unknown absolute configuration

s-BuLi-(–)-sparteine also deprotonates **395** enantioselectively (though now with selectivity for the pro-*R* proton[80]) to form an organolithium **396** which cyclises to **397** in 72% yield.[172,80] The result of cyclising the racemic stannane **398** again proves that the enantioselective step is the deprotonation and that the (–)-sparteine plays no part in determining the stereochemical outcome during the cyclisation step.

Cl, Ph, N, Boc, 395 → s-BuLi, (–)-sparteine, toluene → Cl, Li, Ph, N, Boc, 396 → Ph, N, Boc, 397 — 72% yield, 96% ee

Cl, Me_3Sn, Ph, N, Boc, 398 → s-BuLi, (–)-sparteine, toluene → Cl, Li, Ph, N, Boc → Ph, N, Boc, (±)-397 — 44% yield

Similar results were obtained with the benzylamine-derivative **399**, which can be lithiated by *s*-BuLi-(–)-sparteine to lead to a methylated product in good ee via a configurationally stable organolithium.[82] The reaction of **400** proves again that the presence of (–)-sparteine during the electrophilic substitution step plays no part in determining the enantioselectivity of this reaction.

Ph, NBoc, OMe, 399 → n-BuLi, (–)-sparteine → Li, Ph, NBoc, OMe → MeOTf → Me, Ph, NBoc, OMe

Me_3Sn, Ph, NBoc, OMe, 400 → n-BuLi, (–)-sparteine → Li, Ph, NBoc, OMe → MeOTf → Me, Ph, NBoc, OMe

Further evidence in favour of this explanation of the mechanism comes from the cyclisation of racemic, deuterated *d*-**395**.[172] Deuterated starting materials are useful tools in this area because the outcome of the often competing stereoselective deprotonation and kinetic isotope effect sheds useful light on the mechanism of the lithiation reactions. The general principle is this: since deuterium can only perturb a mechanism significantly by the primary kinetic isotope effect, any change in the stereochemical outcome of a reaction once deuterium is incorporated indicates that the deprotonation step features some stereoselectivity.

Cl D Ph N Boc *d*-395 —*s*-BuLi, (–)-sparteine→ Ph D N Boc *d*-397 43% yield (88% *d*) 30% ee

We can use the example of *d*-**395** to explain how this works. Suppose that (–)-sparteine **12** were in fact not involved at all in the first, deprotonation step from **395**. The kinetic isotope effect (which can have very high values at –78 °C[162,163]) would ensure that most (though not all) of the deuterium remained in the intermediate organolithium, but this would be the only difference between the deuterated and non-deuterated versions of the reaction. Cyclisation would therefore give the product *d*-**397**, mainly deuterated, but with ee essentially identical to the undeuterated analogue (any difference arising solely from a negligible secondary kinetic isotope effect).

This is clearly not the case for the reaction of *d*-**397**: the ee of the product is considerably lower. There must therefore be enantioselectivity in the deprotonation step. Since the starting material *d*-**395** is racemic, each enantiomer will experience either matched or a mismatched selectivity between the kinetic isotope effect (which favours loss of H) and the *s*-BuLi-(–)-sparteine deprotonation (which favours, in this case loss of the pro-X hydrogen, whether H or D). The effect of this competition is two-fold: firstly, the product contains less deuterium than it would otherwise (50% D if the kinetic isotope effect is 0, but approaching 100% D if the kinetic isotope effect is large). Secondly, the enantioselectivity will be lower (0% if the kinetic isotope effect totally dominates the selectivity; unchanged as the kinetic isotope effect decreases in importance). Either of these two effects indicates asymmetric deprotonation: in the cyclisation of **395**, it is the second which is more significant, with deuteration dropping only to 88%, but ee dropping to 30%. The effect of a primary kinetic isotope effect on the lithiation of a racemic deuterated starting material is illustrated in Scheme 5.4.1.

The perturbation of the reaction on incorporation of deuterium into the starting material provides proof that the deprotonation step of the reaction is enantiodetermining. However, on its own it does not provide evidence of the role of that (–)-sparteine in the stereochemical course of the second, substitution step of the reaction. In this sense, the result with the stannane **398** is complementary since it provides no positive evidence regarding the mechanism of the first step, but does prove that there is no enantioselectivity involved in the second.

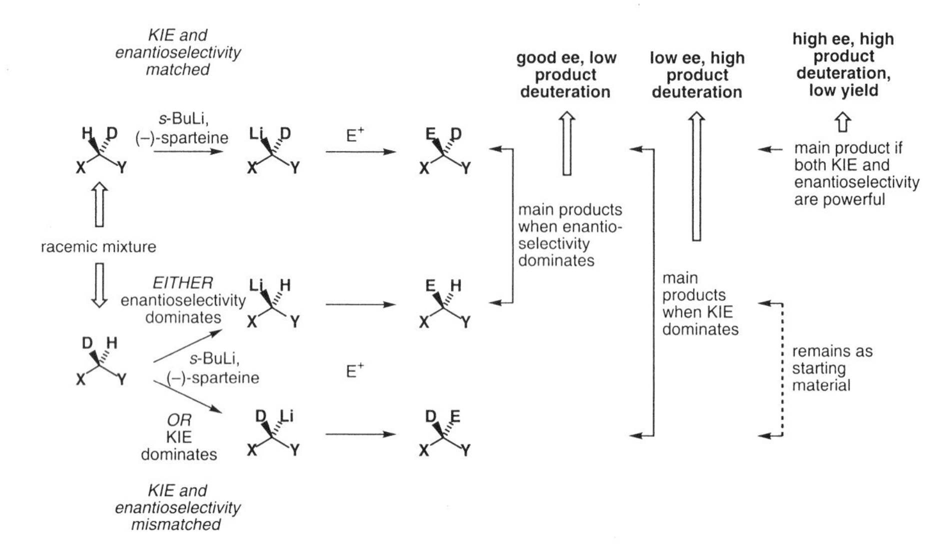

Scheme 5.4.1: The effect of a kinetic isotope effect on an asymmetric α-lithiation

An example from the work of Schlosser demonstrates a reaction in which the intermediate is configurationally unstable, but which nonetheless deuterium substitution indicates is formed by an asymmetric deprotonation. Racemic deuterated *N*-Boc benzylamine derivative **401** is deprotonated enantioselectively in a kinetic resolution (the kinetic isotope effect means only 50% of the starting material is consumed) – as indicated by high ee (85-88%) in the remaining starting material. However, re-formation of enantiomerically enriched organolithium **402** by deprotonating this material with *s*-BuLi–TMEDA returned racemic products **403** on electrophilic quench: the deprotonation is asymmetric but the organolithium is configurationally unstable.[173] Attempted deprotonation of the enantiomerically enriched intermediate leads to no reaction: KIE and enantioselectivity are both powerful and are at loggerheads.

401 — s-BuLi, (–)-sparteine, Et_2O → 402 + (remaining starting material)

CO_2 → 403 + 88% ee, >95% D

s-BuLi, TMEDA, Et_2O → 402 ⇌ 402; CO_2 → 403 (±), 47-63% D

s-BuLi, (–)-sparteine → no deprotonation (mismatched KIE and asymmetric deprotonation selectivity)

Similar results have been obtained for related compounds:[174] for example, **404** is asymmetrically deprotonated by chiral lithium amide bases.[81] Dearomatising cyclisation (section 7.2.4.6) gives products **406** with up to 84% ee, but the intermediate organolithium racemises too fast to be trapped intermolecularly: other electrophiles give **407** in only 0-44% ee.

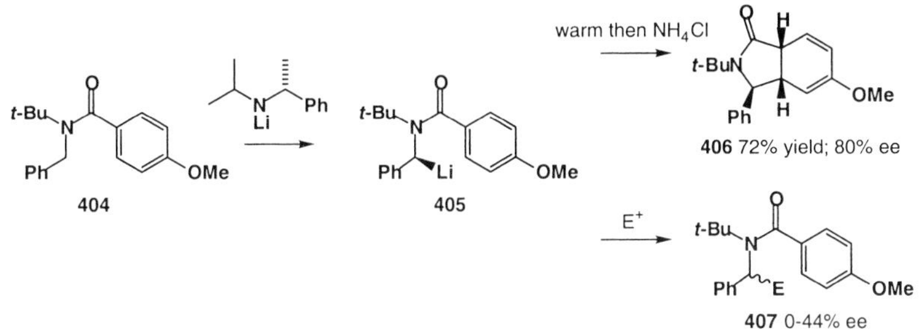

404 → **405** → warm then NH_4Cl → **406** 72% yield; 80% ee; E^+ → **407** 0-44% ee

The related cinnamylamide **408** is deprotonated by *n*-BuLi-(–)-sparteine to give a single enantiomer of the allyllithium species **409** which reacts with various electrophiles to give **410** in high ee.[85,175]

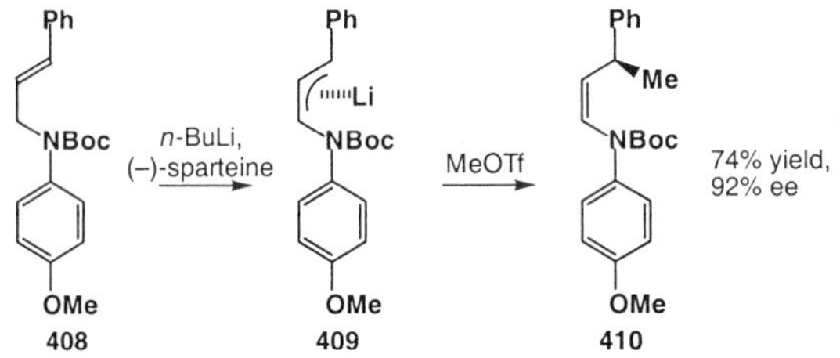

408 → *n*-BuLi, (–)-sparteine → **409** → MeOTf → **410** 74% yield, 92% ee

Deprotonation of *O*-alkyl carbamates may be achieved in an enantioselective manner with *s*-BuLi-(–)-sparteine, and the most effective of these reactions employ the oxazolidinones **411**. The related compounds **412** perform similarly, but have less neat NMR spectra. Enantioselective lithiation of **413**, followed by carboxylation and methylation with diazomethane, generates the protected α-hydroxy acid **414** in >95% ee.[176] Many other electrophiles perform well in the quench step, but not allylic or benzylic halides, which lead to partial racemisation.[177,30]

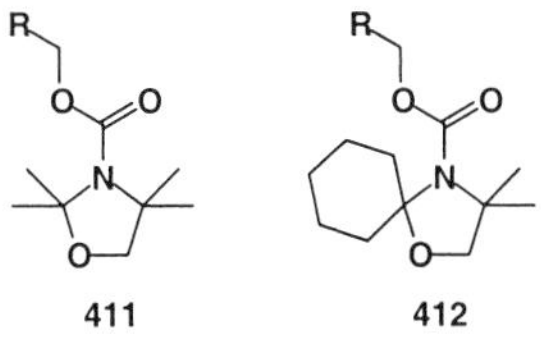

411 **412**

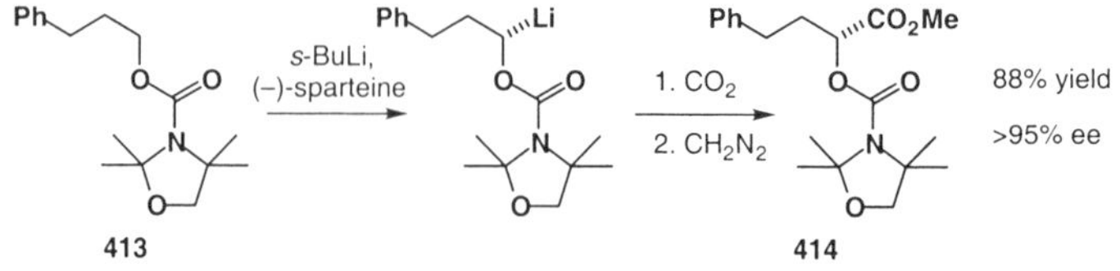

This method, employing methylation by MeI, was used to synthesise pheromones such as (*S*)-sulcatol **416**.[176] The oxazolidinone protecting group features a safety-catch which is released by treatment with acid: intramolecular assistance to hydrolysis of the secondary amide **415** rapidly generates the free alcohol product **416**.

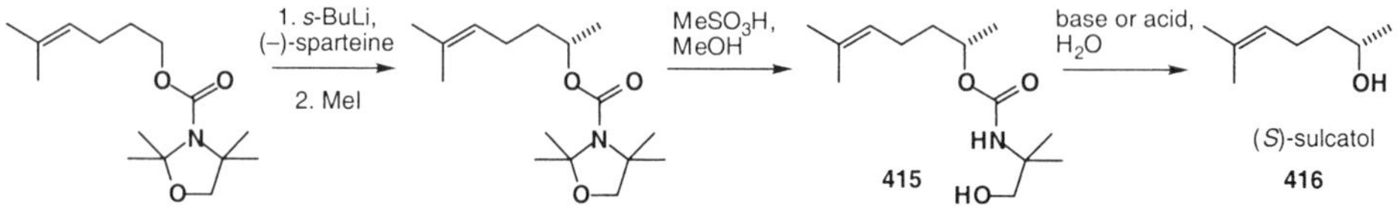

The important commercial compound (*R*)-pantolactone is formed in high yield and ee from the bis-carbamate **417** after carboxylation and lactonisation.[178] Attempts to quench the organolithium with Me_3SiCl failed; instead the silyl chloride acted as a Lewis acid, promoting displacement of the second carbamate and leading to the cyclopropane **420**.[162,179] Further functionalisation of the cyclopropane was possible by stereospecific alkylation of the ring.

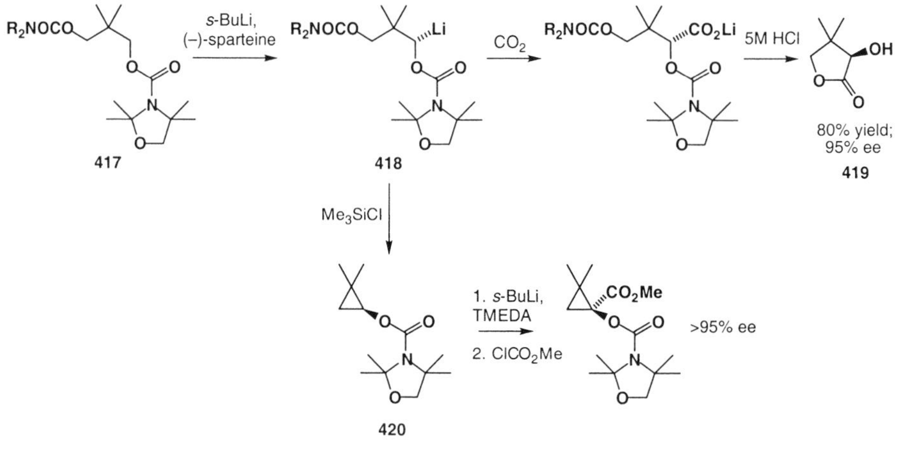

Carbamates containing unsaturated substituents may undergo enantioselective anionic cyclisations (see also section 7.2.4): for example, **421** is deprotonated enantioselectively by *s*-BuLi-(–)-sparteine and gives the cyclopentane **422**.[180]

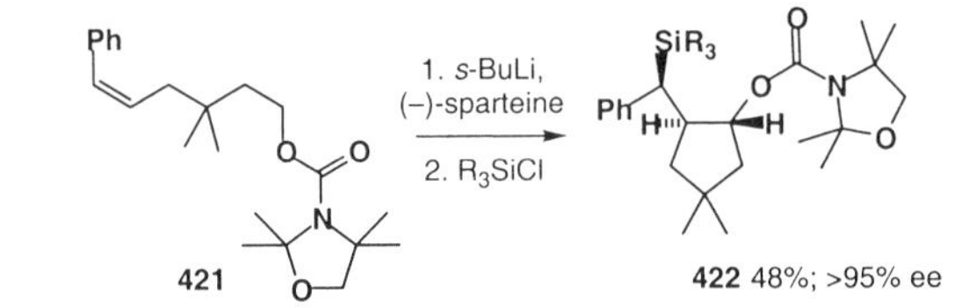

On quenching lithiated carbamates with MeOD, a deuterated carbamate is obtained which cannot be lithiated again by *s*-BuLi-(–)-sparteine, due to a powerful kinetic isotope effect (with a value >70 at –78 °C). Deprotonation can be achieved with *s*-BuLi–TMEDA, and the kinetic isotope effect now ensures that the product retains its deuterium almost entirely.[162] It also ensures that the deprotonation is stereospecific, and because of the configurational stability of the intermediate, a single enantiomer of the product **423** is formed.

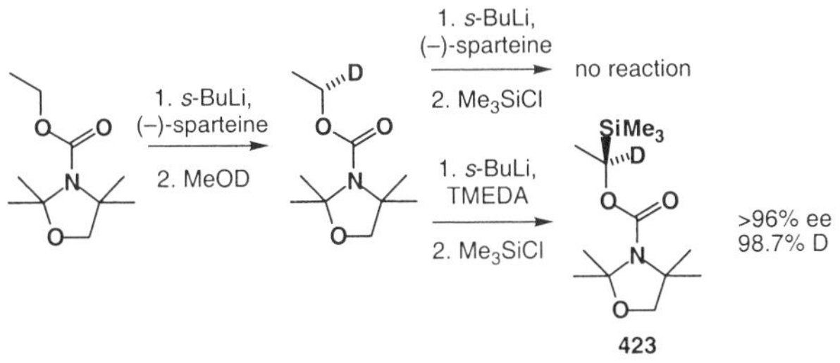

Alkyl carbamates such as these cannot be deprotonated in the absence of sparteine or TMEDA – the reaction is truly an instance of ligand accelerated catalysis (neither can they be deprotonated by *n*-BuLi or *t*-BuLi-(–)-sparteine). Furthermore, if the *s*-BuLi-(–)-sparteine complex is first treated with a non-deprotonatable *O*-isopropyl carbamate, it becomes inactive towards deprotonation of further carbamates: presumably a BuLi–sparteine–carbamate complex is formed irreversibly, and we can deduce that such a complexation is the first step in the deprotonation of other alkylcarbamates.[176]

In the event that the alkylcarbamate contains further lithium-coordinating heteroatoms, enantioselective deprotonation becomes impossible because sparteine is displaced from the lithium in this reactive complex. So, for example, the *N,N*-dimethylamine **424** (R = Me) is lithiated with only 10% ee, which its dibenzyl analogue is lithiated in 97% ee.[162] Most oxygen-containing functional groups do not have this effect.[178,181]

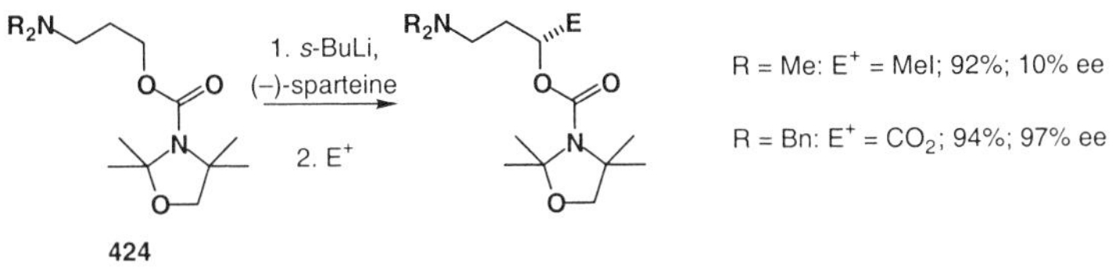

In cases where the deprotonation occurs not at the enantiotopic protons of a methylene group, but at more widely separated enantiotopic positions, configurational stability of the intermediate is not an issue in whether such a reaction can lead to enantiomerically enriched products. For example, *s*-BuLi-(–)-sparteine desymmetrises the epoxides **425** and **427** by lithiation followed by ring opening to a carbene **426** or **428** which inserts into a nearby C–H bond.[182-184]

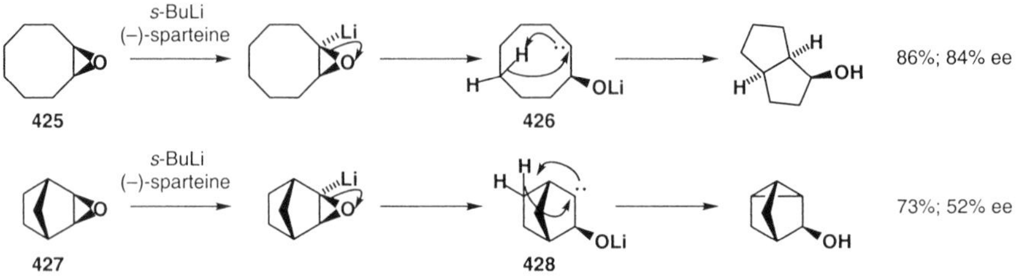

The enantiotopic methyl groups of the phosphine-borane complexes **429** and **430** can similarly be desymmetrised by *n*-BuLi-(–)-sparteine. On coupling and deprotection, valuable chiral diphosphines **431** are formed.[185]

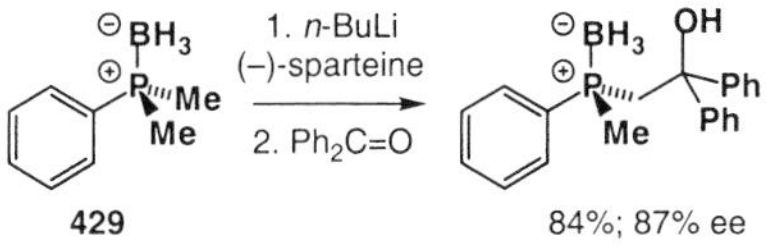

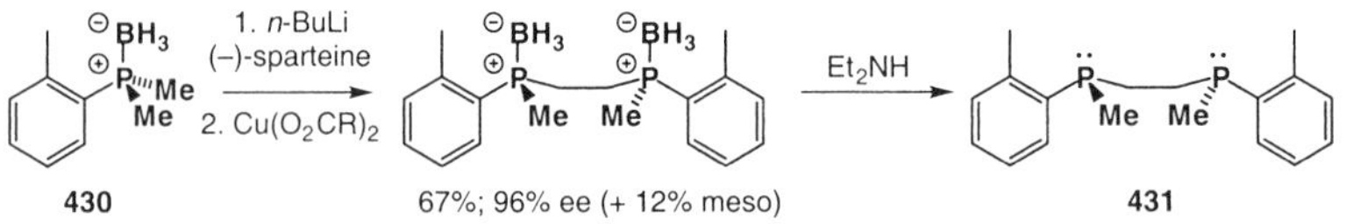

Desymmetrisation of the enantiotopic methyl groups of **432** with a chiral lithium amide base leads to atropisomeric amides in good enantiomeric excess.[186]

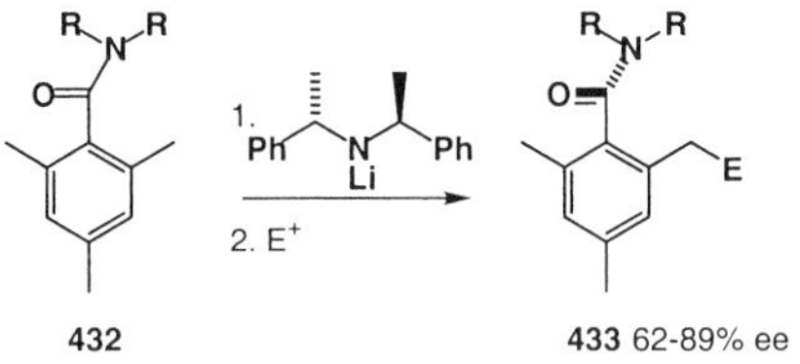

Desymmetrisation by enantioselective ortholithiation has been achieved with ferrocenylcarboxamides **434**,[187] and also (with chiral lithium amide bases) a number of chromium-arene complexes.[188] The chromium arene complex **435**, on treatment with s-BuLi-(–)-sparteine, gives **436** enantioselectively, and reaction with electrophiles leads to **437**. However, further treatment with *t*-BuLi generates the doubly lithiated species **438**, in which the new organolithium centre is more reactive than the old, which still carries the (–)-sparteine ligand. Reaction of **438** with an electrophile followed by protonation therefore gives *ent*-**437**.[189]

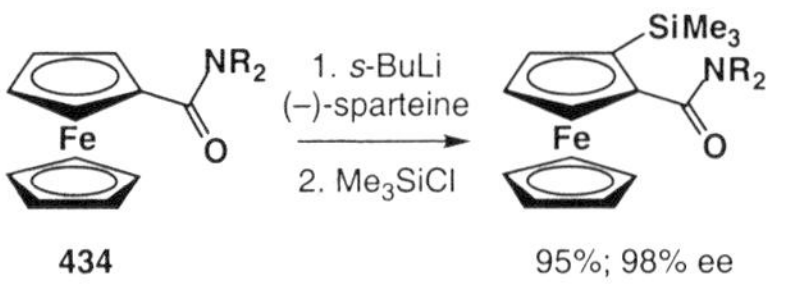

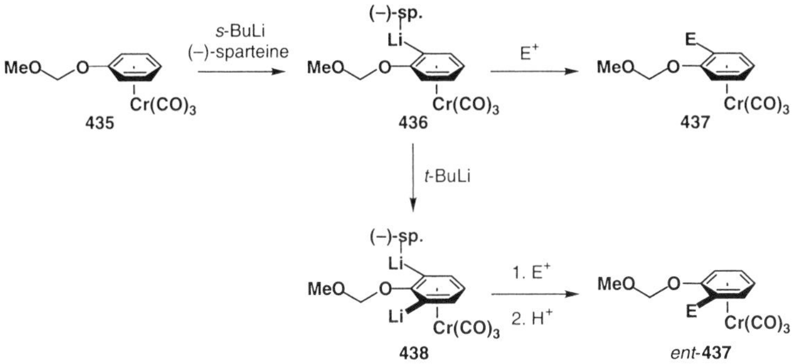

References

1. Fraenkel, G.; Duncan, J. H.; Martin, K.; Wang, J. *J. Am. Chem. Soc.* **1999,** *121*, 10538.
2. Ahlbrecht, H.; Harbach, J.; Hoffmann, R. W.; Ruhland, T. *Liebigs Ann. Chem.* **1995**, 211.
3. Dress, R. K.; Rölle, T.; Hoffmann, R. W. *Chem. Ber.* **1995**, 673.
4. Hoffmann, R. W.; Lanz, J.; Metternich, R.; Tarara, G.; Hoppe, D. *Angew. Chem., Int. Ed. Engl.* **1987,** *26*, 1145.
5. Hoffmann, R. W.; Rühl, T.; Chemla, F.; Zahneisen, T. *Liebigs Ann. Chem.* **1992**, 719.
6. Hoffmann, R. W.; Rühl, T.; Harbach, J. *Liebigs Ann. Chem.* **1992**, 725.
7. Beak, P.; Basu, A.; Gallagher, D. J.; Park, Y. S.; Thayumanavan, S. *Acc. Chem. Res.* **1996,** *29*, 552.
8. Sandström, J. *Dynamic NMR Spectroscopy*; Academic Press: London, 1982.
9. Letsinger, R. L. *J. Am. Chem. Soc.* **1950,** *72*, 4842.
10. Tarbell, D. S.; Weiss, M. *J. Am. Chem. Soc.* **1939,** *61*, 1203.
11. Cram, D. J. *Fundamentals of Carbanion Chemistry*; Academic Press: New York, 1965.
12. Curtin, D. Y.; Koehl, W. J. *J. Am. Chem. Soc.* **1962,** *84*, 1967.
13. Witkanowski, M.; Roberts, J. D. *J. Am. Chem. Soc.* **1966,** *88*, 737.
14. Fraenkel, G.; Dix, D. T.; Carlson, M. *Tetrahedron Lett.* **1968**, 579.
15. Fraenkel, G.; Dix, D. T. *J. Am. Chem. Soc.* **1966,** *88*, 979.
16. Applequist, D. E.; Peterson, A. G. *J. Am. Chem. Soc.* **1961,** *83*, 862.
17. Dewar, M. J. S.; Harris, J. M. *J. Am. Chem. Soc.* **1969,** *91*, 3652.
18. Walborsky, H. M.; Impastato, F. J.; Young, A. E. *J. Am. Chem. Soc.* **1964,** *86*, 3283.
19. Tanaka, K.; Minami, K.; Funaki, I.; Suzuki, H. *Tetrahedron Lett.* **1990,** *31*, 2727.
20. Köbrich, G.; Merkel, J.; Imkampe, K. *Chem. Ber.* **1973,** *106*, 2017.
21. Walborsky, H. M.; Hornyak, F. M. *J. Am. Chem. Soc.* **1955,** *77*, 6026.
22. Walborsky, H. M.; Youssef, A. A.; Motes, J. M. *J. Am. Chem. Soc.* **1962,** *84*, 2465.

23. Walborsky, H. M.; Periasamy, M. P. *J. Am. Chem. Soc.* **1974,** *96*, 3711.
24. Newman-Evans, R. H.; Carpenter, B. K. *Tetrahedron Lett.* **1985,** *26*, 1141.
25. Still, W. C.; Sreekumar, C. *J. Am. Chem. Soc.* **1980,** *102*, 1201.
26. Jephcote, V. S.; Pratt, A. J.; Thomas, E. J. *J. Chem. Soc., Chem. Commun.* **1984**, 800.
27. Chang, P. C.-M.; Chong, J. M. *J. Org. Chem.* **1988,** *53*, 5584.
28. Chong, J. M.; Mar, E. K. *Tetrahedron Lett.* **1990,** *31*, 1981.
29. Matteson, D. S.; Tripathy, P. B.; Sarkar, A.; Sadhu, K. M. *J. Am. Chem. Soc.* **1989,** *111*, 4399.
30. Hoppe, D.; Hintze, F.; Tebben, P. *Angew. Chem., Int. Ed. Engl.* **1990,** *29*, 1422.
31. Sommerfeld, P.; Hoppe, D. *Synlett* **1992**, 764.
32. Wright, A.; West, R. *J. Am. Chem. Soc.* **1974,** *96*, 3227.
33. Lindermann, R. J.; Ghannam, A. *J. Am. Chem. Soc.* **1990,** *112*, 2396.
34. Hoffmann, R. W.; Brückner, R. *Angew. Chem., Int. Ed. Engl.* **1992,** *31*, 647.
35. Tomooka, K.; Igarishi, T.; Watanabe, M.; Nakai, T. *Tetrahedron Lett.* **1992,** *33*, 5795.
36. Tomooka, K.; Igarishi, T.; Kishi, N.; Nakai, T. *Tetrahedron Lett.* **1999,** *40*, 6257.
37. Sawyer, J. S.; Macdonald, T. L.; McGarvey, G. J. *J. Am. Chem. Soc.* **1984,** *106*, 3376.
38. Lesimple, P.; Beau, J.-M.; Sinaÿ, P. *J. Chem. Soc., Chem. Commun.* **1985**, 894.
39. Cohen, T.; Lin, M.-T. *J. Am. Chem. Soc.* **1984,** *106*, 1130.
40. Rychnovsky, S. D. *Chem. Rev.* **1995,** *95*, 2021.
41. Rychnovsky, S. D.; Mikus, D. E. *Tetrahedron Lett.* **1989,** *30*, 3011.
42. Rychnovsky, S. D. *J. Org. Chem.* **1989,** *54*, 4982.
43. Lindermann, R. J.; Griedel, B. D. *J. Org. Chem.* **1991,** *56*, 5491.
44. Hodgson, D. M.; Norsikian, S. L. M. *Org. Lett.* **2001,** *3*, 461.
45. Molander, G. A.; Mautner, K. *J. Org. Chem.* **1989,** *54*, 4042.
46. Eisch, J. J.; Galla, J. E. *J. Am. Chem. Soc.* **1976,** *98*, 4646.
47. Burford, C.; Cooke, F.; Roy, G.; Magnus, P. *Tetrahedron* **1983,** *39*, 867.
48. Mani, N. S.; Townsend, C. A. *J. Org. Chem.* **1997,** *62*, 636.
49. Hoppe, D.; Krämer, T. *Angew. Chem., Int. Ed. Engl.* **1986,** *25*, 160.
50. Krämer, T.; Hoppe, D. *Tetrahedron Lett.* **1987,** *28*, 5149.
51. Hoppe, D.; Carstens, A.; Krämer, T. *Angew. Chem., Int. Ed. Engl.* **1990,** *29*, 1424.
52. Carstens, A.; Hoppe, D. *Tetrahedron* **1994,** *50*, 6097.
53. Derwing, C.; Hoppe, D. *Synthesis* **1996**, 149.
54. Hammerschmidt, F.; Hanninger, A. *Chem. Ber.* **1995,** *128*, 1069.
55. Behrens, K.; Fröhlich, R.; Meyer, O.; Hoppe, D. *Eur. J. Org. Chem.* **1998**, 2397.
56. Hoppe, D.; Zschage, O. *Angew. Chem., Int. Ed. Engl.* **1989,** *28*, 69.
57. Gawley, R. E.; Zhang, Q. *Tetrahedron* **1994,** *50*, 6077.
58. Alezta, V.; Bonin, M.; Micouin, L.; Husson, H.-P. *Tetrahedron Lett.* **2000,** *41*, 651.
59. Häner, R.; Olano, B.; Seebach, D. *Helv. Chim. Acta* **1987,** *70*, 1676.
60. Vedejs, E.; Moss, W. O. *J. Am. Chem. Soc.* **1993,** *115*, 1607.
61. Vedejs, E.; Kendall, J. T. *J. Am. Chem. Soc.* **1997,** *119*, 6941.
62. Rondan, N. G.; Houk, K. N.; Beak, P.; Zajdel, W. J.; Chandrasekhar, J.; Schleyer, P. v. R. *J. Org. Chem.* **1981,** *46*, 4108.

63. Seebach, D.; Wykypiel, W.; Lubosch, W.; Kalinowski, H. O. *Helv. Chim. Acta* **1978,** *61*, 3100.
64. Fraser, R. R.; Grindley, T. B.; Passannanti, S. *Can. J. Chem.* **1975,** *53*, 2473.
65. Gawley, R. E.; Hart, G. C.; Bartolotti, L. J. *J. Org. Chem.* **1989,** *54*, 175.
66. Beak, P.; Kerrick, S. T. *J. Am. Chem. Soc.* **1991,** *113*, 9708.
67. Beak, P.; Kerrick, S. T.; Wu, S.; Chu, J. *J. Am. Chem. Soc.* **1994,** *116*, 3231.
68. Elworthy, T. R.; Meyers, A. I. *Tetrahedron* **1994,** *50*, 6089.
69. Gross, K. M. B.; Jun, Y. M.; Beak, P. *J. Org. Chem.* **1997,** *62*, 7679.
70. Kopach, M. E.; Meyers, A. I. *J. Org. Chem.* **1996,** *61*, 6764.
71. Gawley, R. E.; Zhang, Q. *J. Am. Chem. Soc.* **1993,** *115*, 7515.
72. Coldham, I.; Hufton, R.; Snowden, D. J. *J. Am. Chem. Soc.* **1996,** *118*, 5322.
73. Loewe, M. F.; Boes, M.; Meyers, A. I. *Tetrahedron Lett.* **1985,** *26*, 3295.
74. Meyers, A. I.; Dickman, D. I. *J. Am. Chem. Soc.* **1987,** *109*, 1263.
75. Rein, K.; Goicoechea-Pappas, M.; Anklekar, T. V.; Hart, G. C.; Smith, G. A.; Gawley, R. E. *J. Am. Chem. Soc.* **1989,** *111*, 2211.
76. Meyers, A. I.; Guiles, J.; Warmus, J. S.; Gonzalez, M. A. *Tetrahedron Lett.* **1991,** *32*, 5505.
77. Chong, J. M.; Park, S. B. *J. Org. Chem.* **1992,** *57*, 2220.
78. Pearson, W. H.; Lindbeck, A. C.; Kampf, J. W. *J. Am. Chem. Soc.* **1993,** *115*, 2622.
79. Pearson, W. H.; Lindbeck, A. C. *J. Am. Chem. Soc.* **1991,** *113*, 8546.
80. Faibish, N. C.; Park, Y. S.; Lee, S.; Beak, P. *J. Am. Chem. Soc.* **1997,** *119*, 11561.
81. Clayden, J.; Menet, C. J.; Mansfield, D. J. *J. Chem. Soc., Chem. Commun.* **2002**, 38.
82. Park, Y. S.; Boys, M. L.; Beak, P. *J. Am. Chem. Soc.* **1996,** *118*, 3757.
83. Park, Y. S.; Beak, P. *J. Org. Chem.* **1997,** *62*, 1574.
84. Hara, O.; Ho, M.; Hamada, Y. *Tetrahedron Lett.* **1998,** *39*, 5537.
85. Weisenburger, G. A.; Beak, P. *J. Am. Chem. Soc.* **1996,** *118*, 12218.
86. Boche, G.; Opel, A.; Marsch, M.; Harms, K.; Haller, F.; Lohrenz, J. C. W.; Thümmier, C.; Kock, W. *Chem. Ber.* **1992,** *125*, 2265.
87. Boche, G.; Marsch, M.; Harbach, J.; Harms, K.; Ledig, B.; Schubert, F.; Lohrenz, J. C. W.; Ahlbrecht, H. *Chem. Ber.* **1993,** *126*, 1887.
88. Ahlbrecht, H.; Harbach, J.; Hauck, T.; Kalinowski, H.-O. *Chem. Ber.* **1992,** *125*, 1753.
89. Köbrich, G. *Angew. Chem., Int. Ed. Engl.* **1967,** *6*, 41.
90. Schmidt, A.; Köbrich, G.; Hoffmann, R. W. *Chem. Ber.* **1991,** *124*, 1253.
91. Seyferth, D.; Lambert, R. L. *J. Organomet. Chem.* **1973,** *55*, C53.
92. Seyferth, D.; Lambert, R. L.; Massol, M. *J. Organomet. Chem.* **1975,** *88*, 255.
93. Warner, P. M.; Cheng, J.-C.; Koszewski, N. J. *Tetrahedron Lett.* **1985,** *26*, 5371.
94. Kitakani, K.; Hiyama, T.; Nozaki, H. *J. Am. Chem. Soc.* **1975,** *97*, 949.
95. Kitakani, K.; Hiyama, T.; Nozaki, H. *Bull. Chem. Soc. Jap.* **1977,** *50*, 3288.
96. Harada, T.; Katsuhira, T.; Hattori, K. *J. Org. Chem.* **1993,** *58*, 2958.
97. Taylor, P. C. G.; Hobbs, W. E.; Jaquet, M. *J. Org. Chem.* **1971,** *36*, 369.
98. Baird, M. S.; Baxter, A. G. W. *J. Chem. Soc., Perkin Trans. 1* **1979**, 2317.
99. Köbrich, G.; Trapp, H. *Chem. Ber.* **1966,** *99*, 670.

100. Hoffmann, R. W.; Ruhland, T.; Bewersdorf, M. *J. Chem. Soc., Chem. Commun.* **1991**, 195.
101. Hoffmann, R. W.; Bewersdorf, M. *Chem. Ber.* **1991,** *124*, 1249.
102. Hoffmann, R. W.; Bewersdorf, M.; Ditrich, K.; Krüger, K.; Stürmer, R. *Angew. Chem., Int. Ed. Engl.* **1988,** *27*, 1176.
103. Hoffmann, R. W.; Julius, M.; Chemla, F.; Ruhland, T.; Frenzen, G. *Tetrahedron* **1994,** *50*, 6049.
104. Lutz, G. P.; Wallin, A. P.; Kerrick, S. T.; Beak, P. *J. Org. Chem.* **1991,** *56*, 4938.
105. McDougal, P. G.; Condon, B. D.; Lafosse, M. D.; Lauro, A. M.; VanDerveer, D. *Tetrahedron Lett.* **1988,** *29*, 2547.
106. Reich, H. J.; Bowe, M. D. *J. Am. Chem. Soc.* **1990,** *112*, 8994.
107. Ritter, R. H.; Cohen, T. *J. Am. Chem. Soc.* **1986,** *108*, 3718.
108. Brickmann, K.; Brückner, R. *Chem. Ber.* **1993,** *126*, 1227.
109. Hoffmann, R. W.; Dress, R. K.; Ruhland, T.; Wenzel, A. *Chem. Ber.* **1995**, 861.
110. Hoffmann, R. W.; Koberstein, R. *J. Chem. Soc., Chem. Commun.* **1999**, 33.
111. Hoffmann, R. W.; Koberstein, R.; Remacle, B.; Krief, A. *J. Chem. Soc., Chem. Commun.* **1997**, 2189.
112. Nakamura, S.; Nakagawa, R.; Watanabe, Y.; Toru, T. *J. Am. Chem. Soc.* **2000,** *122*, 11340.
113. Hoppe, D.; Kaiser, K.; Stratmann, O.; Fröhlich, R. *Angew. Chem., Int. Ed. Engl.* **1997,** *36*, 2784.
114. Marr, F.; Fröhlich, R.; Hoppe, D. *Org. Lett.* **1999,** *1*, 2081.
115. Stratman, O.; Kaiser, B.; Fröhlich, R.; Meyer, O.; Hoppe, D. *Chem. Eur. J.* **2001,** *7*, 423.
116. Boche, G. *Angew. Chem., Int. Ed. Engl.* **1989,** *28*, 277.
117. Gais, H.-J.; Hellmann, G.; Günther, H.; Lopez, F.; Lindner, H. J.; Braun, J. *Angew. Chem., Int. Ed. Engl.* **1989,** *28*, 1025.
118. Gais, H.-J.; Hellmann, G. *J. Am. Chem. Soc.* **1992,** *114*, 4439.
119. Tanikaga, R.; Hamamura, K.; Hosoya, K.; Kaji, A. *J. Chem. Soc., Chem. Commun.* **1988**, 817.
120. Boche, G.; Marsch, M.; Harms, K.; Sheldrick, G. M. *Angew. Chem., Int. Ed. Engl.* **1985,** *24*, 573.
121. Gais, H.-J.; Vollhardt, J.; Hellmann, G.; Paulus, H.; Lindner, H. J. *Tetrahedron Lett.* **1988,** *29*, 1259.
122. Raabe, G.; Gais, H.-J.; Fleischhauer, J. *J. Am. Chem. Soc.* **1996,** *118*, 4622.
123. Reich, H. J.; Dykstra, R. R. *Angew. Chem., Int. Ed. Engl.* **1993,** *32*, 1469.
124. Krief, A.; Evrard, G.; Badaoui, E.; De Beys, V.; Dieden, R. *Tetrahedron Lett.* **1989,** *30*, 5635.
125. Hoffmann, R. W.; Bewersdorf, M. *Tetrahedron Lett.* **1990,** *31*, 67.
126. Hoffmann, R. W.; Julius, M.; Olfmann, K. *Tetrahedron Lett.* **1990,** *31*, 7419.
127. Ruhland, T.; Dress, R. K.; Hoffmann, R. W. *Angew. Chem., Int. Ed. Engl.* **1993,** *32*, 1467.
128. Denmark, S. E.; Dorow, R. L. *J. Org. Chem.* **1990,** *55*, 5926.

129. O'Brien, P.; Warren, S. *Tetrahedron Lett.* **1995,** *36*, 8473.
130. Fraenkel, G.; Winchester, W. R.; Williard, P. G. *Organometallics* **1989,** *8*, 2308.
131. Zarges, W.; Marsch, M.; Harms, K.; Koch, W.; Frenking, G.; Boche, G. *Chem. Ber.* **1991,** *124*, 543.
132. Almena, J.; Foubelo, F.; Yus, M. *J. Org. Chem.* **1994,** *59*, 3210.
133. Kato, T.; Marumoto, S.; Sato, T.; Kuwajima, I. *Synlett* **1990**, 671.
134. Marsch, M.; Harms, K.; Zschage, O.; Hoppe, D.; Boche, G. *Angew. Chem., Int. Ed. Engl.* **1991,** *103*, 321.
135. Krief, A.; Hobe, M.; Dumont, W.; Badaoui, E.; Guittet, E.; Evrard, G. *Tetrahedron Lett.* **1992,** *33*, 3381.
136. Krief, A.; Hobe, M. *Tetrahedron Lett.* **1992,** *33*, 6527.
137. Krief, A.; Hobe, M. *Tetrahedron Lett.* **1992,** *33*, 6529.
138. Thayumanavan, S.; Basu, A.; Beak, P. *J. Am. Chem. Soc.* **1997,** *119*, 8209.
139. Clayden, J.; Pink, J. H. *Tetrahedron Lett.* **1997,** *38*, 2565.
140. Clayden, J.; Helliwell, M.; Pink, J. H.; Westlund, N. *J. Am. Chem. Soc.* **2001,** *123*, 12449.
141. Peoples, P. R.; Grutzner, J. B. *J. Am. Chem. Soc.* **1980,** *102*, 4709.
142. Paquette, L. A.; Ra, C. S. *J. Org. Chem.* **1988,** *53*, 4978.
143. Hoell, D.; Lex, J.; Müllen, K. *Angew. Chem., Int. Ed. Engl.* **1983,** *22*, 243.
144. Bousbaa, J.; Ooms, F.; Krief, A. *Tetrahedron Lett.* **1997,** *38*, 7625.
145. Seyferth, D.; Vaughan, L. G. *J. Am. Chem. Soc.* **1964,** *86*, 883.
146. Curtin, D. Y.; Crump, J. W. *J. Am. Chem. Soc.* **1958,** *80*, 1922.
147. Panek, E. J.; Neff, B. L.; Chu, H.; Panek, M. G. *J. Am. Chem. Soc.* **1975,** *97*, 3996.
148. Knorr, R.; Lattke, E. *Tetrahedron Lett.* **1977**, 3969.
149. Negishi, E.-i.; Boardman, L. D.; Tour, J. M.; Sawada, H.; Rand, C. L. *J. Am. Chem. Soc.* **1983,** *105*, 6344.
150. Reich, H. J.; Borst, J. P.; Coplien, M. B.; Phillips, N. H. *J. Am. Chem. Soc.* **1992,** *114*, 6577.
151. Nesmeyanov, A. N.; Borisov, A. E. *Tetrahedron* **1957,** *1*, 158.
152. Sawyer, J. S.; Kucerovy, A.; Macdonald, T. L.; McGarvey, G. J. *J. Am. Chem. Soc.* **1988,** *110*, 842.
153. Hoffmann, R.; Minkin, V. I.; Carpenter, B. K. *Bull. Soc. Chim. Fr.* **1996,** *133*, 117.
154. Tomooka, K.; Komine, N.; Nakai, T. *Tetrahedron Lett.* **1997,** *38*, 8939.
155. Kitching, W.; Olszowy, H. A.; Harvey, K. A. *J. Org. Chem.* **1982,** *47*, 1893.
156. Corey, E. J.; Eckrich, T. M. *Tetrahedron Lett.* **1984,** *25*, 2415.
157. Basu, A.; Beak, P. *J. Am. Chem. Soc.* **1996,** *118*, 1575.
158. Thayumanavan, S.; Lee, S.; Liu, C.; Beak, P. *J. Am. Chem. Soc.* **1994,** *116*, 9755.
159. Clayden, J.; Pink, J. H. *Tetrahedron Lett.* **1997,** *38*, 2561.
160. Neumann, H.; Seebach, D. *Chem. Ber.* **1978,** *111*, 2785.
161. Hammerschmidt, F.; Hanninger, A.; Völlenkle, H. *Chem. Eur. J.* **1997,** *3*, 1728.
162. Hoppe, D.; Paetow, M.; Hintze, F. *Angew. Chem., Int. Ed. Engl.* **1993,** *32*, 394.
163. Kaiser, B.; Hoppe, D. *Angew. Chem., Int. Ed. Engl.* **1995,** *34*, 323.

164. Clayden, J.; Kenworthy, M. N.; Youssef, L. H.; Helliwell, M. *Tetrahedron Lett.* **2000,** *41*, 5171.
165. Quesnelle, C.; Iihama, T.; Aubert, T.; Perrier, H.; Snieckus, V. *Tetrahedron Lett.* **1992,** *33*, 2625.
166. Beak, P.; Tse, A.; Hawkins, J.; Chen, C.-W.; Mills, S. *Tetrahedron* **1983,** *39*, 1983.
167. Pratt, S. A.; Goble, M. P.; Mulvaney, M. J.; Wuts, P. G. M. *Tetrahedron Lett.* **2000,** *41*, 3559.
168. Clayden, J.; Lund, A.; Youssef, L. H. *Org. Lett.* **2001,** *3*, 4133.
169. Wiberg, K. B.; Bailey, W. F. *J. Am. Chem. Soc.* **2001,** *123*, 8231.
170. Dieter, R. K.; Topping, C. M.; Chandupatla, K. R.; Lu, K. *J. Am. Chem. Soc.* **2001,** *123*, 5132.
171. Kobayashi, S.; Shiraishi, N.; Lam, W. W.-L.; Manabe, K. *Tetrahedron Lett.* **2001,** *42*, 7303.
172. Wu, S.; Lee, S.; Beak, P. *J. Am. Chem. Soc.* **1996,** *118*, 715.
173. Schlosser, M.; Limat, D. *J. Am. Chem. Soc.* **1995,** *117*, 12342.
174. Barberis, C.; Voyer, N.; Roby, J.; Chénard, S.; Tremblay, M.; Labrie, P. *Tetrahedron* **2001,** *57*, 2065.
175. Serino, C.; Stehle, N.; Park, Y. S.; Florio, S.; Beak, P. *J. Org. Chem.* **1999,** *64*, 1160.
176. Hoppe, D.; Hense, T. *Angew. Chem., Int. Ed. Engl.* **1997,** *36*, 2282.
177. Hintze, F.; Hoppe, D. *Synthesis* **1992**, 1216.
178. Paetow, M.; Ahrens, H.; Hoppe, D. *Tetrahedron Lett.* **1992,** *33*, 5323.
179. Paetow, M.; Kotthaus, M.; Grehl, M.; Fröhlich, R.; Hoppe, D. *Synlett* **1994**, 1034.
180. Woltering, M. J.; Fröhlich, R.; Hoppe, D. *Angew. Chem., Int. Ed. Engl.* **1997,** *36*, 1764.
181. Schwerdtfeger, J.; Hoppe, D. *Angew. Chem., Int. Ed. Engl.* **1992,** *31*, 1505.
182. Hodgson, D. M.; Lee, G. P. *J. Chem. Soc., Chem. Commun.* **1996**, 1015.
183. Hodgson, D. M.; Wisedale, R. *Tetrahedron Asymmetry* **1997,** *8*, 1275.
184. Hodgson, D. M.; Lee, G. P. *Tetrahedron Asymmetry* **1997,** *8*, 2303.
185. Muci, R. A.; Campos, K. R.; Evans, D. A. *J. Am. Chem. Soc.* **1995,** *117*, 9075.
186. Clayden, J.; Johnson, P.; Pink, J. H. *J. Chem. Soc., Perkin Trans. 1* **2001**, 371.
187. Tsukazaki, M.; Tinkl, M.; Roglans, A.; Chapell, B. J.; Taylor, N. J.; Snieckus, V. *J. Am. Chem. Soc.* **1996,** *118*, 685.
188. O'Brien, P. *J. Chem. Soc., Perkin Trans. 1* **1998**, 1439.
189. Tan, Y.-L.; Widdowson, D. A.; Wilhelm, R. *Synlett* **2001**, 1632.

CHAPTER 6

Stereoselective and Stereospecific Substitution Reactions of Organolithiums

6.1 Stereospecific reactions of organolithium compounds

6.1.1 Introduction

In this section, we shall discuss the stereospecific[1] reaction of organolithiums with electrophiles. In general, this reaction proceeds with retention of stereochemistry – for non-stabilised (i.e. non-benzylic or allylic – we shall use this term in this sense throughout this section) organolithiums there are only two specific and one general exception to this rule. Early theoretical studies[2] nonetheless showed that both retentive and invertive electrophilic substitutions were energetically feasible. For stabilised organolithiums – compounds in which the lithium finds itself in an allylic or benzylic position – both retention and inversion of stereochemistry are indeed observed, and we shall devote much of this section to a discussion of these reactions. In a few cases, radicals intervene, and a mixture of retentive and invertive products is formed.

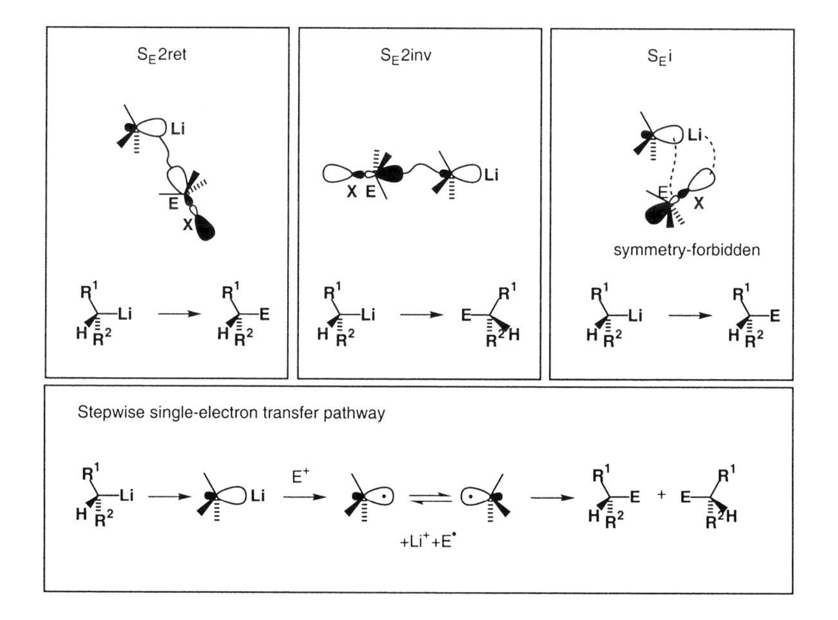

Scheme 6.1.1 Frontier orbitals and electrophilic substitution

Possible pathways for electrophilic substitution are summarised in scheme 6.1.1. Gawley[3] has proposed the terms "S_E2ret" and "S_E2inv" as a convenient means of distinguishing between the two steric courses of the reaction, and we shall employ these terms to indicate the concerted formation of a new C–E bond and breakage of the old C–Li and E–X bonds. Note that both S_E2ret and S_E2inv are allowed by orbital symmetry, and both require inversion at the electrophilic centre. In neither case is there an interaction between the lithium cation and the leaving group. Such a transition state would lead to retention at the electrophilic centre – S_Ei – and is forbidden by orbital symmetry. A further possibility, which operates surprisingly rarely, is substitution with complete loss of stereospecificity: this may happen through a non-concerted S_E1-type mechanism involving single electron transfer.

Clearly, in this section we shall deal only with organolithiums displaying at least some configurational stability. Loss of stereospecificity is obviously possible if the organolithium itself is not configurationally stable.

6.1.2 Vinyllithiums

The simple case of double bond stereospecificity in the reactions of vinyllithiums was studied extensively in the 1950's and 1960's: in every case, the reaction proceeds with reliable retention. Seyferth[4] proved this by quenching with electrophiles two vinyllithiums **1** and **2**. The geometry of both the organolithiums and the products was evident from their NMR coupling constants.

J = 22.2 Hz

1

E^+ = CO_2, Me_3SiCl, PhCHO

J = 17.4 Hz

2

Previous work[5,6] had shown that, overall, transmetallation–electrophilic quench via vinyllithiums proceeds with retention or stereochemistry, and Seyferth's work[4] finally proved that each of the two steps in the sequence are retentive.

6.1.3 Non-stabilised alkyllithiums

6.1.3.1 The general rule: retention (S_E2ret)

Non-stabilised organolithiums – that is, organolithiums in which the lithium-bearing centre is tetrahedral – generally react with high stereospecificity. This fact is essential for the use of chiral bases such as *s*-BuLi-(–)-sparteine in the synthesis of enantiomerically enriched compounds by asymmetric deprotonation, as well as for the use of enantiomerically pure stannanes as precursors to organolithium intermediates which themselves react with retention. All available evidence points to reliable retention of stereochemistry[7] – in other words, non-stabilised organolithiums typically react in an $S_E2$ret fashion – but in most cases this has not been unequivocally proved. Frequently the stereochemistry of the organolithium itself can be inferred only from the stereochemistry of a precursor such as a stannane or from the precedented enantioselectivity of a chiral base. So while it is possible to be sure that an entire sequence of organolithium formation / electrophilic substitution proceeds with retention, it is not certain beyond all doubt that each individual step is retentive (scheme 6.1.2). All we can say is that in only one case is there evidence for fully stereospecific inversion of stereochemistry in the reaction of an unstabilised organolithium, and only in one case is the evidence for a tin-lithium exchange proceeding with incomplete retention of stereochemistry. Whether there are any hidden pairs of invertive reactions, still giving overall retention, remains to be seen.

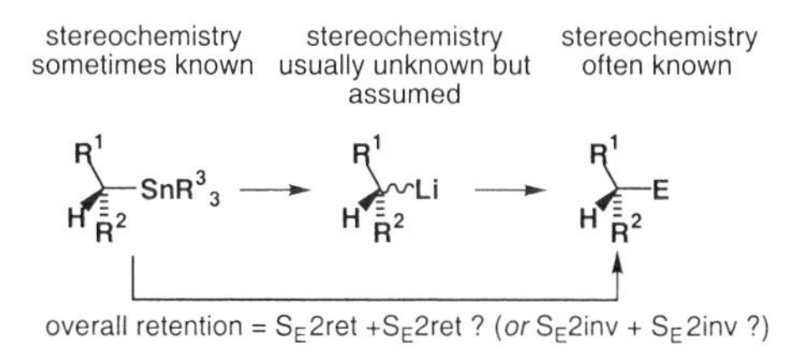

Scheme 6.1.2 Evidence for stereospecificity

The two most important classes of non-stabilised organolithiums readily formed in enantiomerically enriched form are the lithiated *N*-Boc pyrrolidine **3**[8] and α-alkoxyorganolithiums **11**, **13**[9,10] and **15**.[11] Pyrrolidines **3** (unlike their *N*-alkyl protected cousins – see below) react stereospecifically with all electrophiles, including benzophenone, giving products of more or less consistent enantiomeric excess, the value reflecting the enantioselectivity of the deprotonation step.[8] The stereochemistries of the two products **4** (E^+ = Ph_2CO) and **4** (E^+ = CO_2) were proved by comparison with L-proline-derived material, and the other electrophiles were assumed to follow suit. In the unlikely event that Bu_3SnCl, whose tendency to react with inversion with stabilised organolithiums is noted below, reacts with inversion in this case, the stereochemistry of re-formation of the organolithium followed by a second quench would be explicable only by an unprecedented fully invertive tin-lithium exchange.

3 → (E^+, $S_E2$ret) → **4**

E^+	E	ee
Ph_2CO	$-C(OH)Ph_2$	90 (proved *R*)
CO_2	$-CO_2H$	88 (proved *R*)
Me_3SiCl	$-SiMe_3$	96
Me_2SO_4	–Me	94
Bu_3SnCl	$-SnBu_3$	94

(94 → 96: 1. *s*-BuLi 2. Me_3SiCl)

By using an achiral base and a chiral starting material, whose diastereotopic protons H^A and H^B were easily assigned in the ^1H NMR spectrum, Meyers was able to clarify further the sense of stereospecificity of these reactions.[12] Chiral *N*-Boc pyrrolidine **5**, after deprotonation by *s*-BuLi-(–)-sparteine and electrophilic quench, gave methylated and deuterated products **8** and **7** of provable stereochemistry. Re-deprotonation of **7** was prevented by the kinetic isotope effect – so H^B must be the proton removed by the base, and both substitutions must be retentive (and the intermediate organolithium **6** must be configurationally stable). The only alternative explanation – that both deprotonation and substitution are invertive – can be discounted as unnecessarily complicated.[13] On raising the temperature of the organolithium, epimerisation was observed – an indication of the configurational instability of **6** at higher temperature.

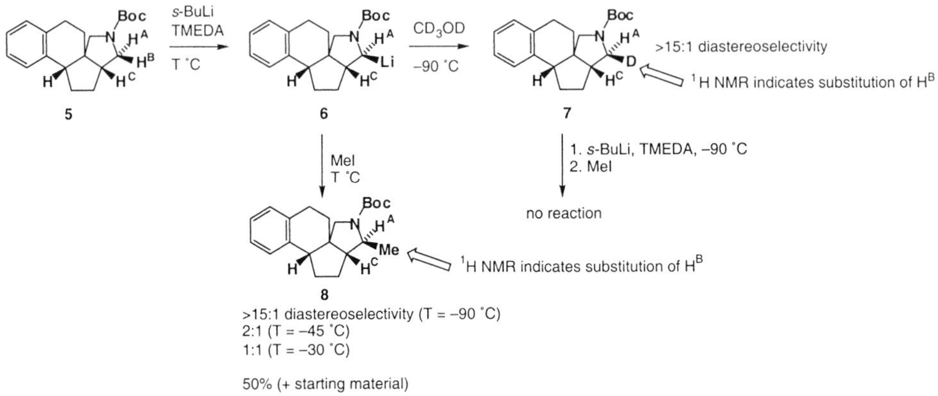

The "enolates" of aziridinecarboxylic esters behave very much like unstabilised organolithiums – they are configurationally stable organolithiums – and as such, not surprisingly react with retention of stereochemistry. Aziridine **9** was proved to react via **10** with retention with benzyl bromide; other electrophiles react stereospecifically, presumably with retention.[14]

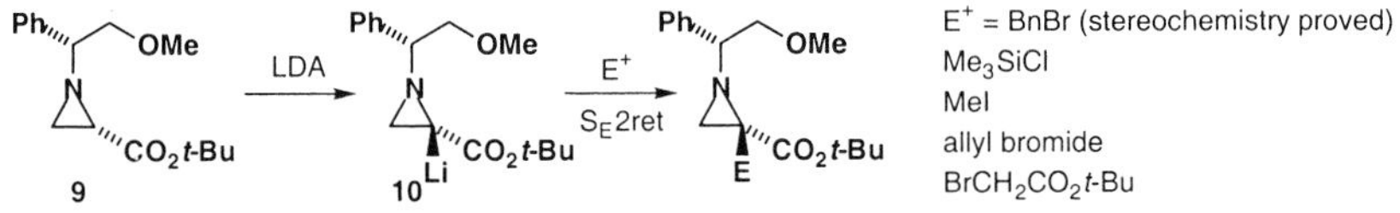

The overall retentive nature of the substitution of α-alkoxyorganolithiums **11** and **13**, formed by a presumably retentive tin-lithium exchange, was proved for the stannylation of **13**, which on transmetallation and re-stannylation, returned the starting diastereoisomer of the stannane **14**.[9] Later work on diastereoisomeric tin-substituted cyclohexanes was able to prove overall retention for a wider range of electrophiles, but still the assignment of $S_E2$ret is based on assumed retention during tin-lithium exchange (see section 5.2.1).[10]

E^+	E
Me_2CO	$-C(OH)Me_2$
Me_3SiCl	$-SiMe_3$
Me_2SO_4	$-Me$
Bu_3SnCl	$-SnBu_3$

Retention in the reactions of **15** is established both from presumed retention in the Sn-Li exchange step of a stannylation-destannylation sequence and by evidence that the *s*-BuLi-(–)-sparteine complex used to make the organolithium reliably removes the pro-*R* proton adjacent to a carbamate (see below for crystallographic evidence involving a similar compound).[11] The stereochemistry of the products **16**, almost all formed essentially in enantiomerically pure form, was proved for the CO_2 adduct and the MeI adduct by comparison with known compounds. The only electrophiles for which incomplete retention of stereochemistry has been observed are the benzylic or allylic halides. These probably react in part by single electron transfer S_E1 mechanisms, rather than by partial S_E2inv.[15] For example, **15** reacts with allyl bromide to give **16** (E = allyl) with only 42% ee.

E^+	E	ee
CO_2	$-CO_2H$	>95 (proved *S*)
MeI	–Me	>95 (proved *S*)
Me_3SiCl	$-SiMe_3$	>95
allyl bromide	–allyl	42
Me_3SnCl	$-SnMe_3$	>95

stereochemistry deducible by analogy

Anionic cyclisations of α-alkoxyorganolithiums similarly proceed with retention, in contrast to their [2,3]-sigmatropic rearrangements (see sections 7.2 and 8.4).

Configurationally stable sulfur-substituted organolithiums probably also react with retention. Hoffmann showed that the retro-anionic cyclisation of **17** and electrophilic substitution with MeI of **18** proceeded with overall retention, indicating either double inversion or double retention.[16] A tin electrophile behaved similarly.

80% ee known configuration

Early work on unfunctionalised organolithiums[17,18] demonstrated that carbonation of, for example, the diastereoisomeric cyclopropyllithiums *cis*- and *trans*-**20** was fully stereospecific, with the overall transformation of *cis*- and *trans*-**19** to **20** and **21** proceeding with retention.[19] Bromination of similar organolithiums is reported to proceed with incomplete stereospecificity, probably due to a competing radical pathway, which can be prevented by carrying out the reaction in pentane.

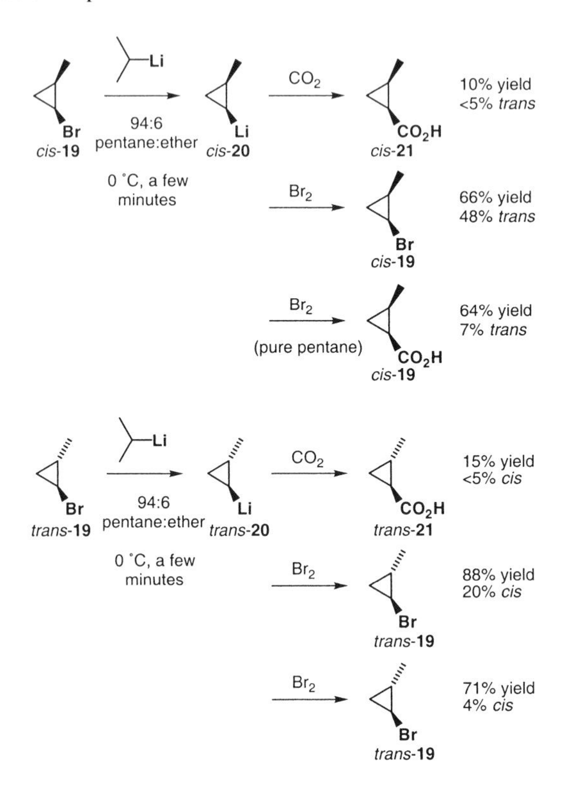

6.1.3.2 *The exception – alkylation of lithiated N-alkyl pyrrolidines and piperidines: inversion (S_E2inv)*

Lithiated *N*-alkyl pyrrolidines or piperidines **22** (in which the Li atom is not stabilised by intramolecular O–Li coordination) similarly react with retention with most electrophiles (all acylating agents, aldehydes and aliphatic ketones[20,21]) and cyclise with retention onto an unactivated alkene[22] (scheme 6.1.3). With electrophiles prone to single electron transfer – benzyl bromide, *t*-butyl bromoacetate, benzophenone – totally racemic products are formed (presumably by S_Ei rather than by competing S_E2inv and S_E2ret). But with other alkylating agents, inversion of stereochemistry occurs – cleanly for lithiated piperidines and with about 20% retention for lithiated pyrrolidines (scheme 6.1.3). It is not clear why these two organolithiums behave so differently from the analogous *N*-Boc compound **3**.

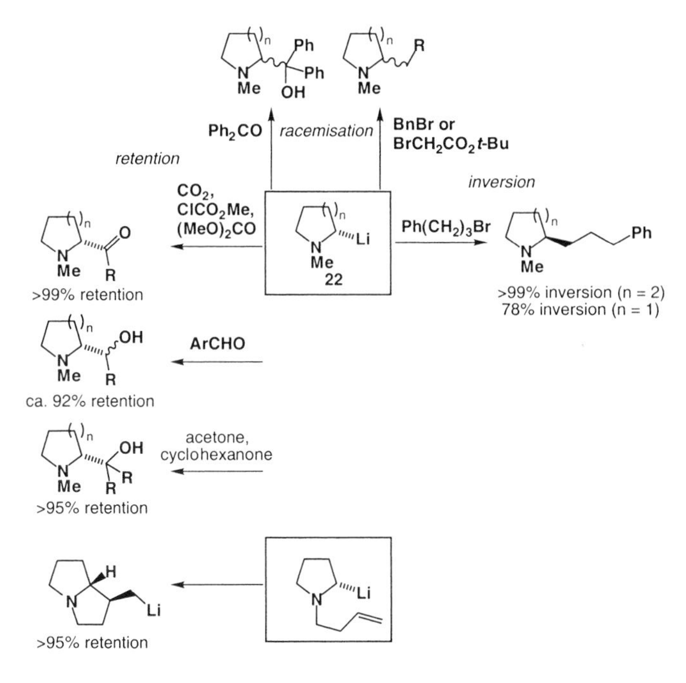

Scheme 6.1.3 Stereospecificity in the substitution of lithiopiperidines and lithiopyrrolidines

6.1.3.3 Rearrangements ([1,2] and [2,3], except Brook rearrangements) of unstabilised organolithiums: inversion (S_E2inv)

The first evidence that the [2,3]-Wittig rearrangement of organolithiums proceeds with inversion followed theoretical work[23] predicting that this should be the case. Cohen used the known axial preference of the formation of organolithiums by reduction of 2-alkylthiopyrans **23** to produce the organolithium **24** diastereoselectively, the vinylic and lithio substituents lying *trans*. After 12 h at –78 °C, a mixture of [1,2]- and [2,3]-Wittig products were produced, both stereoselectively; the [2,3]-product **25** results from inversion at one reacting centre and retention at the other. Since the migration must be suprafacial across the allylic system, the conclusion must be that the reaction proceeds with inversion at the organolithium centre.[24] These rearrangements are discussed in greater detail in Chapter 8.

SPh 23 — Li, DBB → Li 24 ⇌ Li → [Li] — inversion → OH 25 45%

More definitive evidence for inversion comes from an example lacking the diastereoselective bias present in **24**.[25] Stannane **26** was made by an asymmetric acylstannane reduction of known enantioselectivity.[26] Transmetallation of **26** (62% ee) gives **27** which rearranges to **28** (62% ee). The stereochemistry of the product **28**, which was proved by conversion to the naturally occurring ant pheromone **29** of known absolute configuration, demonstrates that the rearrangement proceeds with inversion of stereochemistry, the only assumption being retention of stereochemistry in the tin-lithium exchange step.

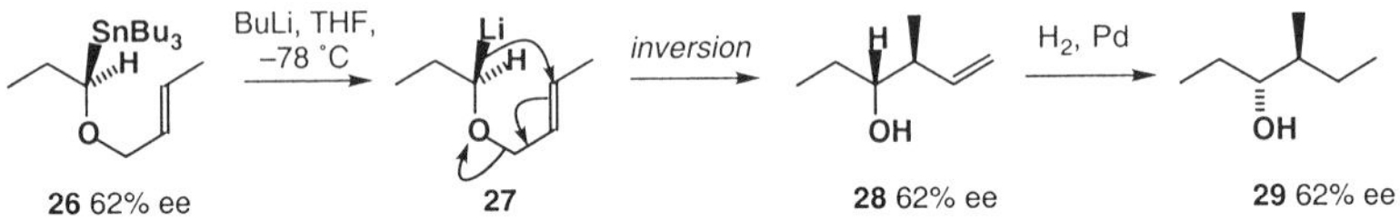

26 62% ee — **27** — **28** 62% ee — **29** 62% ee

[1,2]-Wittig rearrangements proceed with predominant inversion, with radicals intervening in the mechanism.[27] For example, stannane (*R*)-**30** of 88% ee rearranges, on transmetallation with alkyllithiums, to the alcohol (*R*)-**31** of 42% ee, a reaction demonstrating 74% invertive stereospecificity.

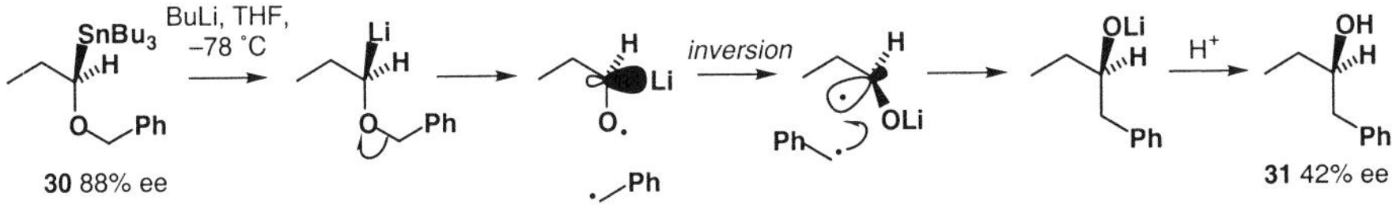

30 88% ee — **31** 42% ee

There is a marked contrast between these rearrangements and two similar classes of reaction. Firstly, anionic cyclisations (see section 7.2) of α-alkoxyorganolithiums such as **32**, which proceed with *retention*.[28] Yields in the cyclisation are poor from **33** without the driving force of methoxide elimination, but the methoxy group has no effect on the stereospecificity of the cyclisation.

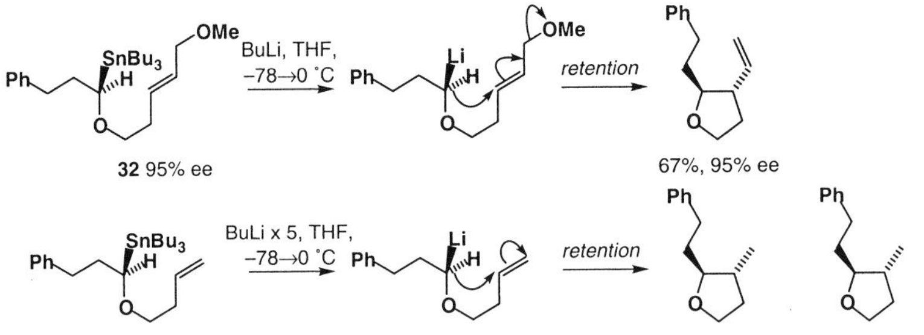

32 95% ee — 67%, 95% ee

33 90% ee — 13%, 90% ee — 2%, 90% ee

Secondly, Brook rearrangements (section 8.2), which proceed with *retention* when unstabilised organolithiums intervene[29-31] (though with *inversion* via stabilised benzylic organolithiums[32]). Enantiomerically pure **34**, for example, gives **35** with 97% ee on transmetallation. Brook rearrangements of **36**, which proceed without cyclopropane ring opening, show that no radicals intervene in this reaction, in contrast with the [1,2]-Wittig rearrangement.[29]

3 x BuLi, THF, −78 °C

H^+

34

35 91% yield, 97% ee

36

6.1.4 Stabilised alkyllithiums: retention ($S_E2$ret) or inversion ($S_E2$inv)?

6.1.4.1 Benzyllithiums

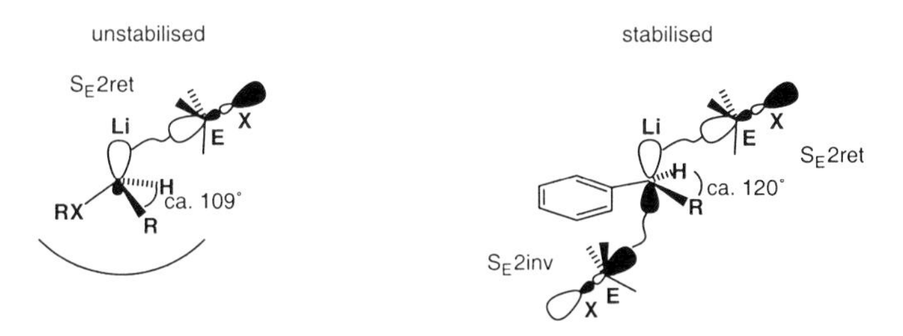

Scheme 6.1.4 Frontier orbitals in the substitution of benzyllithiums

With stabilised (conjugated) – and in particular benzylic – organolithiums, the situation is far from clear. In general, an adjacent π-system gives the C–Li bond more p-character, increasing the planarity of the organolithium and opening up to a greater extent, both electronically and sterically, possibilities for attack on the rear lobe of the C–Li σ bond (scheme 6.1.4). Hoppe carried out a thorough study of the stereochemical course of the reaction of a range of acylating agents[33,34] and other electrophiles[35,34] with the configurationally stable tertiary benzylic organolithium **37**, and scheme 6.1.5 summarises these results. The stereochemistry of the organolithium is deduced from presumed retention during deprotonation, and, except where indicated, the stereochemistry of the products was proved by comparison with known compounds.

In general, retention of configuration is seen when the electrophile has a leaving group capable of coordination to the lithium counterion – MeO^- or AcO^-. Inversion of configuration is seen with halide or cyanide leaving groups, or on addition to "heterocumulenes" such as CO_2 or CS_2. There is some confusion over the course taken by the protonation of **37** with AcOH: early reports[33] of inversion (which were rationalised by assuming the protonating species was the salt of TMEDA and AcOH) have been corrected[36,35] – AcOH protonates with retention. Nonetheless, more recent reports have suggested that deutero-hydrochloric acid (DCl) may protonate with inversion,[37] and there is still some uncertainty in the area.

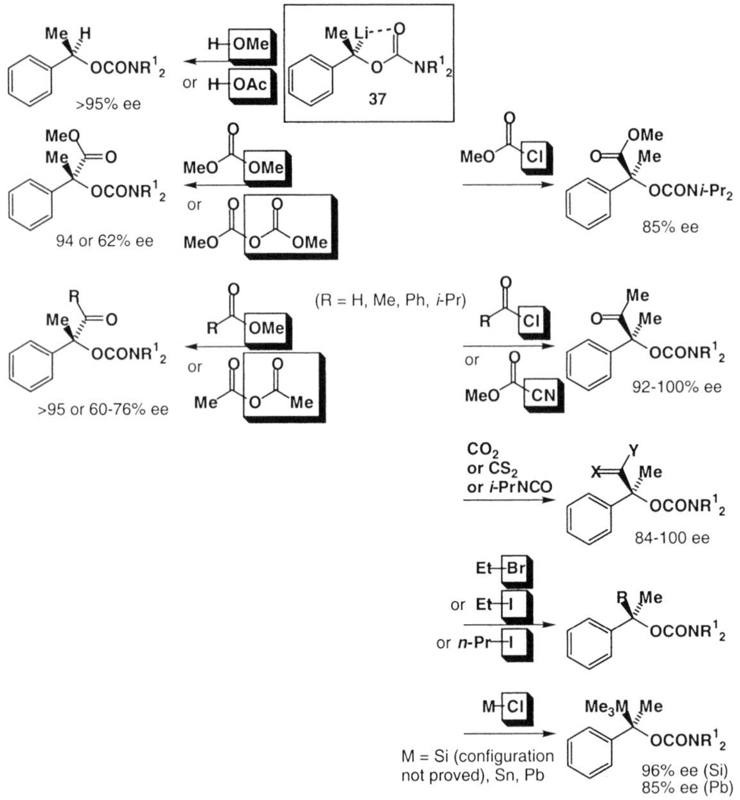

Scheme 6.1.5 Stereospecificity in the electrophilic substitution of an α-carbamoyloxy benzyllithium

Altering the structure of the benzyllithium by introducing a ring creates indanylcarbamate **38**. Its reactions follow a similar pattern to benzyllithium **37** but with a general shift towards retentive substitution (scheme 6.1.6).[38] The principal defectors from inversion with **37** to retention with **38** are CO_2 and MeOCOCl.

Following the pattern set by **37**, Bu_3SnCl reacts with inversion with the indanyllithium **38**, but other stannanes behave very differently – Bu_3SnOTf, for example, reacts with retention! The stereospecificity of the reaction depends strongly on the stannane's alkyl and leaving groups. It is impossible to disentangle steric and electronic effects in this work,[39,38] and there is certainly a solvent-dependence to stereospecificity. The tetralinyllithium **39** reacts in a similar manner: inversion with menthyldimethyltin chloride was proved by X-ray crystallography; Me_3SnCl reacts with a moderate degree of inversion, while protonation proceeds with retention.[39]

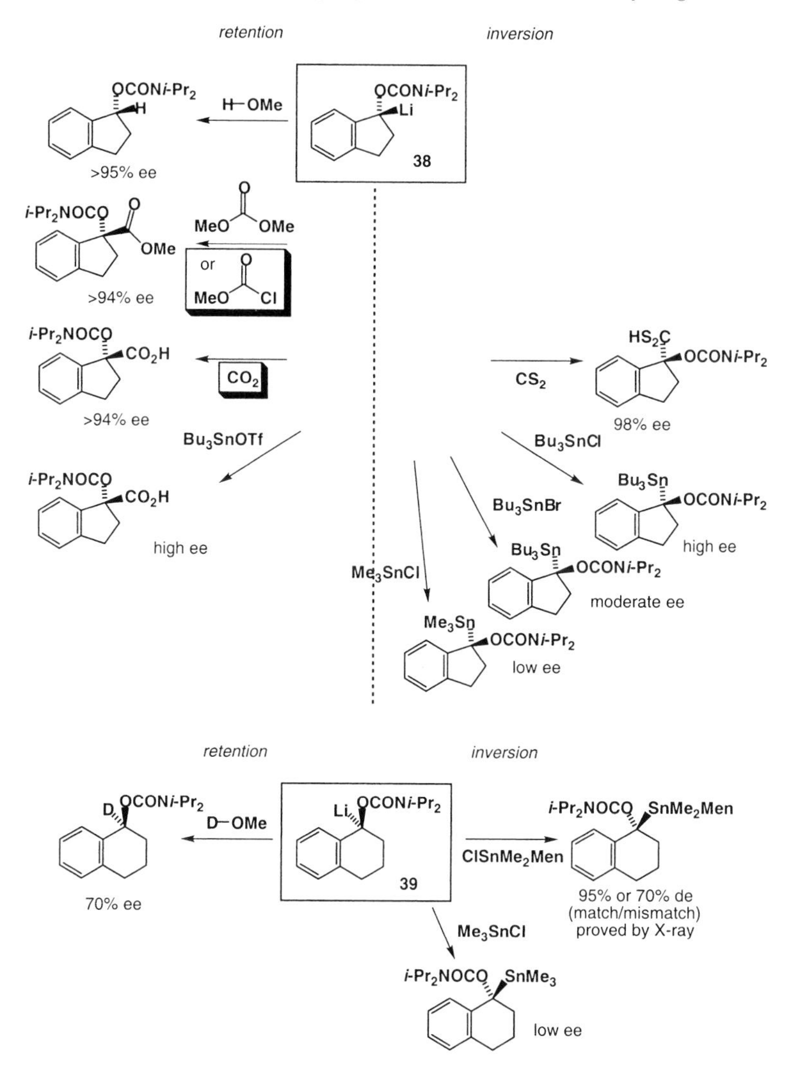

Scheme 6.1.6 Stereospecificity in the electrophilic substitution of cyclic α-carbamoyloxy benzyllithiums

The reason why some electrophiles (in particular CO_2 and MeOCOCl) react with opposite stereospecificity with **37** and with **38** was probed by Hoppe,[38] who calculated (MOPAC98) that **37** was somewhat more planar than **38**. Cyclic **38** appeared to require over 40 kJ mol^{-1} more energy than acyclic **37** to pass through a planar transition state and produce products with inversion, which is in accord with its greater preference for retentive substitution.

The reactions of the lithiated hindered ester **40** follow a very similar pattern to these cyclic benzyllithiums.[36] A key difference is that dimethylcarbonate and methyl chloroformate both react with **40** with retention. AcOH and MeOH still protonate with retention, while stannylation and silylation proceed with inversion (scheme 6.1.7).

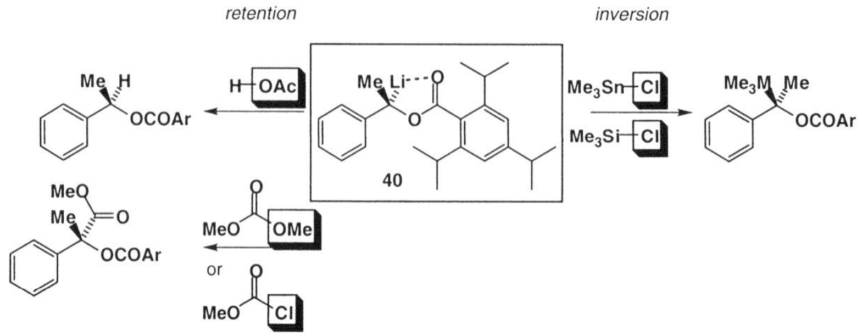

Scheme 6.1.7 Stereospecificity in the electrophilic substitution of an α-acyloxy benzyllithium

X-ray evidence was used to provide unequivocal proof of the invertive electrophilic substitution with trialkyltin halides.[40] Starting with (*R*) or (*S*)-**41**, lithiation (*s*-BuLi, toluene, Et_2O, –78 °C) gave organolithiums **40** which each reacted with menthyldimethyltin chloride to give different diastereoisomers of the stannane **42**, of which **42a** was crystalline. Transmetallation back to the organolithiums **40** must be retentive, because when these organolithiums were deuterated they each gave the enantiomers of the two original starting materials.

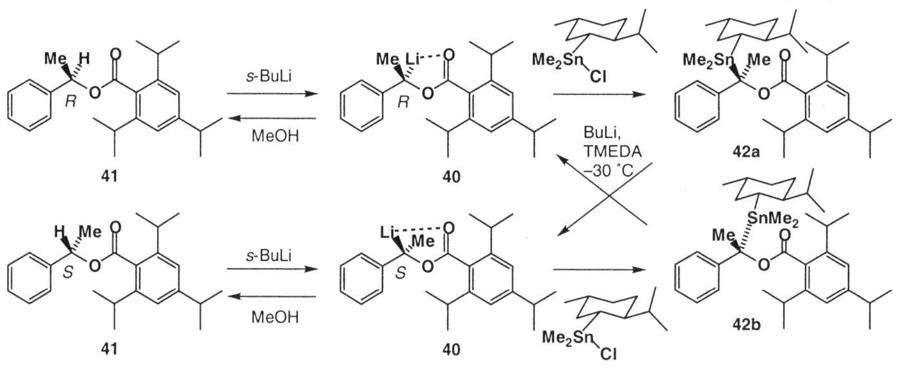

Beak published a similar study of the stereospecificity of the reactions of electrophiles with **43**.[41] He reached a similar conclusion: that fast-reacting, non-coordinating electrophiles react with inversion of configuration (scheme 6.1.8). As far as the results go, they mimic exactly the behaviour of **40**: methyl chloroformate, which reacted with **37** with inversion and with **40** with retention, also reacts with **43** with retention.

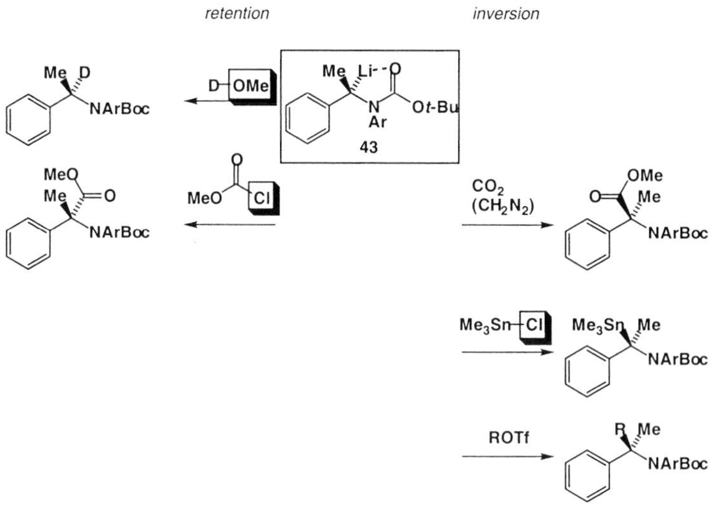

Scheme 6.1.8 Stereospecificity in the electrophilic substitution of an α-amido benzyllithium

Similar results were obtained with the secondary benzyllithium **44**, which reacted with MeOD with retention and with MeOTf with inversion. A similar cyclisation with departure of chloride also proceeds with inversion,[41] but rearrangement reaction described below has a similar intermediate **59** which reacts with retention.

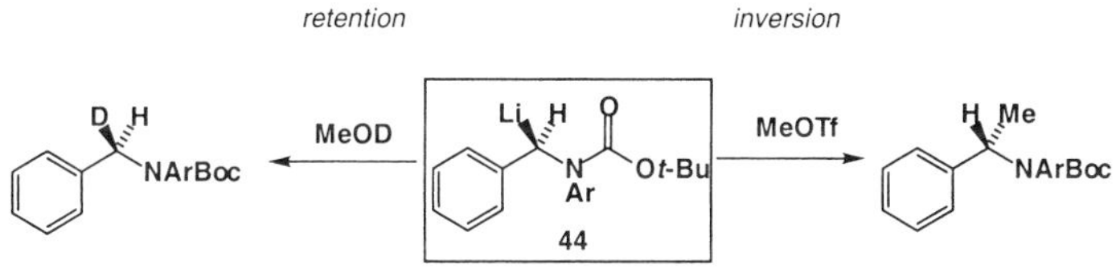

Beak has also reported the stereochemical behaviour of non-α-heterosubstituted organolithiums. For the organolithium **45**, which is configurationally stable at low temperatures, the results are straightforward: all electrophiles react with inversion.[42] Evidence for this was provided by a simple pair of sequences in which **45** reacts either directly with an electrophile or is first converted to a stannane which can then be substituted, via tin-lithium exchange, with the same electrophile (scheme 6.1.9). Opposite enantiomers of the products were produced. Given the generally reliable assumption that tin-lithium exchange is retentive, at least one of the two electrophilic quenches in the second scheme must go with inversion. Since each product from the reaction had been proved independently to have the same absolute configuration, all electrophilic substitutions of **45** must go with the same sense: inversion.

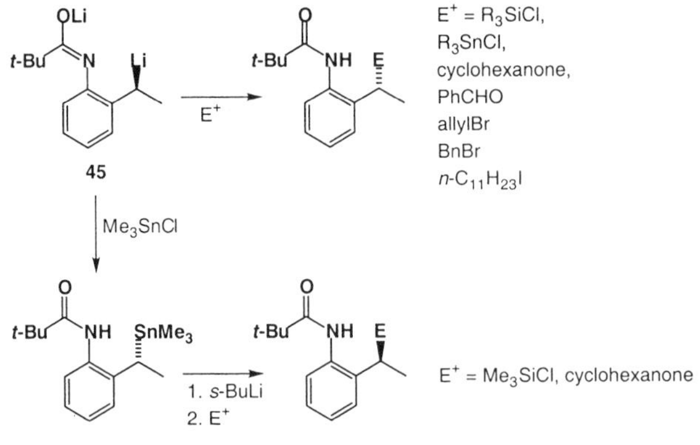

Scheme 6.1.9 Invertive electrophilic substitutions of a non-heterosubstituted benzyllithium

In contrast, **46** demonstrates divergent behaviour with different electrophiles (scheme 6.1.10).[43] The situation here is complicated by the fact that the stereochemical outcome is the result of a dynamic kinetic resolution of two interconverting diastereoisomeric organolithium complexes **46a** and **46b** (see section 6.2). It is not possible to be sure whether the different stereochemical outcomes represent retentive/invertive reactions or whether they represent halides and tosylates reacting at different rates with diastereoisomeric substrates.

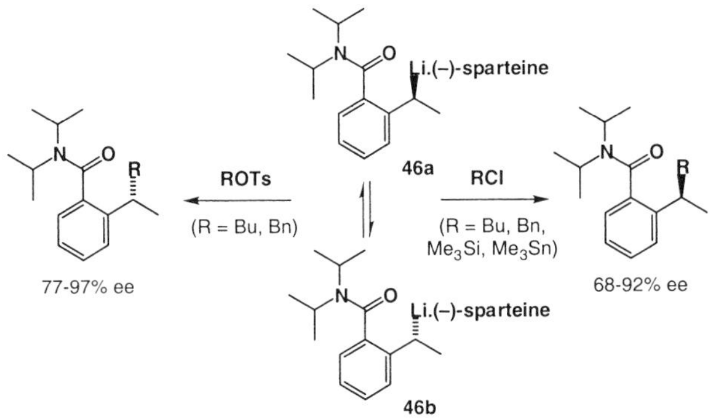

Scheme 6.1.10 Stereospecific electrophilic substitutions of a non-heterosubstituted benzyllithium

Some indication that the reactions of **46** with electrophiles are probably not determined entirely by the retentive or invertive nature of the substitution step comes from the reactions of the structurally similar though configurationally stable organolithiums **47a** and **47b**[44,45] (scheme 6.1.11). Organolithium **47a** is formed on diastereoselective deprotonation of the naphthamide using *s*-BuLi,[46] and reacts with most electrophiles (alkyl halides *and* methyl tosylate,[44] silyl halides,[44,47,48] ketones,[44] aldehydes[49] and imines[49,50]) with retention but with trialkyltin halides with inversion.[44] Its atropisomeric diastereoisomer **47b** also reacts with EtI with retention, but at a much lower level of stereospecificity.[46]

If **47a** reacts with selectivity corresponding to **46** then ROTs and Me_3SnCl must be choosing to react more rapidly with the opposite diastereoisomer of **46** from the one chosen by RCl.

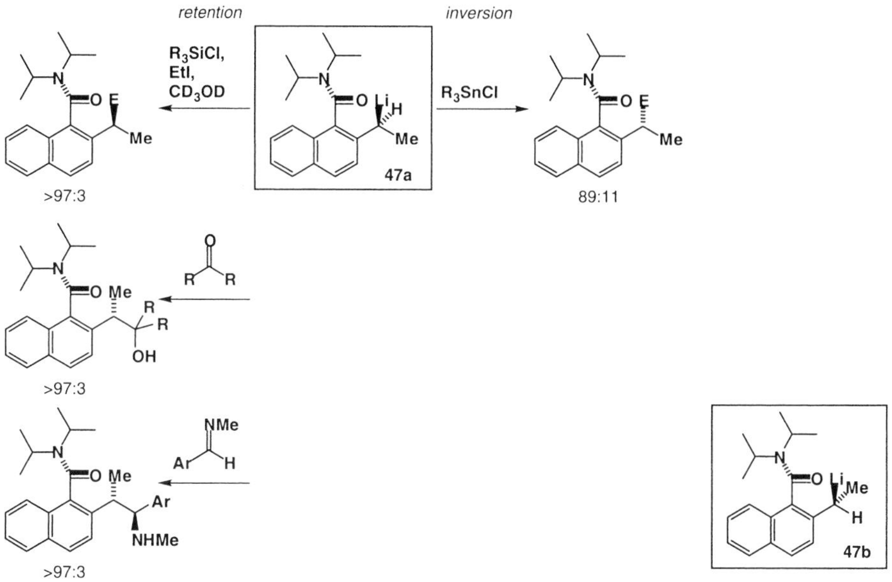

Scheme 6.1.11 Stereospecific electrophilic substitutions of a non-heterosubstituted naphthyllithium

Non-heterosubstituted organolithiums usually have low configurational stability, but the very rapid cyclisation of **49** means that – assuming tin-lithium exchange of **48** goes with retention – invertive substitution can be proved.[51] Inversion at both nucleophilic and electrophilic centres is a common feature of cyclopropane-forming reactions.[7]

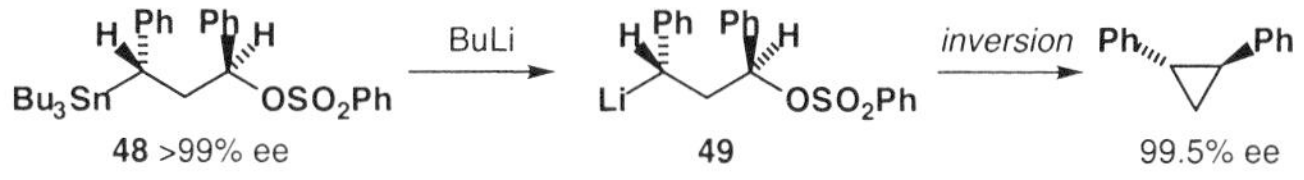

The indenyllithium **50** is one of few stereochemically defined organolithiums whose configuration is known with absolute certainty: in this case, from the X-ray crystal structure of its (–)-sparteine complex. It can therefore be shown to react with all acylating agents with retention of configuration[52] (scheme 6.1.12). Only with acetone is there some loss of *regio*selectivity.

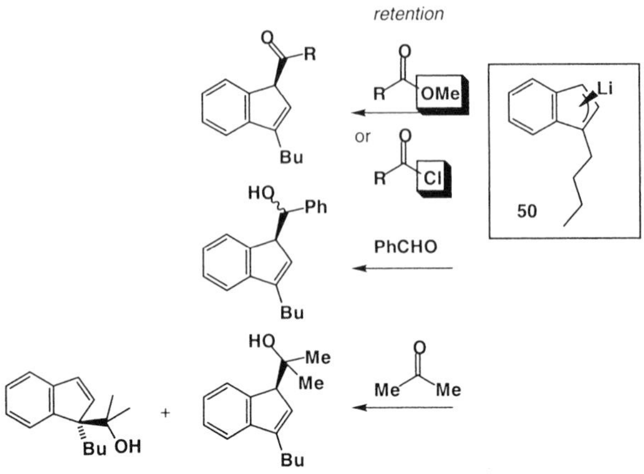

Scheme 6.1.12 Stereospecific electrophilic substitutions of an indenyllithium

6.1.4.2 Allyllithiums

Allyllithium **51** reacts with metallic electrophiles ($Ti(Oi\text{-}Pr)_4$ and R_3SnCl) with inversion[53] (scheme 6.1.13). The stereochemistry of the starting organolithium was deduced by X-ray crystal structure of the related lithiated allylsilane **52**.[54] The enantiomeric excesses of allylstannanes **54** were slightly lower than those of the allyltitaniums **53**, and the regiochemistry was dependent on the stannane's alkyl groups. With regard to stereospecificity, allyllithium **51** behaves like benzyllithium **37**.

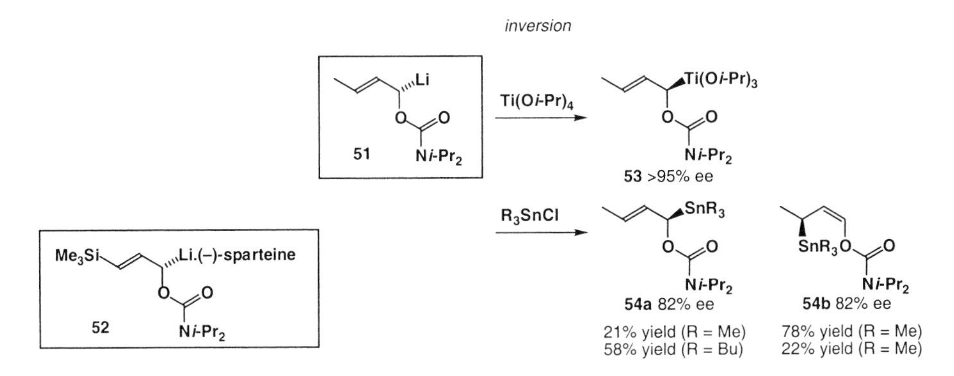

Scheme 6.1.13 Stereospecific electrophilic substitutions of an allyllithium

Cinnamyllithium **55** also stannylates with inversion: it also reacts with almost all other electrophiles with inversion[55] (scheme 6.1.14). There is a regiochemical question with **55** as well, but only CO_2 reacts non-regioselectively, and of the remaining electrophiles, all react at the α position except alkylating agents. Lack of consistent enantioselectivity (and the apparent reaction with methyl tosylate with retention) can be explained by a "negative kinetic resolution" of the configurationally unstable organolithium–sparteine complex – it appears that the minor organolithium **55b** (forming about 10% of the mixture) is *more* reactive with alkylating agents than the major organolithium **55a**, reducing or even inverting the enantioselectivity of alkylation reactions.

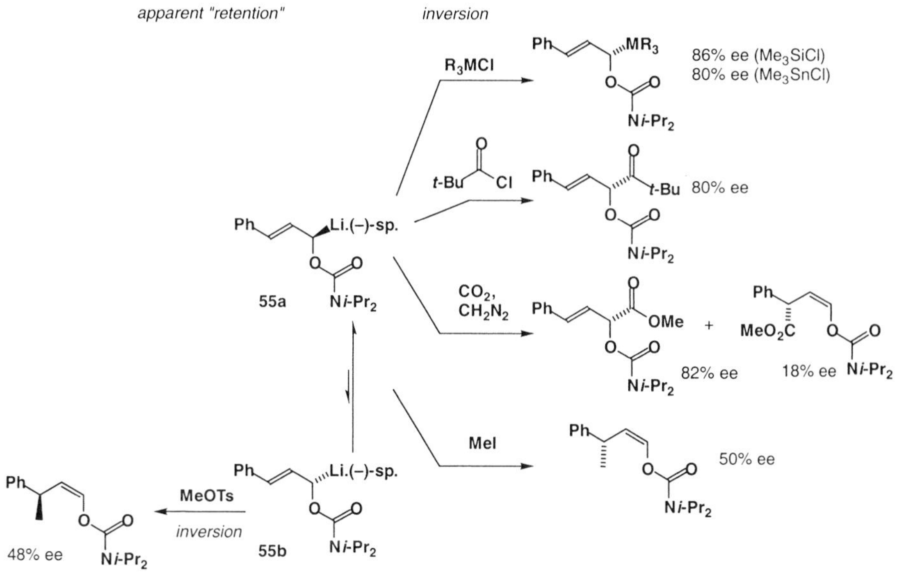

Scheme 6.1.14 Stereospecific electrophilic substitutions of a cinnamyllithium

6.1.4.3 Rearrangements of stabilised organolithiums

As early as 1974 it was known that the Brook rearrangement of benzylic organolithiums such as **56** proceeded with inversion.[32] Brook rearrangements of non-benzylic organolithiums proceed with retention (see also section 8.2).

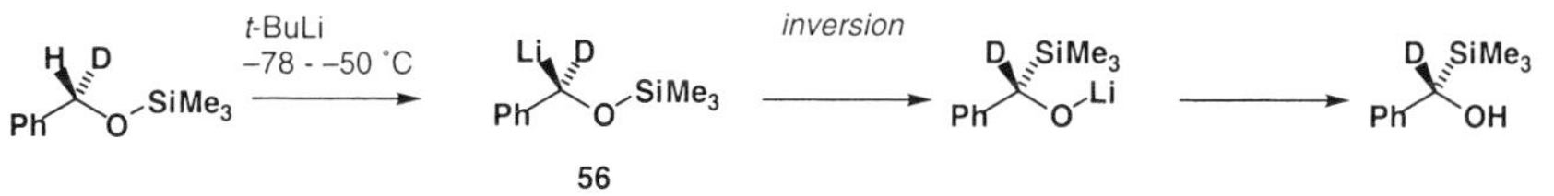

By contrast, the similar phosphate-phosphonate rearrangement proceeds with retention, whether it proceeds via a tertiary[56] or a secondary[57] organolithium **57** or **58**, as does the related amide-ketone rearrangement of **59**.[58] These retentive rearrangements presumably involve C=O–Li or P=O–Li coordination not possible in the Brook rearrangement.

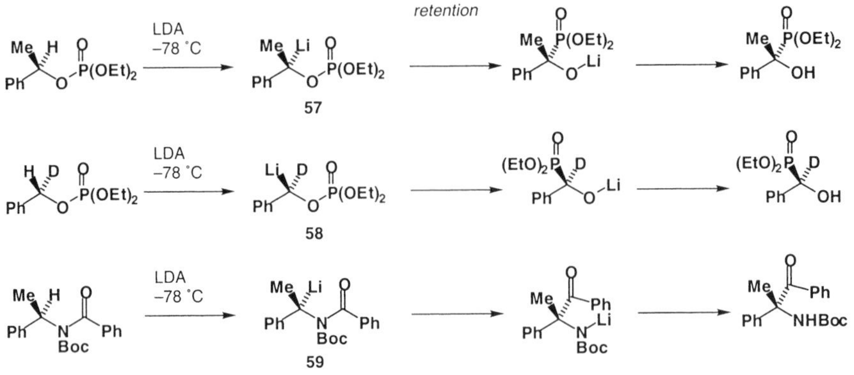

6.2 Stereoselective substitution in the presence of chiral ligands

6.2.1 Introduction: Mechanisms

The use of chiral ligands – in particular, the diamine (–)-sparteine – to complex with lithium in the course of asymmetric reactions is not a simple story.[59,15] A chiral ligand may influence the stereochemical course of either the formation or the reaction of an organolithium – or both, or neither. Add to this the possibility of configurational instability in the organolithium, where a chiral ligand may control both the rate of equilibration and the equilibrium constant between the two organolithium epimers, and detailed mechanisms of enantioselective organolithium reactions start to take on the potential for baroque complexity. Since 1985, reactions illustrating almost all possible types of interaction between chiral ligands and organolithiums have been identified. The simplest type of organolithium reactions involving chiral ligands is the *formation* of configurationally stable organolithiums by enantioselective deprotonation, and these reactions have already been discussed in section 5.4. In the discussion which follows, we shall concentrate on the *reactions* of organolithiums, showing the ways in which chiral ligands can endue organolithium reactions with enantioselectivity, and giving examples of each type of reaction.

6.2.2 Chiral ligands

By far the most important of the chiral ligands used to control the enantioselectivity of organolithium reactions is (–)-sparteine **60**, first used in 1968 to direct the addition of ethylmagnesium bromide to benzaldehyde in 22% ee.[60,61] In its ground state, (–)-sparteine exists in conformation **60a**, but only slightly higher in energy is conformation **60b**, which can function as a bidentate ligand.[62,63] It is extracted from the seeds of a variety of legumes, such as Scotch broom, and is commercially available either as the free base or as the sulfate pentahydrate.

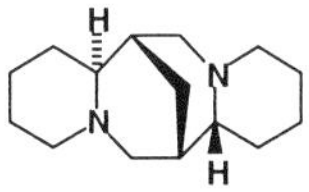

(–)-sparteine **60** 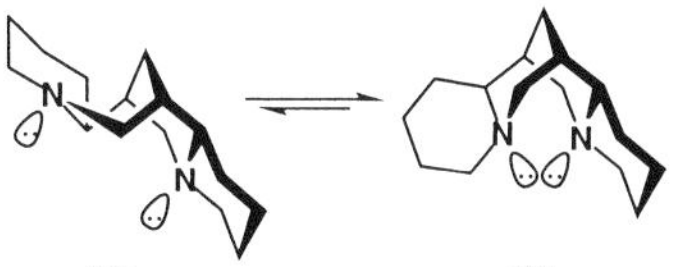 **60a** **60b**

Its enantiomer, (+)-sparteine, is also a natural product, but is more easily obtained by resolution and reduction of the racemic lactam **61**[64] (obtained from the bitter lupin *Lupinus albus*). Nonetheless, (+)-**60** is much less readily available than (–)-**60**, and its general use in synthesis is impractical at present. Sparteine's two diastereoisomers α-isosparteine **64** and β-isosparteine **65** also occur naturally, though they are more conveniently obtained by isomerisation of (–)-sparteine **60** via the enamines **62** and **63**.[65] Although both diastereoisomers display the folkloric advantage of C_2-symmetry, in all comparisons (–)-sparteine performs better than **64** or **65**.

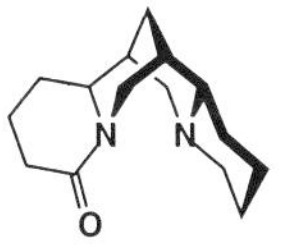

61 **60** **62** **63** (–)-α-isosparteine **64** (–)-β-isosparteine **65**

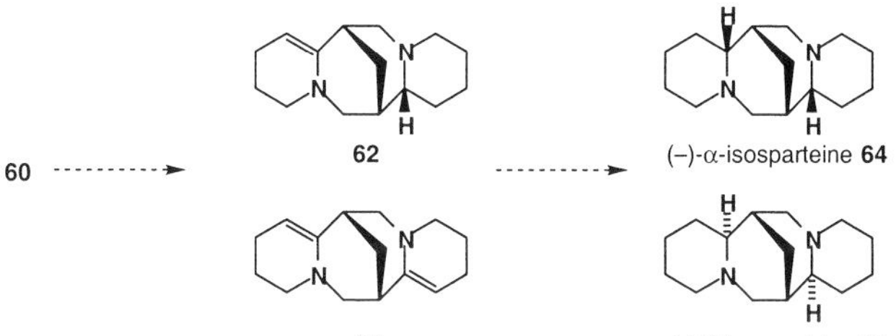

In fact, in an extensive survey carried out by Beak,[66] (–)-sparteine out-performed a wide range of likely contenders in an asymmetric deprotonation reaction. The only other ligands approaching its effectiveness were the proline-based ligand **66** and the bicyclic **67**, whose central bispidine ring system mimics the bicyclic core of sparteine itself. Di-*n*-butylbispidine **68** has on occasion been used as a powerful achiral lithium-coordinating ligand which out-performs TMEDA. More recent work[67-70] has uncovered a further class of powerful potential alternatives to sparteine – bis-oxazolines such as **69**.

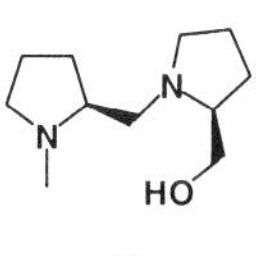

66 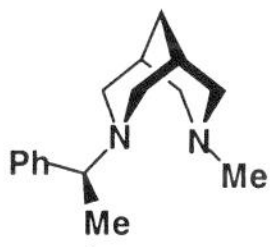 **67**  **68** 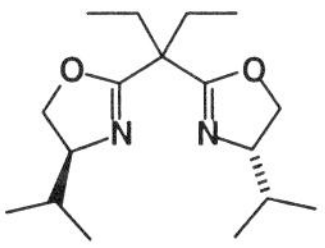 **69**

Further development of (–)-sparteine's remarkable success story is marred by two unfortunate aspects of (–)-sparteine chemistry. Firstly, as mentioned above, its (+)-enantiomer is not readily available. And secondly, it is difficult to make small modifications to the structure of (–)-sparteine without resorting to a lengthy total synthesis of the tetracyclic ring system. Recently, progress towards solving both of these problems has been made in the form of the synthesis of either enantiomer of the simplified sparteine-like ligand **70**.[71] Early results suggest that **70** can perform as well as (–)-sparteine in asymmetric deprotonation reactions.

70

6.2.3 Enantioselective deprotonation

Chiral ligands can promote the enantioselective *formation* of chiral organolithiums by asymmetric deprotonation. This mechanism was discussed in section 5.4.

6.2.4 Enantioselective substitution

The lithiation and silylation of the amide **71** is enantioselective for a very different reason. When the intermediate organolithium is made from the racemic stannane **72**, it still gives the product **74** in good enantioselectivity provided the electrophile reacts in the presence of (–)-sparteine.[72] The reaction must therefore be an enantioselective substitution. Furthermore, reaction of the deuterated analogue **75** gives a result which is not consistent with asymmetric deprotonation: yield, deuterium incorporation and product ee are all high.

Ph NHMe 71; Bu_3Sn Ph NHMe 72 → *s*-BuLi, –78 °C, THF → Li···OLi Ph NMe 73 → Me_3SiCl, (–)-sparteine → Me_3Si Ph NHMe 74

72%, 82% ee from **71**
48%, 78% ee from **72**

D Ph NHMe 75 → *s*-BuLi, –78 °C, THF → D Li···OLi Ph NMe → Me_3SiCl, (–)-sparteine → Me_3Si D Ph NHMe

75%, 90% ee, 94% D

The mechanism by which organolithium **73** reacts enantioselectively is illustrated in scheme 6.2.1. For a reaction in which (–)-sparteine determines the enantioselectivity in the substitution step, and not the deprotonation step, deuterium incorporation will simply reflect the magnitude of the kinetic isotope effect. The product's ee will be the same irrespective of whether or not the starting material is deuterated (contrast the reactions of *d*-**395** in section 5.4, summarised in scheme 5.4.1).

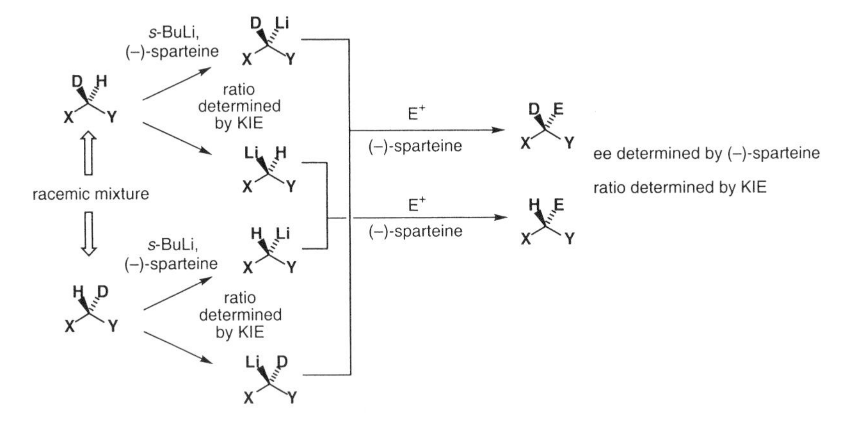

Scheme 6.2.1 Asymmetric substitution in the presence of (–)-sparteine.

6.2.5 Configurational stability, stereospecificity, and dynamic resolutions

Configurational stability is closely linked to the mechanism by which an asymmetric organolithium reaction takes place. For an asymmetric deprotonation to control the enantioselectivity of a substitution reaction, the intermediate organolithium must be configurationally stable. If the electrophile is added in a subsequent step (as in the reaction of **387** in section 5.4) that configurational stability must extend to the macroscopic timescale. For the relatively rare case of an intramolecular quench (**395**, section 5.4, is an example), the configurational stability need extend only to the microscopic timescale. The electrophile must also, of course, react with the organolithium with well-defined stereospecificity. Configurational stability and stereospecificity are discussed in detail in sections 5.1 and 5.2.

For an asymmetric substitution reaction, the intermediate organolithium must usually be *configurationally unstable*, or the ligand will be unable to direct the course of the reaction (an exception might be a hypothetical reaction in which the ligand allows the two organolithium enantiomers to react with opposite stereospecificity). This means that, in the presence of a chiral ligand such as (–)-sparteine, the two organolithium stereoisomers will be present in unequal proportions, because they form interconverting diastereoisomeric complexes. Whether or not this ratio of complexes is relevant to the subsequent substitution step of the reaction depends on the degree of conformational stability in the intermediate. Two possibilities emerge: in one, the organolithium is configurationally unstable on the timescale of its addition to an electrophile – in other words, it does not even have microscopic configurational stability. The product ratio will reflect the respective rates of reaction of the two diastereoisomeric complexes, and not their ratio. This situation is illustrated diagrammatically in figure 6.2.1. Effectively, the organolithium intermediate undergoes a *dynamic kinetic resolution*. In the other, the rate of interconversion of the organolithium stereoisomers is slow on the timescale of the addition of the organolithiums to an electrophile – in other words, it has microscopic (but not macroscopic) configurational stability. The product ratio now reflects the relative proportions of the two diastereoisomeric organolithium complexes, whose equilibrium, for good enantioselectivity, must lie heavily to one side. This is still a dynamic resolution, because the initial racemic organolithium can be made to give a single product enantiomer, but it is under thermodynamic, and not kinetic control: such reactions are known as dynamic thermodynamic resolutions,[73] and are represented schematically in figure 6.2.2.

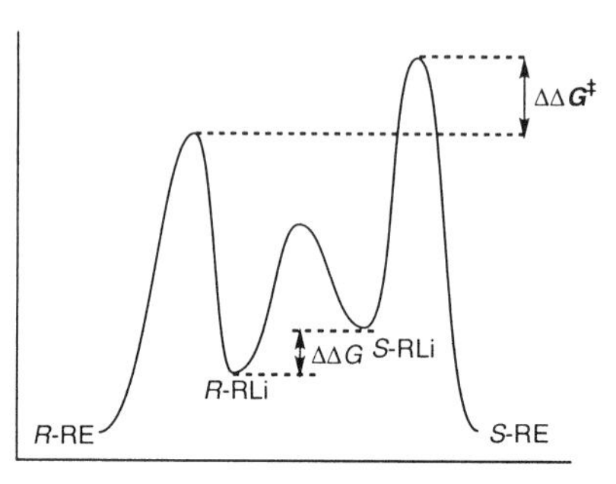

Figure 6.2.1 Dynamic kinetic resolution

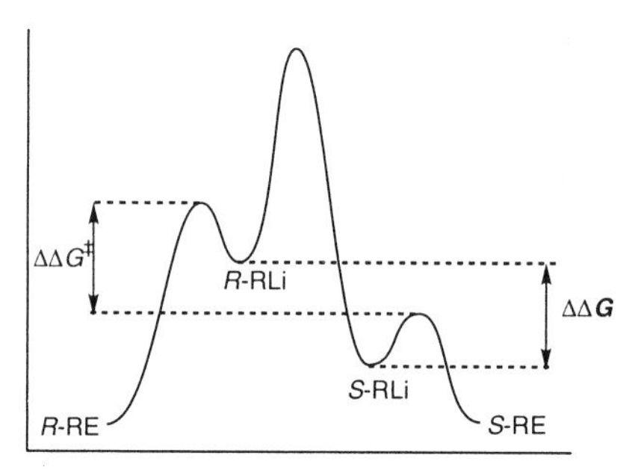

Figure 6.2.2 Dynamic thermodynamic resolution

The difference between these two mechanisms is the basis of the Hoffmann test, discussed in section 5.1.1.

6.2.6 Dynamic thermodynamic resolution[73]

The 2-ethylanilide **76** may be laterally lithiated with *s*-BuLi and quenched with Me_3SiCl to give the silane **78**. The product is formed in 82% ee if the racemic intermediate organolithium **77** is first treated with (–)-sparteine at –25 °C and then Me_3SiCl at –78 °C: the reaction is an asymmetric substitution, with (–)-sparteine governing the enantioselectivity of the electrophilic quench step. But how does it do this?

s-BuLi, –25 °C

1. (–)-sparteine, –25 °C or –78 °C
2. Me_3SiCl, –78 °C

76 **77** **78**

63%, 82% ee ((–)-sparteine added at –25 °C)
52%, 21% ee ((–)-sparteine added at –78 °C)

A key to the answer comes from changing the temperature at which the organolithium is treated with (–)-sparteine: adding (–)-sparteine at –78 °C *reduces* the ee of **78** to only 21%. (–)-Sparteine is able to exert its effect more efficiently at higher temperatures, exactly as would be expected if it were controlling the equilibration of a pair of organolithiums (which might be fast at –25 °C but slow at –78 °C), but in contrast to what would be expected if it were controlling the rates of reaction of two organolithiums.

Two reactions of the partially enantiomerically enriched stannane **79** also point to this explanation of the result. Stannane **79** has 66% ee, so transmetallation probably (see section 5.2) gives an organolithium **77** with 66% ee. If the transmetallation is carried out at –78 °C, the product **78** of electrophilic quench does indeed have an ee close to 66%, suggesting that at this temperature (–)-sparteine is unable to perturb the ratio of the two enantiomeric organolithiums. However, if the mixture of organolithium **77** and (–)-sparteine is warmed to –25 °C for 2 h before being cooled again to –78 °C and quenched, the other enantiomer of the product **78** is formed in 85% ee! (–)-Sparteine has succeeded in more than reversing the ratio of the organolithiums and hence more than inverting the ee of the product.

s-BuLi, –25 °C

1. (–)-sparteine, –25 °C or –78 °C
2. Me_3SiCl, –78 °C

79 66% ee **77** presumably 66% ee **78**

(*R*)-**78**, 85% ee (–25 °C)
(*S*)-**78**, 62% ee (–78 °C)

The mechanism of stereocontrol in the reactions of **77** is summarised in scheme 6.2.2.

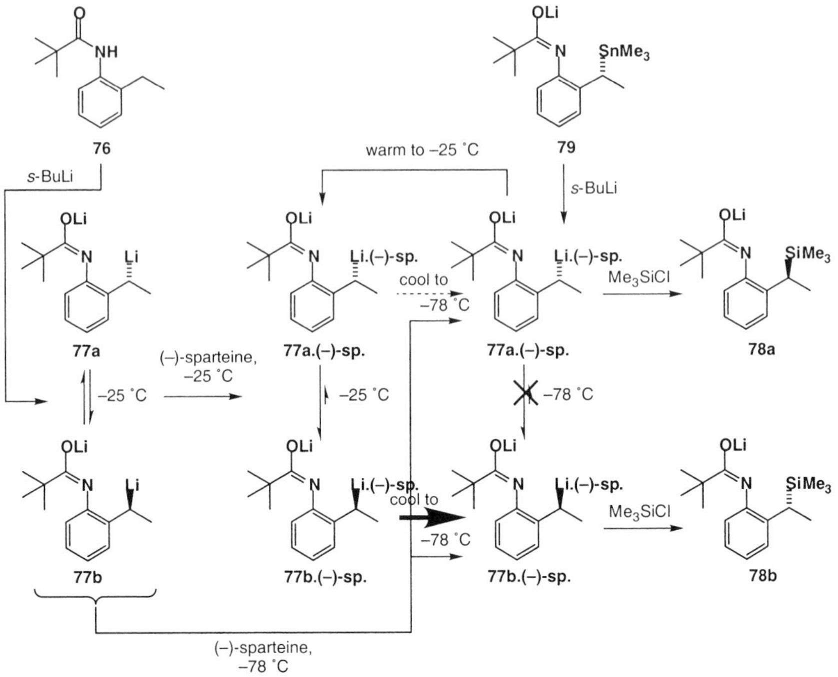

Scheme 6.2.2 Asymmetric substitution by dynamic thermodynamic resolution.

All of the enantioselectivities described here arise on treatment of the non-racemising intermediate organolithium **77** with an excess of electrophile, such that both enantiomers react to completion. But since the (–)-sparteine complexes of the two enantiomers of **77** are diastereoisomeric, it should in principle be possible to separate them kinetically by using a deficit of electrophile, such that only the more reactive of the two diastereoisomeric complexes is quenched. In such a case, ee should increase. And it does.[74] The 21% ee obtained on treatment of the unequilibrated **77**- complex with a full equivalent of Me_3SiCl at –78 °C can be improved to 82% ee when only 0.1 equiv. Me_3SiCl is used: in other words, the intermediate organolithiums **77**-(–)-sparteine in this case are present in about a 60:40 ratio (total reaction gives 21% ee), but the major diastereoisomeric complex reacts much faster than the minor, allowing the formation of one product in high enantiomeric excess. An even better result is obtained if the organolithiums **77** are first allowed to equilibrate and *then* the faster-reacting one (which is evidently also the major one in this case) is allowed to react selectively with a deficit of electrophile: the ee improves to 98%. This modification is illustrated in scheme 6.2.3.

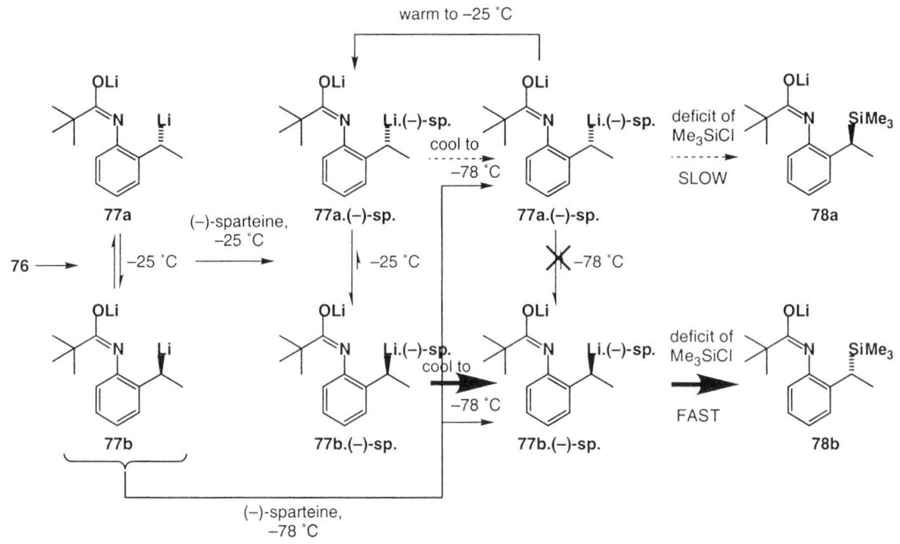

Scheme 6.2.3 Asymmetric substitution by dynamic thermodynamic resolution with a deficit of electrophile.

Yields in the reactions with 0.1 equivalents of electrophile are necessarily low, but high yields and high ee's can be obtained if a pair of equilibrated organolithiums are first cooled, treated with a deficit of electrophile, and then the remainder is re-equilibrated again, and finally treated with further electrophile, as shown below.

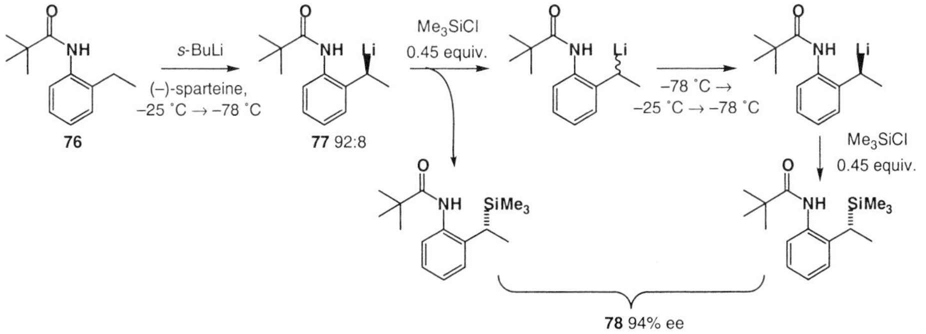

The detailed mechanism of a similar asymmetric silylation of **80** to give **81** is not yet known.[15]

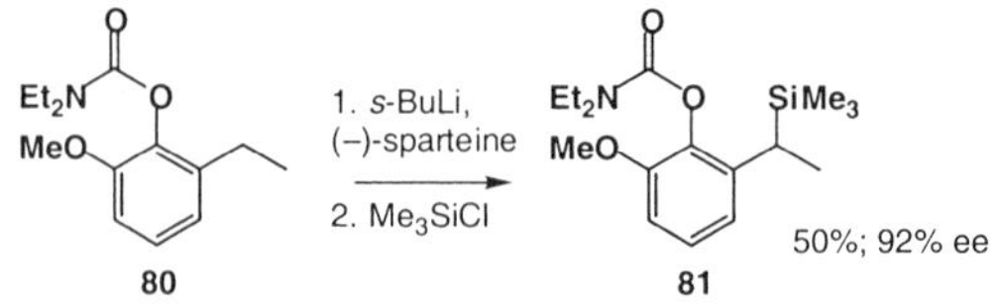

The change of ee of a product with varying amounts of electrophile is effectively a variant of the Hoffmann test, for which Beak coined the term "poor man's Hoffmann test", because it does not require an enantiomerically pure electrophile. The use of this test to prove configurational stability on the microscopic timescale is described in section 5.1.1.

In some cases, crystallisation of the intermediate organolithium-(–)-sparteine complexes is necessary to force their equilibration to the major diastereoisomer.[54] This type of dynamic thermodynamic resolution was involved in one of the very first effective uses of (–)-sparteine in organolithium chemistry. Hoppe showed in 1988 that the carbamate **82** could be deprotonated in the presence of (–)-sparteine, and that when the product **83** was transmetallated with titanium and then added to an aldehyde, a homoaldol product **85** was formed in 83% ee. It became apparent that acceptable enantiomeric excesses were obtained only when the intermediate organolithium-(–)-sparteine complex was allowed to crystallise: the complexes **83a** and **83b** interconvert in solution, but one diastereoisomer crystallises preferentially, leading to a dynamic resolution of the organolithium.

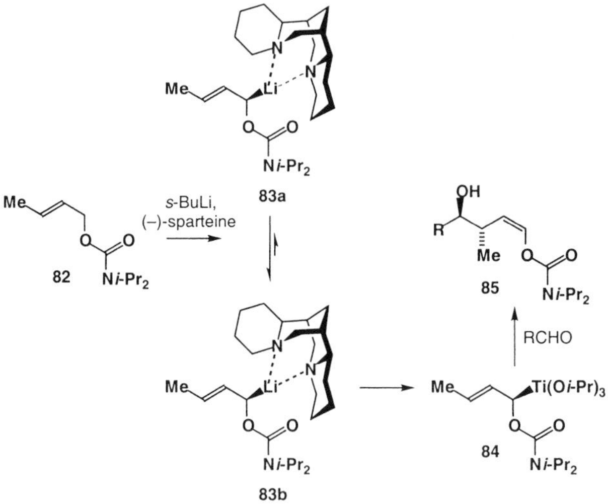

A similar effect controls the lithiation and substitution of the benzylic carbamate **86**.[33,35] Lithiation of **86** with *s*-BuLi-(–)-sparteine in ether gives low enantiomeric excesses, but when the lithiation is carried out in hexane, a solvent in which the intermediate complex is not soluble, the enantiomeric excess of the product **88** increases to 82%. Even higher enantiomeric excesses are obtained if the intermediate suspension of organolithium-(–)-

sparteine complex **87** is filtered: the crystals lead to the product in 90% ee, while the mother liquors give a product with only 38% ee.

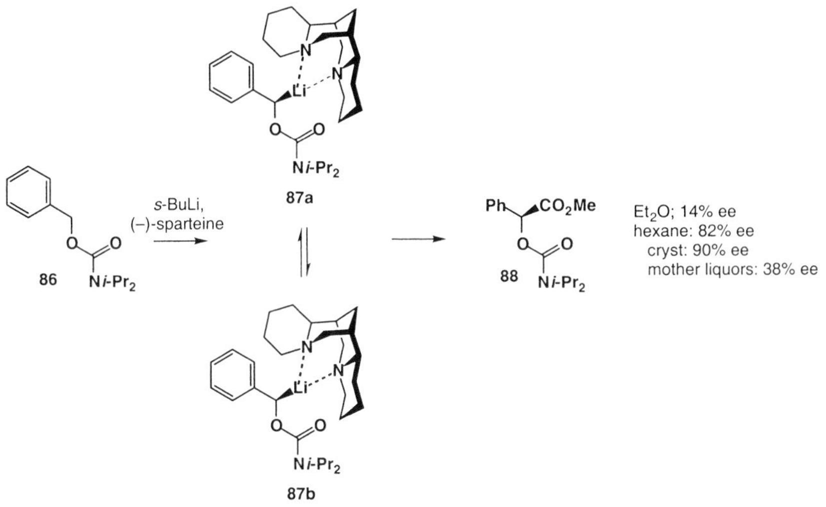

The formation of diastereoisomerically pure complexes of **90** with (–)-sparteine is also controlled by crystallisation. Treatment of the indene **89** with BuLi and (–)-sparteine in ether gives, on warming, a yellow precipitate which reacts with carbonyl electrophiles to provide the products **91** typically with good regioselectivity and >95% ee.[52] An X-ray crystal structure proved the stereochemistry of the intermediate complex to be that shown as **90b**, and hence proved the stereochemical course of the substitution (see section 6.1). The complex is readily decomposed by THF, in the presence of which it rearranges to a racemic η^1 allyllithium.

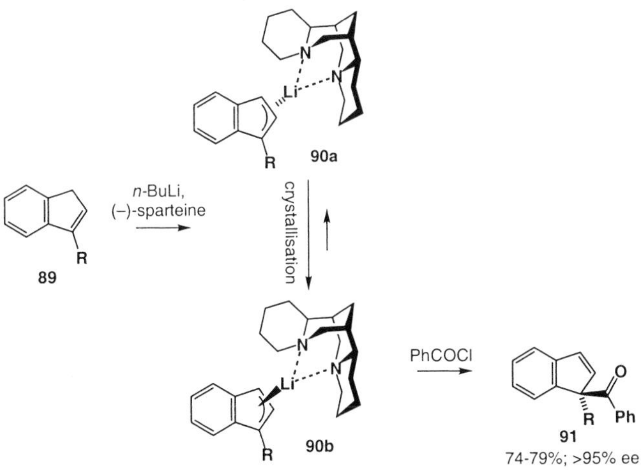

Section 5.4 describes the enantioselective deprotonation of a deuterated benzylamine derivative **431** to give a configurationally unstable organolithium **432**.[75] Over a period of minutes, in the presence of (–)-sparteine, the organolithium **432** gave products of increasing ee as the organolithiums equilibrated under thermodynamic control to the more stable **432**-(–)-sparteine complex, a process that could be accelerated by the precipitation of the complex.

6.2.7 Dynamic kinetic resolution

The organolithium **92** is not configurationally stable on the microscopic timescale: (–)-sparteine may be able to perturb an equilibrating mixture of **92a** and **92b** in favour of one enantiomer or the other, but this can have no effect on the ratio of products formed on electrophilic quench. However, treating **92** with electrophiles in the presence of (–)-sparteine still gives products with good enantiomeric excesses in reactions which must amount to dynamic kinetic resolutions.[43] As expected, the enantiomeric excess of the product does not change during the course of the reaction.

s-BuLi
92a
R–X
(–)-sparteine
FAST
major product:
X = OTs electrophiles
reacting with retention
major product:
X = Cl electrophiles
reacting with inversion
92b
R–X
(–)-sparteine
SLOW
minor product

6.2.8 Summary: Mechanisms of asymmetric functionalisation with (–)-sparteine

Scheme 6.2.4 summarises the possible mechanisms by which (–)-sparteine induces asymmetry into organolithium reactions, highlighting organolithiums whose reactions typify of each type of asymmetric induction.

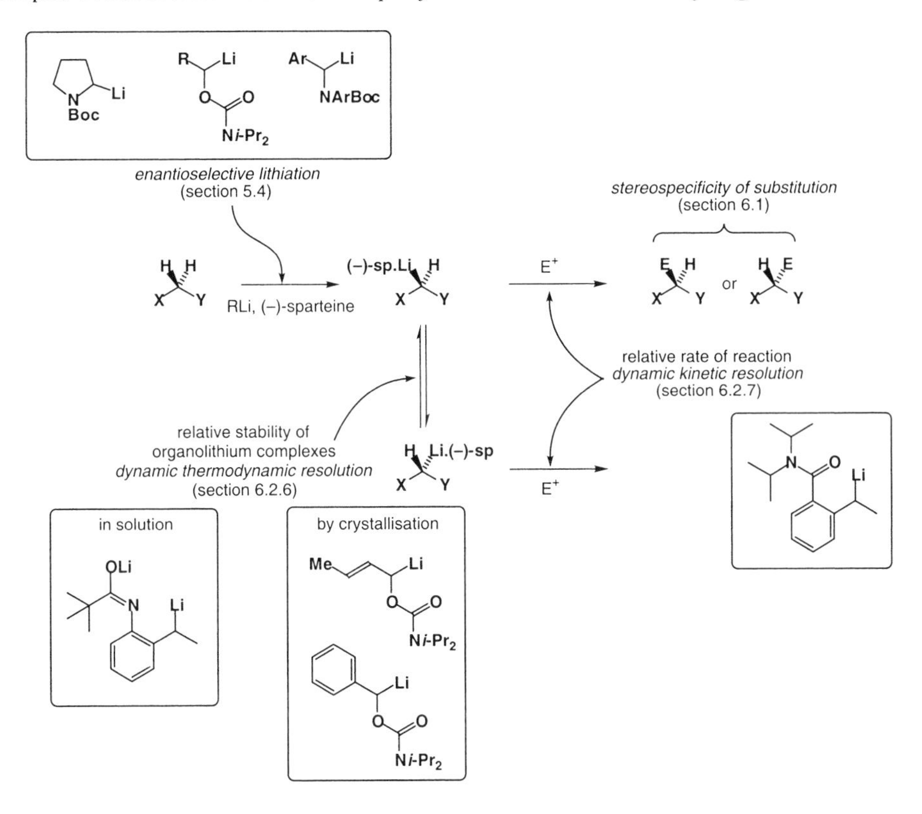

Scheme 6.2.4 Asymmetric functionalisation with (–)-sparteine

References

1. Zimmerman, H. E.; Singer, L.; Thyagarajan, B. S. *J. Am. Chem. Soc.* **1959,** *81*, 108.
2. Jemming, E. J.; Chandrasekhar, J.; Schleyer, P. v. R. *J. Am. Chem. Soc.* **1979,** *101*, 527.
3. Gawley, R. E. *Tetrahedron Lett.* **1999,** *40*, 4297.
4. Seyferth, D.; Vaughan, L. G. *J. Am. Chem. Soc.* **1964,** *86*, 883.
5. Nesmeyanov, A. N.; Borisov, A. E. *Tetrahedron* **1957,** *1*, 158.
6. Curtin, D. Y.; Crump, J. W. *J. Am. Chem. Soc.* **1958,** *80*, 1922.
7. Fleming, I.; Rowley, M. *Tetrahedron* **1986,** *42*, 3181.
8. Beak, P.; Kerrick, S. T. *J. Am. Chem. Soc.* **1991,** *113*, 9708.
9. Still, W. C.; Sreekumar, C. *J. Am. Chem. Soc.* **1980,** *102*, 1201.
10. Sawyer, J. S.; Kucerovy, A.; Macdonald, T. L.; McGarvey, G. J. *J. Am. Chem. Soc.* **1988,** *110*, 842.
11. Hoppe, D.; Hintze, F.; Tebben, P. *Angew. Chem., Int. Ed. Engl.* **1990,** *29*, 1422.
12. Kopach, M. E.; Meyers, A. I. *J. Org. Chem.* **1996,** *61*, 6764.
13. Hoffmann, R.; Minkin, V. I.; Carpenter, B. K. *Bull. Soc. Chim. Fr.* **1996,** *133*, 117.
14. Alezta, V.; Bonin, M.; Micouin, L.; Husson, H.-P. *Tetrahedron Lett.* **2000,** *41*, 651.

15. Hoppe, D.; Hense, T. *Angew. Chem., Int. Ed. Engl.* **1997,** *36*, 2282.
16. Hoffmann, R. W.; Koberstein, R. *J. Chem. Soc., Chem. Commun.* **1999**, 33.
17. Walborsky, H. M.; Impastato, F. J.; Young, A. E. *J. Am. Chem. Soc.* **1964,** *86*, 3283.
18. Letsinger, R. L. *J. Am. Chem. Soc.* **1950,** *72*, 4842.
19. Applequist, D. E.; Peterson, A. G. *J. Am. Chem. Soc.* **1961,** *83*, 862.
20. Gawley, R. E.; Low, E.; Zhang, Q.; Harris, R. *J. Am. Chem. Soc.* **2000,** *122*, 3344.
21. Gawley, R. E.; Zhang, Q. *J. Org. Chem.* **1995,** *60*, 5763.
22. Coldham, I.; Hufton, R.; Snowden, D. J. *J. Am. Chem. Soc.* **1996,** *118*, 5322.
23. Wu, Y. D.; Houk, K. N.; Marshall, J. A. *J. Org. Chem.* **1990,** *55*, 1421.
24. Verner, E. J.; Cohen, T. *J. Am. Chem. Soc.* **1992,** *114*, 375.
25. Tomooka, K.; Igarishi, T.; Watanabe, M.; Nakai, T. *Tetrahedron Lett.* **1992,** *33*, 5795.
26. Chang, P. C.-M.; Chong, J. M. *J. Org. Chem.* **1988,** *53*, 5584.
27. Tomooka, K.; Igarishi, T.; Nakai, T. *Tetrahedron* **1994,** *50*, 5927.
28. Tomooka, K.; Komine, N.; Nakai, T. *Tetrahedron Lett.* **1997,** *38*, 8939.
29. Lindermann, R. J.; Ghannam, A. *J. Am. Chem. Soc.* **1990,** *112*, 2396.
30. Hudrlik, P. F.; Hudrlik, A. M.; Kulkarni, A. K. *J. Am. Chem. Soc.* **1982,** *104*, 6809.
31. Wilson, S. R.; Hague, M. J.; Misra, R. N. *J. Org. Chem.* **1982,** *47*, 747.
32. Wright, A.; West, R. *J. Am. Chem. Soc.* **1974,** *96*, 3227.
33. Carstens, A.; Hoppe, D. *Tetrahedron* **1994,** *50*, 6097.
34. Hoppe, D.; Carstens, A.; Krämer, T. *Angew. Chem., Int. Ed. Engl.* **1990,** *29*, 1424.
35. Derwing, C.; Hoppe, D. *Synthesis* **1996**, 149.
36. Hammerschmidt, F.; Hanninger, A. *Chem. Ber.* **1995,** *128*, 1069.
37. Norsikian, S.; Marek, I.; Klein, S.; Poisson, J.-F.; Normant, J.-F. *Chem. Eur. J.* **1999,** *5*, 2055.
38. Derwing, C.; Frank, H.; Hoppe, D. *Eur. J. Org. Chem.* **1999**, 3519.
39. Hammerschmidt, F.; Hanninger, A.; Simov, B. P.; Völlenkle, H.; Werner, A. *Eur. J. Org. Chem.* **1999**, 3511.
40. Hammerschmidt, F.; Hanninger, A.; Völlenkle, H. *Chem. Eur. J.* **1997,** *3*, 1728.
41. Faibish, N. C.; Park, Y. S.; Lee, S.; Beak, P. *J. Am. Chem. Soc.* **1997,** *119*, 11561.
42. Basu, A.; Beak, P. *J. Am. Chem. Soc.* **1996,** *118*, 1575.
43. Thayumanavan, S.; Lee, S.; Liu, C.; Beak, P. *J. Am. Chem. Soc.* **1994,** *116*, 9755.
44. Clayden, J.; Pink, J. H. *Tetrahedron Lett.* **1997,** *38*, 2565.
45. Clayden, J.; Helliwell, M.; Pink, J. H.; Westlund, N. *J. Am. Chem. Soc.* **2001,** *123*, 12449.
46. Clayden, J.; Pink, J. H. *Tetrahedron Lett.* **1997,** *38*, 2561.
47. Clayden, J.; Pink, J. H.; Yasin, S. A. *Tetrahedron Lett.* **1998,** *39*, 105.
48. Clayden, J.; Kenworthy, M. N.; Youssef, L. H.; Helliwell, M. *Tetrahedron Lett.* **2000,** *41*, 5171.
49. Clayden, J.; Darbyshire, M.; Pink, J. H.; Westlund, N.; Wilson, F. X. *Tetrahedron Lett.* **1998,** *38*, 8587.
50. Clayden, J.; Westlund, N.; Wilson, F. X. *Tetrahedron Lett.* **1999,** *40*, 3331.
51. Krief, A.; Hobe, M. *Tetrahedron Lett.* **1992,** *33*, 6529.

52. Hoppe, I.; Marsch, M.; Harms, K.; Boche, G.; Hoppe, D. *Angew. Chem., Int. Ed. Engl.* **1995,** *34*, 2158.
53. Paulsen, H.; Graeve, C.; Hoppe, D. *Synthesis* **1996**, 141.
54. Marsch, M.; Harms, K.; Zschage, O.; Hoppe, D.; Boche, G. *Angew. Chem., Int. Ed. Engl.* **1991,** *30*, 321.
55. Behrens, K.; Fröhlich, R.; Meyer, O.; Hoppe, D. *Eur. J. Org. Chem.* **1998**, 2397.
56. Hammerschmidt, F.; Völlenkle, H. *Liebigs Ann. Chem.* **1986**, 2653.
57. Hammerschmidt, F.; Hanninger, A. *Chem. Ber.* **1995,** *128*, 823.
58. Hara, O.; Ho, M.; Hamada, Y. *Tetrahedron Lett.* **1998,** *39*, 5537.
59. Beak, P.; Basu, A.; Gallagher, D. J.; Park, Y. S.; Thayumanavan, S. *Acc. Chem. Res.* **1996,** *29*, 552.
60. Nozaki, H.; Aratani, T.; Toraya, T. *Tetrahedron Lett.* **1968**, 4097.
61. Nozaki, H.; Aratani, T.; Noyori, R. *Tetrahedron Lett.* **1968**, 2087.
62. Bohlmann, F.; Schumann, D.; Arndt, C. *Tetrahedron Lett.* **1965**, 2705.
63. Wiewiorowski, M.; Edwards, O. E.; Bratek-Wiewiorowska, M. D. *Can. J. Chem.* **1967,** *45*, 1447.
64. Clemo, G. R.; Raper, R.; Short, W. S. *J. Chem. Soc.* **1949**, 663.
65. Okamoto, Y.; Suzuki, K.; Kitayama, T.; Yuki, H.; Kageyama, K.; Miki, K.; Tanaka, N.; Kasai, N. *J. Am. Chem. Soc.* **1982,** *104*, 4618.
66. Gallagher, D. J.; Wu, S.; Nikolic, N. A.; Beak, P. *J. Org. Chem.* **1995,** *60*, 8148.
67. Kang, J.; Cho, W. O.; Cho, H. *Tetrahedron Asymmetry* **1994,** *5*, 1347.
68. Tomooka, K.; Komine, N.; Nakai, T. *Tetrahedron Lett.* **1998,** *39*, 5513.
69. Komine, N.; Wang, C.-F.; Tomooka, K.; Nakai, T. *Tetrahedron Lett.* **1999,** *40*, 6809.
70. Hodgson, D. M.; Cameron, I. D.; Christlieb, M.; Green, R.; Lee, G. P.; Robinson, L. A. *J. Chem. Soc., Perkin Trans. 1* **2001**, 2161.
71. Harrison, J. R.; O'Brien, P.; Pater, D. W.; Smith, N. M. *J. Chem. Soc., Chem. Commun.* **2001**, 1203.
72. Beak, P.; Du, H. *J. Am. Chem. Soc.* **1993,** *115*, 2516.
73. Beak, P.; Anderson, D. R.; Curtis, M. D.; Laumer, J. M.; Pippel, D. J.; Weisenburger, G. A. *Acc. Chem. Res.* **2000,** *33*, 715.
74. Basu, A.; Gallagher, D. J.; Beak, P. *J. Org. Chem.* **1996,** *61*, 5718.
75. Schlosser, M.; Limat, D. *J. Am. Chem. Soc.* **1995,** *117*, 12342.

CHAPTER 7

Regio- and Stereoselective Addition Reactions of Organolithiums

The addition of organolithiums to polarised C=X bonds is one of the most widely used ways of making C–C bonds, and (excepting some unusual intramolecular cases) will not be discussed in this book other than to say it is a reliable and successful reaction. With a few exceptions,[1-3] stereoselectivity is not a general feature of organolithium addition reactions to C=O π bonds. Much of this chapter will concern controlled addition of organolithiums to C=C π bonds: after an overview of carbolithiation, we shall review the development of intramolecular carbolithiation, or anionic cyclisation.

7.1 Intermolecular addition to π bonds: Carbolithiation

The addition of an organolithium to an unactivated, non-polarised alkene to form a new C–C bond suffers from problems in its application as a viable synthetic method.[4] The product is also an organolithium, meaning that there is the potential for the formation of at least two new C–C bonds during the reaction – and indeed cascade carbolithiation reactions in the guise of anionic polymerisation[5,6] were one of the first organolithium reactions to be exploited. A major task facing a chemist trying to control carbolithiation is therefore to provide stabilisation for the product organolithium to prevent the formation of polymers by its continued reaction. The alternative solution – to make the carbolithiation intramolecular – is discussed in section 7.2.4.

7.1.1 Carbolithiation of simple alkenes

Secondary organolithiums **1** add to ethylene about one million times as fast as primary organolithiums (tertiary organolithiums even faster still),[4] so it is easy to control the addition of *t*-BuLi, *s*-BuLi, *i*-PrLi etc. to give a primary organolithium product **2**. The reaction is quantitative in Et_2O at –25 °C.[7,8]

$CH_2{=}CH_2$, Et_2O, –25 °C

1 → **2** generally quantitative

With unactivated terminal alkenes such as **4**, the reaction still works, but requires forcing conditions (80 °C, 48 h) and gives only low yields because side reactions such as β-elimination and allylic deprotonation compete. The only organolithium formed is the primary one (**5**), with the alkyl group attacking the secondary carbon – in contrast, organomagnesium and organoaluminium reagents add to the primary carbon to generate a secondary organometallic.

4

3 Li

80 °C, 48 h

Li 5 30%

Strained alkenes may react moderately well even with primary organolithiums - norbornene, for example, gives the *exo* adduct **6** with *n*-BuLi,[9] though the reaction is vastly improved if a coordinating group is incorporated (see below).

n-BuLi

TMEDA, hexane
25 °C, 36 h

Li 6

7.1.2 Carbolithiation of conjugated alkenes and alkynes

All other carbolithiations require the product organolithium to be stabilised, either by conjugation or by coordination. Activation of the starting material by TMEDA, DABCO or (–)-sparteine is also often advantageous. One of the simplest alkene acceptors of carbolithiation, which has surprisingly remained unexploited till recently, is styrene. In THF, styrene polymerises readily on reaction with organolithiums, but in ether the reaction is quite controllable and gives good yields of the stabilised, benzylic organolithium **7**.[10,11] The organolithium product can be quenched to yield arylalkylcarboxylic acids such as **8**, and the addition can be used to initiate cyclisation[12] or rearrangement reactions.[13]

RLi, –78 °C, Et_2O

Li R 7

R = *n*-, *s*-, *t*-BuLi etc.

1. BuLi, –78 °C, Et_2O

2. CO_2

CO_2H

8 75%

Butadiene reacts similarly: the allyllithium product **9** is much less reactive, though a second addition can be forced at 25 °C.[14]

t-BuLi, pentane, Et_2O, 20 °C

Li

9 95%

25 °C, 8 h

Li

+

Li

Carbolithiation of terminal dienes such as **11** is promoted by (–)-sparteine **10**, irrespective of any asymmetric induction which may result. With sparteine, organolithiums even add to dienols such as **12**.[15]

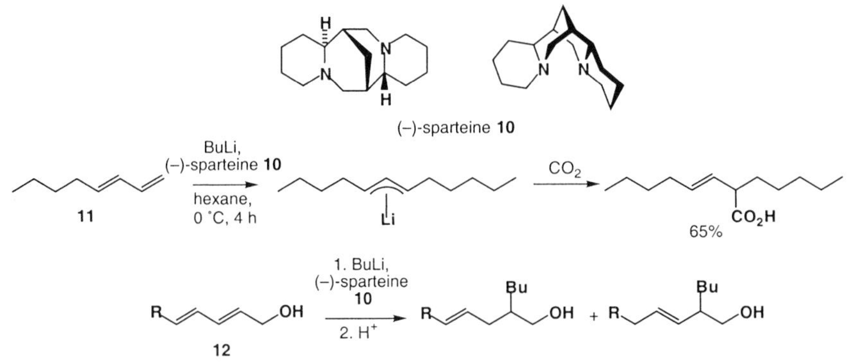

(–)-sparteine **10**

BuLi, (–)-sparteine **10**, hexane, 0 °C, 4 h; **11**; Li; CO_2; CO_2H 65%

12; 1. BuLi, (–)-sparteine **10**; 2. H^+; Bu, OH + Bu, OH

Other "internal" dienes are not reactive towards carbolithiation, but enynes are. Although the yields are variable, the products are valuable allenyllithiums such as **13**.[16]

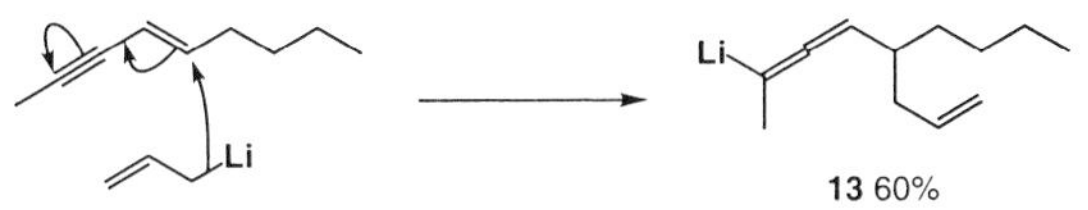

Li; **13** 60%

In general, alkynes cannot be carbolithiated under the usual conditions because they are deprotonated at the alkyne carbon (if terminal) or at propargylic positions. Diphenylacetylene **14** can do neither of these, and it carbolithiates to give **15** in which the lithioalkene turns out to be a powerful ortholithiation-directing group (presumably through coordination in the TMEDA-containing aggregate) to give **16**.[17,18]

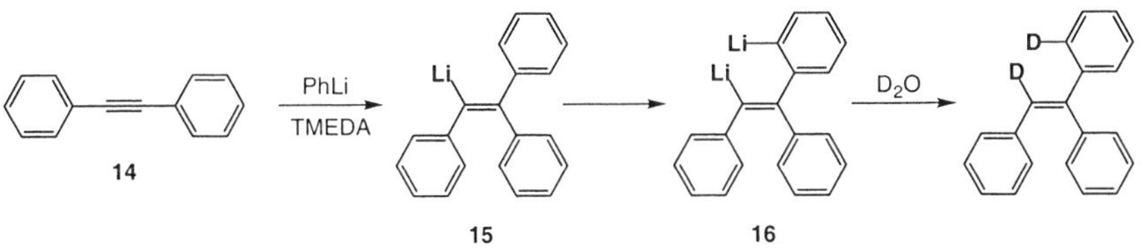

14; PhLi, TMEDA; **15**; **16**; D_2O

It has been reported that carbolithiation of ether-substituted alkynes such as **17** is catalysed by iron(III), a method which has allowed the remarkable stereoselective construction of tetrasubstituted double bonds such as that of **18**.[19]

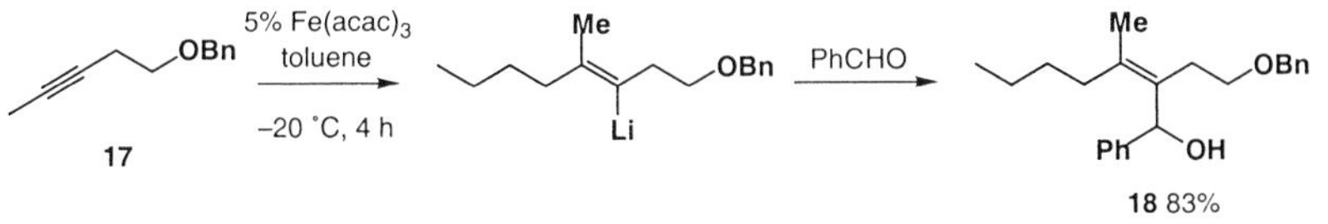

17; BuLi x 3, 5% $Fe(acac)_3$, toluene, –20 °C, 4 h; Me, Li, OBn; PhCHO; **18** 83%

Conjugated alkenes which are part of aromatic rings are occasionally carbolithiated, but unless they are stabilised by further conjugation the products are unstable. For example, organolithiums will add to naphthalene, but the product eliminates lithium hydride to give the rearomatised compound **19**.[20] Dearomatised products are obtained if the naphthalene bears an electron-withdrawing group such as an oxazoline substituent.[21]

RLi

19

7.1.3 Carbolithiation of functionalised alkenes

By far the most successful of all carbolithiations are those in which coordination to a heteroatom accelerates the reaction and stabilises the product. For example, modifying two of the reactions above by incorporating a coordinating oxygen atom improves rates and yields dramatically:[16,22]

20 BuLi 21 80%

22 *i*-PrLi petroleum ether −20 °C 23 87-94%

24 RLi 25 quant.

In contrast with unreactive, unfunctionalised terminal alkenes, allylic and homoallylic ethers (**22**, **24**) and alcohols (**20**) from which the product organolithiums (**21**, **23**, **25**) can be chelated in a (preferably) five-membered, oxygen-containing ring, carbolithiate rapidly and cleanly.[23] Coordination overrides any preference for the lithium to be bonded to the primary carbon, but cannot overcome the unfavourability of forming a tertiary organolithium – **26** gives **27**, but **28** cannot be carbolithiated. Coordination to sulfur in similar thioethers **29** works too.

26 *i*-PrLi petroleum ether, Et_2O, −20 °C 27 89%

28

29 *i*-PrLi petroleum ether, Et_2O, −25 °C 50%

With allylic alcohols, the organolithium adds to the secondary carbon when this allows formation of a five-membered chelate (**31**, **33**) and of a stabilised or primary organolithium. Allyl alcohol itself requires the presence of TMEDA,[24,25] while isopropyllithium adds to cinnamyl alcohol **32** even without TMEDA.[26]

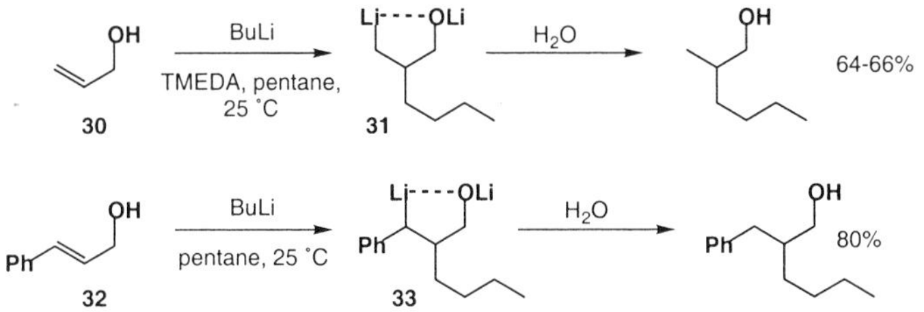

30 **31** **32** **33**

When the alkene is neither terminal nor conjugated, addition appears to occur the other way round – to give a β-alkoxylithium such as **34** – which eliminates Li_2O under the conditions of the reaction.[26] Cyclic allylic alcohols similarly give alkenes by elimination, and geminally disubstituted allylic alcohols such as **35** are simply unreactive.[25]

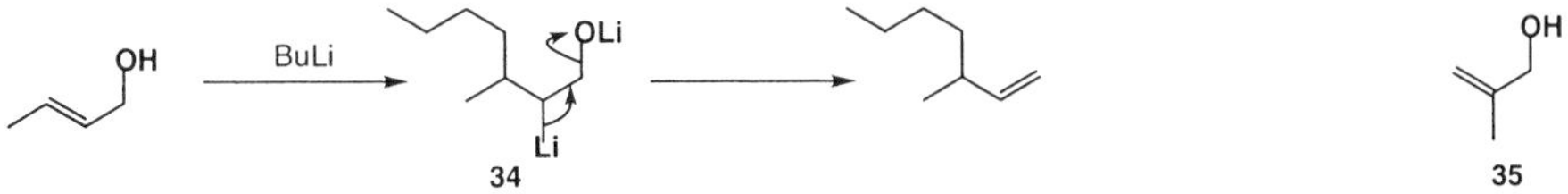

34 **35**

Coordination to O means that the additions are highly stereoselective when the hydroxyl group is at a stereogenic centre,[26,25] and the sense of the selectivity can be rationalised by a transition state approximating to **36**.

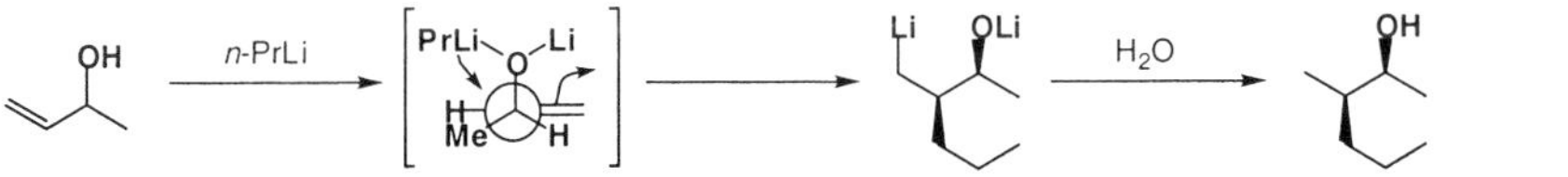

36 >98% 1 diastereoisomer

Stereoselectivity also arises in additions to cinnamyl alcohol **32** when the product organolithium is quenched with an electrophile. Single diastereoisomers of both **38** and **39** are obtained.[27-29] Since non-heterosubstituted benzyllithiums are typically configurationally unstable, the selectivity presumably arises from the preferred direction of attack on a more or less planar or rapidly epimerising organolithium **37**. Represented as shown below, attack from the top face yields the observed stereoselectivity.

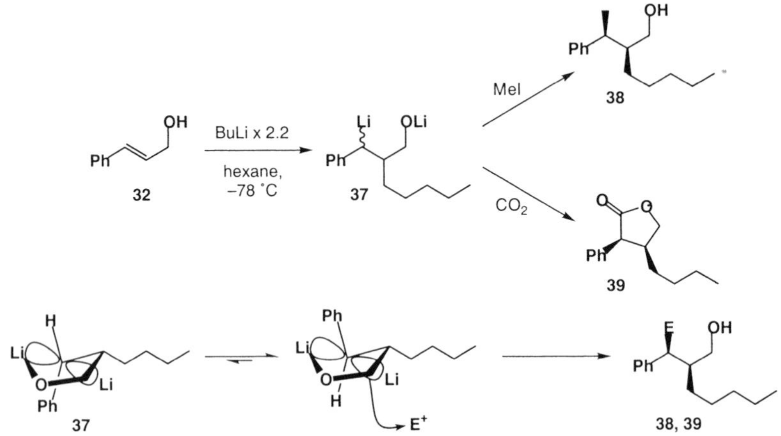

Interestingly, **40**, the phenylthio-substituted analogue of **32**, reacts with opposite diastereoselectivity. Presumably, the "anion" **41** is now very much pyramidal, coordinated to both lithium ions on the same face, and reacting with retention, as expected for an organolithium at tetrahedral carbon (see chapter 6).[27-29]

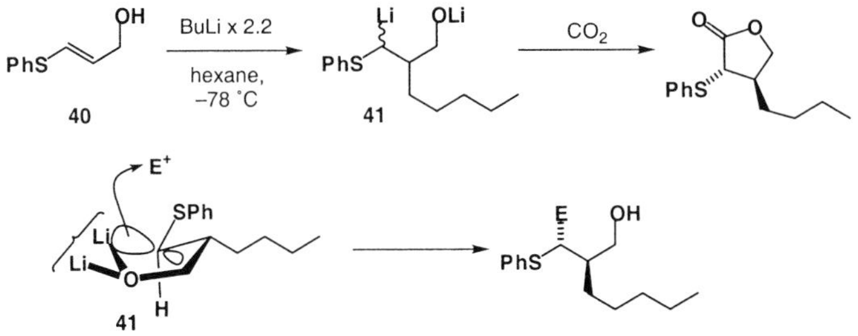

An alkoxy-substituted cinnamyl system **44** is transiently generated by the treatment of cinnamaldehyde **42** with lithiated trimethylethylenediamine **43**, and this protected version of cinnamaldehyde undergoes carbolithiation with good diastereoselectivity in favour, like **32**, of the *syn* product **45**.[30]

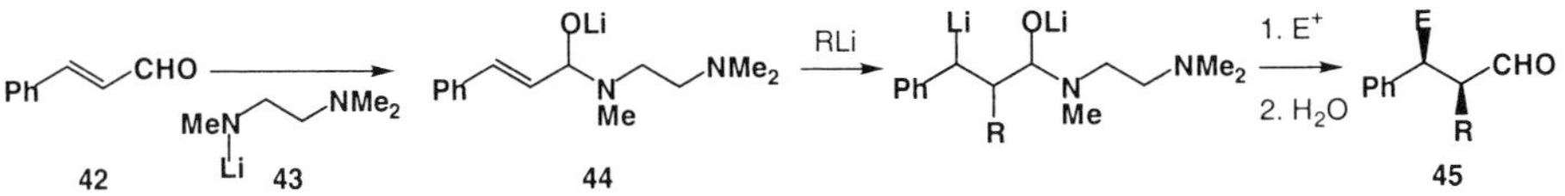

Overall the reaction achieves an anti-Michael addition across the double bond of **42**, an umpolung reactivity pattern also observed in some unsaturated amides (see below).[31]

Cinnamyl ethers are carbolithiated to give intermediates similar to **37**, but without the additional lithium cation of the alkoxide substituent they presumably exist in a structure closer to **46**. The products of deuteration and carboxylation of this organolithium have *anti* stereochemistry, consistent with electrophilic substitution with retention.[32] The stereochemistry of alkylation was unfortunately not determined.

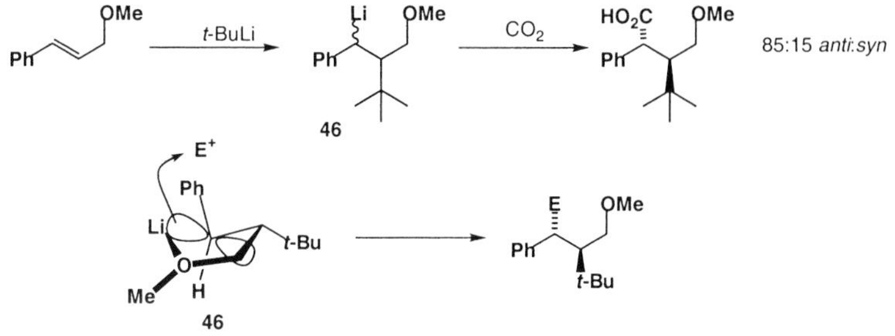

A similar study[33] of the cinnamyl amine **47** revealed that electrophilic quench with zinc bromide, which leads to lithium-zinc exchange, proceeds with inversion. Deuteration of the organolithium **48a** formed by addition of BuLi to **47** gave the *anti* deuterated compound **49a**, presumably by retentive reaction with DCl. Evidence that the stereoselectivity is a result of thermodynamic control over the ratio of two diastereoisomeric complexed organolithiums **48a** and **48b** is provided by the fact that stereoselectivity increases at higher temperature. When the organolithium **48a** is transmetallated with zinc bromide and then immediately deuterated, the other diastereoisomer **49b** is formed. The intermediate organozinc **50b** is formed under kinetic control, because on standing it slowly isomerises to the more stable isomer **50a** which is deuterated to give **49a**.

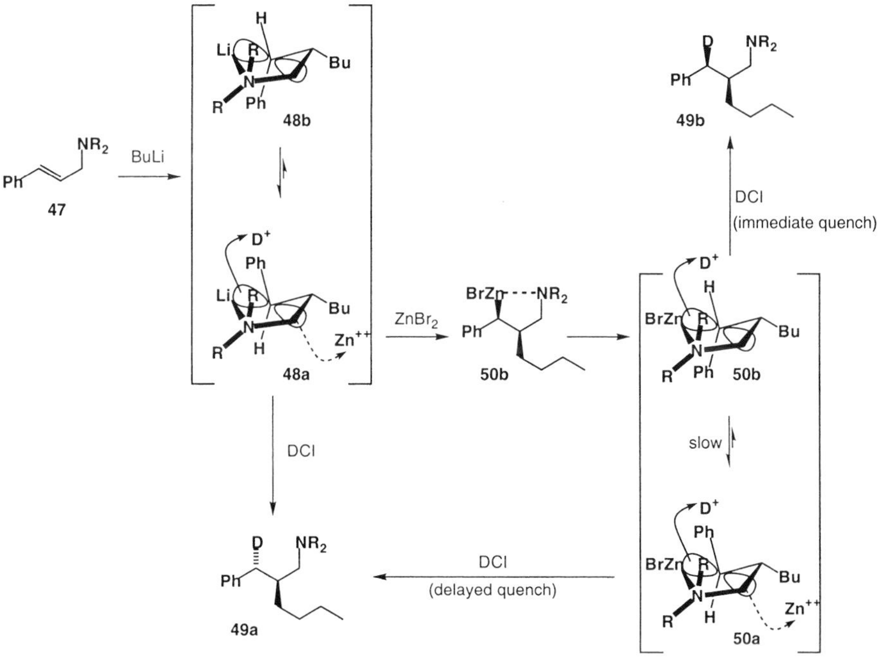

The electron-rich oxygen atom of a secondary amide derivative of cinnamic acid can be as effective as that of a cinnamyl ether or alcohol in directing *anti*-Michael additions to some amide Michael acceptors. For example, BuLi adds to the amide **51** to give a 90:10 ratio of anti-Michael:Michael addition products **52**:**53**.[31] With the alkyne **54** the effect is even more pronounced: MeLi gives solely the anti-Michael product **55**. These reactions are however very sensitive to starting material and reagent structure.

Me$_3$Si NHMe 51 RLi; Li OLi Me$_3$Si NMe R; H_2O; Me$_3$Si NHMe R anti-Michael **52**; 90:10 (R = Bu)
R OLi Me$_3$Si NMe Li; H_2O; Me$_3$Si NHMe Michael **53**

Me$_3$Si NHMe 54 RLi; Li- - -OLi Me$_3$Si NMe R; H_2O; Me$_3$Si NHMe R anti-Michael **55**; sole product (R = Me)

Umpolung reactivity is also evident in the addition of an organolithium to a vinyl carbamate **56**.[34] The product **57** can be quenched with electrophiles at low temperature, but rearranges to **58** by a [1,2]-acyl transfer on warming.

NEt$_2$ O O Ph 56; RLi; NEt$_2$ O O Li R Ph 57; E^+; NEt$_2$ O O E R Ph; warm to 20 °C; NEt$_2$ O OH Ph R 58

Carbolithiation of a trifluoromethyl-substituted alkene **59** produces an unstable intermediate **60** which eliminates lithium fluoride to give gem-difluoroalkenes **61**.[35]

F_3C Ph 59; RLi; [F_3C Li R Ph] 60; F F R Ph 61

7.1.4 Enantioselective carbolithiation

(–)-Sparteine **10** is a powerful promoter of carbolithiation, and can lead to good levels of asymmetric induction in the addition step. Successful enantioselective carbolithiations have all involved functionalised styrenic double bonds. Potentially, two new stereogenic centres are formed in most reactions of this type, with enantioselectivity determined on formation of the first. Even if no stereogenic centre is formed on addition of the alkyllithium, enantioselectivity may still result at the quench step: for example, the enantioselective carbolithiation–electrophilic quench of **62** gives **63**. The 72% ee presumably represents the result of an enantioselective substitution of a configurationally unstable organolithium (see section 6.2).[36]

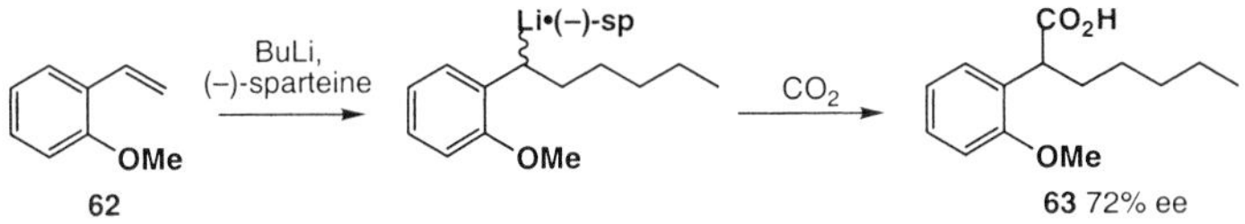

(–)-Sparteine is such a powerful activator of organolithiums that it promotes carbolithiation even of unfunctionalised 1,2-disubstituted styrenic double bonds, giving for example **64**.[37]

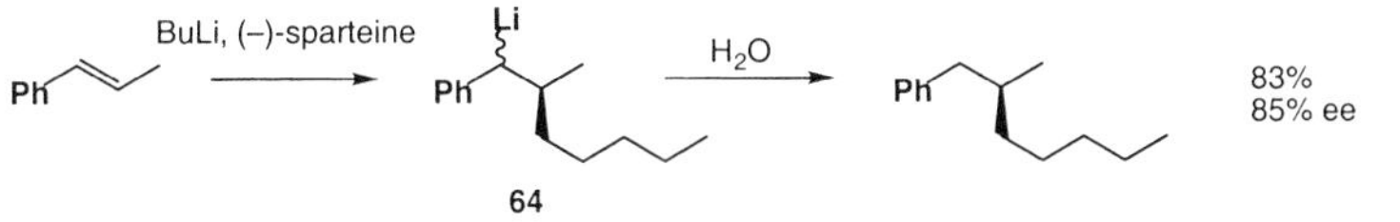

The addition of BuLi to cinnamyl alcohol **32** in the presence of (–)-sparteine gives **65**, which can be protonated and oxidised to give the carboxylic acid **66** in 80% ee.[38,39]

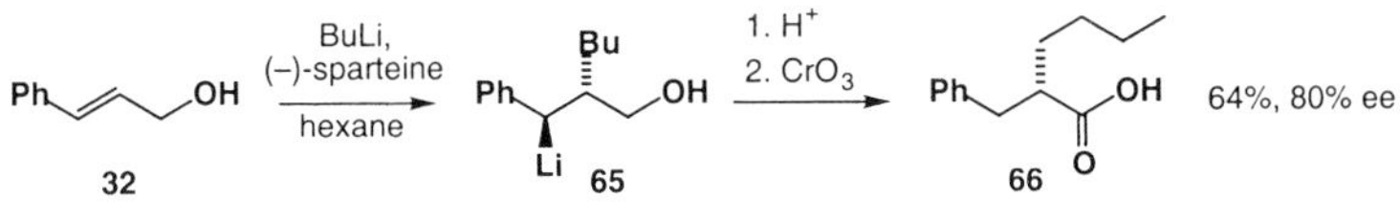

Enantioselective carbolithiation of cinnamaldehyde **42** can be carried out by a modification of the method described above, replacing lithiotrimethylethylenediamine **43** with a chiral diamine derivative.[40]

By protecting cinnamyl alcohol **37** as an acetal, ee is improved further, and new modes of reactivity are opened up. At –50 °C, **67** carbolithiates in the presence of catalytic (–)-sparteine and leads to the alcohol **69** in 90-95% ee. On warming to 20 °C, however, **68** undergoes a cyclisation with substitution at the C–O bond to give the cyclopropane **70**, also in >90% ee.[41]

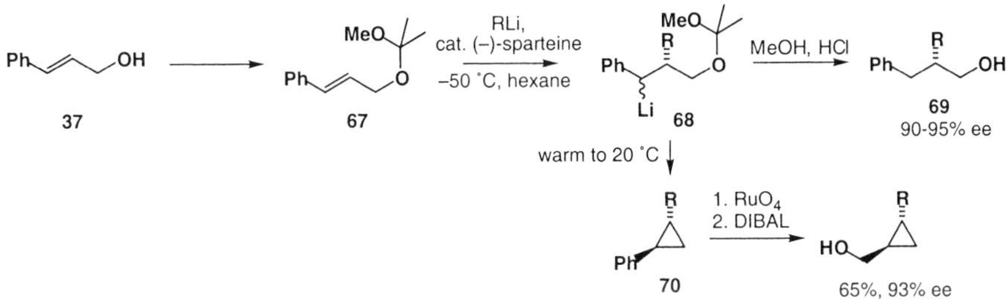

7.2 Intramolecular addition and substitution reactions: anionic cyclisation

Attack of an organolithium on an electrophile within the same molecule is only controllable when the electrophile is stable to the method used to make the organolithium. This has generally meant that cyclisation reactions of organolithiums (for which we shall use the term "anionic cyclisation") begin with a transmetallation or halogen-metal exchange. Even so, organolithiums have now been successfully cyclised onto hindered carbonyl compounds, nitriles, halides, selenides, tosylates, alkynes, activated and unactivated alkenes and aromatic rings.

The first cyclisations to be put to synthetic use were those of aryl lithiums onto carbonyl compounds, imines and epoxides. These are known as "Parham cyclisations", and the method for transforming an aryl bromide to an aryllithium the "Parham protocol", after W. E. Parham, who developed the reaction. We will survey the use of Parham cyclisations in synthesis, before assessing intramolecular attack of other electrophiles. The most important of these are the alkenes, and the usefulness of anionic cyclisations onto unactivated double bonds compares very favourably with radical cyclisations, particularly with regard to stereochemical control.

7.2.1 Anionic cyclisations onto carbonyl compounds and derivatives

7.2.1.1 Cyclisations of aryllithiums – Parham cyclisations[42]

Although halogen-metal exchange reactions of aryl bromides are extremely fast, chemoselectivity in the reaction of butyllithium with a carbonyl-containing aryl bromide is possible only at low temperature, and only if the carbonyl group is sterically or electronically deactivated. Lithium carboxylates, formed *in situ* from the acid by the addition of an extra equivalent of alkyllithium, work quite well here – treating the acids **71**, for example, with two equivalents of BuLi at –100 °C gives good yields of indanone and tetralone **73** and **74** via the organolithiums **72**.[43]

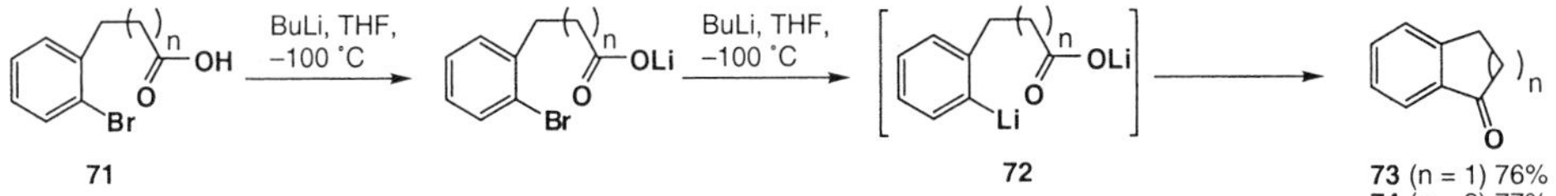

BuLi, THF, –100 °C; BuLi, THF, –100 °C

71 **72**

73 (n = 1) 76%
74 (n = 2) 77%

This regioselective synthesis of benzannelated rings, which, unlike its Friedel Crafts equivalent, is not perturbed by the nature of other directing groups carried by the ring, has been used in several syntheses. Indanone **76** from cyclisation of **75** is an intermediate in a synthesis of nanaomycin A,[44] and tetralone **77** is an intermediate in a synthetic study towards daunomycin.[45]

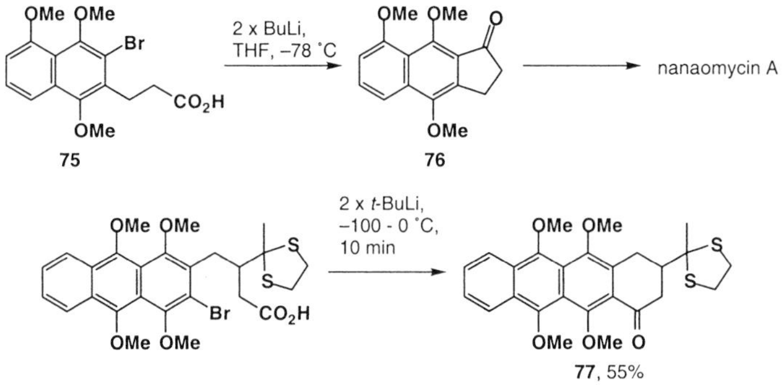

Introducing the carboxylate trap by carbonating a second organolithium site within the molecule leads to a neat synthesis of the tricyclic **78**:[46]

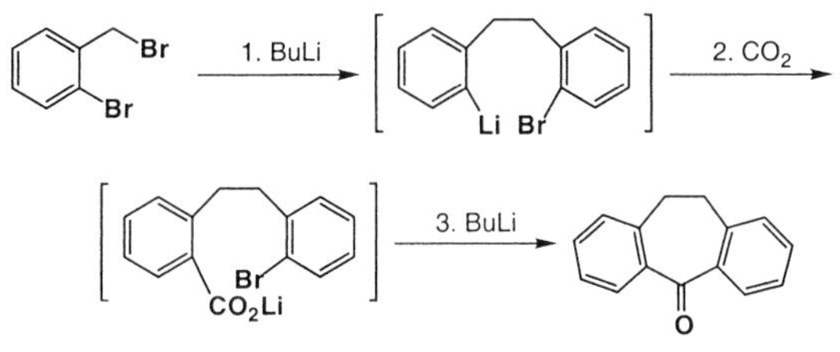

Diethylamide[43] **79** and *N,N*-dimethyl carbamate[47,48] **81** are also sufficiently unreactive to be able to fend off direct attack by butyllithiums.

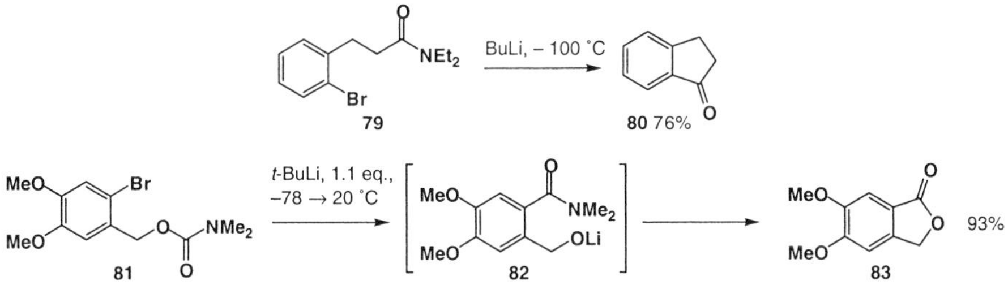

Amides cyclise directly to ketones such as **80**; the carbamates on the other hand pass through a cyclic transition state which collapses to an amide **82** (an "anionic *ortho*-Fries rearrangement" – section 2.3.2.1.4) which lactonises to **83**.

N-Alkylimides, and the sodium (but not lithium) anions of *N*-unsubstituted imides (**84**), behave similarly, and the cyclisations of **84** and **86** provide a stereoselective (with regard to the enamine double bond) route to cyclic acyl enamines **85** and **87** – again, effectively an N $\rightarrow$ C acyl transfer.[49]

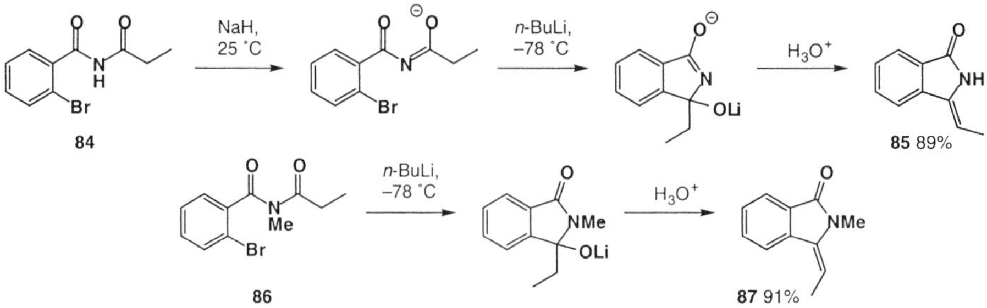

We should include the somewhat similar cyclisation onto a phosphine oxide here, even though no carbonyl group is involved. In this strange reaction, deprotonation of **88** α to phosphorus protects this position while halogen-metal exchange generates an aryllithium which displaces PhLi from phosphorus to give **89**.[50]

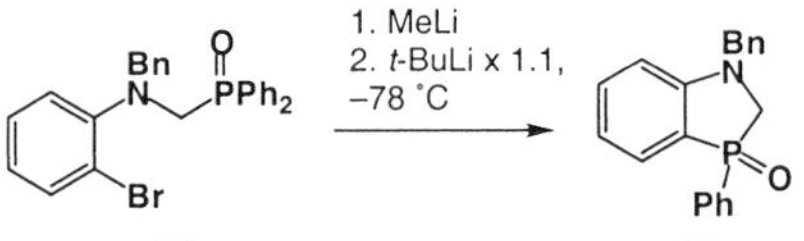

7.2.1.2 Cyclisations of alkenyllithiums

Vinyl iodides and vinyl bromides undergo halogen-metal exchange with *n*- or *t*-BuLi and can be cyclised onto carbonyl groups.[51] A useful study has shown that even ketones are suitable as electrophilic traps – direct attack on the carbonyl group of **90** to give **92** accounted for only 11% of the product, with the majority being **93** arising from halogen-metal exchange to **91** followed by cyclisation. Under these conditions, the corresponding aldehyde gave about 50:50 direct addition:cyclisation.[52,53] *t*-BuLi was less effective in these reactions, probably because it slows down the transmetallation step.

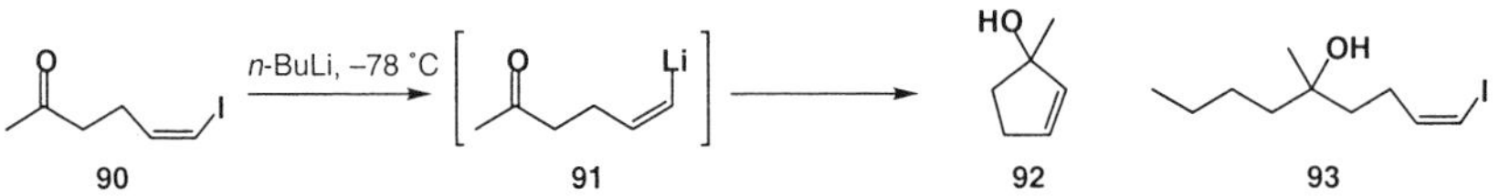

With metallic lithium, the problem of alkyllithium addition to the electrophile is avoided, and provided the reacting substituents were arranged *cis* on the existing five-membered ring, the ketone **94** could be cyclised to a single isomer of the bicyclic 5,5-system **95** in 95% yield.[54] The *trans*-fused system **97** formed in only 44% yield from **96**, as a mixture of diastereoisomers – the major product was then simple reduction to the uncyclised alkene **98**.

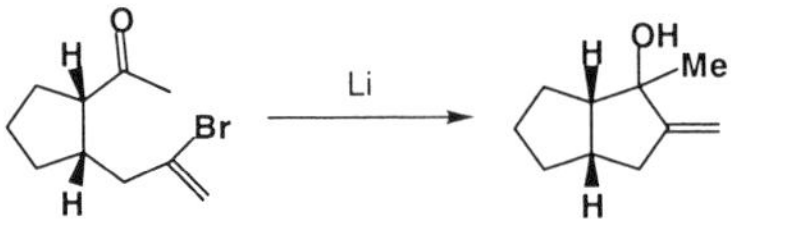

Li

94

95 95%
single diastereoisomer

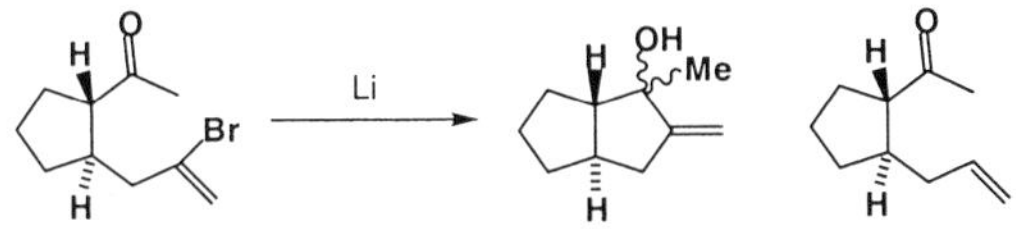

Li

96

97 44%
two diastereoisomers

98 48%

Not surprisingly, amides function well as traps for vinyl iodide or vinyl bromide-generated alkenyllithiums. Reduction of the products from lactams **99** and **101** gives valuable intermediates **100** and **102** for alkaloid synthesis:[55,56]

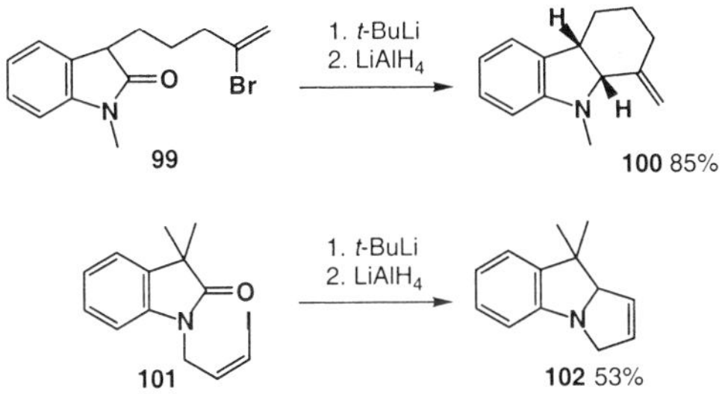

1. *t*-BuLi
2. $LiAlH_4$

99

100 85%

1. *t*-BuLi
2. $LiAlH_4$

101

102 53%

In an unconventional cyclopentenone synthesis, Negishi cyclised vinyllithiums derived from the 1-iodo-1-silyl alkenes **103** onto preformed lithium carboxylate salts.[57] He later found[58] that amides **104** function in this reaction rather better than the carboxylate salts, and that the silyl substituent is not necessary for cyclisation. Nitriles, on the other hand, fail to cyclise.

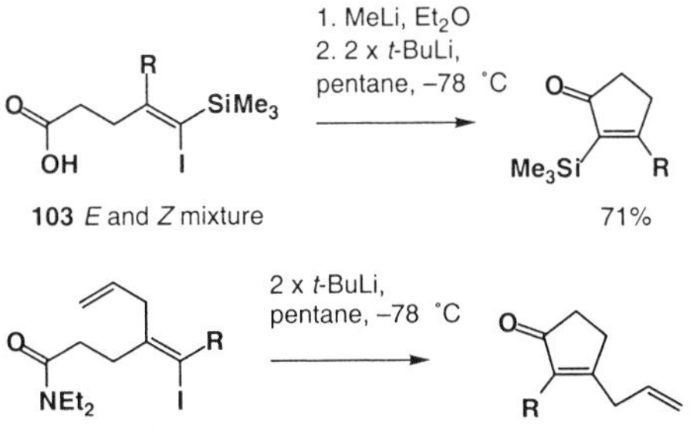

1. MeLi, Et_2O
2. 2 x *t*-BuLi,
pentane, –78 °C

103 *E* and *Z* mixture

71%

2 x *t*-BuLi,
pentane, –78 °C

104

71%

7.2.1.3 Cyclisations of alkyllithiums

The relatively slow transmetallations of alkyl halides means that generally only alkyl iodides can be used as precursors for cyclisation onto any but the least reactive of carbonyl groups. The resulting alkyllithiums will cyclise onto nitriles, amides, esters, and, in some cases ketones. Direct addition to the C=O group is slower with *t*-BuLi than with *n*-BuLi, but so is the transmetallation, and *n*-BuLi is generally the better choice except with ketones.

Cyclisation of iodonitrile **105**[59] to **107** was an early example of an anionic cyclisation. Slow transmetallation of the choronitrile **108** allowed the BuLi to attack the nitrile instead, leading to a quite different cyclisation product **109**. The successful iodine-lithium exchange of primary alkyl iodide **105** to give **106** even with *n*-BuLi must be driven by the cyclisation itself.

BuLi, Et_2O, 0 °C, 30 min
105 → [106] → [ ] → 107

BuLi, Et_2O, 0 °C, 30 min
108 → [ ] → 109

The iodoamide **110** and iodoester **111** both cyclise to cyclopentanone, but to avoid further reaction in the case of the ester it was necessary to trap the tetrahedral intermediate with Me_3SiCl. Six-membered rings formed from homologous starting materials only in very low yield.[52]

110 → (*n*-BuLi) → 71%

111 → (1. *n*-BuLi, −110 °C; 2. Me_3SiCl; 3. H_2O) → 69%

The ester cyclisation was more reliable with more hindered starting materials **112** and **114**, making the bicyclic products **113** and **115** in acceptable yield.[60]

112 → (*t*-BuLi x 2, Et_2O, pentane, −100 → −50 °C) → **113** 67%

114 → (*t*-BuLi x 2, Et_2O, pentane, −100 → −50 °C) → **115** 66%

Ketones **116** gave moderate yields of cyclised product **117**, with direct attack by the alkyllithium forming the main by-products **118**.[52]

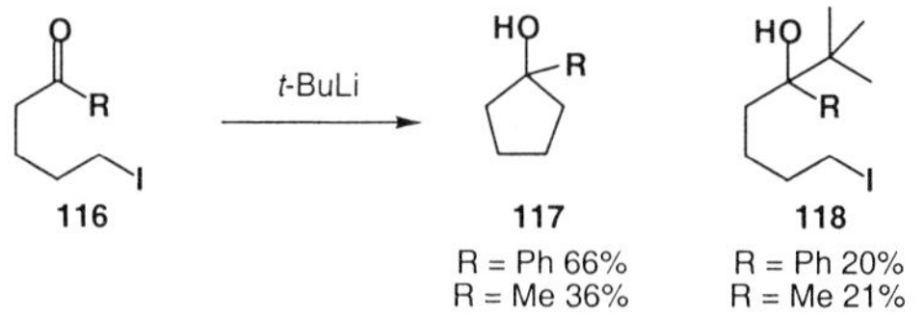

116 → (t-BuLi) → 117 + 118

117: R = Ph 66%; R = Me 36%

118: R = Ph 20%; R = Me 21%

α-Amino alkyllithiums cyclise onto both amides (**119**) and carboxylates (**121**) in reasonable yield to give pyrrolidin-3-ones **120**.[61]

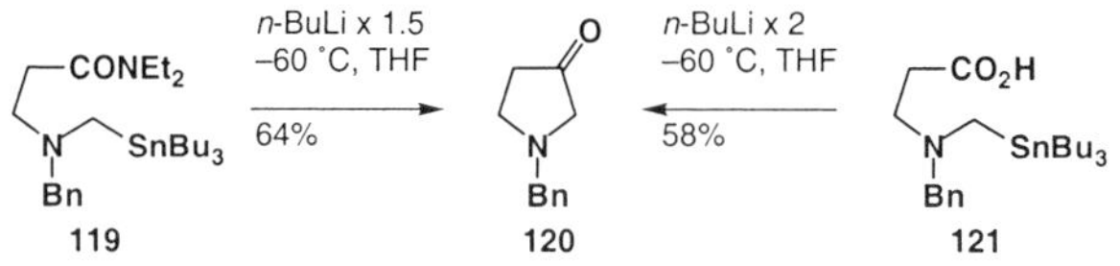

119 → (n-BuLi x 1.5, −60 °C, THF, 64%) → 120 ← (n-BuLi x 2, −60 °C, THF, 58%) ← 121

Comparable reactions have been made diastereoselective by the incorporation of a chiral substituent at nitrogen.[62]

There is one example, unique for several reasons, of the formation of a four-membered ring by anionic cyclisation onto an oxazoline. Attempted oxazoline-directed lithiation of the styrene **122** gave, instead, the cyclobutane **124** via addition of the alkyllithium to give a benzylic organolithium **123** which cyclises stereoselectively.[63] The initial intermolecular carbolithiation proceeds remarkably easily – no additives (such as TMEDA) are required, even with MeLi.

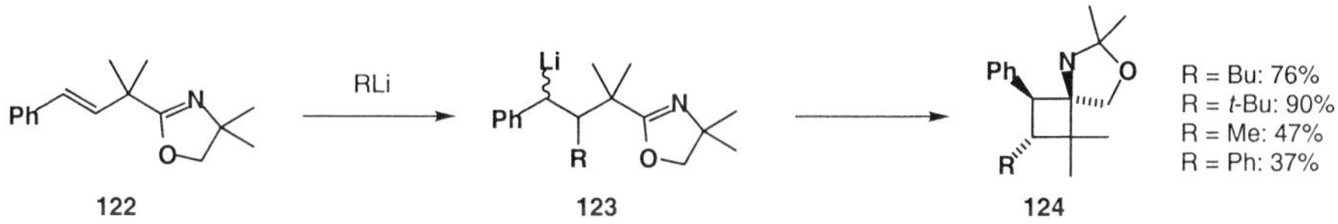

122 → (RLi) → 123 → 124

R = Bu: 76%
R = t-Bu: 90%
R = Me: 47%
R = Ph: 37%

7.2.1.4 Cyclisations of alkynyllithiums

The acidity of alkynyl protons means that alkynyllithium cyclisations are rather trivial to devise – the starting alkyne can be deprotonated with LDA or LiHMDS and cyclised even onto aldehydes. An example from Kende's laboratory demonstrates the use of an alkynyllithium cyclisation in the synthesis of an enediyne **125**:[64]

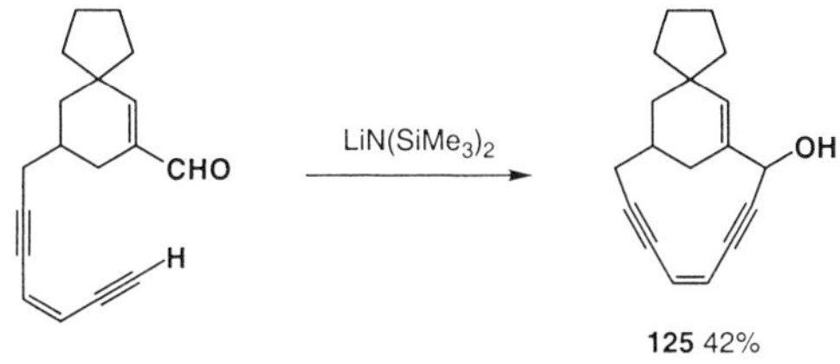

→ ($LiN(SiMe_3)_2$) →

125 42%

7.2.2 Anionic cyclisations onto epoxides

Aryl and alkyllithiums can be made to cyclise onto epoxides: a Parham-style cyclisation is possible using an aryl bromide **126**, provided the temperature is kept low (–100 °C) to prevent the epoxide reacting with BuLi.[65]

BuLi, –100 °C
126 → 79%

Under these conditions, the smaller of the two possible rings is formed, and yields are good provided the end of the epoxide nearer the organolithium does not carry another substituent. However, activating the epoxide with a Lewis acid ($MgBr_2$) both increases the scope of the reaction (four- and six-membered rings can be formed: **127** gives **128**) and can change its regioselectivity: **130** gives **131** under the normal conditions of the reaction but **129** in the presence of $MgBr_2$.[66]

127 → 128 (BuLi, $MgBr_2$)

130 → 129 (BuLi, $MgBr_2$); 130 → 131 64% (BuLi, –100 °C)

Using the less reactive aryl chlorides in this reaction can give rise to a competitive epoxide lithiation, especially when the epoxide is benzylic: for example, **132** gives **134** via **133**.

132 → [133] → 134 (BuLi, $MgBr_2$)

Both Cooke[67] and Babler[68] made systematic studies of the anionic cyclisation of iodo- and bromoepoxides **135**. In these cases, the preference for 5- over 6-ring formation (exo-cyclisation) could be *increased* from 10:1 to 100:1 by adding magnesium salts, or reversed to 1:4 by adding copper. Transmetallation from both iodo- and bromoepoxides is rapid at –78 °C with either *s*-BuLi or *t*-BuLi (there is insufficient driving force for transmetallation with *n*-BuLi), but the cyclisation takes hours, and best yields were obtained on warming to 0 °C.

135: *s*-BuLi or *t*-BuLi x 2, –78 → 0 °C; 10 : 1 (78 %)
+ Mg^{2+}: 100:1
+ Cu^{+}: 1:4

Exo-cyclisation, to give the smaller of the two possible rings, is still preferred for 4- vs. 5-ring formation, and is susceptible to the substitution pattern of the epoxide, as demonstrated by the examples **136-138**.[67]

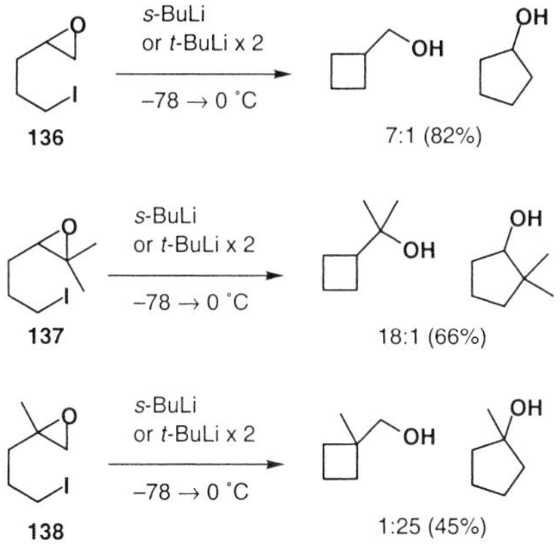

s-BuLi or t-BuLi x 2, −78 → 0 °C

136 7:1 (82%)

137 18:1 (66%)

138 1:25 (45%)

Cyclisation of **139** is not regioselective and gives a 60:40 mixture of six- and five-membered rings, but the same class of reaction starting from **140** allows the synthesis of the remarkable strained cage compound **141**.[69]

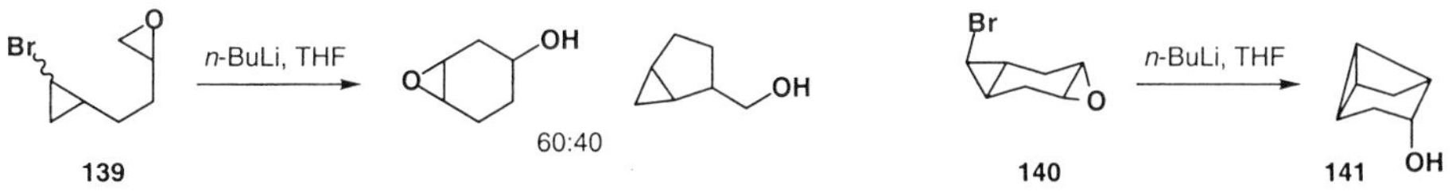

n-BuLi, THF

139 60:40 **140** **141**

Epoxides will also capture organolithiums formed by deprotonation, notably cyclopropyllithiums and ortholithiated tertiary amides. Ortholithiated amides cyclise regioselectively in a valuable route to benzofurans such as **142** and **143** and (less efficiently) benzopyrans **144**.[70]

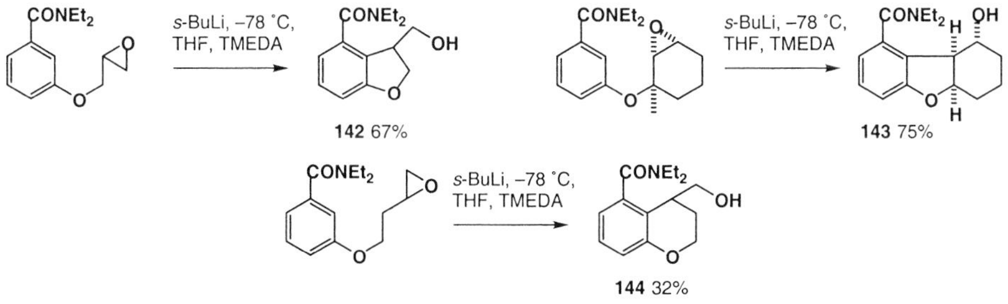

s-BuLi, −78 °C, THF, TMEDA

142 67% **143** 75% **144** 32%

The fulvene **145** offered the opportunity for a sequential inter-intramolecular carbolithiation: the intermediate cyclopentadienyllithium **146** undergoes 3-*exo* cyclisation to give **147**, a precursor to longifolene.[71]

MeLi

longifolene

145 146 147 65%

7.2.3 Anionic cyclisations onto alkyl halides and similar compounds

Compounds containing two halogen atoms will undergo halogen-metal exchange and cyclisation provided that the cyclisation step is faster than the second halogen-metal exchange. This is the case if both are iodine, or if one is iodine and the other bromine, or if one is a vinyl or aryl bromide or iodide.

Cyclisation of a compound containing two bromine atoms is successful provided one of the bromines is an aryl bromide and the other is a non-benzylic alkyl bromide.[46] A type of controlled intramolecular Wurtz coupling then takes place, the aryl bromide transmetallating fastest. Four-, five-, six- and seven-membered rings can be formed from **148**, **149** and **150**.[72]

n-BuLi, –100 °C

148

R = Bu, 86%
R = H, 68%

n-BuLi, –100 °C

149

R = Me, 46%
R = H, 78%

n-BuLi, –100 °C

150

With diiodides, the reaction is successful even if both are primary alkyl iodides.[73,74] For small (three-, four- and five-membered rings) the cyclisation is the fastest step, and the second transmetallation does not get the chance to take place: cyclopropane **151**, cyclobutane **152** and cyclopentane **153** all form in high yield from their respective diiodides. Six- and seven-membered rings fail to form: presumably the slower cyclisation fails to compete with the second transmetallation.

t-BuLi, ether, pentane, –23 °C

151 93%

t-BuLi, ether, pentane, –23 °C

152 87%

t-BuLi, ether, pentane, –23 °C

153 98%

With a vinyl iodide–alkyl bromide precursor such as **154** the discrepancy between the rates of the two transmetallations means that it is even possible to trap the uncyclised mono-transmetallated vinyl lithium at –78 °C, giving **155**. Warming to 0 °C promotes cyclisation to **156**, and 5-, 6-[53] (**157**) and even 3-membered[75] rings (**158**) form successfully.

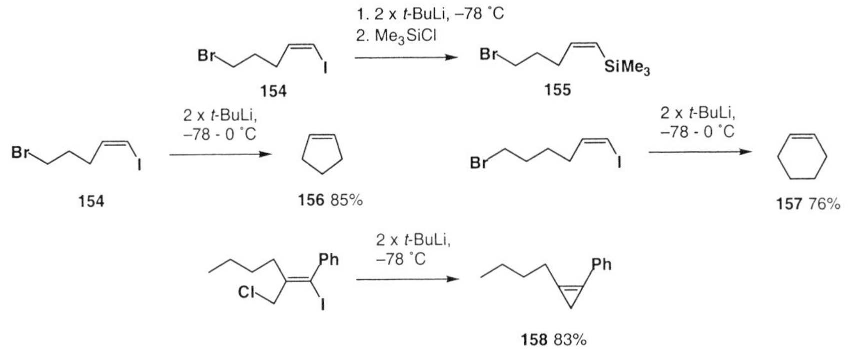

1. 2 x *t*-BuLi, –78 °C
2. Me_3SiCl

154 → **155**

2 x *t*-BuLi, –78 - 0 °C

154 → **156** 85%

2 x *t*-BuLi, –78 - 0 °C → **157** 76%

2 x *t*-BuLi, –78 °C → **158** 83%

For cyclisation, simple vinyl iodides need to be *Z*, or polymerisation occurs. However, a silyl substituent geminal to the lithium atom promotes causes geometrical instability in the vinyllithium, and even *E* 1-iodo-1-silyl alkenes **159** and **160** cyclise successfully:[76]

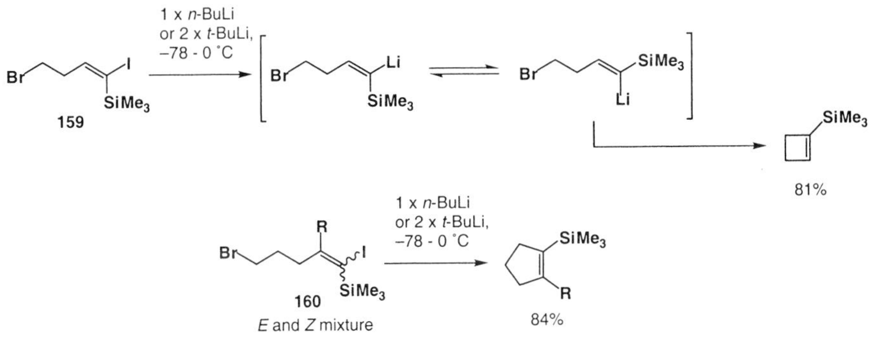

1 x *n*-BuLi or 2 x *t*-BuLi, –78 - 0 °C

159 → 81%

1 x *n*-BuLi or 2 x *t*-BuLi, –78 - 0 °C

160 *E* and *Z* mixture → 84%

Chemoselectivity is also achievable by transmetallating a vinyl stannane and closing onto an alkyl chloride (**161**) or (better) bromide (**162**). Attempted formation of the six-membered ring **163** with the chloride analogue of **162** led to (retro-Brook) silyl migration from O to C.[77]

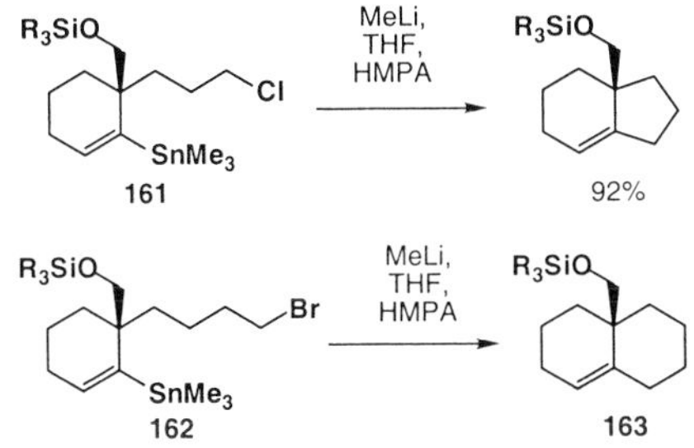

MeLi, THF, HMPA

161 → 92%

MeLi, THF, HMPA

162 → **163**

Making the vinyllithium by a Shapiro reaction (section 8.1) from **164** bypasses the chemoselectivity problem altogether, and provides a useful cyclisation route to exo-methylene derivatives of five-, six- and seven-membered rings **165**.[78] The main by-products **166** are the result of protonating the vinyllithium.

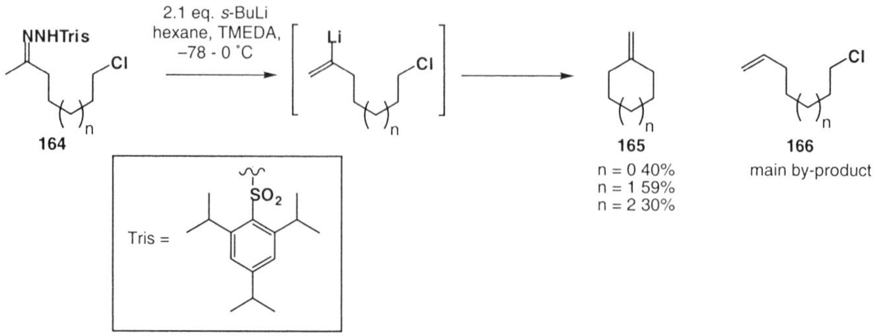

Krief has made selenide **168** containing a mesylate leaving group by ring-opening of propylene oxide.[79] Selenium-lithium exchange (with two equivalents of BuLi because the first deprotonates the mesylate group) gives cyclopropane **169** effectively by overall replacement of the oxygen atom of the epoxide with the alkylidene group of **167**.

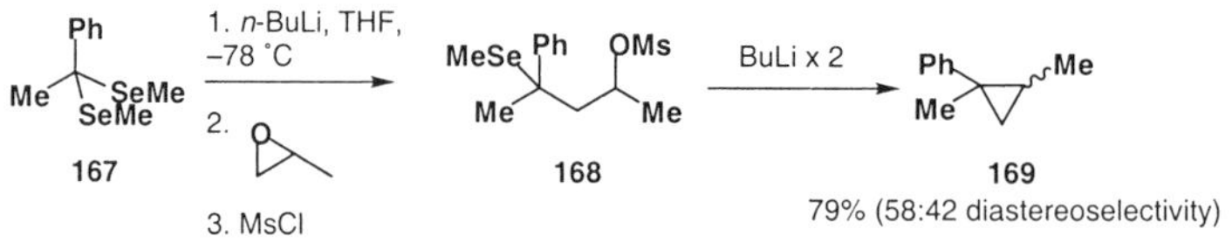

An alternative, and this time stereoselective (though the origin of the stereoselectivity is unclear) selenium-based route to cyclopropanes again starts with Se–Li exchange of **167**, followed by addition to a styrene, and then organolithium cyclisation of **170** with selenide as a leaving group, effectively shortening the route above by cutting out the epoxide and combining two steps into one pot.

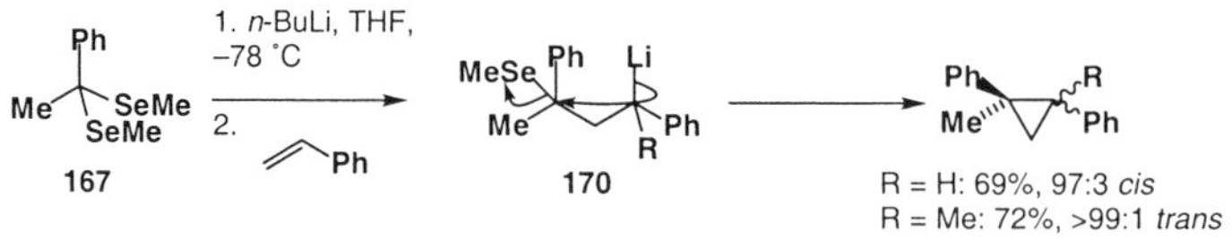

Overall, the transformations are equivalent to carbene additions to the styrenes. However, a carbene *mechanism* can be ruled out since the only alkenes which are successful are those carrying anion-stabilising groups.

These reactions can also be performed using tin-lithium exchange, and have been used to investigate the stereospecificity of the substitution reactions involved and the configurational stability of the intermediates (see section 5.1).[80-82]

Cyclisation with substitution of chloride was made enantioselective by Beak[83] in a remarkable reaction initiated by an asymmetric deprotonation of **171** (see section 5.4 for a discussion), followed by cyclisation to **172**. Yields and enantiomeric excesses were highly solvent-dependent.

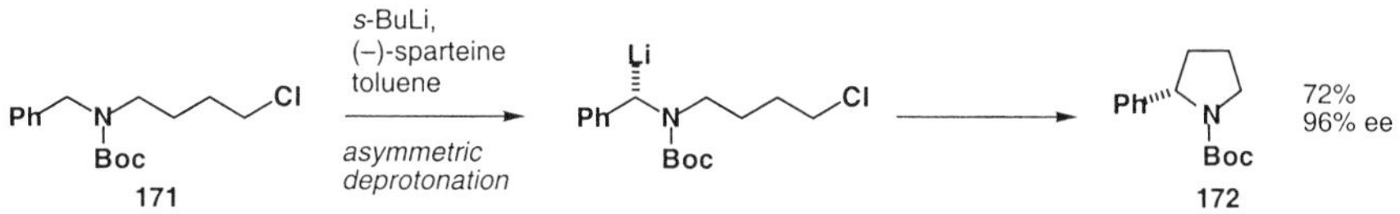

7.2.4 Anionic cyclisations onto alkenes and alkynes

7.2.4.1 Cyclisation onto activated alkenes

Electron-poor alkenes are in general too susceptible to direct attack by alkyllithiums to be useful as traps in cyclisation reactions. For example, the unsaturated *t*-butyl esters **173** cyclise successfully to give four- or five-membered rings **174**, but six-membered rings form in only very low yield, with the major side reaction being direct attack of BuLi on the unsaturated ester.[84]

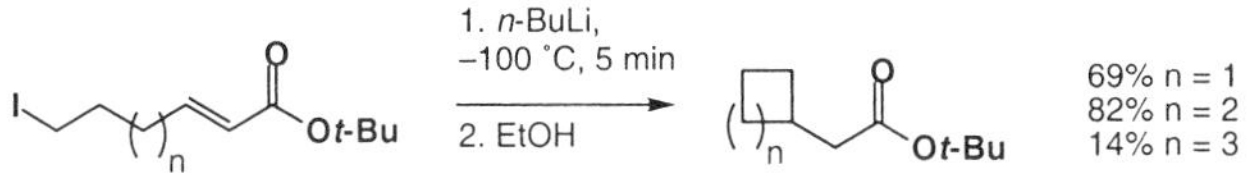

Unsaturated ketones and amides are not successful traps for alkyllithiums, and the only other activated alkene to have successfully trapped an alkyllithium is the unsaturated phosphorane **175**.[85] This trap works remarkably well with both vinyl and alkyl iodides, giving good yields of ring sizes 3-6, and is not perturbed even by a second substituent on the Michael acceptor, as in **176**.

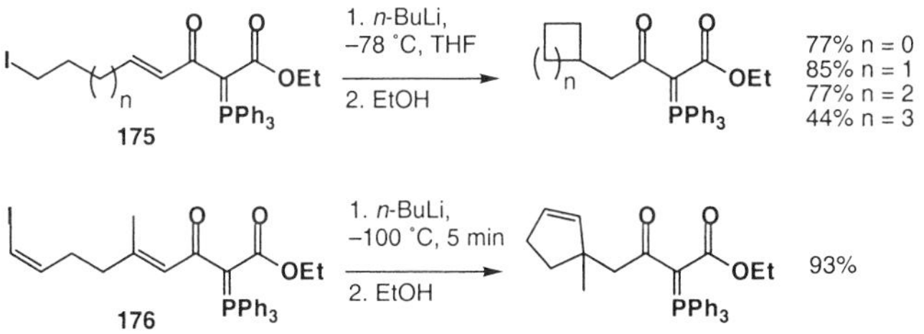

Cyclisations of alkyllithiums onto *E* unsaturated phosphoranes or *t*-butyl esters are less stereoselective than the corresponding cyclisations onto unactivated alkenes, presumably because the additional stabilisation allows the transition state to become "looser".[86] With *Z* enoates **177**, however, *trans* selectivity is high because of congestion in the transition state leading to the *cis* isomer of **178**. With an alkoxy substituent (**177**, R = OMe), the stereoselectivity reverses, possibly due to Li-coordination.

E or Z

1. *n*-BuLi, –100 °C, THF
2. EtOH

177 → 178

E, R = Et: 90%, 3:1 *trans*:*cis*
Z, R = Et: *trans* only
E, R = OMe: 46%, 1.5:1 *cis*:*trans*

The faster halogen-metal exchange and controlled conformation in vinyl and aryl systems meant that aryllithiums and vinyllithiums can be cyclised remarkably efficiently onto cyclic vinylogous amides to give **179** and **180**. **180** was used in a synthesis of (+)-indolizidine 209D.[87]

n-BuLi, THF, –78 °C

179 91%

1. *t*-BuLi, THF, –78 °C
2. (Tf_2N-pyridine, Cl)

80%

H_2, Pt/C

180 79%

(+)-indolizidine 209D

7.2.4.2 Cyclisation onto unactivated alkenes

During the 1950's and 1960's it became evident that organolithiums would add to unactivated alkenes, especially when the reaction was driven by release of ring strain or by conversion of a secondary or tertiary organolithium to a primary one. We discussed the intermolecular additions to alkenes in section 7.1. The 1960's and early 1970's saw plenty of examples of intramolecular attack of a variety of organometallics[88] – particularly organomagnesiums[89-91] – onto unactivated alkenes.[92] These cyclisations (of organo-Al, Mg, Li, Zn and, later,[93] Na compounds) were assumed to be promoted by metal–alkene complexation[94] (and therefore required metals bearing empty orbitals), though at this stage the degree to which the cyclisations were radical-mediated was not clear. Organozinc cyclisations are again becoming an important cyclisation method.[95]

The first clear demonstration of cyclisation onto an alkene involving an organolithium came in 1968, when a Russian group showed that formation of secondary alkyllithium **182** from iodide **181** led to the *cis*-dimethylcyclopentane (5:1 mixture with *trans*) in a matter of hours at room temperature[96,97] – a reaction that was already known for Grignard reagents.[92] The stereoselectivity of the reaction (it gave mainly *cis*-**183**) now suggests that the cyclisation itself involves radicals, a point discussed below.

181 → 1. Li, 0 °C, Et_2O → [182] → 2. 20 °C, 3. H_2O → 183 (5:1)

Shortly afterwards they showed that even primary alkyllithiums **184** and **185** would cyclise, the driving force being simply exchange of a π for a σ bond. Both five- and six-membered rings could be formed (though 6-ring formation only occurred in the presence of TMEDA and would not reach completion) and the cyclisation appeared to be irreversible since in the 6-ring case no ring-opened product was generated when the cyclised organolithium **186** was formed by another route.[98,99]

Br → 1. Li, 0 °C, Et_2O → [184] → 2. 20 °C, 4 h → Li → 3. H_2O

I → 1. Li, 0 °C, Et_2O → [185] → 2. TMEDA, 2 eq., 20 °C, 4 h → 186 → 3. H_2O

The rate of cyclisation is highly solvent-dependent: hexenyllithium **184** for example takes (at 25 °C) 8 days to cyclise in pentane, 96 h in benzene and less than an hour in ether.[100-102]

Shortly afterwards, Lansbury[103] managed to cyclise the *endo* isomer of the mixture of stereoisomeric norbornenes **188a** and **188b**. Reduction of the benzylic ether **187** gave the organolithium from both isomers, but only the *endo* isomer **188b** can cyclise, giving **189a** and **189b**.

187 (OMe, Ph) → Li, Ph_2O → [188a, 188b] ; 188a → H_2O ; 188b → [Li, Ph] → H_2O → 189a 48%, 189b 25%

When this reaction was repeated on an oxa-analogue **190**, with the organolithium now being formed by deprotonation, the *endo* isomer **191b** again cyclised, to give **193**, while the *exo* isomer **191a** underwent a [1,2]-Wittig rearrangement to **192**.

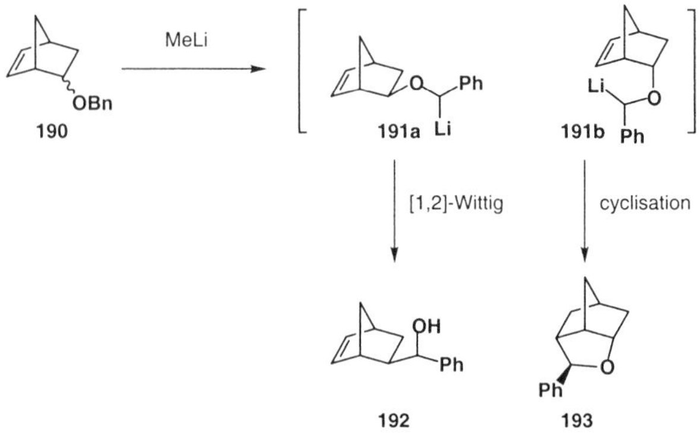

MeLi
190
191a
191b
[1,2]-Wittig
cyclisation
192
193

The benzyl ether **194** also cyclises, and interestingly forms in addition to the expected **195** a by-product **196** resulting from attack of the organolithium on the aromatic ring.[104]

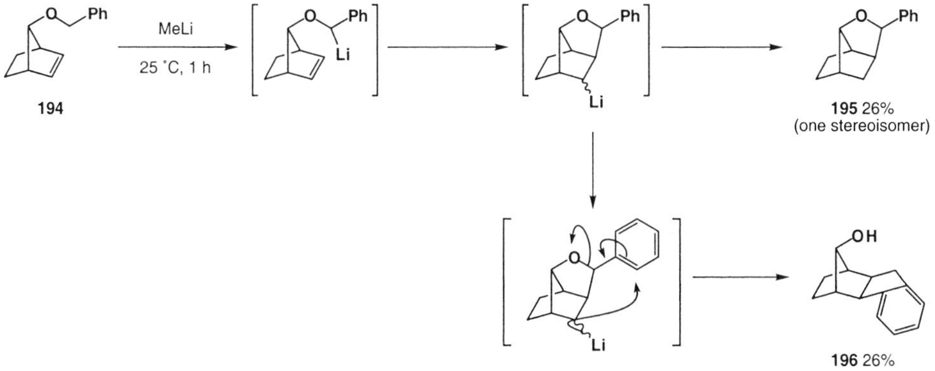

MeLi
25 °C, 1 h
194
195 26%
(one stereoisomer)
196 26%

Without careful design of the starting material, it is not usually possible to form three- and four-membered rings by organolithium cyclisation: no cyclised product was obtained when butenyllithium **198** and pentenyllithium **200** were formed from organomercury compounds.[100] Furthermore, while cyclopentylmethyllithium and cyclohexylmethyllithium **186** are stable to ring opening (see above), early work[105,106] had already shown that this was not the case for cyclopropylmethyllithium **197** and cyclobutylmethyllithium **199**: both undergo ring opening in a matter of hours at –70 °C or minutes at –20 °C.

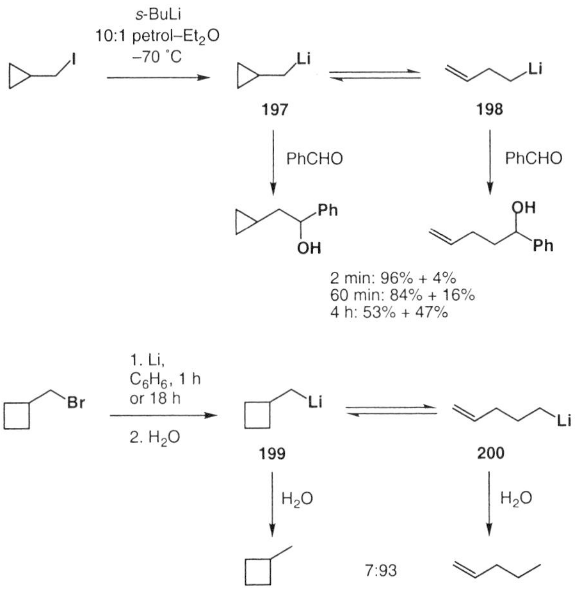

s-BuLi
10:1 petrol–Et_2O
−70 °C

197 198

PhCHO PhCHO

2 min: 96% + 4%
60 min: 84% + 16%
4 h: 53% + 47%

1. Li, C_6H_6, 1 h or 18 h
2. H_2O

199 200

H_2O H_2O

7:93

There are nonetheless just a few isolated examples of three- or four-membered ring forming anionic cyclisations: curiously it was in fact a three- and a four-membered ring forming reaction which provided the first two pieces of evidence that organolithiums do indeed cyclise onto unactivated double bonds. In 1960, Wittig[107] showed that *n*-BuLi not only added to one of the double bonds of norbornadiene but that the major product of the reaction arose from organolithium cyclisation of the intermediate **201** onto the other double bond. This odd reaction is presumably made possible by the strain already inherent in the bicyclic system. After protonation, the major product was **202**.

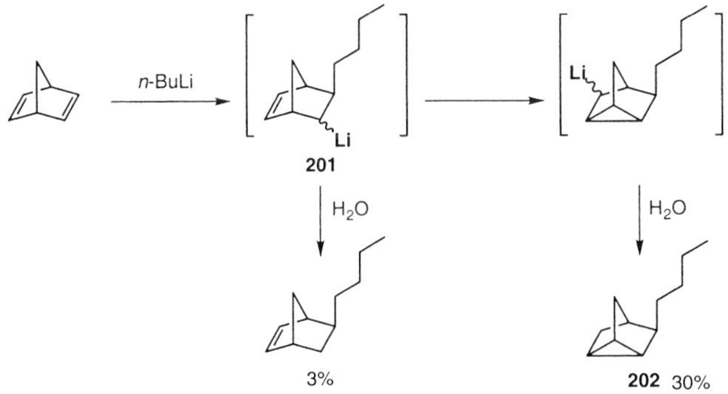

n-BuLi

201

H_2O H_2O

3% **202** 30%

The allyl norbornenyl ether **203**, like its benzyl cousin, exhibits unusual cyclisation chemistry, including a four-membered ring formation, on lithiation.[104] Both a four-membered ring and a five-membered ring (formed by a 5-*endo-trig* cyclisation![108]) are generated in the products **204** and **205**. The oxygen atom must be directing these reactions: the *endo*-methyl group of the four-membered ring product **205** can only reasonably be explained by coordination to oxygen.

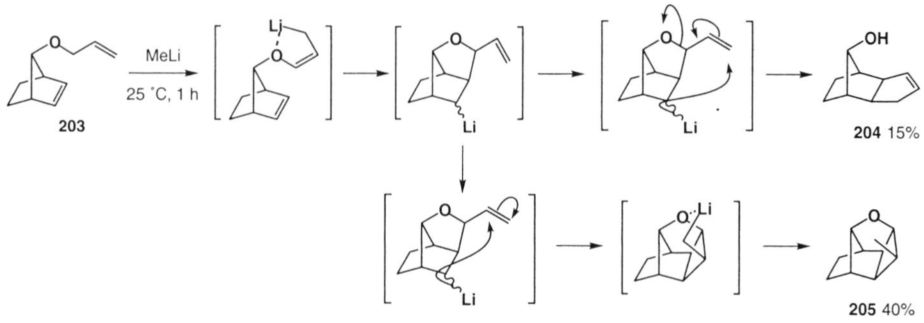

From simple, acyclic organolithium starting materials, three- and four-membered rings can usually be formed only transiently as intermediates. Their existence can be deduced in rearrangements of secondary to primary organolithiums such as **206** to **207**[105] or **209** to **210**.[109] The deuterium labelling in **208** demonstrated that three-membered ring formation must be faster than four (only the non-deuterated alkene cyclised). The fact that the organolithium **209** could be trapped with D_2O prior to cyclisation indicates that these reactions are not radical mediated.

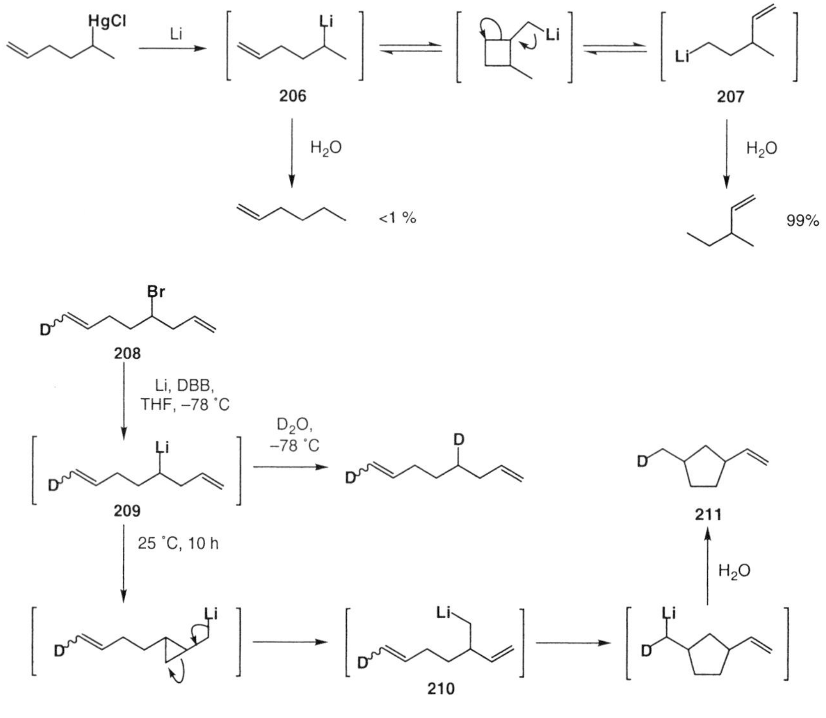

Similar reversible three-membered ring forming reactions, driven by the formation of a primary from a tertiary alkyllithium, are involved in the reactions below.[110] The tertiary organolithium **212** can be trapped directly with an electrophile at –78 °C, but after 2 h at –40 °C has rearranged to the primary organolithium **213**. This rearrangement can be used in ring-contractions (converting **214** to **215**) and expansions (converting **216** to **217**).

Li, DBB, –78 °C, hexane, THF; –40 °C, 2 h; RCHO; SPh; Li; **212**; **213**; R OH 87%; R OH 70%

214 SPh; 1. Li, DBB, –78 - –40 °C, hexane, THF 2. RCHO; **215** R OH; **216** SPh; 1. Li, DBB, –78 - –40 °C, hexane, THF 2. H$^+$; **217** 79%

A three-membered ring intermediate has been trapped in one isolated case. Stopping the rearrangement of the organolithium derived from **218** after 5 min at –78 °C gave 43% of the cyclopropane **220**; at –40 °C **221** became the sole product.

PhS HO **218** 17%; 1. Li, DBB, –78 °C, 5 min 2. H$^+$; HO **219** 17%; HO **220** 43%; HO **221** 40%

As discussed later, others have noted formation of three- and four-membered ring products from phenyl-, silyl or alkynyl-substituted alkenes, which generate more stabilised product organolithiums, or from compounds bearing allylic leaving groups.[111]

After a lull of some ten years, Bailey's paper,[112] in 1985 – essentially on the mechanism of the organolithium cyclisations – reignited interest in anionic cyclisations of organolithiums onto alkenes as a synthetic method for forming carbocycles. This was a period when radical cyclisations, and radical cascade reactions, were very much to the fore – Curran published his now classic[113] synthesis of hirsutene in 1986 – and the cyclisation of 1-hexenyl systems was being used as a probe for the participation of radicals in a mechanism. However, the work described above had already shown that 1-hexenyl anions could also cyclise, albeit at a slower rate.[114] Bailey was first of all concerned with showing that the halogen-metal exchange of primary alkyl iodides with *t*-BuLi at –78 °C did not proceed via radical intermediates, as had been suggested by Ashby, on the basis that cyclisation products could be isolated from this reaction.[115] This matter was discussed in section 3.2.1, and Bailey proved, by NMR, that the cyclisation of the organolithium **184** was slower than the cyclisation of the corresponding radical by 8 to 10 orders of magnitude.[112] Ashby's cyclisation products turned out to be due to the slow reaction of organolithiums with water at this temperature, which allowed them to persist in the reaction mixture and cyclise as the temperature was raised.[116]

t-BuLi, – 78 °C
pentane–ether
Li
184
Li

Alkyl bromides, on the other hand, undergo halogen-metal exchange via single electron transfer processes, and cyclisations of alkyl bromide-derived organolithiums may proceed with significant contribution from radical pathways.[117]

Parallel work on aryllithium and vinyllithium cyclisation onto alkenes highlighted further differences between radical and organolithium cyclisations. Halogen-metal exchange of aryl bromide **222** gave an organolithium **223** which cyclised very slowly indeed in THF at –78 °C (no cyclisation was detectable after 120 min), but with a useful rate at 23 °C (92% yield of the indane **226** after an hour).[118] Cyclisation in ether was slower (but gave less protonated material **225** as a by-product), but rates in Et_2O–TMEDA were comparable to those in THF.

Br
222
2 x n-BuLi
–78 °C, THF
Li
223
23 °C
Li
224
1. Li, Np,
–78 °C, 1 h
2. D_2O
46% **225** + 52% **226**
D_2O
D
225
only product after 2 h at –78 °C
D_2O
D
226
92% after 1 h at 23 °C

Formation of the organolithium with lithium-naphthalene in THF, by contrast, gave 52% cyclisation product **226** after 1 h at –78 °C, conditions under which the cyclisation of organolithium **223** is only very slow. This halogen-metal exchange presumably does proceed to some degree via a single electron transfer, and the cyclisation is a radical one; lack of cyclisation at –78 °C using *n*-BuLi suggests that radicals do not intervene in this case and the cyclisation at 23 °C is a true anionic cyclisation. Similar experiments have been carried out with organosodium cyclisations – radicals intervene in naphthalene-mediated bromine-sodium exchanges but not when an arene promoter is absent.[119] The stereochemical differences between radical and anionic cyclisation are discussed below.

Both aryllithium and vinyllithium cyclisations onto alkenes are successful despite the associated increase in organolithium basicity, and the first vinyllithium cyclisation was demonstrated by Chamberlin, who cyclised the Shapiro-derived **227** onto a terminal alkene to give a single stereoisomer (>50:1 stereoselectivity) of the bicyclic 5-*exo* product **229** and less than 2% of the 6-endo product.[120] The product organolithium **228** could be trapped with other electrophiles too, provided these were added after only a few minutes at 0 °C: longer than this, and the product is protonated by the THF solvent. There is no mechanistic ambiguity here: this is certainly an anionic cyclisation, but it is interesting to note the contrast in regio- and stereoselectivity with a radical cyclisation of **230**, which gives a 1:1 mixture of 6,5- and [6,6-fused products **231** and **232**, of which the 6,5-product **231** is a 3:1 mixture of diastereoisomers.

NNHTris
t-BuLi x 2
THF, 0 °C
Li
227
Li
H
228
$BrCH_2CH_2Br$
Br
229
61%

Br
Bu_3SnH
230
H
231
1:1
232

With the radical controversy solved, Bailey and others set about the detailed investigation of anionic cyclisation as a synthetic method for the formation of five-membered carbocyclic rings, and we shall divide the remainder of this section, which covers work published between 1987 and 2000, into the main classes of compounds produced by the cyclisation: cyclopentanes, furans, pyrrolidines, and polycyclic products. It turns out that organolithiums are unique among the alkali metals in their ability to undergo anionic cyclisation to give cyclopentanes, particularly where primary organometallics are concerned.[121]

7.2.4.2.1 Cyclopentanes

Optimum conditions for anionic cyclisation starting from iodoalkanes start with a halogen-metal exchange at –78 °C using two equivalents of *t*-BuLi in 3:2 pentane:ether.[122] These conditions ensure that the exchange process does not generate radicals which, as we shall note, can lessen the stereoselectivity of the cyclisation. Alkyl bromides give messy reactions, and alkyl chlorides are inert under these conditions. Even with two equivalents of *t*-BuLi, some (a few percent yield) uncyclised, protonated products are produced, apparently by elimination of HI from the *t*-BuI by-product. Cyclisation is very slow at –78 °C (it has a half-life of over 100 days), but on warming to room temperature, at which temperature the half-life is 5 minutes, cyclisation is soon complete. Excessive reaction times are undesirable if the cyclised cyclopentylmethyllithium is to be trapped with electrophiles other than H^+, forming the products **233**, since the intermediate organolithium is slowly protonated by ether. This is an important reason for avoiding THF as a solvent for anionic cyclisations to be followed by functionalisation of the product organolithium. Similar conditions promote cyclisation to the cyclohexane **234**.

1. 2 x *t*-BuLi, – 78 °C 3:2 pentane–ether; 2. 25 °C, 1 h; 3. E^+ → **233**

E = OH ($E^+ = O_2$); 78%
E = H (E^+ = MeOH); 89%
E = CO_2H ($E^+ = CO_2$); 54%

1. 2 x *t*-BuLi, – 78 °C 3:2 pentane–ether; 2. 25 °C, 2 h; 3. E^+ → **234** 68%

Cyclisations of ring-containing starting materials to form bicyclic products **235** and **236** are rather slower, but an effective way to increase the rate of any anionic cyclisation is to add TMEDA: **235** is formed in only 15% yield after 4 h at 25 °C without TMEDA, but 62% with.

1. 2 x *t*-BuLi, 2 x TMEDA – 78 °C; 3:2 pentane–ether; 2. 25 °C, 4 h; 3. MeOH → **235** 62%

1. 2 x *t*-BuLi, 2 x TMEDA – 78 °C; 3:2 pentane–ether; 2. 25 °C, 4 h; 3. MeOH → **236** 94%

Transmetallation of **237** produces an organolithium **238** which requires TMEDA and warming to cyclise, giving products **239**.[123] An alternative way of making these bicyclic products, which does not require TMEDA, is to start with the methylene cyclohexane **240**[124]). Additives such as THF or TMEDA are also necessary for the formation of the **241**,[125] and the relative efficacy of THF, TMEDA or PMDTA is illustrated by cyclisations to **242** involving the formation of a quaternary carbon atom.[125]

237 —*t*-BuLi, ether, pentane, –78 °C→ **238** —1. TMEDA (2 equiv.) 2. warm to 20 °C 3. E^+→ **239** ←1. *t*-BuLi, ether, pentane, –78 – +20 °C 2. E^+— **240**

238 —MeOH, –78 °C→

E = H 80%
E = CH_2OH 68%
E = CHO 65%
E = $SiMe_3$ 61%

(yields from **237**)

1. *t*-BuLi, ether, pentane, –78 °C; 2. additive; 3. warm to 20 °C; 4. MeOH → **241**

additive:

none: 26% (25:1 *endo*:*exo*)
THF: 85% (50:1 *endo*:*exo*)
TMEDA: 91% (50:1 *endo*:*exo*)

1. *t*-BuLi, ether, pentane, –78 °C; 2. additive; 3. warm to 20 °C; 4. MeOH → **242**

additive:

none: 10%
THF: 70%
TMEDA: 77%
PMDTA: 45%

Cuparene **244** contains two adjacent quaternary centres, and by adding TMEDA Bailey was able to make it in 76% yield from **243**.[126] Although the standard conditions for halogen-metal exchange employ two equivalents of *t*-BuLi, in this case it turned out that the second equivalent added directly to the styrene double bond.[127] One equivalent turned out to give acceptable yields.

t-BuLi x 1.2, pentane, Et_2O, –78 – 0 °C

243 → **244** cuparene, 76%

243 → (Bu_3SnH, AIBN) → **245** 90%

In general, the most successful cyclisations are 5-*exo*, starting and ending with a primary organolithium. 6-*Exo* cyclisations (for example, the formation of **234**) are slower, though they will give an acceptable yield in the presence of TMEDA. 6-*Endo* anionic cyclisations almost never occur – the preference for 5-*exo* over 6-*endo* with organolithiums is much more so than with radicals, which, for example, give a 90% yield of the 6-ring product **245** from **243**, avoiding the formation of two adjacent quaternary carbons. The preference for *exo* over *endo* cyclisation arises from the Li–|| interaction present in the transition state of the cyclisation, as discussed below.

Cyclisation of the organolithium **247** formed from the selenide **246** is unfavourable because it would generate a three-membered ring. But in the presence of ethylene, an intermolecular carbolithiation provides an unusual route to a primary organolithium, giving the cyclopentane **248**.[128]

246 → (BuLi, –78 °C, THF) → **247** → (=) → → (1. warm 2. H^+) → **248**

Also possible, but requiring careful optimisation, are reactions which start with a secondary organolithium. Such reactions are very fast even at –78 °C, but the formation of the organolithium turns out to be harder to control than the formation of a primary organolithium under these conditions, producing by-products due to elimination and Wurtz-type coupling. Unlike with primary alkyl iodides, halogen-metal exchange using *t*-BuLi and secondary alkyl iodides can involve the participation of radicals[116] (this is certainly the case in the absence of ether), and this may contribute to the unpredictable nature of these reactions.

It is possible to form, and cyclise, tertiary alkyllithiums, provided they are benzylic or allylic, by starting with a selenide. Krief has used selenium acetals to construct the starting materials **249** and **252**, and on treatment with *n*-BuLi an extremely rapid (less than 20 min even at –110 °C; effectively instantaneous at –78 °C) selenium-lithium exchange ensues to give tertiary organolithiums **250** and **253**. Cyclisation to give **251** or **254** takes half an hour at –78 °C, and has >20:1 stereoselectivity in favour of the *trans* product for **251**[129] and 9:1 for **254**.[130] As usual, cyclisation is slower in ether, but stereoselectivity in this system turns out to be highly solvent dependent as well: changing to Et_2O or pentane completely reverses the selectivity of the cyclisation of **250** to yield **251** as a 98:2 *cis*:*trans* mixture of diastereoisomers. Trapping with Me_2SO_4 gave the ethyl-substituted compound **251** (E = Me): bizarrely, trapping with MeI led to nucleophilic substitution at iodine, generating **251** (E = I) and, presumably, methyllithium.[131]

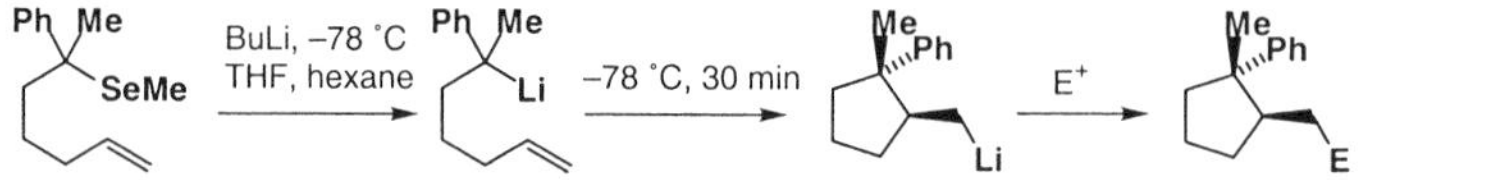

249 **250** **251** >95:5 stereoselectivity

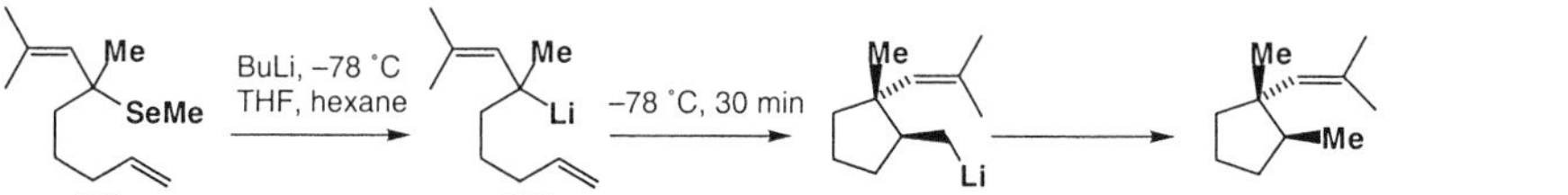

252 **253** **254** 54%, >90:10 stereoselectivity

The analogous sulfide can also be used as the starting material, but higher temperatures (–30 °C) are necessary and the reaction is less stereoselective.[132] Replacement of the exocyclic methyl group with OMe also leads to very rapid cyclisation[133] – the only way of trapping the intermediate **256** before it cyclised was by using an *in situ* quench with Me_3SiCl. This reaction can be made catalytic in *t*-BuLi. A comparable secondary alkyllithium stabilised by being both benzylic and having an adjacent oxygen substituent failed to cyclise, and instead underwent [1,2]-Wittig rearrangement.[134] Tertiary benzylic organolithiums α to sulfur fail to cyclise.[132]

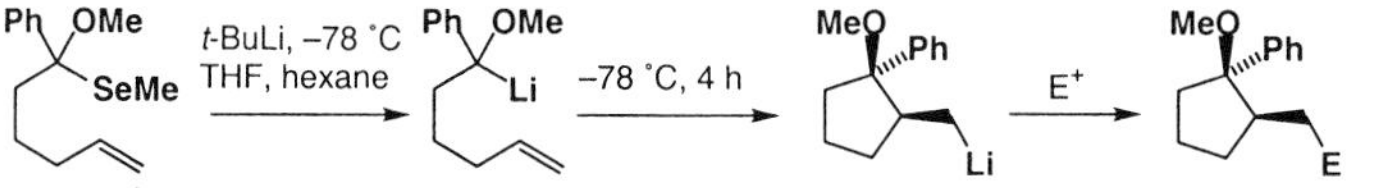

E = H; 81%
E = $SiMe_3$; 84%

255 **256** >95:5 stereoselectivity

Similar tertiary organolithiums have also been formed, and cyclised (but less stereoselectively) by decarboxylation:[135]

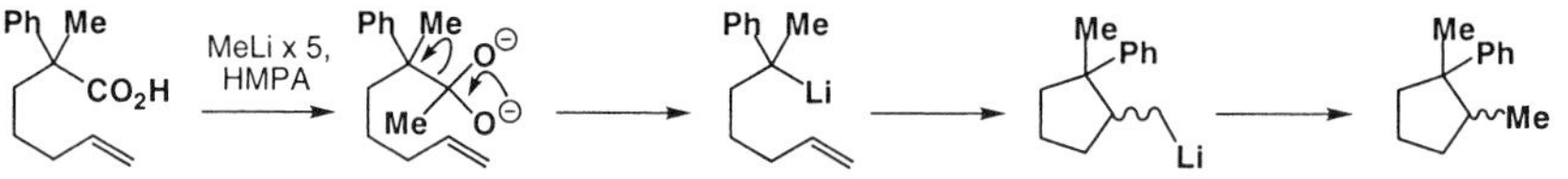

71% 3:1 *cis*:*trans*

When the alkene trap is 1,1-disubstituted, cyclisation of a tertiary benzylic organolithium gives a product with two adjacent quaternary centres. Krief applied this type of cyclisation to the synthesis of cuparene **244**[136] using a different disconnection from the one used by Bailey (above). Though it is irrelevant to the synthesis of cuparene, the cyclisation of **257** to **258** is also stereoselective and produces a single stereoisomer of **259** on carbonation of the cyclised organolithium. The tertiary organolithium is too basic to cyclise in THF and in this solvent gives products including **260**, which arise by intermolecular attack of the organolithium on ethylene produced by decomposition of THF.

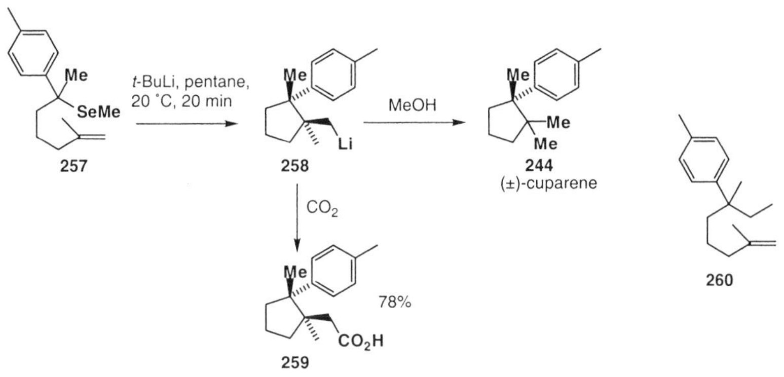

Vinyllithium cyclisations have already been mentioned, and generate a useful array of functionalised products – for example **262** from **261**.[137] They are also highly stereoselective, as discussed below.

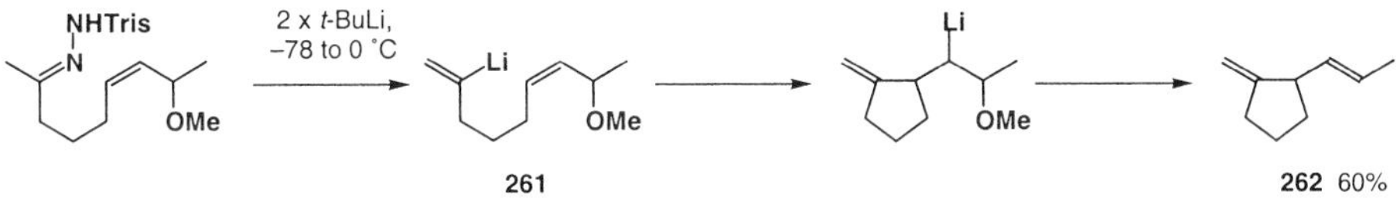

A general rule of anionic cyclisation onto an alkene is that although (with some caveats) the starting organolithium may be primary, secondary or tertiary, the product organolithium must always be primary, or the cyclisation fails.[122,137] (In other words, the alkene trap may only be monosubstituted or 1,1-disubstituted). The need for the alkene trap to be unsubstituted at one end is a key point of divergence from radical cyclisations: the formation of tertiary radicals by cyclisation onto trisubstituted alkenes is common. Exceptions to this rule include compounds in which the product organolithium has the opportunity to undergo an elimination reaction: for example, the cyclisation of **261**, where the trap is an allylic ether.[137] This useful trick has been used a number of times for the synthesis of alkene containing products, particularly vinyl tetrahydrofurans and pyrrolidines, as described below. Cyclisation onto a cyclic allylic ether provided one of the very first demonstrations of an organolithium cyclisation directed towards the total synthesis of a natural product. Lautens showed that iodides **263** and **266** could be cyclised to the polycyclic products **265** and **268** respectively (via organolithiums **264** and **267**) in excellent yield.[138] The fact that the elimination of the ether in this case is a 5-*endo-trig* ring opening seems to be inconsequential.

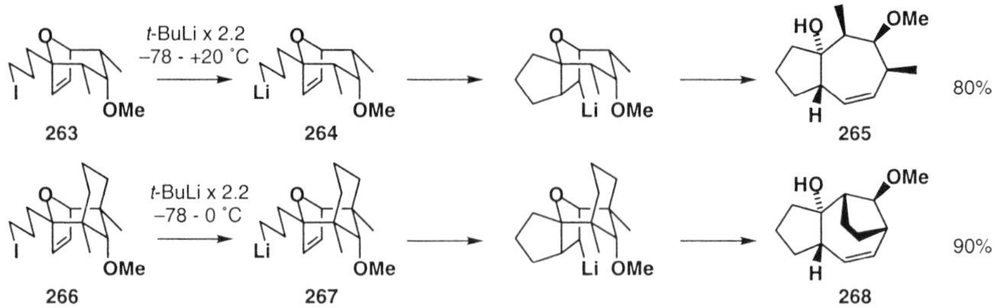

Cyclopropanes can function similarly, with the closure of the larger ring being driven by ring opening of the smaller.[139] Unlike the allylic ethers, the forward reaction in these cases is rather sluggish, and the cyclisation of **269** proceeds only at room temperature in the presence of TMEDA: as usual in such cases, competing protonation by solvent (Et_2O) precludes high yields in electrophilic quenching reactions other than simple protonation.

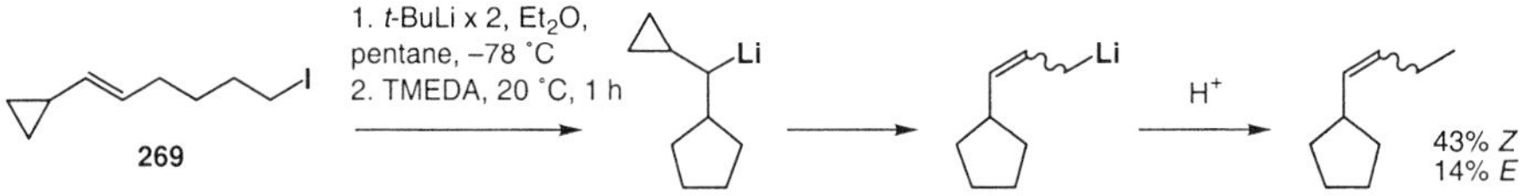

Anionic cyclisation can be used to form cyclopropanes when the product cyclopropylmethyllithium can undergo an elimination reaction.[140,111] The simple primary organolithium **270** requires the presence of TMEDA to cyclise, but gives 88% of vinylcyclopropane. **271** cyclises with moderate stereoselectivity.

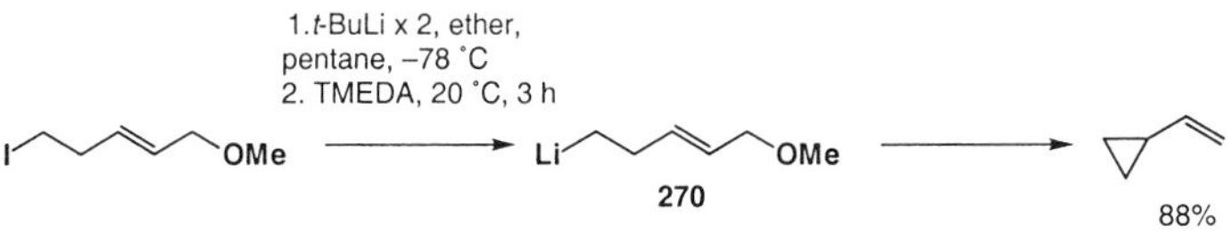

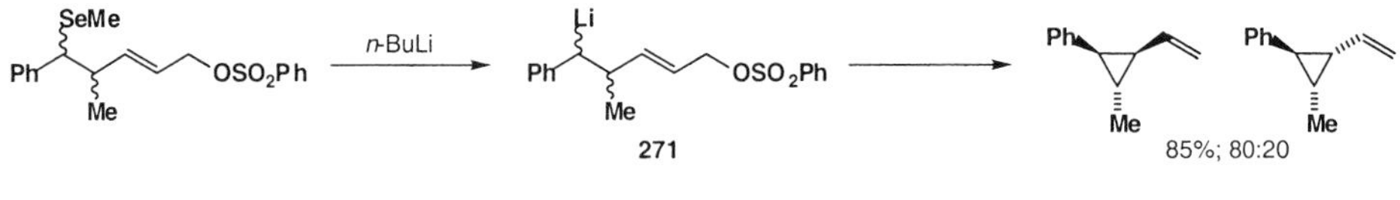

Further exceptions to the general failure of disubstituted alkenes to act as organolithium traps are those compounds where the product organolithium is stabilised by an adjacent conjugating or electron-withdrawing group. Activated alkenes such as unsaturated esters or acylphosphoranes were discussed above, but phenyl,[141,139] silyl,[139] arylthio,[142,143] alkynyl,[144] and alkenyl[145] groups are sufficiently stabilising to allow cyclisations of organolithiums to occur onto styrenes, vinylsilanes, vinylsulfides and enynes **272**. Phenyl and silyl *alkynes* are also good organolithium traps (see below).

t-BuLi x 2, Et_2O, pentane, −78 - 0 or +20 °C; E^+

272

95% (R = Ph, E = H)
72% (R = Ph, E = Me_2COH)
84% (R = Ph, E = $SiMe_3$)
84% (R = $SiMe_3$, E = H)
70% (R = $SiMe_3$, E = Me_2COH)
78% (R = $SiMe_3$, E = $SiMe_3$)

While the additions to these disubstituted alkenes **272** are in fact faster than the additions to their monosubstituted analogues (and additions to *E* alkenes are faster than additions to *Z* alkenes), the product organolithiums are nonetheless rather unstable and, especially with R = $SiMe_3$, tend to deprotonate solvent. In these cases the electrophilic quenches are therefore best carried out at 0 °C.

Since the sulfide substituents can be removed reductively, the cyclisations of **273** and **276** are synthetically equivalent to cyclisations onto disubstituted double bonds, giving compounds such as **275**.[142,143] Like the corresponding cyclisations onto monosubstituted alkenes they are highly stereoselective, with the sense of stereoselectivity being solvent-dependent: the products **274** and **277** are *trans* as shown for a cyclisation conducted in THF: in pentane complete *cis* selectivity is obtained. Similar cyclisations onto vinyl sulfides have been used to explore the stereochemical course of the anionic cyclisation reaction, and are discussed below.[146]

273 → BuLi, −78 °C, THF, hexane → −78 °C, 30 min → **274** → 1. MeI 2. RaNi, 93% → **275**

276 → BuLi, −78 °C, THF, hexane → −78 °C, 30 min → **277** (68%) → RaNi, 87% → **275**

Benzylic tertiary organolithiums generated by selenium-lithium exchange can in fact be forced to undergo additions in low yield even to 1,2-disubstituted alkenes without stabilising substituents, and in THF or ether **278** gives mixtures of 5-*exo* and 6-*endo* products **279** and **280**.[147] In pentane an unusual cyclisation onto an aromatic ring occurs (see below). Attempts to cyclise onto a trisubstituted double bond led merely to ethyl-containing products **281** resulting from addition to ethylene generated by decomposition of THF.

278 **279** **280** **281**

Cyclisation onto an enyne generates a propargylic organolithium that is sufficiently stabilised that it can be used to form three- and four-membered rings.[144] A problem is that the propargyllithium reacts non-regioselectively giving a mixture of allene and alkyne products.

This can be overcome by transmetallating to the propargylzinc, which reacts regioselectively to give only alkyne products. In this way, cyclopropane **282**, cyclobutane **283** and cyclopentane **284** can all be formed in good yield and with high diastereoselectivity.

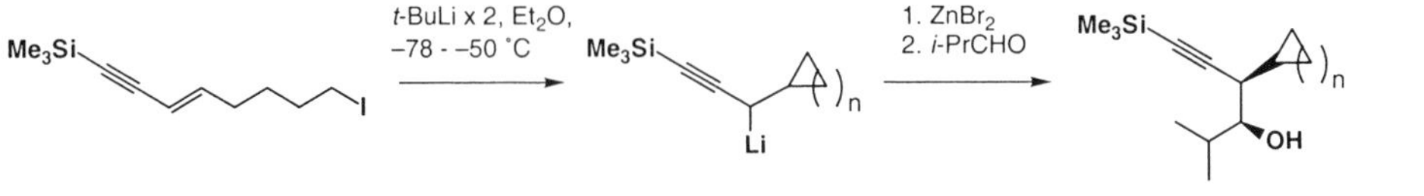

Me_3Si ... I → (t-BuLi x 2, Et_2O, −78 - −50 °C) → Me_3Si ... Li → (1. $ZnBr_2$ 2. i-PrCHO) → Me_3Si ... OH

282 n = 1 28% (98:2 dias)
283 n = 2 55% (98:2 dias)
284 n = 3 72% (99:1 dias)

The stabilising effect of the alkynyl, phenyl and silyl substituents in these reactions means they are perhaps coming close to Cooke's cyclisations onto activated double bonds, which do indeed give three- and four-membered ring products in good yield. Cooke himself showed that there is little difference between the two when he cyclised both **285** and **286** in high yield.[145]

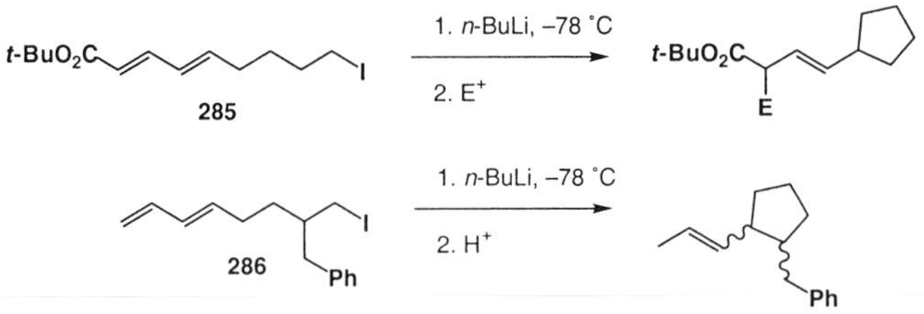

t-BuO_2C ... I (**285**) → (1. n-BuLi, −78 °C 2. E^+) → t-BuO_2C ... E

286 ... I, Ph → (1. n-BuLi, −78 °C 2. H^+) → ... Ph

The cyclisation of α-carbamoyloxy substituted organolithiums formed by deprotonation, leading to carbamate-substituted cyclopentanes,[141] is discussed below.

7.2.4.2.2 Cascade reactions

If the product organolithium is provided with a second, appropriately sited alkene (or some other electrophile), it can be encouraged to undergo an anionic "cascade" cyclisation. To date, this strategy has not been widely applied, and the first examples, starting from iodides **287** and **288**, were published by Bailey in 1989.[148] The inability to cyclise organolithiums onto anything but terminal alkenes limits the structural types that can be made by this method, but unlike with radical reactions it is a straightforward matter to trap the final organolithium intermolecularly – with acetone in the case shown here.

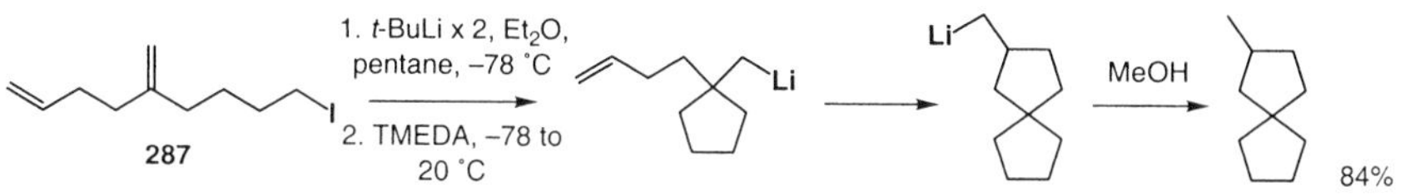

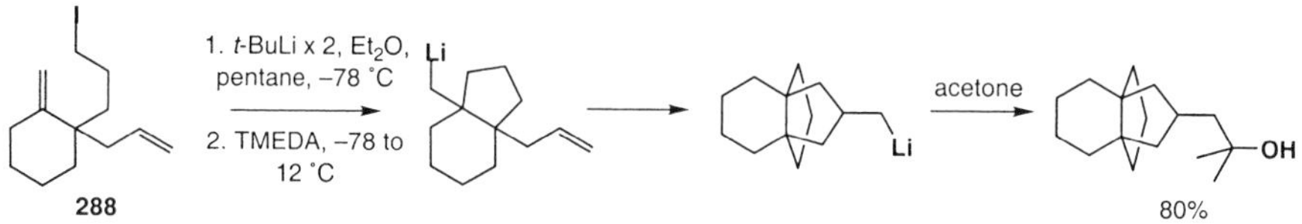

Anionic cascade cyclisations can be stereoselective (just like their simpler counterparts: see below). The cyclisation of **289**, for example, gives solely the *trans* ring-junction product **292** (stereochemistry is determined in the first cyclisation to **290**) despite the fact that the *cis* 5,5-fused system is nearly 30 kJ mol^{-1} more stable.[149] This argues against the reversibility of the cyclisation. Because the cyclisation of **290** is rather slow, trapping the product organolithium **291** with electrophiles other than a proton can pose problems, and considerable amounts of material protonated by the ether solvent are obtained. This can be overcome by decreasing the proportion of ether from 40% to 10% of the solvent volume.[150]

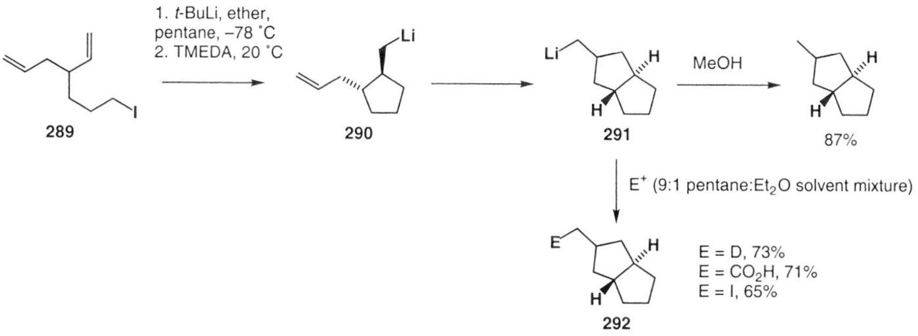

Krief has applied selenium chemistry to some anionic cascade cyclisations.[128] For example, **293** can be cyclised in two successive 5-*exo* reactions to give a mixture of the stereoisomers of **297**. If **294** is warmed to 0 °C, an alternative sort of tandem process occurs: after cyclisation onto the alkene, the organolithium **295** undergoes an intramolecular displacement, stereoselectively generating the 5,3-fused system of **296**. Similar intramolecular cyclopropanations (of α-bromo organolithiums) have been described by Hoffmann[151] but are probably mediated by carbenes rather than a sequential cyclisation–substitution sequence.

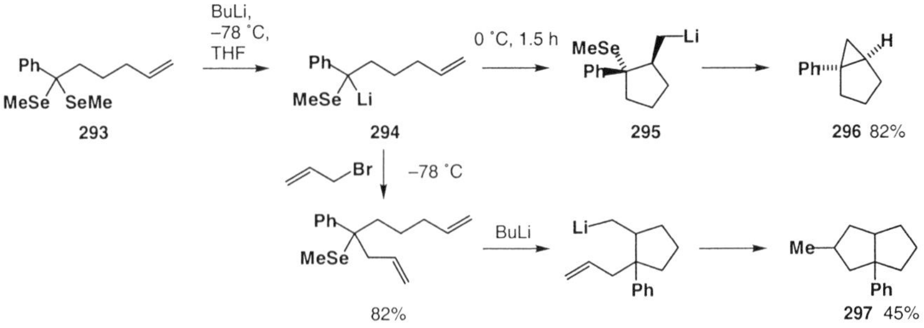

Cyclisation of **298** in pentane leads to a cascade of two very curious anionic cyclisations in remarkably good yield: a usually highly unfavourable addition to a 1,2-disubstituted alkene, followed by attack on an aromatic ring with subsequent rearomatisation to give **299**.[147]

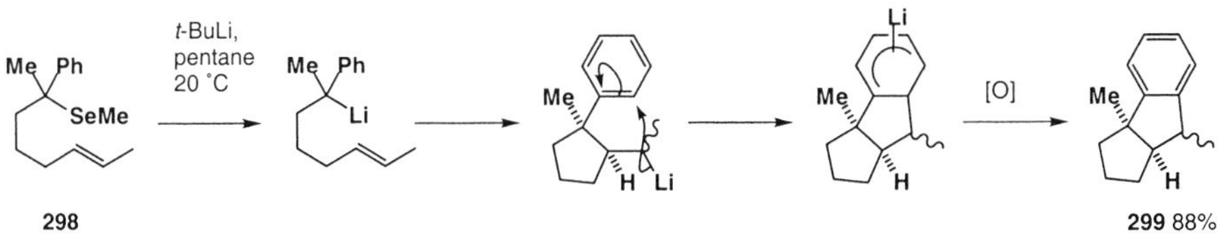

7.2.4.2.3 Tetrahydrofurans

The formation of tetrahydrofurans by cyclisation of alkoxy-substituted organolithiums was pioneered by Broka.[134] He used tin-lithium exchange of **300**, **302** and **304** to generate organolithiums which cyclised onto monosubstituted alkenes or allylic ethers to give products **301**, **303** and **305**. The reaction is highly (>10:1) stereoselective, favouring the *cis* isomers. A lack of success with simple disubstituted alkene traps demonstrated that radicals are not involved in the reaction. This general feature of anionic cyclisations – that simple alkene traps must be monosubstituted – was overcome in **302** and **304** by introducing an allylic leaving group in the usual way.

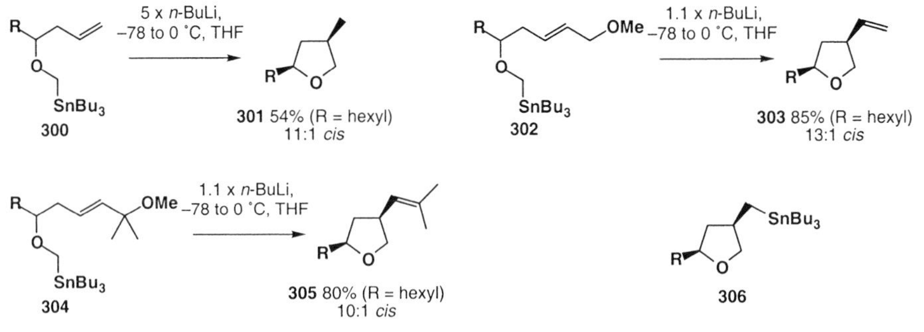

A problem can arise in anionic cyclisations when the organolithium is formed by tin-lithium exchanges such as these. The product organolithium may become stannylated by nucleophilic substitution at tin of the Bu_4Sn produced by the reaction. In the cyclisation of **300**, for example, formation of the stannylated product **306** could only be completely avoided if 5 equivalents of *n*-BuLi were used in the reaction. This can be turned to advantage, because it means the cyclisation can be made catalytic in alkyllithium, as described below in a synthesis of stannylated pyrrolidines. Cyclisations starting with selenium-lithium exchange may behave similarly, giving selenenylated by-products unless a large excess of BuLi is employed.

In a study of synthetic routes to the phorbol class of compounds, Lautens cyclised the stannane **307** (with 5 equiv MeLi, to avoid stannane reincorporation) to the 7,5-fused ring system **308**.[138] This reaction also works in the synthesis of analogous saturated five-membered nitrogen and sulfur heterocycles (see below).

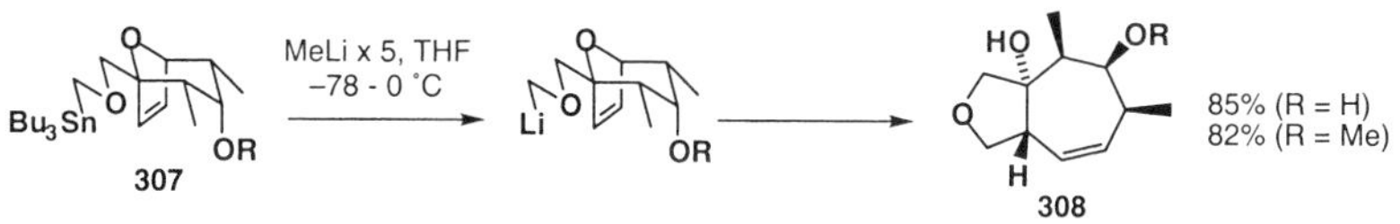
Bu3Sn
OR
307
MeLi x 5, THF
−78 - 0 °C
Li
OR
HO
OR
H
308
85% (R = H)
82% (R = Me)

α-Alkoxy organolithiums can also be made from sulfides using Cohen's lithium naphthalene reduction method. This allows the formation of tetrahydrofurans **310** and **312** from monothioacetals **309** and **311** – as with the stannanes **302** and **304**, yields are best when the trap is an allylic ether.[152]

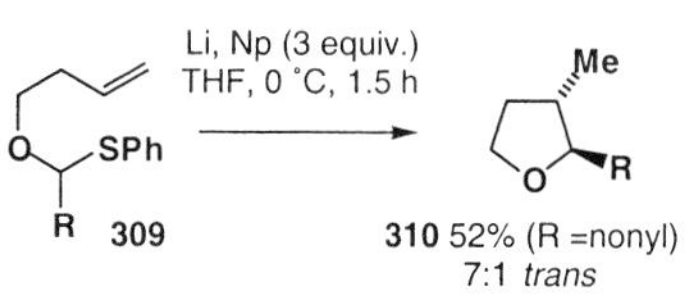
SPh
R 309
Li, Np (3 equiv.)
THF, 0 °C, 1.5 h
Me
R
310 52% (R =nonyl)
7:1 trans

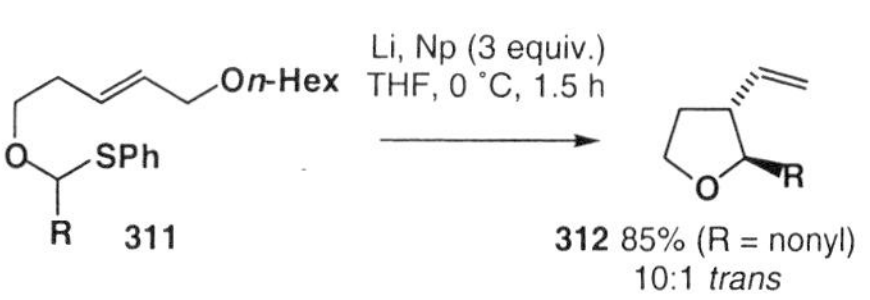
On-Hex
SPh
R 311
Li, Np (3 equiv.)
THF, 0 °C, 1.5 h
R
312 85% (R = nonyl)
10:1 trans

Since the reduction is believed to proceed via a radical mechanism, it seems plausible that these cyclisations are in fact radical-mediated. However, strong evidence against this comes from the stereoselectivity of the reaction: the *trans* selectivity of these cyclisations contrasts powerfully with the *cis* selectivity of the corresponding radical reactions.

Comparable secondary α-alkoxy organolithiums were made from stannanes **313** and **314** and cyclised to tetrahydrofurans during a study of the stereospecificity of the cyclisation reaction with regard to the lithium-bearing centre.[153] The stereochemical aspects of these reactions are discussed below.

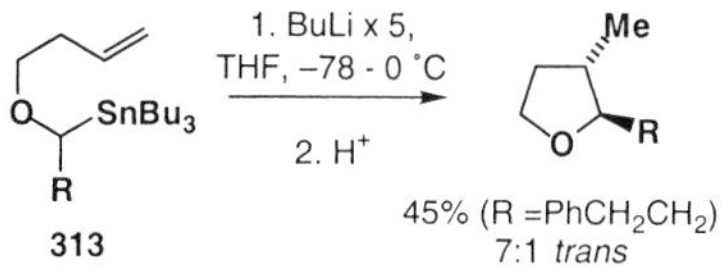
SnBu3
R
313
1. BuLi x 5,
THF, −78 - 0 °C
2. H+
Me
R
45% (R =PhCH2CH2)
7:1 trans

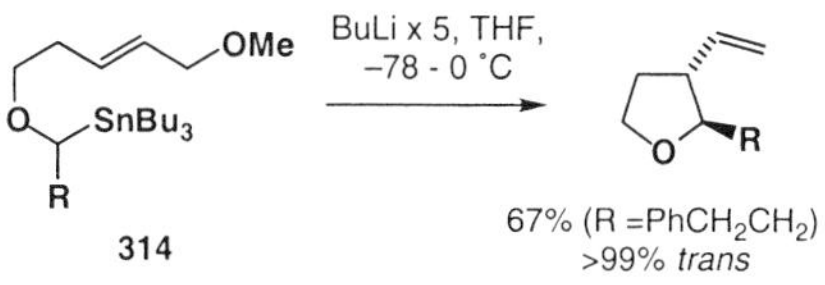
OMe
SnBu3
R
314
BuLi x 5, THF,
−78 - 0 °C
R
67% (R =PhCH2CH2)
>99% trans

Cyclisation onto the enol ether **315** transiently generates a tetrahydrofuran **316**, but this undergoes rapid elimination to **317** – a transformation which is overall equivalent to a [1,4]-Wittig rearrangement.[154]

t-BuLi, pentane, Et_2O, –78 °C

315 316 317 95%

7.2.4.2.4 Pyrrolidines

The same Sn–Li and S–Li exchange methods work equally well in the synthesis of pyrrolidines from α-amino organolithiums. Broka's first example,[152] the cyclisation of **318** (published in 1989) has been followed up by a considerable amount of work by Coldham, who used tin-lithium exchange of **319** to generate an α-amino organolithium which cyclises to the pyrrolidinylmethyllithium **320**.[155] This product attacks the tetramethyltin by-product of the transmetallation to give the stannane **321** as the final product – best yields of the stannane (which can subsequently be further functionalised) are obtained when a further two equivalents of tetramethyltin are added at the start of the reaction.[156] The stannane-forming step regenerates MeLi, and it is therefore possible to do this cyclisation using only 0.4 equiv. MeLi – the first example of an anionic cyclisation catalytic in alkyllithium. The same method can be used to form an azetidine **322**.

Li, Np (3 equiv.) THF, 0 °C, 1.5 h

318 >56% (R = hexyl) 6:1 *cis*

MeLi x 2 or MeLi x 0.4, 2 x Me_4Sn, THF; Me_4Sn; + MeLi

319 320 321 70-72%

MeLi, THF

322 35%

Reincorporation of the tin can be avoided by using a tri*butyl*stannane precursor.[156] This means that the product organolithium can be trapped with other electrophiles, provided that the solvent is 10:1 hexane-ether, and not THF, which otherwise protonates the product organolithium.[157] The hexane-ether mixture slows down the tin-lithium exchange reaction dramatically, and it proceeds only at room temperature. Cyclisation of **323** thus leads to the synthesis of a simple GABA uptake inhibitor **324**. *N*-acyl pyrrolidines cannot be made directly from α-amido organolithiums, as these fail to cyclise, possibly because they are too well stabilised by O–Li coordination.[138]

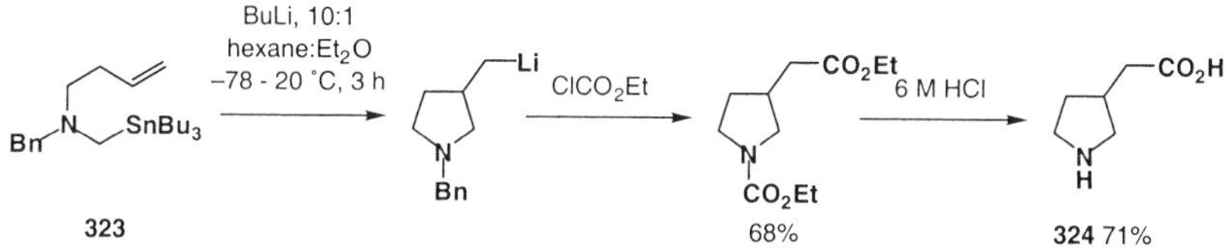

Cyclisations onto allylic ethers such as **325** give vinyl-substituted pyrrolidines in good yield.[61] Enantioselective versions of similar syntheses of pyrrolidines are discussed below.

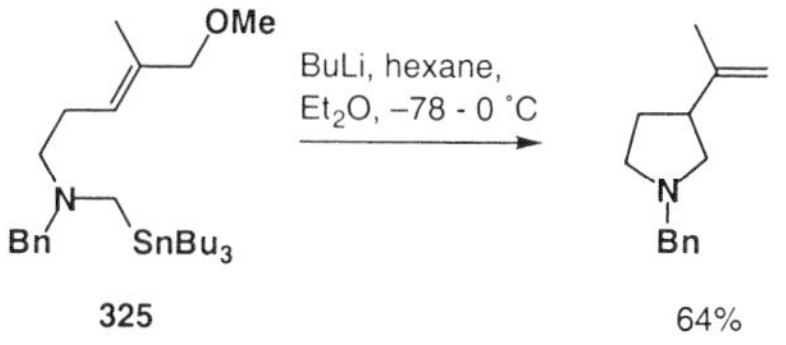

Heteroatom-containing rings can in principle be made by siting the heteroatom other than α to the organolithium, but the problem arises that β and γ heteroatoms will be prone to elimination from the organolithium starting material or from the cyclised product respectively. This is not a problem with an aryllithium, and both Liebeskind[158] and Bailey[159] simultaneously published a pyrrolidine-forming reaction from β-amino substituted arylbromides **326**, from which the products are dihydroindoles **327**.

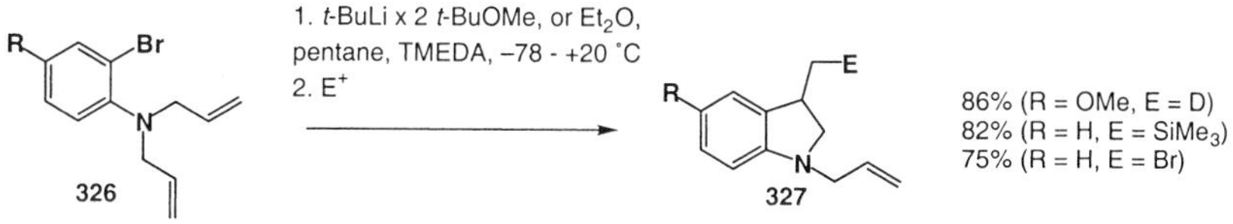

With an appropriately sited fluoro substituent, elimination to a benzyne **328**, followed by addition of butyllithium, generates an alkyl-substituted organolithium **329** which cyclises to give differently substituted dihydroindoles **330**.[160]

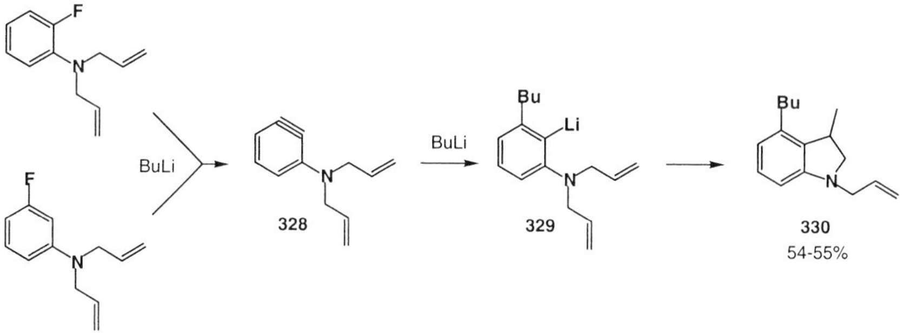

Neither is elimination a problem for the vinyllithium **331**, which survives long enough to give the pyrrolidine **332**.[161]

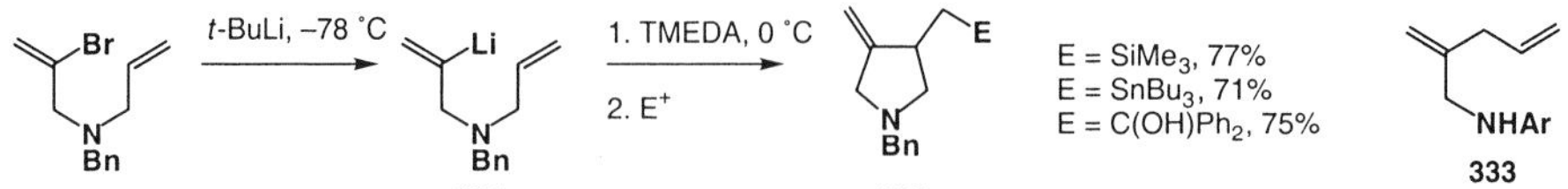

Br, N, Bn — *t*-BuLi, –78 °C → Li, N, Bn **331** — 1. TMEDA, 0 °C; 2. E^+ → E, N, Bn **332**

E = $SiMe_3$, 77%
E = $SnBu_3$, 71%
E = $C(OH)Ph_2$, 75%

NHAr **333**

An *N-alkyl* protecting group is crucial to the regioselectivity of this reaction: with *N-aryl* groups products **333** are obtained, arising from 6-*endo* cyclisation followed by elimination.

7.2.4.2.5 Tetrahydrothiophenes

Only one example is known of the cyclisation of an α-thioorganolithium to give a tetrahydrothiophene,[138] a cyclisation of **334** analogous to that of **307**.

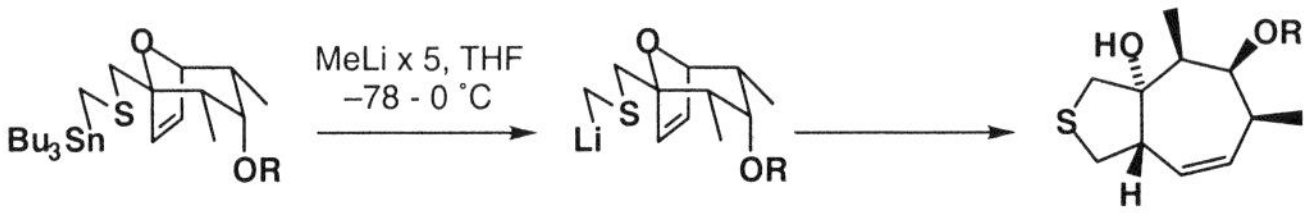

Bu_3Sn, S, O, OR **334** — MeLi x 5, THF, –78 - 0 °C → Li, S, O, OR → HO, OR, S, H 88%

7.2.4.2.6 Stereoselectivity and mechanism

starting material	*cis* product	*trans* product	ratio (organolithium cyclisation)[a]	ratio (radical cyclisation)[b]
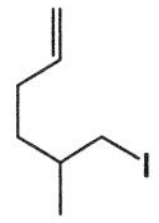	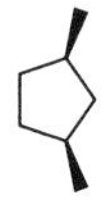	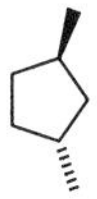	1:10	1:1.8
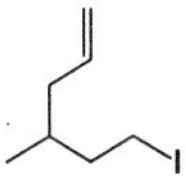	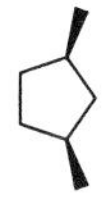	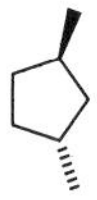	10:1	2.5:1
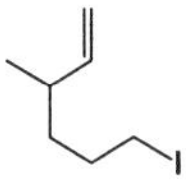	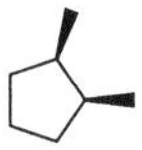		1:12	1:5
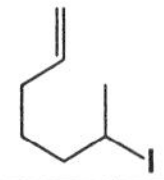	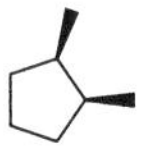		1.7:1	5:1

[a]Organolithium cyclisation conditions: 1. *t*-BuLi x 2, Et_2O, pentane, –78 °C then 20 °C, 1 h; 2. MeOH
[b]Radical cyclisation conditions: Bu_3SnH, PhH, Δ.

Table 7.2.1 Stereoselectivity in organolithium and radical cyclisations

Anionic cyclisations often exhibit high stereoselectivity with regard to stereogenic centres within the newly formed ring, and frequently give stereochemical results which contrast with those obtained from the analogous radical cyclisations, which are usually much less stereoselective. The stereochemistry of the cyclisation to give variously substituted cyclopentanes, along with the stereoselectivity of the cyclisation of the corresponding radicals,[162,163] is illustrated in Table 7.2.1.[125] Cyclisations giving disubstituted tetrahydrofurans and pyrrolidines[164] display similar selectivities. The secondary organolithium is the only example to cyclise less stereoselectively than its radical analogue.[116]

The presence of Lewis-basic additives may alter stereoselectivity (PMDTA, for example, increases the stereoselectivities of the first three organolithium cyclisations in the Table to 1:23, 17:1 and 1:34 respectively), and, as discussed before, also increases the rate of the reaction. The cyclisation to pyrrolidine **335** could be made >25:1 stereoselective in THF because the transmetallation can be carried out at lower temperature: with hexane–ether, the solvent mixture necessary if the product is to be trapped with electrophiles other than H, transmetallation occurs only at 20 °C, and cyclisation at this temperature demonstrates *cis* selectivity similar to that of the analogous cyclopentane.[164]

1. *t*-BuLi x 2, Et_2O, hexane, –78 - +20 °C; 2. MeOH → **335**

R = Me: 78%; 7:1 *cis*:*trans*
R = *i*-Pr: 74%; 6:1 *cis*:*trans*

The *cis* selectivity of the cyclisation of vinyl substituted **336** means it can undergo a second cyclisation, and allows the one-step synthesis of the bicyclic structures **337**.[150]

336 → 1. *t*-BuLi x 2, Et_2O, pentane, –78 °C; 2. TMEDA, warm → → E^+ → **337**

E = H, 89%
E = CO_2H, 65%
E = Br, 75%
E = $SiMe_3$, 81%

In each case, stereoselectivity can be explained by assuming that the reaction proceeds through a chair-like transition state, in which there are interactions between the two ends of the C–Li and C=C bonds, and in which the substituents all occupy pseudo-equatorial positions (structures **338**, **339** and **340**). The same transition state model suffices to explain the stereoselectivities of the furan and pyrrolidine forming reactions above. Stereoselectivity in the cyclisations of the selenide-derived tertiary organolithiums would arise from a conformation with a precedented pseudo-axial phenyl ring.

338 **339** **340**

Chamberlin proposed similar chair-like transition states **342** and **345** to account for the stereoselectivity of the cyclisations of vinyllithiums such as **341** and **344**,[137] and again a striking contrast with similar radical cyclisations is evident. The radicals **343** and **346** cyclise less stereoselectively and less regioselectively than the organolithiums. The difficulty of forming a four-membered C–C–C–Li ring accounts for the unfavourability of *endo* anionic cyclisations.

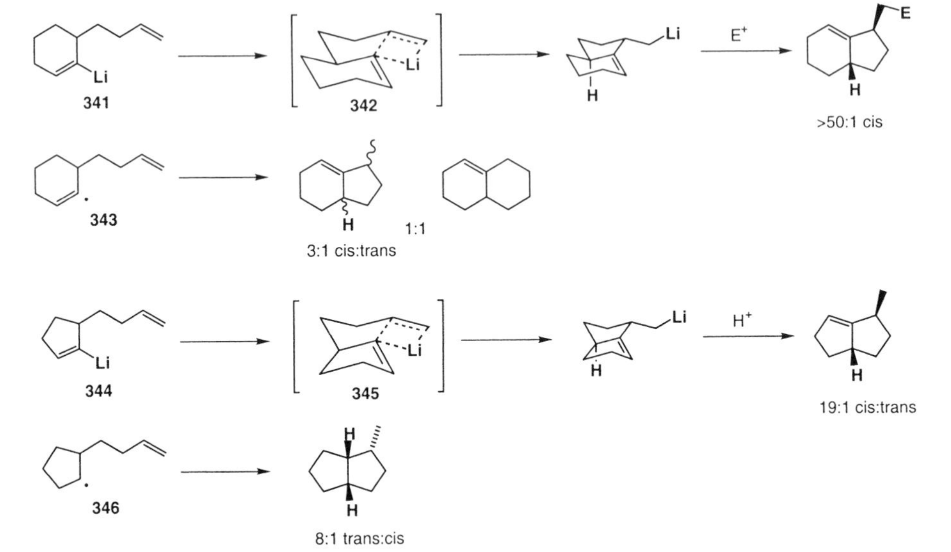

Li
341
342
Li
Li
H
E⁺
E
H
>50:1 cis
343
H
3:1 cis:trans
1:1
Li
344
345
Li
Li
H
H⁺
H
19:1 cis:trans
346
H
H
8:1 trans:cis

A moderately stereoselective vinyllithium cyclisation has been used to make laurene **349** from a vinyl bromide **347**.[165] The transition state **348** leading to the major isomer has a pseudo-axial tolyl group and a pseudo-equatorial methyl – in accordance with precedent for similar cyclohexyl systems.

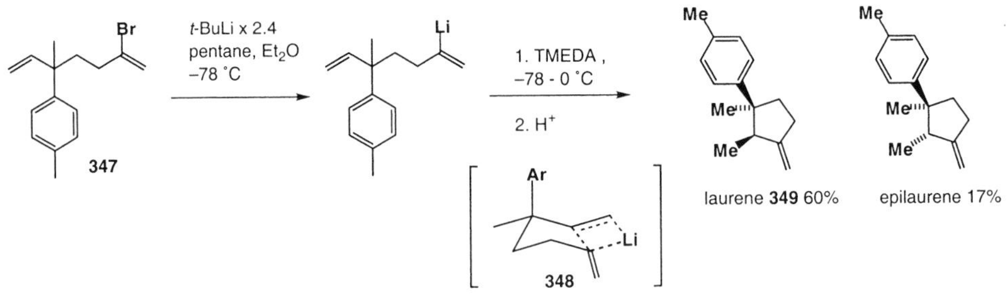

Br
347
t-BuLi x 2.4
pentane, Et_2O
–78 °C
Li
1. TMEDA ,
–78 - 0 °C
2. H^+
Ar
Li
348
Me
Me
Me
laurene **349** 60%
Me
Me
Me
epilaurene 17%

The key to the high stereoselectivity of these reactions, relative to their radical counterparts, is their late, product-like transition state.[152] As early as 1974, it was suggested by Oliver that cyclisations of organolithiums onto unactivated alkenes was promoted by an interaction between the lithium atom and the C=C double bond. There is now plenty of evidence that this is the case: in 1991, *ab initio* calculations by Bailey and coworkers[125] showed that not only is there a sound theoretical basis for treating structures such as **338-340** as reasonable transition states for the reaction, but that a structure **350** (in which there is a clear Li–π(C=C) interaction, but a rather distant C–C interaction of 3.35 Å for the new C–C bond) is an intermediate along the reaction pathway for cyclisation of the simple hexenyllithium **184**. The Li–(C=C) interaction was calculated to be worth over 40 kJ mol^{-1} of stabilisation energy relative to linear 5-hexenyl-1-lithium.

Li

350

These theoretical studies were given real credibility when Hoffmann showed, in 1995, that the Li–(C=C) interaction can be observed by heteronuclear coupling in the NMR spectrum of 5-hexenyl-1-lithium **184**.[166]

X M

351 **352** *cis*-**353** *trans*-**353**

351, X =	conditions	**352**, M =	ratio *cis*:*trans* **353**
I	*t*-BuLi x 2, pentane Et_2O	Li	1.7:1
I	*t*-BuLi x 2, pentane	Li	4:1
Br	*t*-BuLi x 2, pentane Et_2O	Li	5:1
Br	*t*-BuLi x 2, pentane	Li	6:1
Cl	Na	Na	<1:1
Cl	Na, naphthalene	Na	<1:1
Cl	Na, *t*-$BuNH_2$	Na	no reaction
Cl	Na, naphthalene, *t*-$BuNH_2$	Na	4:1

Table 7.2.2 Conditions and stereoselectivity in anionic cyclisations

The stereochemical contrast between radical and anionic cyclisations provides a useful test of mechanism which has been used on a number of occasions to elucidate the course of reactions whose mechanism is ambiguous. For example the results in table 7.2.2 show that the anionic cyclisation of **352** under certain conditions proceeds without a radical component. However, in pentane alone the *cis:trans* ratio of **353** produced from **351** (X = I) increases to 4:1, which has been taken as evidence that cyclisation proceeds partly by a radical mechanism under these conditions. The ratio further increases to 5 or 6:1 with a secondary alkyl bromide as starting material under either set of conditions, suggesting that alkyl bromides do not undergo the clean, non-radical mediated halogen-metal exchange that iodides do in pentane–ether.

The same type of reasoning was used to investigate some organosodium reactions: the anionic cyclisation of the organosodium **352** gives <1:1 *cis:trans*-**353** and cannot therefore go via a radical mechanism.[119] *t*-BuNH_2 rapidly protonates organosodiums, and carrying out the reaction starting with the alkyl chloride gave no cyclised product. However, making the organosodium using sodium naphthalenide in the presence of *t*-BuNH_2 did give some cyclised product – and the source of this product was evident from its stereochemical composition: with 4:1 *cis:trans*-**353** it must be formed by radical cyclisation. It can be deduced that sodium naphthalenide promotes Cl–Na exchange via a radical process, while with sodium metal no radical intermediates are involved.

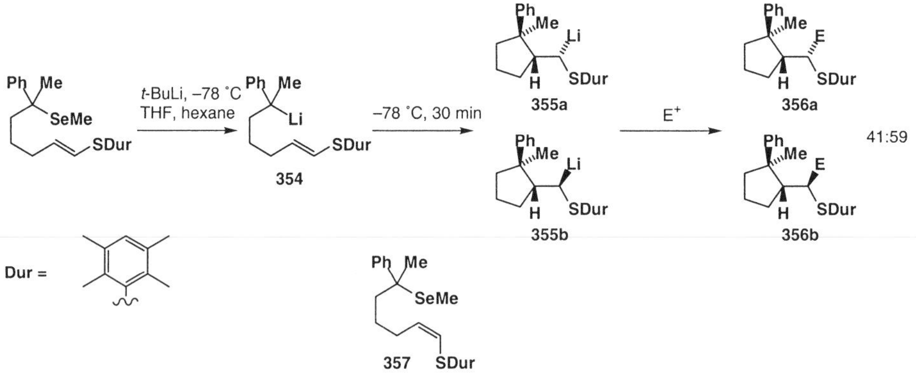

The proposed concerted *syn* addition of carbon and lithium to the double bond has consequences for a new lithium-bearing chiral centre too, and it should be possible to test this with an appropriate substrate. The problem is, of course, that anionic cyclisations usually fail if the product would be a secondary organolithium, making it hard to devise a cyclisation generating a lithium-bearing chiral centre. This problem was solved by Hoffmann in 1997,[146,167] who built on Krief's observation that cyclisation onto vinylsulfides is successful, even when the product is a secondary or even a tertiary alkyllithium. There is still a problem with vinyl *phenyl*sulfides, because α-phenylthioorganolithiums are configurationally unstable. However, cyclisation onto vinyl *duryl*sulfide **354** gives a configurationally stable (at –105 ˚C) organolithium **355** which can be trapped with electrophiles. The 1:1 mixture of products **356** indicates that the cyclisation in the presence of THF is *non*-stereospecific: it was not possible to test the stereospecificity in the absence of THF because THF (or another coordinating solvent) is essential for the reaction to proceed at temperatures low enough to ensure configurational stability in the organolithium product. When the *Z*-vinyl sulfide **357** was used as the starting material, the reaction did however appear to be stereospecific, though this result does not prove that the mechanism is concerted. These results may be usefully compared with the demonstrably stereospecific addition of organolithiums to *alkynes* discussed below.

Control of absolute stereoselectivity in anionic cyclisations is still a developing area, and several approaches can be envisaged – among them are: inclusion of an exocyclic stereogenic centre of defined configuration in the starting material to direct the cyclisation (a "chiral auxiliary" approach), the use of a configurationally defined organolithium starting material generated by stereospecific tin-lithium exchange, or the use of a ligand such as (–)-sparteine to direct the course of the cyclisation.

Exocyclic stereogenic centres have so far had little success. Krief succeeded in cyclising **358** with some degree of stereoselectivity,[168] but the "auxiliary" centre could only be removed by a long and inelegant series of steps.

MeSe Ph Me OH
n-BuLi
Me OH Me Ph
358
90%; 90:10 diastereoselectivity

The nitrogen atom of a forming pyrrolidine provides a point of attachment for an auxiliary, but the α-methylbenzyl group of **359** induced only 3:1 selectivity in the new ring of **360**.

N $SnBu_3$ Me Ph
1. *n*-BuLi, THF
2. MeOH
Me N Me Ph
Me N Me Ph
359
360a **360b**
78%, 74:26

Much more successful in a pyrrolidine synthesis was the use of a stereochemically defined organolithium **362** formed by tin-lithium exchange from an almost enantiomerically pure stannane **361**, itself a product of Beak's sparteine lithiation chemistry. Despite the high temperature (20 °C) required for tin-lithium exchange in hexane–ether, it is nonetheless possible to carry out a stereospecific cyclisation via an organolithium which is configurationally stable, even at 20 °C, on the timescale of the cyclisation.[169] The cyclisation proceeds with retention and gives the alkaloid pseudoheliotridane **363** in 87% yield and with no loss of enantiomeric excess.

N $SnBu_3$
BuLi x 2, 10:1
hexane:Et_2O
–78 - 20 °C, 1 h
N Li
then MeOH
N Me H
361 94% ee
362
363 pseudoheliotridane, 87%
94% ee

Cyclisation of a stereodefined organolithium derived from enantiomerically enriched α-alkoxy stannane **364** gave a tetrahydrofuran **365** with no loss of enantiomeric excess and with overall retention.[153]

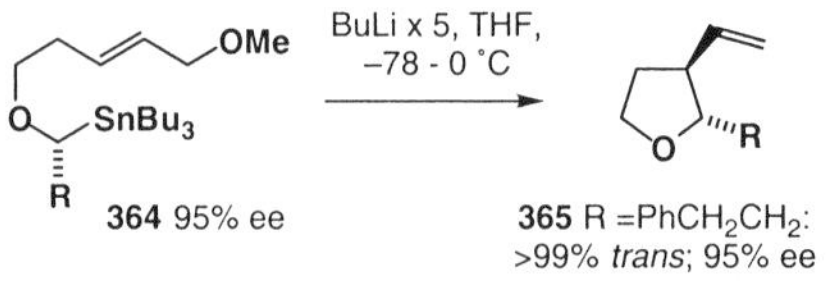

BuLi x 5, THF, −78 - 0 °C

364 95% ee

365 R =$PhCH_2CH_2$: >99% *trans*; 95% ee

Asymmetric deprotonation in the presence of (–)-sparteine could in principle give the configurationally defined organolithiums required for comparable stereoselective cyclisations. In the pyrrolidine series, sparteine induces only low levels (<30%) of enantiomeric excess in the cyclisation of α-amino organolithiums,[164] but it is rather more effective with α-carbamoyloxy substituents. Deprotonation of **366** using *s*-BuLi-(–)-sparteine gave an organolithium **367** which cyclised to give products **368** essentially as single enantiomers, and also single diastereoisomers with respect to the three new stereogenic centres.[141]

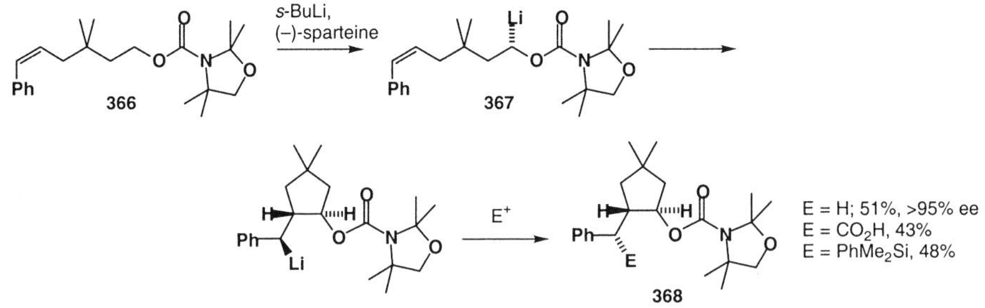

s-BuLi, (–)-sparteine

366 **367** **368**

E = H; 51%, >95% ee
E = CO_2H, 43%
E = $PhMe_2Si$, 48%

Cyclisations of achiral organolithiums in which (–)-sparteine governs the facial selectivity of the attack on the alkene can also give good enantioselectivity. Transmetallation of **369** generates an organolithium which cyclises to give **370** in high ee.[170,171]

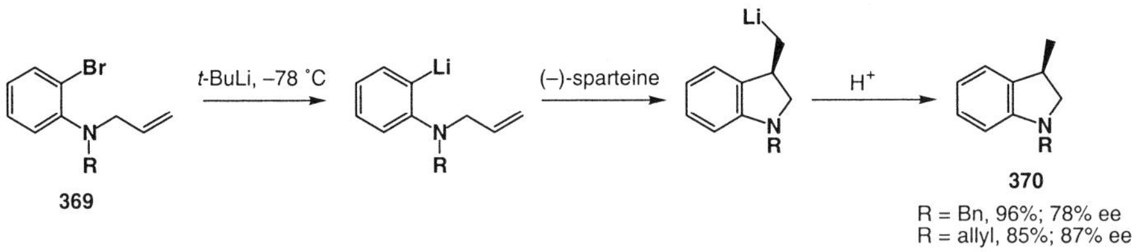

t-BuLi, −78 °C; (–)-sparteine; H^+

369 **370**

R = Bn, 96%; 78% ee
R = allyl, 85%; 87% ee

A dearomatising asymmetric cyclisation initiated by deprotonation with a chiral lithium amide base is discussed in section 5.4.

7.2.4.3 Anionic cyclisation onto allenes

Because the product benefits from allylic stabilisation, cyclisations onto allenes are favourable even when four-membered rings are being formed. For example, the allenyl iodide **371** transmetallates at –78 °C to give a trappable organolithium **372**. After 30 min at 20 °C, **372** has cyclised to the cyclobutenylmethyllithium **373**, which can be trapped with electrophiles with some degree of regioselectivity.[172] This reaction contrasts with the corresponding reaction of organomagnesium compounds, which give preferentially cyclopropyl products by cyclisation onto the proximal double bond.

With the similar compound **374**, something else entirely happens which was itself the first of a new class of dearomatising anionic cyclisations (see below). *Exo*-cyclisation to give a cyclobutene is disfavoured by the gem-dimethyl group, but ought to be favoured by the sulfonyl substituent. However, the reaction proceeds by an unusual intramolecular sulfone-lithium exchange, followed by attack on the phenyl ring of the sulfonyl group by the resulting allenyllithium **375** to give the cyclohexadiene **376**.[172]

7.2.4.4 Anionic cyclisation onto alkynes

The first alkyne cyclisations, from **377**, **379** and **381**, predate the early alkene cyclisations by a couple of years: these three date from 1966[173] and 1967,[174] and illustrate the favourability of both *exo* and *endo-dig* cyclisation. All three generate "benzylic" vinyllithiums (**378**, **380** and **382**), and both aryl (**377**, **379**) and alkyl halides (**381**) are successful starting materials. Similar organomagnesium cyclisations were described at about the same time.[175] However, it is not clear in these reactions how much of the product is due to participation of radicals in the mechanism – alkylbromides undergo halogen-metal exchange with alkyllithiums via radical intermediates (chapter 3).[176] If it really is an anionic cyclisation, cyclisation to **378** is remarkable in being *endo*. *Endo-dig* anionic cyclisations are discussed below.

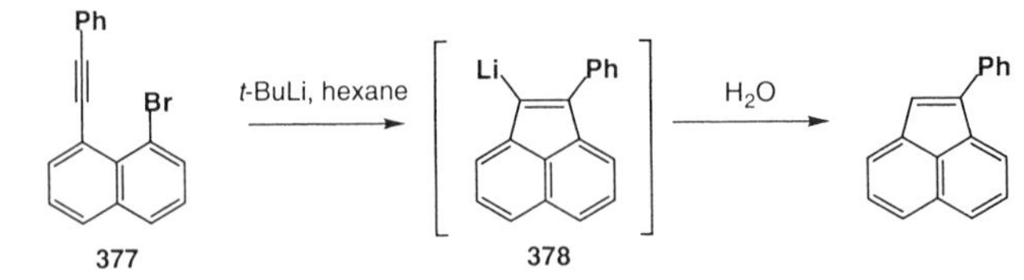

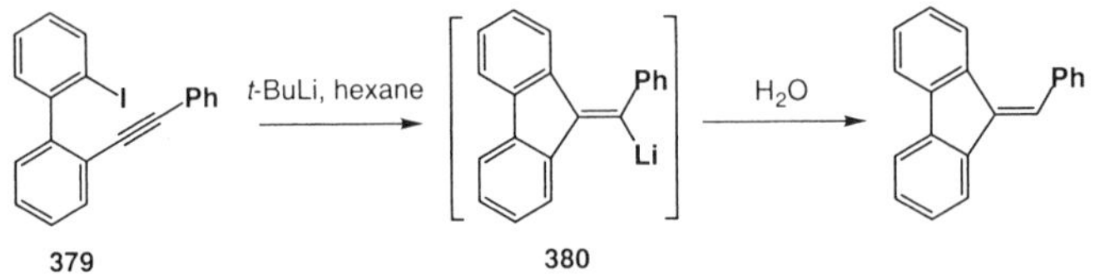

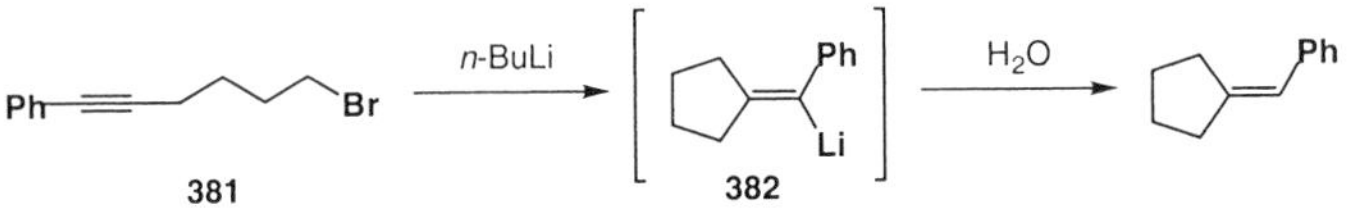

Bailey addressed the question of radical participation in 1989,[177] and showed that replacement of bromine in **381** by iodine in **383** allowed a clean, non-radical-mediated halogen-lithium exchange at –78 °C to give a trappable organolithium. Only on warming did the alkyllithium **384** cyclise (the phenylacetylene **384** had a half-life of 6 min at –50.6 °C, and replacing Ph with *n*-Bu leads to a million-fold reduction in rate to a half life of 7 min at +28.8 °C[178]), to give a vinyllithium **385** which could itself be trapped with electrophiles.[179]

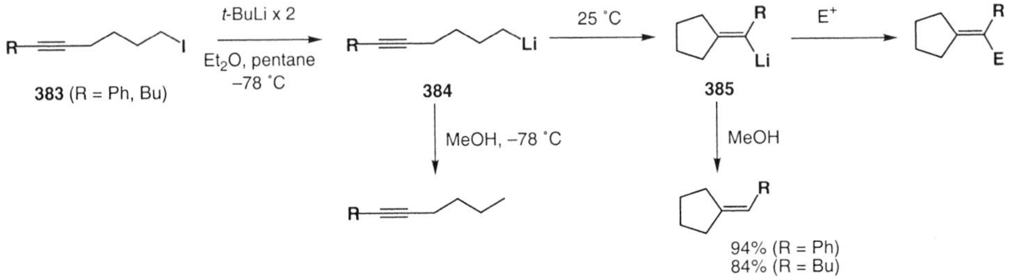

The alkyne addition – effectively an intramolecular carbolithiation – turned out to be a *syn*-stereospecific process, and **386** gave the *Z*-vinyl lithium **387** and hence the *Z*-alkene **388**, with no trace of the *E* isomer. With the more hindered, and less geometrically stable aryl substituted vinyllithium **390a**, **391a** was formed almost exclusively at –78 °C, but quenching the reaction after warming to room temperature gave largely **391b** through isomerisation to **390b**.[179]

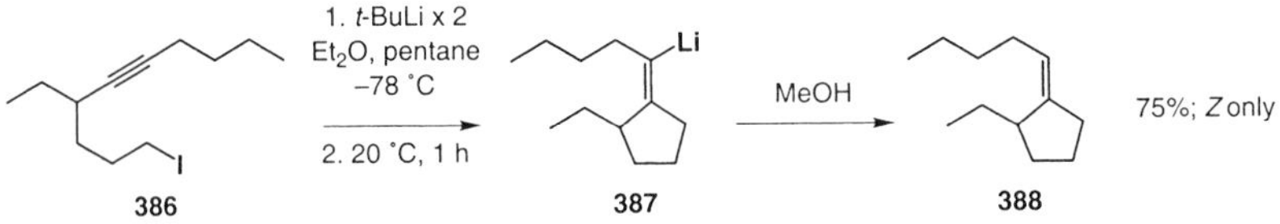

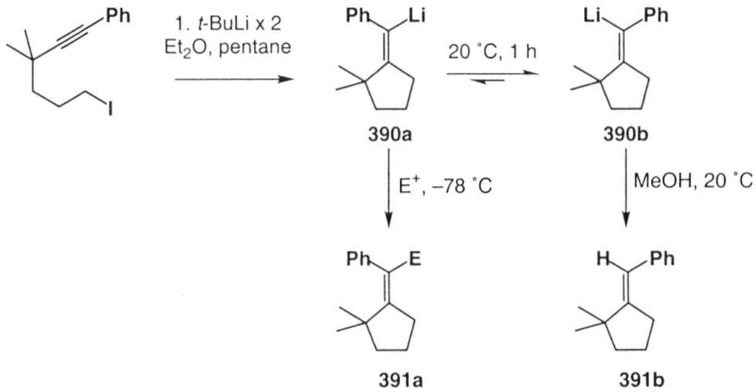

390a **390b**

391a E = H, 88%; E = CO_2H, 76%

391b 70% + 30% **391a**

5-*Exo* cyclisations of vinyllithiums onto phenyl-substituted alkynes **392** and **394** are also *syn* stereospecific, and give the sort of stereodefined dienes **393** and **395** of value in Diels Alder reactions.[180] 6-*Exo* cyclisations are also possible but are much slower, though they still give geometrically pure products (in contrast to 6-*exo* cyclisations onto silyl alkynes: see below). Vinyllithiums cyclise onto alkyl-substituted alkynes (such as **396**) only in the presence of TMEDA, and they do so very slowly. Nonetheless, a single geometrical isomer of the product **397** is obtained.

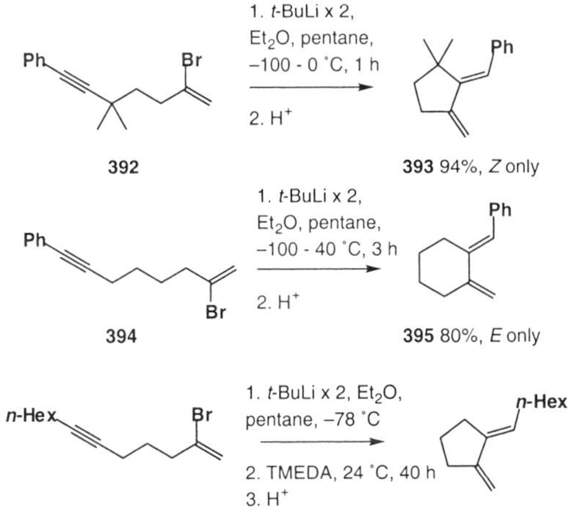

392 → **393** 94%, *Z* only

394 → **395** 80%, *E* only

396 → **397** 76%, *E* only

The cyclisation of alkynylsilane **398** to produce an α-silyl vinyllithium **400** was uncovered during an attempt to add the organolithium **399** to enone **401** as part of a natural product synthesis.[181] Organolithium **399** is unstable and cyclises to **400** before adding to the enone to give **402**.

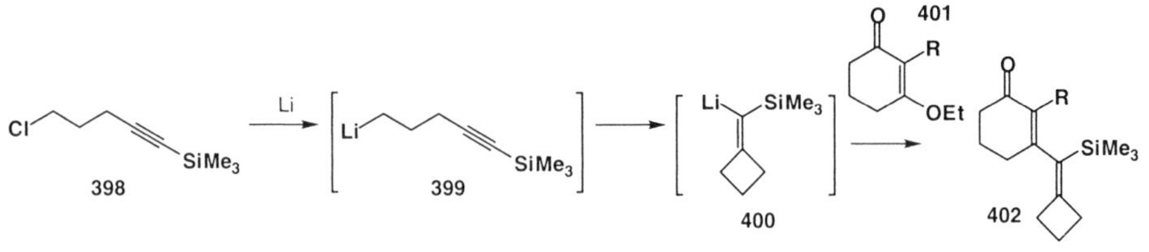

Negishi[182] generalised this cyclisation to show that trialkylsilylalkynes will trap alkyl- (**403**, **404**), vinyl- (**405**, **406**), allenyl- (**407**) and aryllithiums (**408**) intramolecularly. Piers[183] has cyclised a cyclopropyllithium onto an alkynylsilane (**409** – phenyl and alkyl substituted alkynes performed less well), and Coldham[61] found that the α-amino organolithium cyclises onto a silyl alkyne (**410**) though in low yield.

A silyl or aryl anion-stabilising group at the terminus of the alkyne is essential for rapid *exo-dig* cyclisations other than those of alkyllithiums:[177,178] aryl and vinyllithiums will cyclise onto alkynes bearing alkyl substituents only very slowly even in the presence of TMEDA.[180] The allenyllithium cyclisation from **407** is a remarkable example of a cyclisation initiated by a *deprotonation*.

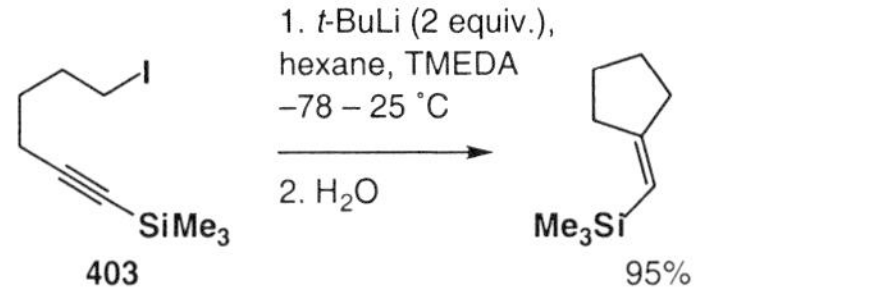

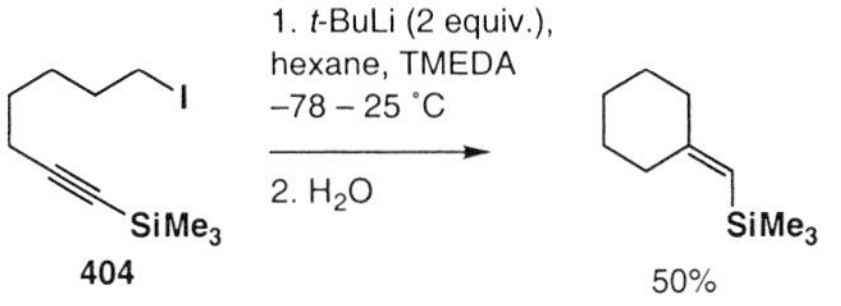

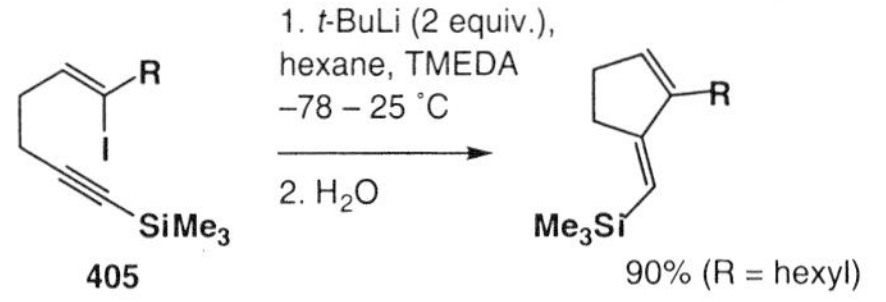

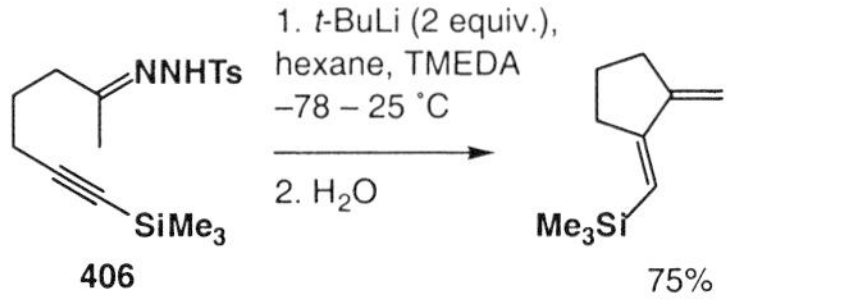

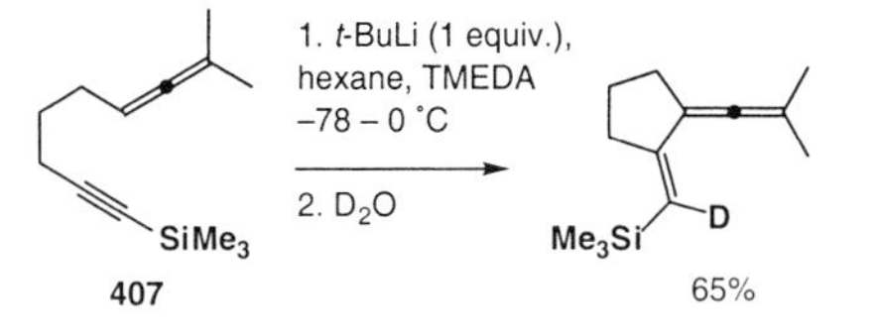

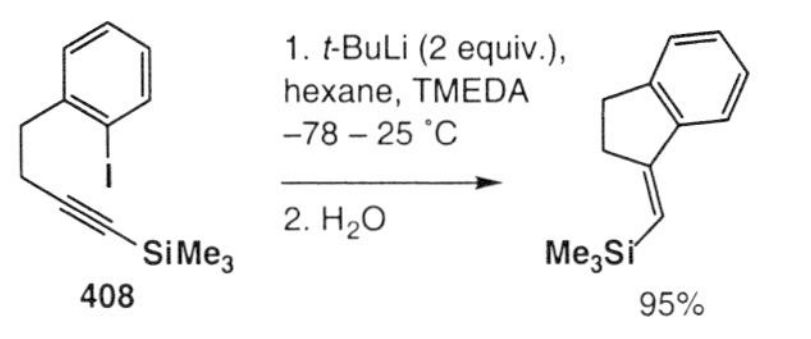

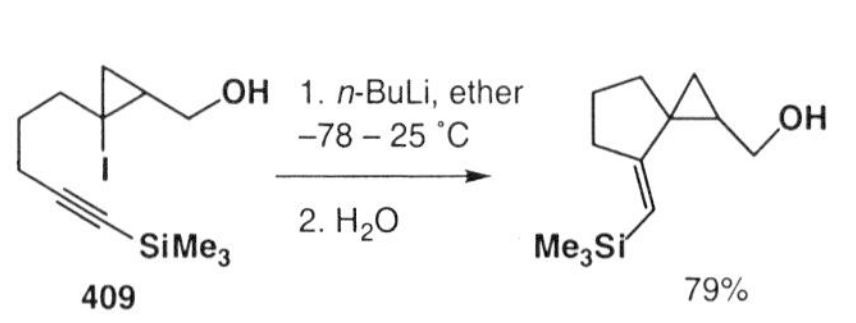

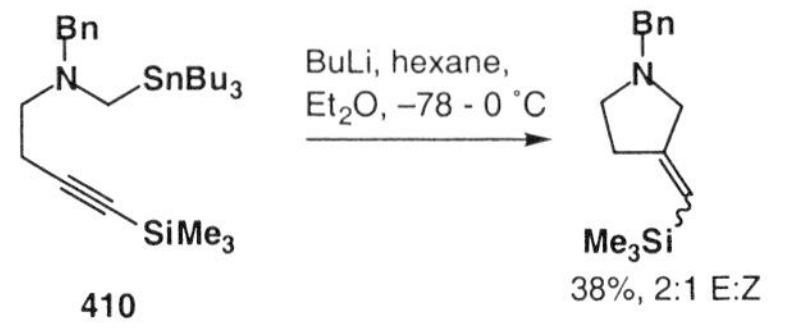

The product silicon-bearing vinyllithiums behaved rather like the phenyl-substituted vinyllithium and an initially formed hindered *Z* isomer will isomerise to *E* at room temperature.[179] A mixture of stereoisomers was obtained in the cyclisation of **410**.[61] Only alkyl substituted vinyllithiums are indefinitely geometrically stable (see section 5.1.12).

With silyl or phenyl stabilising substituents (but again not with alkyl) it was also possible to make four and six-membered rings by *exo-dig* cyclisations.[179] Unsubstituted four-membered rings **412** formed in high yield from **411**, and with unsymmetrical rings it was possible to obtain geometrically pure compounds **414** provided the reactions were carried out at temperatures high enough for cyclisation to occur (generally –50 to –20 °C) but low enough to ensure geometrical stability in the product vinyllithium **413**. Six-membered rings such as **415** could only be formed in high yield when the propargylic positions of the alkyne were blocked, but required such high temperatures (20 °C) that isomerisation of the first-formed *Z*-vinyllithium to *E* was unavoidable.[180]

t-BuLi x 2; Et_2O, pentane, –78 °C; 25 °C; MeOH
411
412
93% (R = Ph)
89% (R = $SiMe_3$)
0% (R = Bu)

t-BuLi x 2; Et_2O, pentane, –78 °C; –50 - –20 °C; H^+
413
414
93% (R = Ph)
90% (R = $SiMe_3$)

t-BuLi x 2; Et_2O, pentane, –78 – +20 °C; 20 °C; E^+
415
81% (R = Ph, E = MeCHOH)
84% (R = $SiMe_3$, E = H)

Cascade reactions using alkyne cyclisations have not been explored, and the isolated example of **416** is unique in a number of ways.[145] For a start, 5-*exo-dig* cyclisation onto the alkyne to give **417** must be sufficiently fast that it competes with halogen-metal exchange of the second iodine atom, which instead traps the product vinyllithium **417**, giving the bicyclic product **418** in 47% yield.

n-BuLi, –78 °C
416
417
418 47%

Endo-trig cyclisations are rare for organolithiums, but *endo-dig* cyclisations are in general more favourable, and 5-*endo-dig* anionic cyclisations do occur. One possible example is the cyclisation of **377**, but a more convincing one is shown below.[184] Evidently the *endo-dig* cyclisation onto an ynol ether is so favourable that even the stabilised sulfone anion **419** cyclises. Primary sulfones such as **421** also cyclise, but give products **420** arising from rearrangement of the vinyl lithium **422** to the more stable lithiated allyl sulfone **423**.

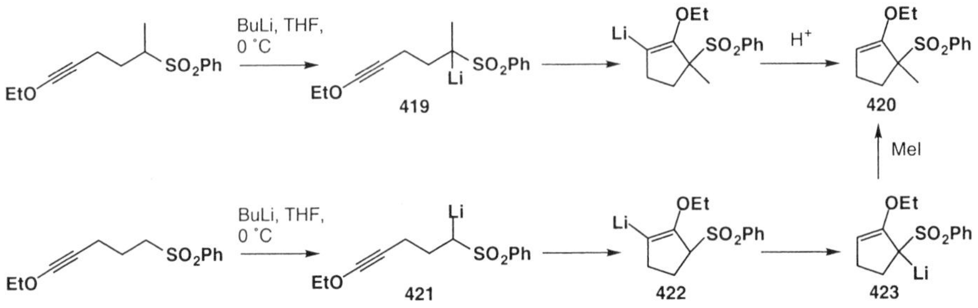

BuLi, THF, 0 °C — SO_2Ph — EtO — **419** — Li — OEt — H^+ — **420** — MeI — **421** — **422** — **423**

With a sulfone capable of giving 5-*exo-dig* vs. 6-*endo-dig*, or 6-*exo-dig* vs. 7-*endo-dig* cyclisations, mixtures are obtained, with *exo* cyclisation dominating. With alkynyl sulfides, solely *exo* products are obtained.

It is impossible for the *endo-dig* cyclisations to proceed via the usual *syn*-carbolithiation mechanism because this would place a *trans* double bond in the ring. Unlike *exo-dig* cyclisations onto alkynylsilanes, however, the product of the 5-*exo-dig* cyclisation of **424** appears to arise by an unexplained *trans* carbolithiation, perhaps due to O–Li coordination.

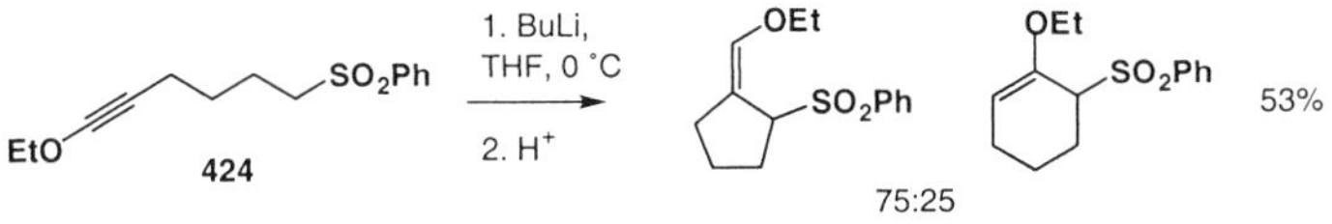

EtO — SO_2Ph — **424** — 1. BuLi, THF, 0 °C; 2. H^+ — OEt — 75:25 — 53%

Anionic cyclisations of yet more stabilised carbon nucleophiles such as enolates fall outside the scope of this book.[185]

A clever synthesis of benzofurans, benzothiophenes and indoles starts with the trifluoroethyl ethers **425** and uses a 5-*endo-dig* cyclisation. Four equivalents of *n*-BuLi perform two eliminations, one substitution, and one lithiation to generate aryllithium **426**, which undergoes 5-*endo-dig* cyclisation. 3-Butylbenzofuran is generated in 40% yield.[186]

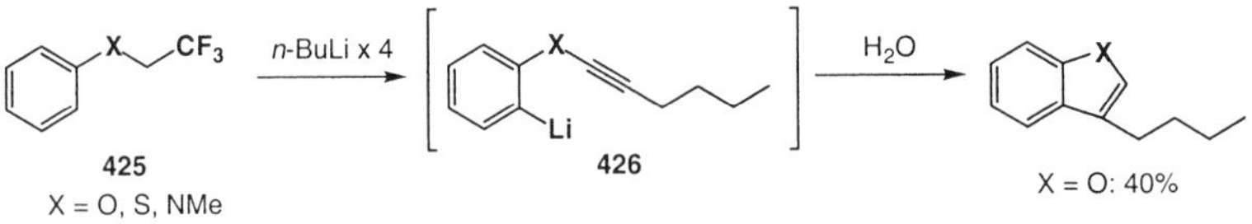

X — CF_3 — **425** — X = O, S, NMe — *n*-BuLi x 4 — Li — **426** — H_2O — X = O: 40%

7.2.4.6 Anionic cyclisation onto aromatic rings

A few isolated examples[104,172,187,147] prior to 1998 showed that organolithiums **427** tethered to aromatic rings can on occasion cyclise, apparently by nucleophilic addition of the organolithium to the π-system of the aromatic ring. In some cases (for example, the conversion of **194** to **196** or **298** to **299**),[104,147] the organolithium **428** rearomatises to yield a new benzo-fused compound **429**; in others (for example, conversion of **374** to **376**, or **431** to **432** below),[172,187] protonation yields a stable, dearomatised bicyclic product **430** or an isomer.

R X Li 427 → 428 → 429 / 430

431 —base→ … —H⁺→ 432

Since 1998, however, it has become clear that in the case of aromatic amides, dearomatising anionic cyclisation is a common pattern of reactivity in *N*-benzyl benzamides and in *N*-benzyl naphthamides.[188-190] Amides **433** and **437**, for example, on treatment with strong base (early reports indicated *t*-BuLi in the presence of HMPA or DMPU at –78 °C, but later[190] it became clear that LDA at 0-20 °C will do) leads first to lithiation, in some cases directly α to nitrogen, and in others firstly *ortho* to the amide[191] – in which case an *ortho*-α anion translocation is required in order for cyclisation to proceed. Dearomatising cyclisation leads to the enolates **435** and **439**, which can be quenched, usually stereoselectively, with electrophiles to give dearomatised products **436** and **440**.[188] Analogous cyclisations of sulfonamides[192] and phosphonamides[193] have also been reported.

433 —*t*-BuLi or LDA→ 434 —> 0 °C→ 435 —E⁺→ 436

E = H: 82%
E = Bn: 72%
E = CH(OH)Ph: 81%

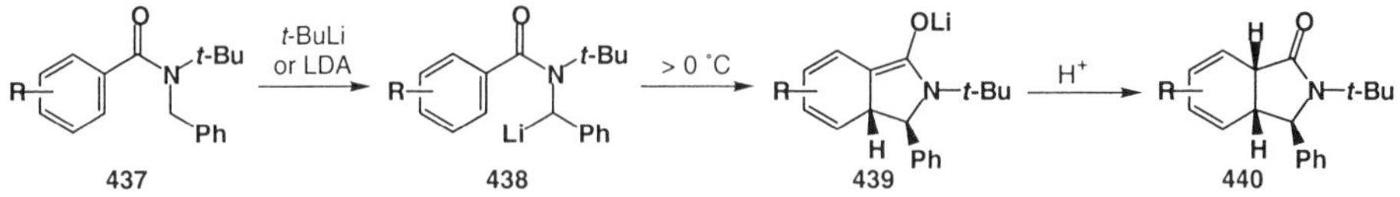

The enol ether products (**442**) of cyclising methoxy-substituted amides (**437**, R = OMe and **441**) may be hydrolysed cleanly to enones such as **443**,[194] and these have proven to be versatile intermediates for the stereoselective synthesis of a range of important heterocycles, principally in the kainoid family of amino acids.[195-197] The general strategy, illustrated below in the synthesis of kainic acid **444** itself,[195] is to retain the new five-membered ring of **443**, but to sacrifice the six-membered ring, while exploiting the *cis* ring junction as a means of ensuring the characteristic kainoid *cis* stereochemistry of the substituents at C2 and C3. Naphthamide starting materials are readily applicable to the synthesis of aryl-substituted kainoids related to acromelic acid.[196] Amide **441** carries the acid-labile but base stable cumyl protecting group,[198] which is proving useful in a number of organolithium reactions.

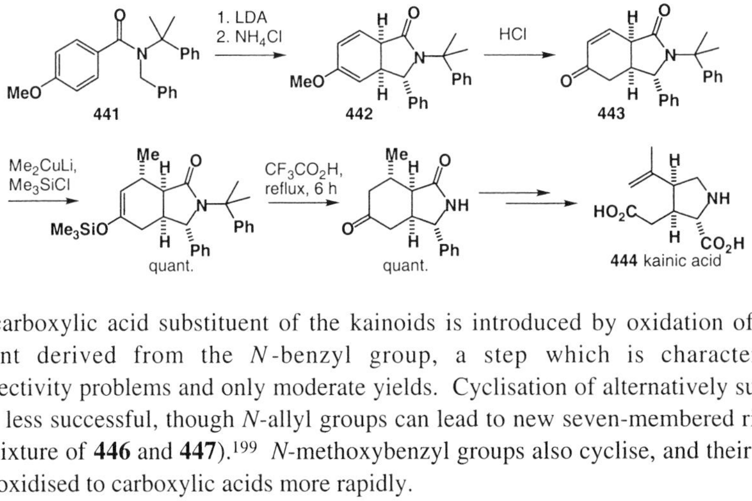

The C1 carboxylic acid substituent of the kainoids is introduced by oxidation of the aryl substituent derived from the *N*-benzyl group, a step which is characterised by chemoselectivity problems and only moderate yields. Cyclisation of alternatively substituted amides is less successful, though *N*-allyl groups can lead to new seven-membered rings (**445** gives a mixture of **446** and **447**).[199] *N*-methoxybenzyl groups also cyclise, and their aromatic rings are oxidised to carboxylic acids more rapidly.

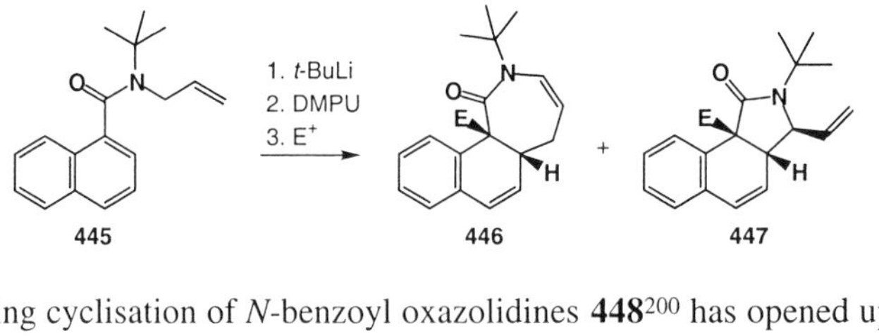

The dearomatising cyclisation of *N*-benzoyl oxazolidines **448**[200] has opened up the possibility that non-benzylated starting materials may prove valuable in syntheses based on dearomatising cyclisation, and also reveals some deeper mechanistic aspects of the reaction. Amide **448** is ortholithiated by *t*-BuLi to give **449**, as indicated by the fact that deuteration after only 30 min gives **453**. However, with time or on warming, **449** cyclises to **452** and gives a product **453** whose congested *cis* stereochemistry was proved by X-ray crystallography. This result may most reasonably be explained by rationalising the cyclisation as a new example of disrotatory electrocyclic ring closure of **450** (best represented as **451**). It is possible therefore that dearomatising anionic cyclisations of amides in general are in fact pericyclic in nature.

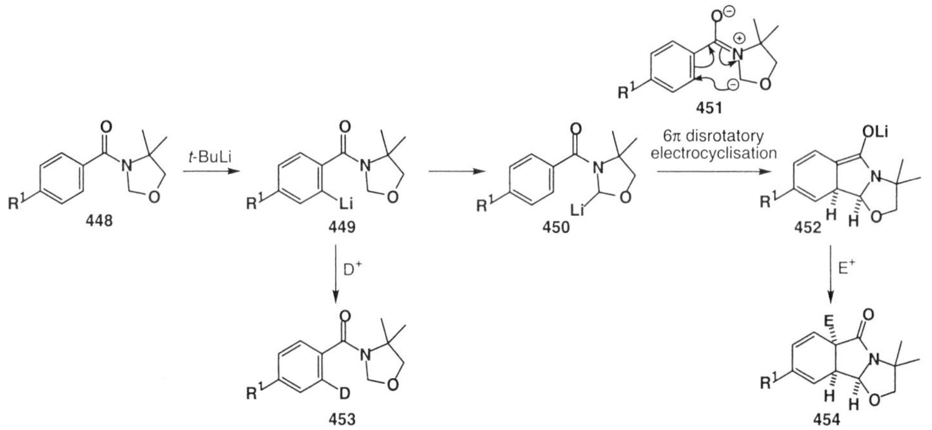

It is clear, nonetheless, that successful dearomatising cyclisations are not confined to systems which can undergo electrocyclic ring closure: **456** for example cyclises to **457**, which may be protonated or alkylated stereoselectively.[201] Without the oxazoline substituent, similar reactions lead to rearomatisation.

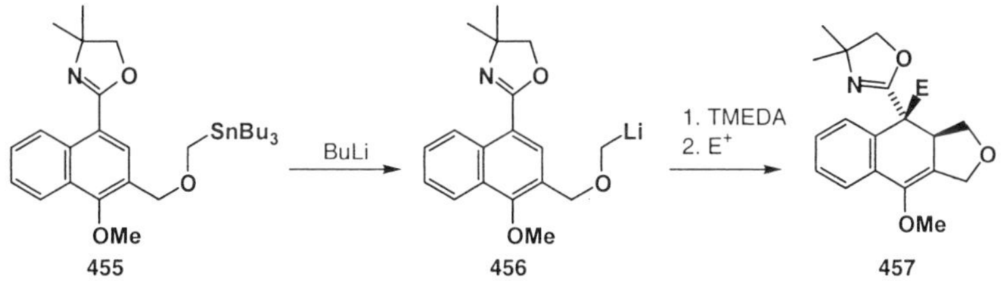

Stereoselectivity in dearomatising cyclisations may be controlled by a number of factors, including rotational restriction in the organolithium intermediates[202,203] and coordination to an exocyclic chiral auxiliary.[197] Most usefully, by employing a chiral lithium amide base, it is possible to lithiate **441** enantioselectively (see section 5.4 for similar reactions) and promote a cyclisation to **442** with >80% ee.[204]

References

1. Corruble, A.; Valnot, J.-Y.; Maddaluno, J.; Prigent, Y.; Davoust, D.; Duhamel, P. *J. Am. Chem. Soc.* **1997,** *119*, 10042; Corruble, A.; Valnot, J.-Y.; Maddaluno, J.; Duhamel, P. *J. Org. Chem.* **1998,** *63*, 8266
2. Smyj, R. P.; Chong, J. M. *Org. Lett.*, **2001**, *3*, 2903

3. Denmark, S. E. *J. Am. Chem. Soc.* **1994,** *116*, 8792.
4. Knochel, P. In *Comprehensive Organic Synthesis;* Trost, B. M.; Fleming, I. Eds.; Pergamon: Oxford, 1990; Vol. 4; chapter 4.4.
5. Waack, R.; Doran, M. A. *J. Org. Chem.* **1967,** *32*, 3395.
6. Morton, M. *Anionic Polymerisation: Principles and Practice*; Academic Press: New York, 1983.
7. Bartlett, P. D.; Tauber, S. J.; Weber, W. P. *J. Am. Chem. Soc.* **1969,** *91*, 6362.
8. Schmitz, R. F.; de Kanter, F. F.; Schakel, M.; Klumpp, G. W. *Tetrahedron* **1994,** *50*, 5933.
9. Richey, H. G.; Wilkins, C. W.; Benson, R. M. *J. Org. Chem.* **1980,** *40*, 5052.
10. Wei, X.; Taylor, R. J. K. *J. Chem. Soc., Chem. Commun.* **1996**, 187.
11. Wei, X.; Johnson, P.; Taylor, R. J. K. *J. Chem. Soc., Perkin Trans. 1* **2000**, 1109.
12. Wei, X.; Taylor, R. J. K. *Tetrahedron Lett.* **1997,** *38*, 6467.
13. Wei, X.; Taylor, R. J. K. *Tetrahedron Lett.* **1996,** *37*, 4209.
14. Glaze, W. H.; Hanicek, J. E.; Moore, M. L.; Chaudhuri, J. *J. Organomet. Chem.* **1972,** *44*, 39.
15. Norsikian, S.; Baudry, M.; Normant, J.-F. *Tetrahedron Lett.* **2000,** *41*, 6575.
16. Miginiac, L. *J. Organomet. Chem.* **1982,** *238*, 235.
17. Mulvaney, J. E.; Gardlund, Z. G.; Gardlund, S. L. *J. Am. Chem. Soc.* **1963,** *85*, 3897.
18. Mulvaney, J. E.; North, D. J. *J. Org. Chem.* **1969,** *24*, 1936.
19. Hojo, M.; Murakami, Y.; Aihara, H.; Sakuragi, R.; Baba, Y.; Hosomi, A. *Angew. Chem., Int. Ed. Engl.* **2001,** *40*, 621.
20. Eppley, R. L.; Dixon, J. A. *J. Am. Chem. Soc.* **1968,** *90*, 1606.
21. Gant, T. G.; Meyers, A. I. *Tetrahedron* **1994,** *50*, 2297.
22. Klumpp, G. W. *Rec. Trav. Chim. Pays-Bas* **1986,** *105*, 1.
23. Veefkind, A. H.; Bickelhaupt, F.; Klumpp, G. W. *Rec. Trav. Chim. Pays-Bas* **1969,** *88*, 1058.
24. Crandall, J. K.; Roja, A. C. *Org. Synth.* **1988,** *6*, 786.
25. Crandall, J. K.; Clark, A. C. *J. Org. Chem.* **1972,** *31*, 4236.
26. Felkin, H.; Swierczewski, G.; Tambuté, A. *Tetrahedron Lett.* **1969**, 707.
27. Kato, T.; Marumoto, S.; Sato, T.; Kuwajima, I. *Synlett* **1990**, 671.
28. Marumoto, S.; Kuwajima, I. *Chem. Lett.* **1992**, 1421.
29. Marumoto, S.; Kuwajima, I. *J. Am. Chem. Soc.* **1993,** *115*, 9021.
30. Brémand, N.; Normant, J.-F.; Mangeney, P. *Synlett* **2000,** *4*, 532.
31. Klumpp, G. W.; Mierop, A. J. C.; Vrielink, J. J.; Bringman, K.; Schafel, M. *J. Am. Chem. Soc.* **1985,** *107*, 6740.
32. Muck-Lichtenfeld, C.; Ahlbrecht, H. *Tetrahedron* **1999,** *44*, 2609.
33. Klein, S.; Marek, I.; Normant, J.-F. *J. Org. Chem.* **1994,** *59*, 2925.
34. Superchi, S.; Sotomayor, N.; Miao, G.; Joseph, B.; Campbell, M. G.; Snieckus, V. *Tetrahedron Lett.* **1996,** *37*, 6061.
35. Bégué, J.-P.; Bonnet-Delpon, D.; Bouvet, D.; Rock, M. H. *J. Org. Chem.* **1996,** *61*, 9111.
36. Wei, X.; Taylor, R. J. K. *Tetrahedron Asymmetry* **1997,** *8*, 665.

37. Norsikian, S.; Marek, I.; Normant, J.-F. *Tetrahedron Lett.* **1997,** *38*, 7523.
38. Klein, S.; Marek, I.; Poisson, J.-F.; Normant, J.-F. *J. Am. Chem. Soc.* **1995,** *117*, 8853.
39. Norsikian, S.; Marek, I.; Klein, S.; Poisson, J.-F.; Normant, J.-F. *Chem. Eur. J.* **1999,** *5*, 2055.
40. Brémand, N.; Mangeney, P.; Normant, J. F. *Tetrahedron Lett.* **2001,** *42*, 1883.
41. Norsikian, S.; Marek, I.; POisson, J.-F.; Normant, J.-F. *J. Org. Chem.* **1997,** *62*, 4898.
42. Parham, W. E.; Bradsher, C. K. *Acc. Chem. Res.* **1982,** *15*, 300.
43. Parham, W. E.; Jones, L. D.; Sayed, Y. A. *J. Org. Chem.* **1975,** *40*, 2394.
44. Kometani, T.; Takeuchi, Y.; Yoshii, E. *J. Chem. Soc., Perkin Trans. 1* **1981**, 1197.
45. Boatman, R. J.; Whitlock, B. J.; Whitlock, H. W. *J. Am. Chem. Soc.* **1977,** *99*, 4823.
46. Parham, W. E.; Jones, L. D.; Sayed, Y. A. *J. Org. Chem.* **1976,** *41*, 1184.
47. Lamas, C.; Castedo, L.; Dominguez, D. *Tetrahedron Lett.* **1990,** *31*, 6247.
48. Paleo, M. R.; Lamas, C.; Castedo, L.; Dominguez, D. *J. Org. Chem.* **1992,** *57*, 2029.
49. Hendi, M. S.; Natalie, K. J.; Hendi, S. B.; Campbell, J. A.; Greenwood, T. D.; Wolfe, J. F. *Tetrahedron Lett.* **1989,** *30*, 275.
50. Couture, A.; Deniau, E.; Grandclaudon, P. *J. Chem. Soc., Chem. Commun.* **1994**, 1329.
51. Piers, E.; Walker, S. D.; Armbrust, R. *J. Chem. Soc., Perkin Trans. 1* **2000**.
52. Cooke, M. P.; Houpis, I. N. *Tetrahedron Lett.* **1985,** *26*, 4987.
53. Boardman, L. D.; Bagheri, V.; Sawada, H.; Negishi, E.-i. *J. Am. Chem. Soc.* **1984,** *106*, 6105.
54. Trost, B. M.; Coppola, B. P. *J. Am. Chem. Soc.* **1982,** *104*, 6879.
55. McCarthy, C.; Jones, K. *J. Chem. Soc., Chem. Commun.* **1989**, 1717.
56. Storey, J. M. D.; McCarthy, C.; Jones, K. *J. Chem. Soc., Chem. Commun.* **1991**, 892.
57. Negishi, E.-i.; Miller, J. A. *J. Am. Chem. Soc.* **1983,** *105*, 6761.
58. Sawada, H.; Webb, M.; Stoll, A. T.; Negishi, E.-i. *Tetrahedron Lett.* **1986,** *37*, 775.
59. Larchevêque, M.; Debal, A.; Cuvigny, T. *J. Organomet. Chem.* **1975,** *87*, 25.
60. van der Does, T.; Klumpp, G. W.; Schakel, M. *Tetrahedron Lett.* **1986,** *27*, 519.
61. Coldham, I.; Lang-Anderson, M. M. S.; Rathmell, R. E.; Snowden, D. J. *Tetrahedron Lett.* **1997,** *38*, 7621.
62. Baird, M. S.; Huber, F. A. M.; Tverezovsky, V. V.; Bolesov, I. G. *Tetrahedron* **2001,** *57*, 1593.
63. Robinson, R. P.; Cronin, B. J.; Jones, B. P. *Tetrahedron Lett.* **1997,** *38*, 8479.
64. Kende, A. S.; Smith, C. A. *Tetrahedron Lett.* **1988,** *29*, 4217.
65. Bradsher, C. K.; Reames, D. C. *J. Org. Chem.* **1978,** *43*, 3800.
66. Dhowan, K. L.; Gowland, B. D.; Durst, T. *J. Org. Chem.* **1980,** *45*, 922.
67. Cooke, M. P.; Houpis, I. N. *Tetrahedron Lett.* **1985,** *36*, 3643.
68. Babler, J. H.; Bauta, W. E. *Tetrahedron Lett.* **1984,** *25*, 4323.
69. Last, L. A.; Fretz, E. R.; Coates, R. M. *J. Org. Chem.* **1982,** *47*, 3211.
70. Shankaran, K.; Snieckus, V. *J. Org. Chem.* **1984,** *49*, 5022.
71. Lei, B.; Fallis, A. G. *J. Am. Chem. Soc.* **1990,** *112*, 4609.
72. Bradsher, C. K.; Hunt, D. A. *J. Org. Chem.* **1981,** *46*, 4600.
73. Bailey, W. F.; Gagnier, R. P. *Tetrahedron Lett.* **1982,** *23*, 5123.

74. Bailey, W. F.; Gagnier, X.; Patricia, J. J. *J. Org. Chem.* **1984,** *49*, 2092.
75. Stoll, A. T.; Negishi, E.-i. *Tetrahedron Lett.* **1985,** *26*, 5671.
76. Negishi, E.-i.; Boardman, L. D.; Tour, J. M.; Sawada, H.; Rand, C. L. *J. Am. Chem. Soc.* **1983,** *105*, 6344.
77. Piers, E.; Tse, H. L. A. *Tetrahedron Lett.* **1984,** *25*, 3155.
78. Chamberlin, A. R.; Bloom, S. H. *Tetrahedron Lett.* **1984,** *25*, 4901.
79. Krief, A.; Hobe, M. *Synlett* **1992**, 317.
80. Krief, A.; Hobe, M. *Tetrahedron Lett.* **1992,** *33*, 6527.
81. Krief, A.; Hobe, M. *Tetrahedron Lett.* **1992,** *33*, 6529.
82. Krief, A.; Hobe, M.; Dumont, W.; Badaoui, E.; Guittet, E.; Evrard, G. *Tetrahedron Lett.* **1992,** *33*, 3381.
83. Wu, S.; Lee, S.; Beak, P. *J. Am. Chem. Soc.* **1996,** *118*, 715.
84. Cooke, M. P. *J. Org. Chem.* **1984,** *49*, 1144.
85. Cooke, M. P.; Widener, R. K. *J. Org. Chem.* **1987,** *52*, 1381.
86. Cooke, M. P. *J. Org. Chem.* **1992,** *57*, 1495.
87. Comins, D. L.; Zhang, Y.-m. *J. Am. Chem. Soc.* **1996,** *118*, 12248.
88. St Denis, J.; Dolzine, T. W.; Oliver, J. P.; Smart, J. B. *J. Organomet. Chem.* **1974,** *71*, 315.
89. Silver, M. S.; Shafer, P. R.; Norlander, J. E.; Ruchart, C.; Roberts, S. D. *J. Am. Chem. Soc.* **1960,** *82*, 2646.
90. Richey, H. G.; Kossa, W. C. *Tetrahedron Lett.* **1969**, 2313.
91. Hill, E. A. *J. Organomet. Chem.* **1975,** *91*, 123.
92. Richey, H. G.; Rees, T. C. *Tetrahedron Lett.* **1966**, 4297.
93. Garst, J. F.; Hines, J. B. *J. Am. Chem. Soc.* **1984,** *106*, 6443.
94. Oliver, J. P.; Smart, J. B.; Emerson, M. T. *J. Am. Chem. Soc.* **1966,** *88*, 4101.
95. Meyer, C.; Marek, I.; Coutemanche, G.; Normant, J.-F. *Tetrahedron* **1994,** *50*, 11665.
96. Drozd, V. N.; Ustynyuk, Y. A.; Tsel'eva, M. A.; Dmitriev, L. B. *Zh. Obsch. Khim.* **1968,** *38*, 2114.
97. Drozd, V. N.; Ustynyuk, Y. A.; Tsel'eva, M. A.; Dmitriev, L. B. *J. Gen. Chem. USSR* **1968,** *38*, 2047.
98. Drozd, V. N.; Ustynyuk, Y. A.; Tsel'eva, M. A.; Dmitriev, L. B. *Zh. Obsch. Khim.* **1969,** *39*, 1991.
99. Drozd, V. N.; Ustynyuk, Y. A.; Tsel'eva, M. A.; Dmitriev, L. B. *J. Gen. Chem. USSR* **1969,** *39*, 1951.
100. St Denis, J.; Dolzine, T. W.; Oliver, J. P. *J. Am. Chem. Soc.* **1972,** *94*, 8260.
101. Dolzine, T. W.; Oliver, J. P. *J. Organomet. Chem.* **1974,** *78*, 165.
102. Richey, H. G.; Veale, H. S. *J. Am. Chem. Soc.* **1974,** *96*, 2641.
103. Lansbury, P. T.; Caridi, F. *J. Chem. Soc., Chem. Commun.* **1970**, 714.
104. Klumpp, G. W.; Schmitz, R. F. *Tetrahedron Lett.* **1974,** *15*, 2911.
105. Hill, E. A.; Richey, H. G.; Rees, T. C. *J. Org. Chem.* **1963,** *28*, 2161.
106. Lansbury, P. T.; Pattison, V. A.; Clement, W. A.; Sidler, J. D. *J. Am. Chem. Soc.* **1964,** *86*, 2247.
107. Wittig, G.; Offen, J. *Angew. Chemie* **1960,** *72*, 781.

108. Baldwin, J. E. *J. Chem. Soc., Chem. Commun.* **1976**, 734.
109. Wilson, S. E. *Tetrahedron Lett.* **1975,** *16*, 4651.
110. Mudryk, B.; Cohen, T. *J. Am. Chem. Soc.* **1993,** *113*, 3855.
111. Krief, A.; Couty, F. *Tetrahedron Lett.* **1997,** *38*, 8085.
112. Bailey, W. F.; Patricia, J. J.; DelGobbo, V. C.; Jarret, R. M.; Okarma, P. J. *J. Org. Chem.* **1985,** *50*, 1999.
113. Nicolaou, K. C.; Sorensen, E. J. *Classics in Total Synthesis*; VCH: Weinheim, 1996.
114. Lee, K.-W.; San Filippo, S. *Organometallics* **1983,** *2*, 906.
115. Ashby, E. C.; Pham, T. N.; Park, B. *Tetrahedron Lett.* **1985,** *26*, 4691.
116. Ashby, E. C.; Pham, T. N. *J. Org. Chem.* **1987,** *52*, 1291.
117. Bailey, W. F.; Patricia, J. J.; Nurmi, T. T.; Wang, W. *Tetrahedron Lett.* **1986,** *27*, 1861.
118. Ross, G. A.; Koppang, M. D.; Bartak, D. E.; Woolsey, N. F. *J. Am. Chem. Soc.* **1985,** *107*, 6742.
119. Garst, J. F.; Hines, J. B.; Bruhnke, J. D. *Tetrahedron Lett.* **1986,** *27*, 1963.
120. Chamberlin, A. R.; Bloom, S. H. *Tetrahedron Lett.* **1986,** *27*, 551.
121. Bailey, W. F.; Punzalan, E. R. *J. Am. Chem. Soc.* **1994,** *116*, 6577.
122. Bailey, W. F.; Nurmi, T. T.; Patricia, J. J.; Wang, W. *J. Am. Chem. Soc.* **1987,** *109*, 2442.
123. Bailey, W. F.; Khanolkar, A. D. *J. Org. Chem.* **1990,** *55*, 6058.
124. Bailey, W. F.; Khanolkar, A. D. *Organometallics* **1993,** *12*, 239.
125. Bailey, W. F.; Khanolkar, A. D.; Gavaskar, K.; Ovaska, T. V.; Rossi, K.; Thiel, Y.; Wiberg, K. B. *J. Am. Chem. Soc.* **1991,** *113*, 5720.
126. Bailey, W. F.; Khanolkar, A. D. *Tetrahedron* **1991,** *47*, 7727.
127. Fraenkel, G.; Geckle, J. M. *J. Am. Chem. Soc.* **1980,** *102*, 2869.
128. Krief, A.; Barbeaux, P. *Tetrahedron Lett.* **1991,** *32*, 417.
129. Krief, A.; Barbeaux, P. *J. Chem. Soc., Chem. Commun.* **1987**, 1214.
130. Krief, A.; Derouane, D.; Dumont, W. *Synlett* **1992**, 907.
131. Krief, A.; Kenda, B.; Maertens, C.; Remacle, B. *Tetrahedron* **1996,** *52*, 7645.
132. Krief, A.; Kenda, B.; Barbeaux, P. *Tetrahedron Lett.* **1991,** *32*, 2509.
133. Krief, A.; Bousbaa, J. *Tetrahedron Lett.* **1997,** *38*, 6291.
134. Broka, C. A.; Lee, W. J.; Shen, T. *J. Org. Chem.* **1988,** *53*, 1336.
135. Paquette, L. A.; Gilday, J. P.; Maynard, G. D. *J. Org. Chem.* **1989,** *54*, 5044.
136. Krief, A.; Barbeaux, P. *Synlett* **1990**, 511.
137. Chamberlin, A. R.; Bloom, S. H.; Cervini, L. A.; Fotsch, C. H. *J. Am. Chem. Soc.* **1988,** *110*, 4788.
138. Lautens, M.; Kumanovic, S. *J. Am. Chem. Soc.* **1995,** *117*, 1954.
139. Bailey, W. F.; Gavaskar, K. V. *Tetrahedron* **1994,** *50*, 5957.
140. Bailey, W. F.; Tao, Y. *Tetrahedron Lett.* **1997,** *38*, 6157.
141. Woltering, M. J.; Fröhlich, R.; Hoppe, D. *Angew. Chem., Int. Ed. Engl.* **1997,** *36*, 1764.
142. Krief, A.; Kenda, B.; Remacle, B. *Tetrahedron Lett.* **1995,** *36*, 7917.
143. Krief, A.; Kenda, B.; Remacle, B. *Tetrahedron* **1996,** *52*, 7435.

144. Lorthiois, E.; Marek, I.; Normant, J.-F. *Tetrahedron Lett.* **1996,** *37*, 6693.
145. Cooke, M. P.; Huang, J. J. *Synlett* **1997**, 535.
146. Hoffmann, R. W.; Koberstein, R.; Remacle, B.; Krief, A. *J. Chem. Soc., Chem. Commun.* **1997**, 2189.
147. Krief, A.; Kenda, B.; Barbeaux, P.; Guittet, E. *Tetrahedron* **1994,** *50*, 7177.
148. Bailey, W. F.; Rossi, K. *J. Am. Chem. Soc.* **1989,** *111*, 765.
149. Bailey, W. F.; Khanolkar, A. D. *Tetrahedron Lett.* **1990,** *31*, 5993.
150. Bailey, W. F.; Khanolkar, A. D.; Gavaskar, K. V. *J. Am. Chem. Soc.* **1992,** *114*, 8053.
151. Stiasny, H. C.; Hoffmann, R. W. *Chem. Eur. J.* **1995,** *1*, 619.
152. Broka, C. A.; Shen, T. *J. Am. Chem. Soc.* **1989,** *111*, 2981.
153. Tomooka, K.; Komine, N.; Nakai, T. *Tetrahedron Lett.* **1997,** *38*, 8939.
154. Bailey, W. F.; Žarcone, L. M. J. *Tetrahedron Lett.* **1991,** *52*, 4425.
155. Coldham, I. *J. Chem. Soc., Perkin Trans. 1* **1993**, 1275.
156. Coldham, I.; Hufton, R. *Tetrahedron Lett.* **1995,** *36*, 2157.
157. Coldham, I.; Hufton, R. *Tetrahedron* **1996,** *52*, 12541.
158. Zhang, D.; Liebeskind, L. S. *J. Org. Chem.* **1996,** *61*, 2594.
159. Bailey, W. F.; Jiang, X.-L. *J. Org. Chem.* **1996,** *61*, 2596.
160. Bailey, W. F.; Carson, M. W. *Tetrahedron Lett.* **1997,** *38*, 1329.
161. Barluenga, J.; Sanz, R.; Fañanás, F. J. *Tetrahedron Lett.* **1997,** *38*, 2763.
162. Beckwith, A. L. J.; Easton, C. J.; Serelis, A. K. *J. Chem. Soc., Chem. Commun.* **1980**, 482.
163. Beckwith, A. L. J.; Lawrence, T.; Serelis, A. K. *J. Chem. Soc., Chem. Commun.* **1980**, 484.
164. Coldham, I.; Hufton, R.; Rathmell, R. E. *Tetrahedron Lett.* **1997,** *38*, 7617.
165. Bailey, W. F.; Jiang, X.-l.; Mcleod, C. E. *J. Org. Chem.* **1995,** *60*, 7791.
166. Rölle, T.; Hoffmann, R. W. *J. Chem. Soc., Perkin Trans. 2* **1995**, 1953.
167. Hoffmann, R. W.; Koberstein, R.; Harms, K. *J. Chem. Soc., Perkin Trans. 2* **1999**, 183.
168. Krief, A.; Bousbaa, J. *Synlett* **1996**, 1007.
169. Coldham, I.; Hufton, R.; Snowden, D. J. *J. Am. Chem. Soc.* **1996,** *118*, 5322.
170. Bailey, W. F.; Mealy, M. J. *J. Am. Chem. Soc.* **2000,** *122*, 6787.
171. Gil, G. S.; Groth, U. *J. Am. Chem. Soc.* **2000,** *122*, 6789.
172. Crandall, J. K.;Ayers, T. A. *J. Org. Chem.* **1992,** *57*, 2993.
173. Kandil, S. A.; Dessy, R. E. *J. Am. Chem. Soc.* **1966,** *88*, 3027.
174. Ward, H. R. *J. Am. Chem. Soc.* **1967,** *89*, 5517.
175. Richey, H. G.; Rothman, A. M. *Tetrahedron Lett.* **1968**, 1427.
176. Bailey, W. F.; Patricia, J. J. *J. Organomet. Chem.* **1988,** *352*, 1.
177. Bailey, W. F.; Ovaska, T. V.; Leipert, T. K. *Tetrahedron Lett.* **1989,** *30*, 3901.
178. Bailey, W. F.; Ovaska, T. V. *Organometallics* **1990,** *9*, 1694.
179. Bailey, W. F.; Ovaska, T. V. *J. Am. Chem. Soc.* **1993,** *115*, 3083.
180. Bailey, W. F.; Wachter-Jurcsak, N. M.; Pineau, M. R.; Ovaska, T. V.; Warren, R. R.; Lewis, C. E. *J. Org. Chem.* **1996,** *61*, 8216.
181. Koft, E. R.; Smith, A. B. *J. Org. Chem.* **1984,** *49*, 832.

182. Wu, G.; Cederbaum, F. E.; Negishi, E.-i. *Tetrahedron Lett.* **1990,** *31*, 493.
183. Piers, E.; Coish, P. D. G. *Synthesis* **1996**, 502.
184. Funk, R. L.; Bolton, G. L.; Brummond, K. M.; Ellestad, K. E.; Stallman, J. B. *J. Am. Chem. Soc.* **1993,** *115*, 7023.
185. Burke, S. D.; Jung, K. W. *Tetrahedron Lett.* **1994,** *35*, 5837.
186. Johnson, F.; Subramanian, R. *J. Org. Chem.* **1986,** *51*, 5041.
187. Padwa, A.; Filipkowski, M. A.; Kline, D. N.; Murphree, S. S.; Yeske, P. E. *J. Org. Chem.* **1993,** *58*, 2061.
188. Ahmed, A.; Clayden, J.; Rowley, M. *J. Chem. Soc., Chem. Commun.* **1998**, 297.
189. Ahmed, A.; Clayden, J.; Yasin, S. A. *J. Chem. Soc., Chem. Commun.* **1999**, 231.
190. Clayden, J.; Menet, C. J.; Mansfield, D. J. *Org. Lett.* **2000,** *2*, 4229.
191. Ahmed, A.; Clayden, J.; Rowley, M. *Tetrahedron Lett.* **1998,** *39*, 6103.
192. Aggarwal, V. K.; Ferrara, M. *Org. Lett.* **2000,** *2*, 4107.
193. Fernández, I.; Ortiz, F. L.; Tejerina, B.; Granda, S. G. *Org. Lett.* **2001,** *3*, 1339.
194. Clayden, J.; Tchabanenko, K.; Yasin, S. A.; Turnbull, M. D. *Synlett* **2001**, 302.
195. Clayden, J.; Tchabanenko, K. *J. Chem. Soc., Chem. Commun.* **2000**, 317.
196. Ahmed, A.; Bragg, R. A.; Clayden, J.; Tchabanenko, K. *Tetrahedron Lett.* **2001,** *42*, 3407.
197. Bragg, R. A.; Clayden, J.; Bladon, M.; Ichihara, O. *Tetrahedron Lett.* **2001,** *42*, 3411.
198. Metallinos, C.; Nerdinger, S.; Snieckus, V. *Org. Lett.* **1999,** *1*, 1183.
199. Ahmed, A.; Clayden, J.; Rowley, M. *Synlett* **1999**, 1954.
200. Clayden, J.; Purewal, S.; Helliwell, M.; Mantell, S. J. *Angew. Chem., Int. Ed. Engl.* in press.
201. Clayden, J.; Kenworthy, M. N., *Org. Lett.*, in press.
202. Bragg, R. A.; Clayden, J. *Tetrahedron Lett.* **1999,** *40*, 8323.
203. Bragg, R. A.; Clayden, J. *Tetrahedron Lett.* **1999,** *40*, 8327.
204. Clayden, J.; Menet, C. J.; Mansfield, D. J. *J. Chem. Soc., Chem. Commun.* **2002**, 38.

CHAPTER 8

Organolithium Rearrangements

This chapter will briefly detail a small range of common rearrangements of which an organolithium is the starting material or product. The Shapiro reaction and Brook rearrangements (sections 8.1 and 8.2) are methods for generating organolithiums; the Wittig rearrangements (sections 8.3 and 8.4) are rearrangement reactions with organolithium starting materials. Section 7.2 also dealt with rearrangements involving intramolecular attack of an organolithium on π systems.

8.1 Shapiro Reaction[1-3]

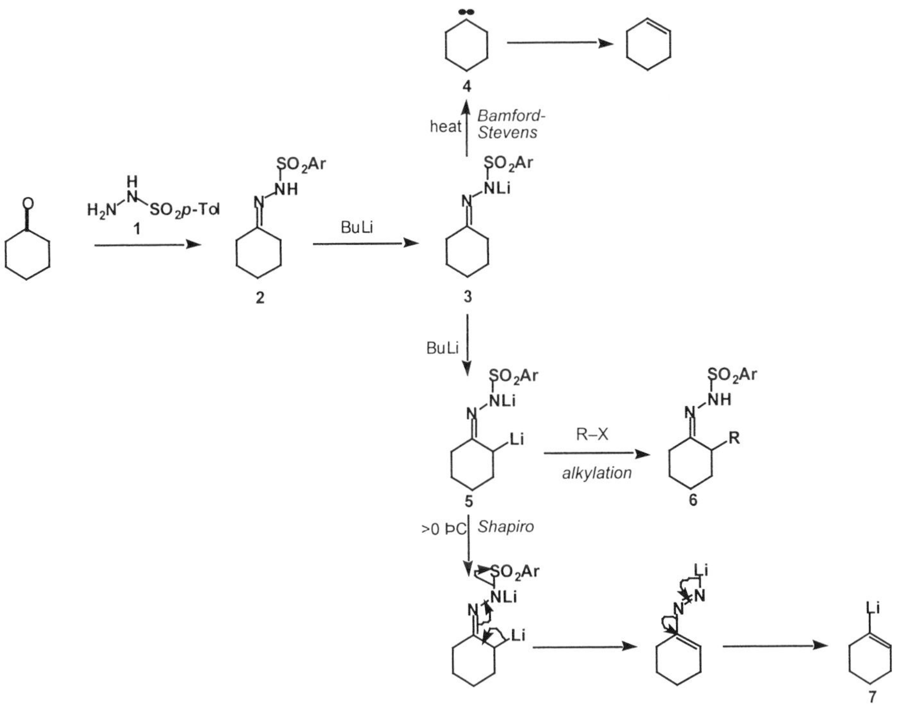

The decomposition of the dianions of arenesulfonylhydrazones, known as the Shapiro reaction,[4] is one of the most reliable ways of making vinyllithium reagents. In the earliest, and still widely used, version of the reaction,[1] a ketone is condensed with *p*-toluenesulfonylhydrazine **1** to yield a crystalline hydrazone such as **2**. Deprotonation of **2** with base (usually BuLi) gives a monoanion **3**, which, if heated, decomposes to a carbene **4** in a reaction known as the aprotic Bamford-Stevens reaction.[5] Alternatively, a second deprotonation with BuLi leads to **5** which collapses between 0 and 25 °C, losing sulfinate and N_2 to yield the vinyllithium **7**. The azaenolate formed by the second deprotonation may be C-alkylated at low temperature (**6**), a useful reaction which increases the structural variety of vinyllithiums which can be made by a subsequent deprotonation.

Both alkylation and Shapiro reactions of unsymmetrical ketones **8** have the potential to form two regioisomeric products. Both depend on the stereochemistry of the hydrazone **9** and on the regiochemistry of the deprotonation. In the presence of TMEDA, deprotonation of either *E*- or *Z*-**9** may occur to give the less basic, less substituted organolithium leading to **10** or **11**,[2] but usually (and always in ether or hydrocarbon solvents) the second deprotonation occurs *syn* to the *N*-sulfonyl substituent due to the kinetic activating effect of the lithiosulfonamide group,[6] and isomers *E*-**9** and *Z*-**9** give Shapiro products **11** and **14** respectively.[2]

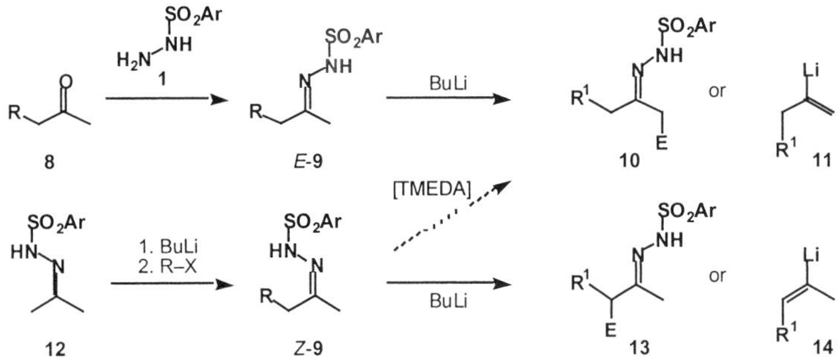

Geometrical stereoselectivity can often be achieved in the condensation of unsymmetrical ketones **8** with tosylhydrazine **1**,[2] and this feature means Shapiro reactions direct from an unsymmetrical ketone **8** via *E*-**9** lead to the less substituted vinyllithium **11**. On the other hand, a sequential alkylation–Shapiro sequence from a starting symmetrical hydrazone **12** will reliably form the more substituted vinyllithium **14** via *Z*-**9** Retention of *Z* stereochemistry in *Z*-**9** is dependent on its re-use almost immediately; on standing, for example, *Z*-**9** (R = vinyl) equilibrates to an 85:15 ratio *E*:*Z*-**9**.[7,8]

With *p*-toluenesulfonylhydrazones, yields of the vinyllithiums can be compromised by their competing protonation by solvent at the temperature required for the decomposition of the anion.[6] Furthermore, *p*-toluenesulfonylhydrazones are partially ortholithiated by BuLi.[9] *p*-Toluenesulfonylhydrazones are therefore suitable only for the direct formation of alkenes by protonation of the vinyllithium.

Usable vinyllithiums are obtained if (a) ortholithiation-resistant hydrazones derived from triisopropylphenylsulfonylhydrazine **15** are used,[9] and (b) if the Shapiro reaction is carried out in a TMEDA–hexane mixture, which avoids protonation of the vinyllithium.[10] For example, **16** forms stereoselectively from **15**, with steric hindrance directing formation of the *E* hydrazone. The dianion of **16** decomposes at 0 °C to yield the vinyllithium **17** which reacts with electrophiles to yield products such as **18**.[11]

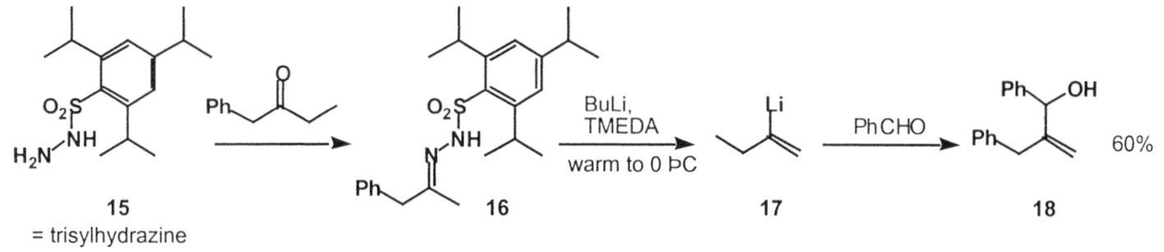

15 = trisylhydrazine

When the second hydrazone deprotonation takes place at a substituted position, Shapiro reaction leads to *Z*-vinyllithiums with high stereoselectivity, and alkylation provides a reliable stereocontrolled route to trisubstituted alkenes.[12] For example, in Corey's synthesis of (±)-ovalicin,[13] the key step was the Shapiro reaction of **19**. Regioselectivity was ensured by sequential alkylation and Shapiro reaction, both taking place *syn* to the *N*-sulfonyl group. The geometrically stable *Z* vinyllithium **20** reacts (as expected) with the ketone **21** with retention of configuration to give the *E* product **22**. Similar methods have been used for the synthesis of *E* vinyl iodides for coupling chemistry.[14]

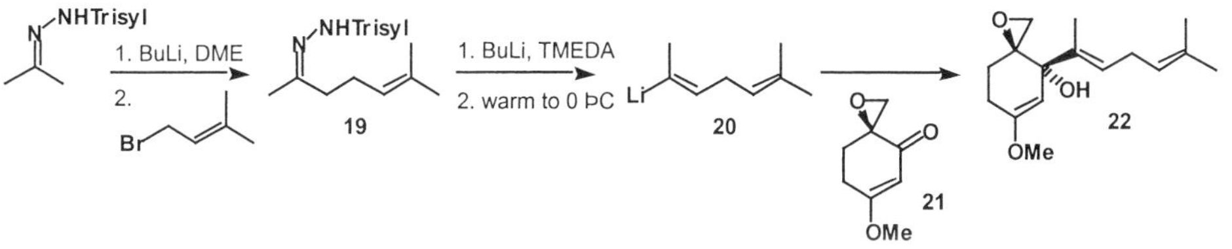

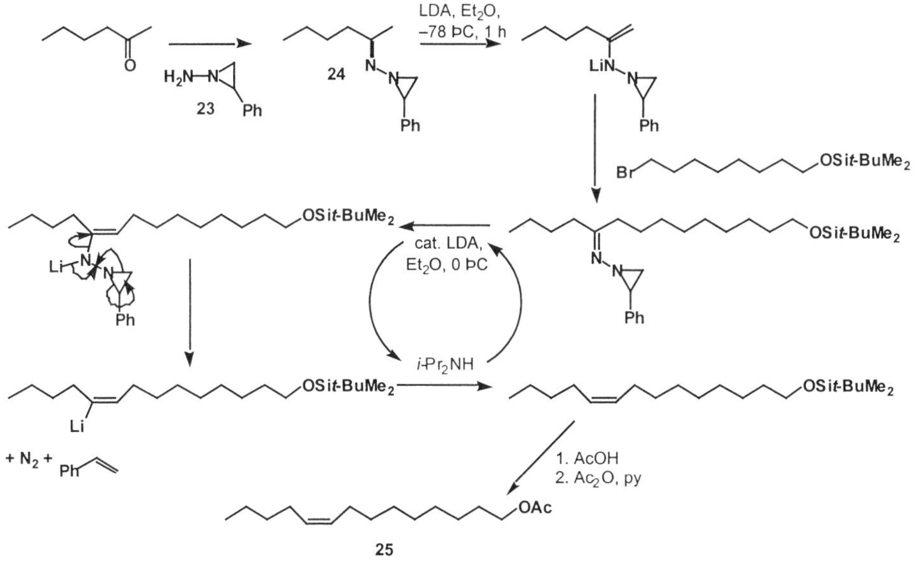

Hydrazones based upon 1-amino-2-phenylaziridine **23** have also been described. Use of excess base leads to the usual Shapiro reaction;[15] however, a unique feature of hydrazones like **24** is that they form alkenes in the presence of only catalytic amounts of LDA.[16] For example, a simple alkylation–Shapiro sequence leads from **24** to **25**, the sex pheromone of the summer fruit tortrix moth (*Adoxophes orana*). Regioselectivity is assured by the *syn*-directing effect of the aziridinyl nitrogen.[17]

The Shapiro reaction has been used as a method for the initiation of anionic cyclisations, as described in section 7.2. Hydrazones derived from **23** are particularly valuable in this area, not only as the source of vinyllithiums but also as carbene equivalents. Nucleophilic addition to **26** promotes the collapse of the product lithioamine to give the organolithium **27**.

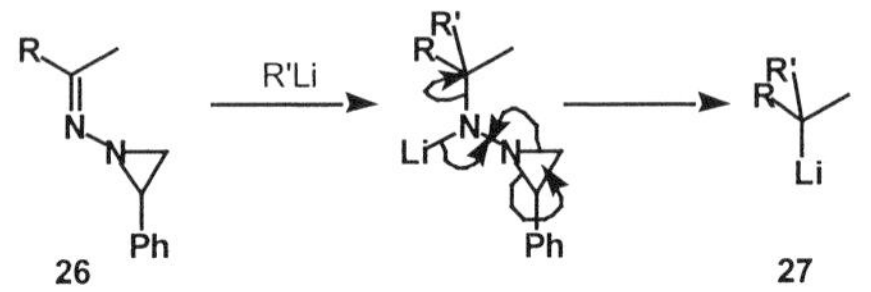

Kim demonstrated the power of this method in a number of spectacular cascade cyclopentane syntheses, such as **29** from **28**.[18]

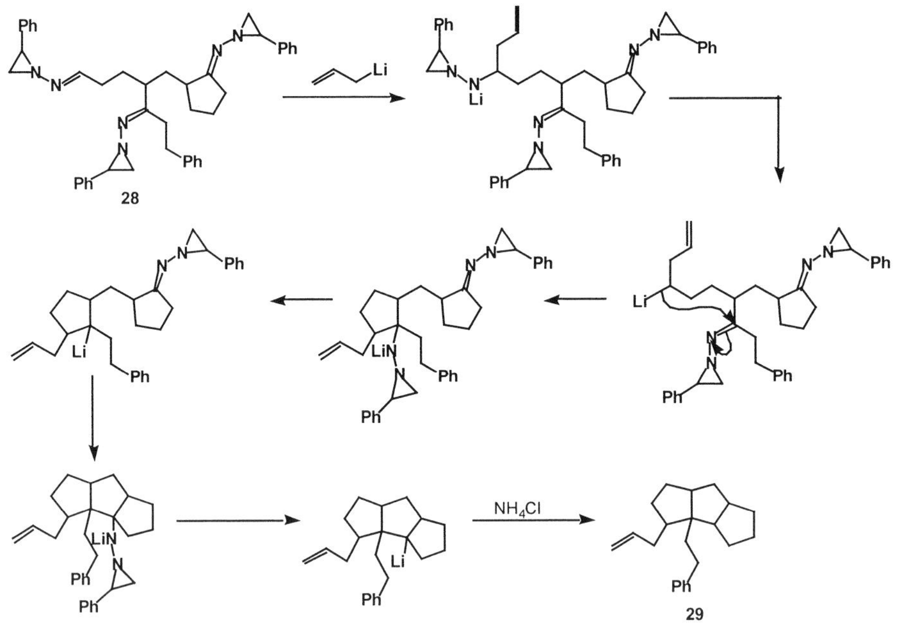

69%, 3:1 ratio of stereoisomers

8.2 Brook Rearrangements[19-21]

Brook rearrangements[19] can be loosely defined as anionic migrations of silicon from carbon to oxygen, converting an oxyanion (typically a lithium alkoxide) **30** or **32** into a carbanion (an organolithium) **31** or **33**. The reverse process – migration of silicon from oxygen to carbon[22]

– is commonly known as the retro-Brook rearrangement.[23] Although Brook's original studies[24-27] encompassed [1,2]-rearrangements, numerous examples of silicon [1,3]- and [1,4]-migrations are now known,[28] and these too fall into the general category of Brook rearrangements.

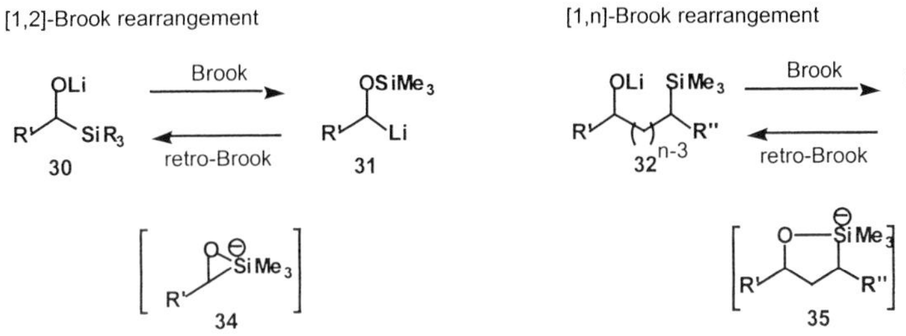

[1,2]-Brook rearrangement

[1,n]-Brook rearrangement

Brook and retro-Brook rearrangements are generally considered to proceed intramolecularly via pentacoordinate silicon-containing intermediates such as **34** and **35**.[29-32] The intramolecularity even of a [1,5]-Brook rearrangement is demonstrated by the fact that *cis*-**36** rearranges to **37**, while *trans*-**36** does not rearrange.

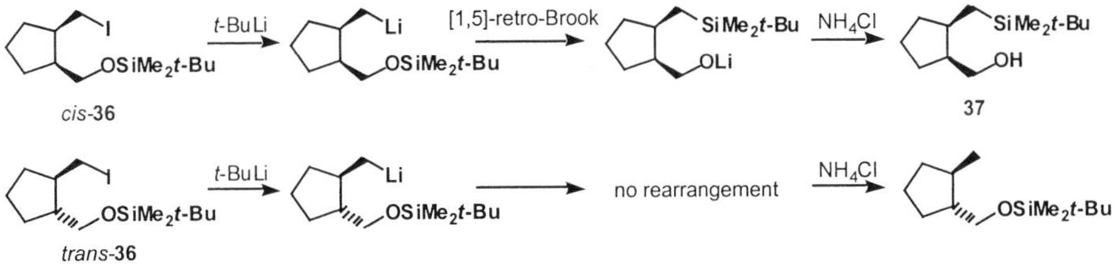

The stereospecificity of the [1,2]-Brook rearrangement depends on whether a benzylic or non-benzylic organolithium intermediate is involved. Brook rearrangements proceed with *retention* when unstabilised organolithiums intervene,[29,33,34] but with *inversion* via stabilised benzylic organolithiums.[35] Enantiomerically pure **38**, for example, gives **39** with 97% ee on transmetallation. Brook rearrangements of **40**, which proceed without cyclopropane ring opening, show that no radicals intervene in this reaction, in contrast with the [1,2]-Wittig rearrangement.[29]

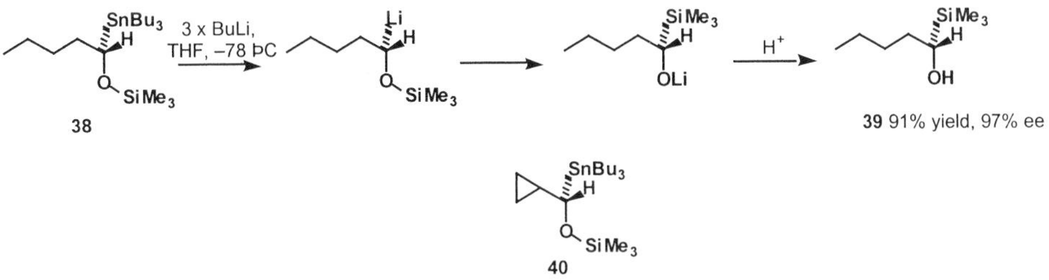

39 91% yield, 97% ee

Brook rearrangements may be carried out with either catalytic or stoichiometric base. With catalytic base, the reaction can be considered an equilibrium between **41** and **42**. The strength of the Si–O bond (about 500-520 kJ mol^{-1}) compared with the Si–C bond (about 310-350 kJ mol^{-1}) means that, provided the anion **33** forms reasonably rapidly (some degree of stabilisation is required), Brook rearrangement (alkoxide formation) is favoured over retro-Brook. Organolithiums **33** may be present as intermediates in the catalytic Brook rearrangement, but their reactivity cannot be exploited under these conditions.

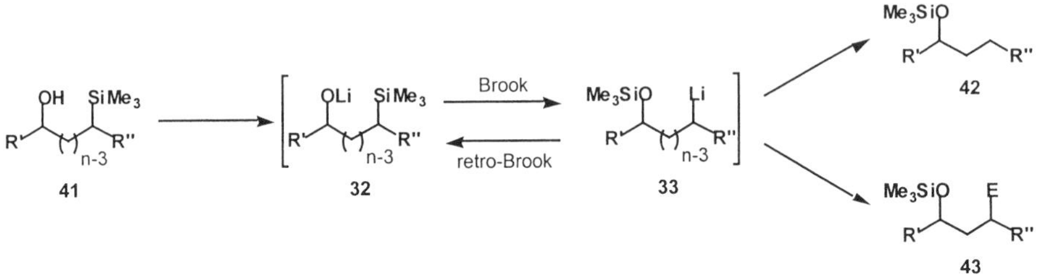

With stoichiometric base, the reaction is an equilibrium between the alkoxide **32** and the "carbanion" (or organometallic) **33**; the organolithium **33** may be formed quantitatively and quenched with electrophiles to give **43**. Stabilisation of the "carbanion" by electron-withdrawing (PhS) or conjugating (vinyl, or phenyl) groups favours Brook rearrangement; retro-Brook rearrangement is favoured by stabilisation of the alkoxide – lithium counterions, for example, give the advantage to the retro-Brook (formation of the organolithium), but this effect can be overcome by the addition of lithium-solvating ligands. Overall, formation of conjugated or α-phenylthio organolithiums by retro-Brook rearrangement in the presence of THF or TMEDA turns out to be a valuable method.

8.2.1 [1,2]-Brook Rearrangements

A particularly effective method for the synthesis of silyloxy-substituted allyllithiums entails the addition of a vinyllithium **44** to an acylsilane **45**.[36,37] The alkoxide **46** thus generated undergoes [1,2]-Brook rearrangement, generating the stabilised allyllithium **47** in the expected *Z* configuration. Alkylation γ to oxygen (**47** is a homoenolate equivalent) generates a silyl enol ether **48**. Careful control is required in this reaction to avoid by-products resulting from the readdition of the product to the starting acylsilane. Alkynyllithiums undergo similar reactions to generate allenyllithiums;[37] functionalised alkynyllithiums generate silyloxycumulenes.[38,39]

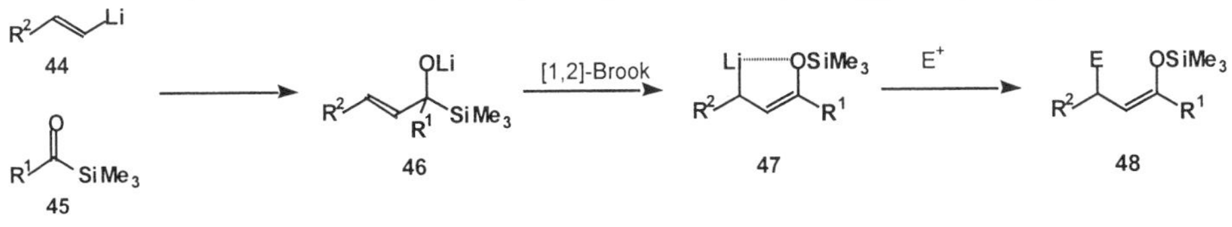

Phenyllithium (and lithiophosphites) also bring with them sufficient stabilisation of the organolithium for Brook rearrangement to occur readily.[40] With acylsilane **49**, intramolecular Michael addition leads to cyclic structures **50**.

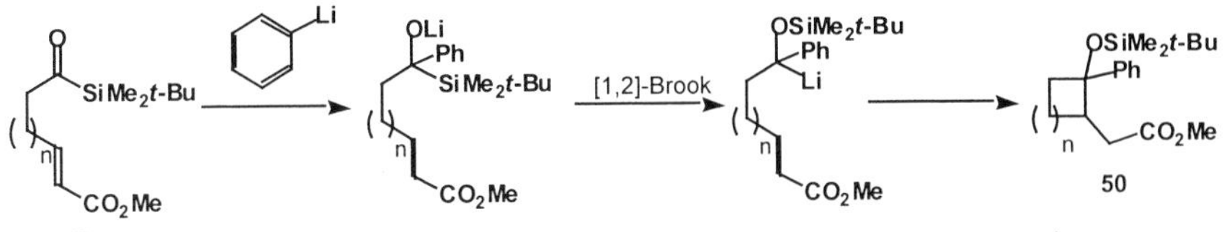

An alternative disconnection of the alkoxide requires the addition of a silyllithium reagent to an enone. Addition of stoichiometric base to the alcohol **51** produces an alkoxide **52**, but no evidence of Brook rearrangement to generate **53** was found on protonation of the product. However, alkoxide **52** must exist in equilibrium with some of the organolithium **53**, since alkylation with a soft electrophile (MeI) produced **54**.[41] The equilibrium concentration of the organolithium **53** is lessened in this case by the impossibility of O–Li coordination.

Nucleophiles other than organolithiums can be employed if the acylsilane component carries a sufficiently[42] anion-stabilising group, such as an α,β-unsaturated acylsilane bearing a β-silyl or β-phenylsulfanyl group. For example, **55** reacts with the lithium enolate **56** to give **57**, which then cyclises to **58**.[43,44] This synthesis of cyclopentanones has been applied to the synthesis of some biologically active marine prostanoids.[45]

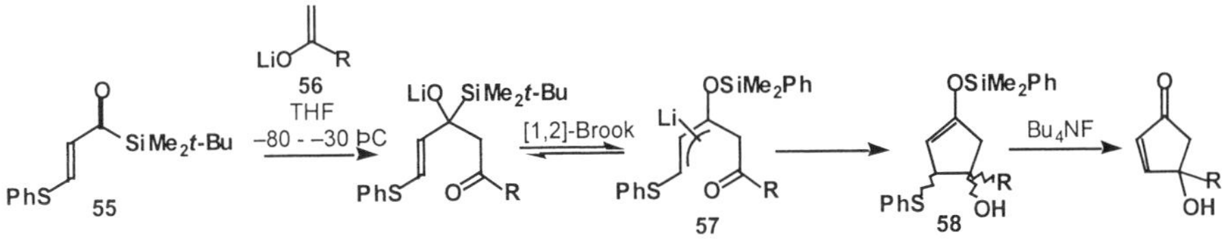

If both components are based upon α,β-unsaturated carbonyl systems, seven-membered rings such as **59** can be formed,[46-49] another reaction applicable in synthesis, this time to the cyanthin ring system.[50]

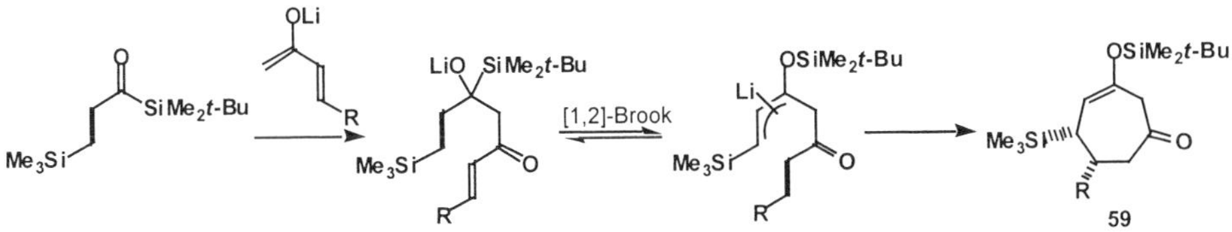

8.2.2 [1,3]-Brook Rearrangements

[1,3]-Brook rearrangements are in competition with Peterson elimination, making their use to generate organolithiums rare. One example is the double alkylation of the lithiated silyldichloromethane **60**, which reacts with aldehydes in THF to generate **61**. HMPA promotes the rearrangement to **62**, allowing a different electrophile to be added in the second organolithium quench step, giving such compounds as **63**.[51,52] The organolithiums in more remote ([1,3] and [1,4]) Brook rearrangements need considerably greater stabilisation for the forward reaction to be preferred than in [1,2]-rearrangements.

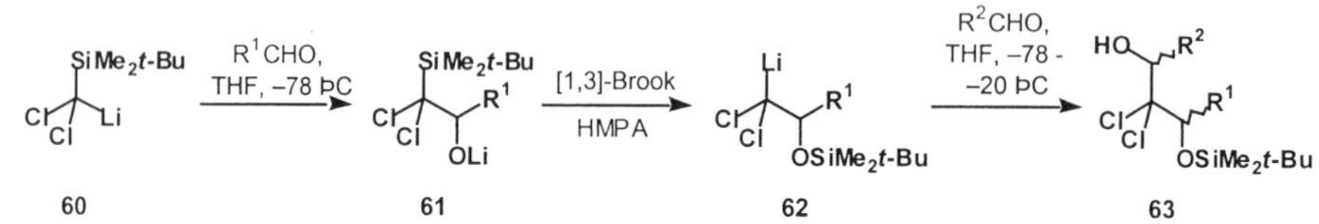

8.2.3 [1,4]-Brook Rearrangements

[1,4]-Brook rearrangements are precipitated by the addition of an α-lithiated silane to an epoxide. The same nucleophile **60,** for example, gives **65** on addition to **64**, with the rate of Brook rearrangement to **66** again under the control of HMPA.[51,52]

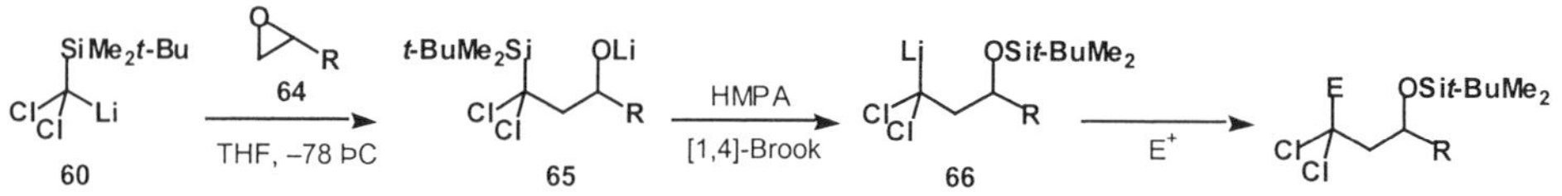

Being able to use HMPA to promote [1,4]-Brook rearrangements is at the heart of the versatility of **67** as an equivalent for the synthon **68**.[53-55] The first alkylation is followed by addition of the second electrophile in the presence of HMPA, allowing the construction of molecules such as **69**, a spiroketal component of the spongistatin antitumour agents.[54]

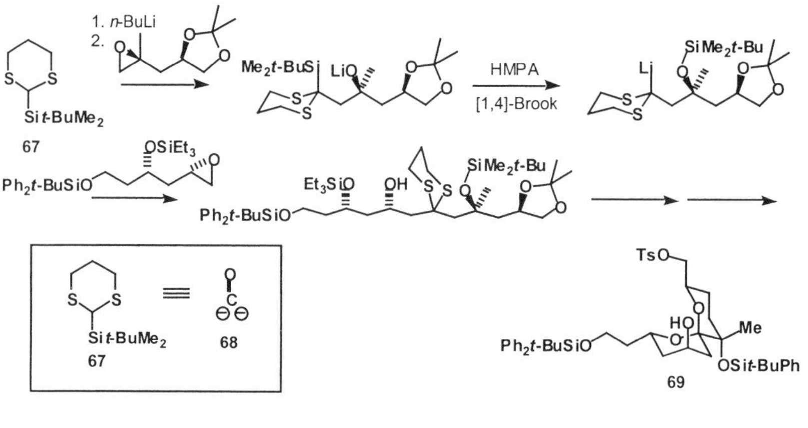

Related compound **70** gives **72** and hence **73** on addition to **71** in a useful cyclopentane synthesis.[56,57] The mannitol-derived bis-epoxide **72** led to a mixture of **73** and **74**, and **74** was used in the synthesis of the carbasugar *epi*-validatol **75**.[58]

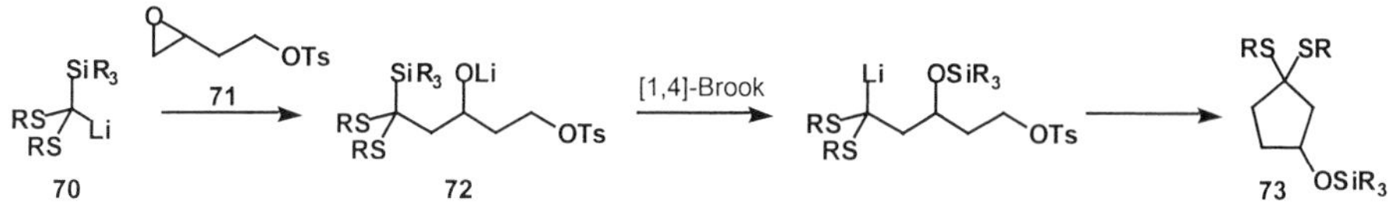

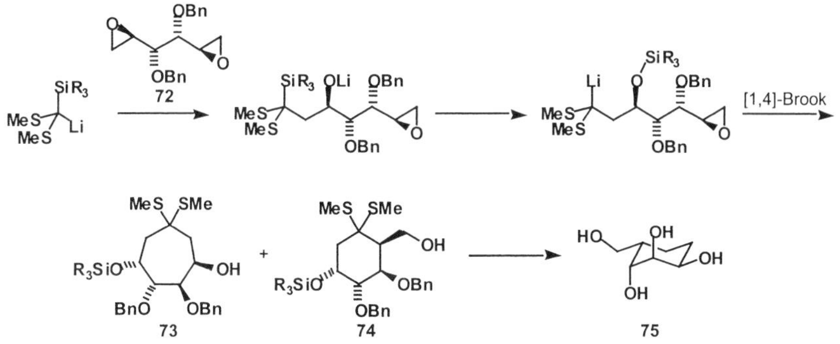

A [1,4]-Brook rearrangement to give an aryllithium product occurs with the arenetricarbonyl chromium complex **76**. The chirality of the starting material controls the diastereoselective addition of methyllithium, giving **77** as a single diastereoisomer.[59]

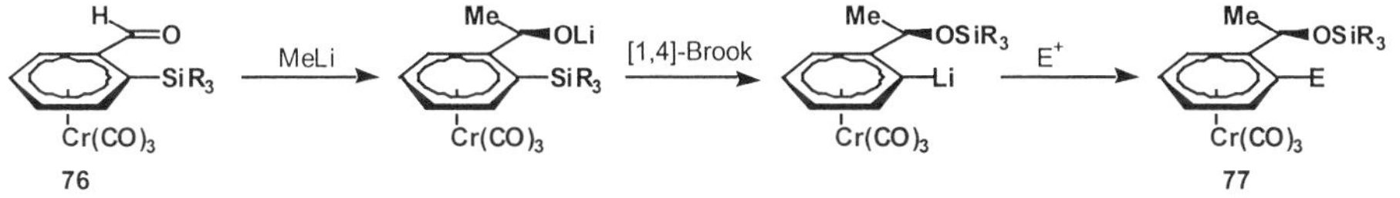

8.2.4 [1,4]-Retro-Brook rearrangements

[1,4]-Retro-Brook rearrangement turns out to be a relatively common (and not always expected) fate of organolithiums containing silyl ether protecting groups and lacking the stabilisation required to favour the forward Brook rearrangement. So, for example, in contrast with the reaction of **76** above, halogen-metal exchange of **78** gives **79**,[60] and attempted superbase lithiation of **80** gave significant amounts of **81**.[61]

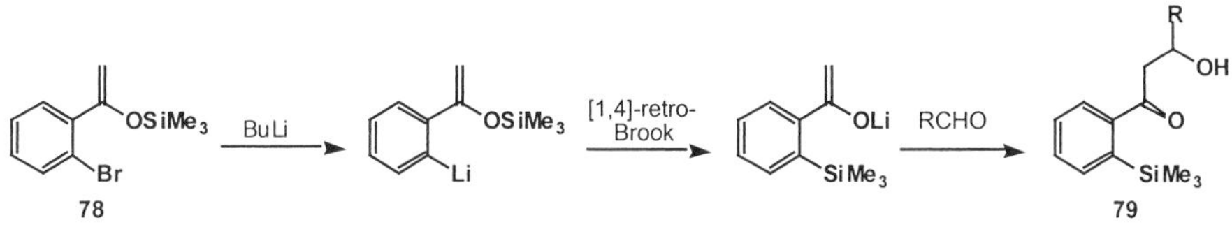

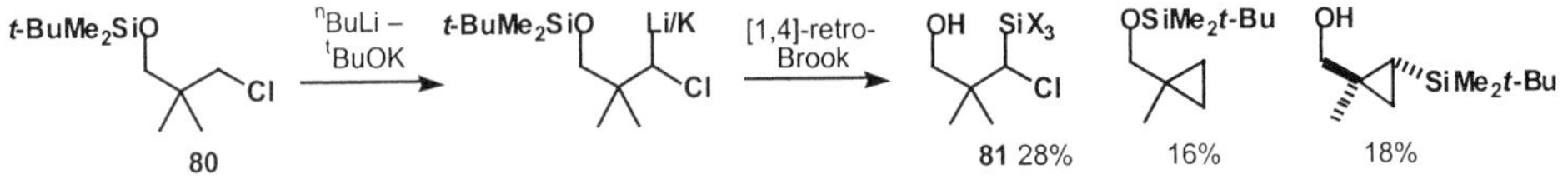

Retro-Brook rearrangements onto vinyl and cyclopropyllithiums **82** and **84** have been used to make vinylsilanes[62] and cyclopropylsilanes[28] **83** and **85**.

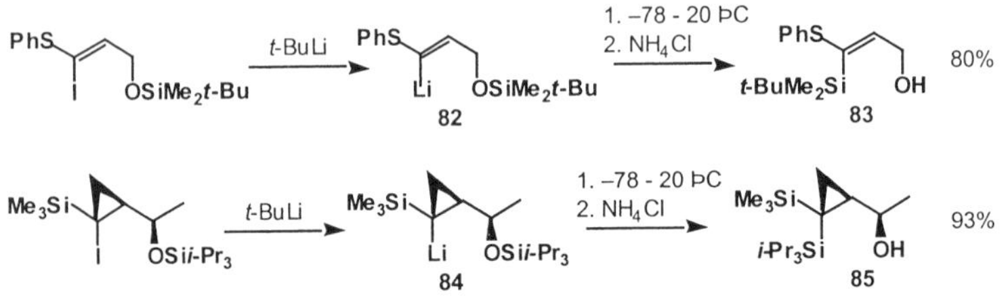

Benzyl and allylsilanes **87** and **89** can be formed by reductive lithiation and [1,4]-retro-Brook rearrangement of benzyl and allylsulfides **86** and **88** containing silyl ethers.[63] The reaction, by virtue of the intermediacy of a configurationally unstable benzyl or allyllithium, is stereoselective but not stereospecific. In the latter case, intramolecularity assures the regioselective formation of **89**.

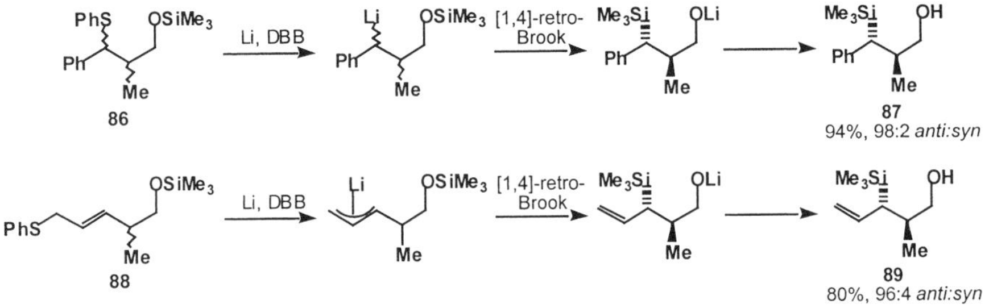

8.3 [1,2]-Wittig Rearrangements[64-66]

The rearrangement of α-alkoxyorganolithiums to alkoxides by deprotonation of an ether and 1,2-migration of an alkyl group from oxygen to carbon was first published by Wittig in 1942,[67] and is now known as the [1,2]-Wittig rearrangement. Vinylogous variants, such as the [2,3]-Wittig rearrangement, have subsequently come to light: some are dealt with in section 8.4.

8.3.1 Mechanism and scope

A general [1,2]-Wittig rearrangement of ether **90** is shown below; mechanistic studies[65,68-70] of the reaction are consistent with formation of a radical pair **94** which rearranges by lithium migration to oxygen to form a product radical pair **95** which then recombines to yield **92**.

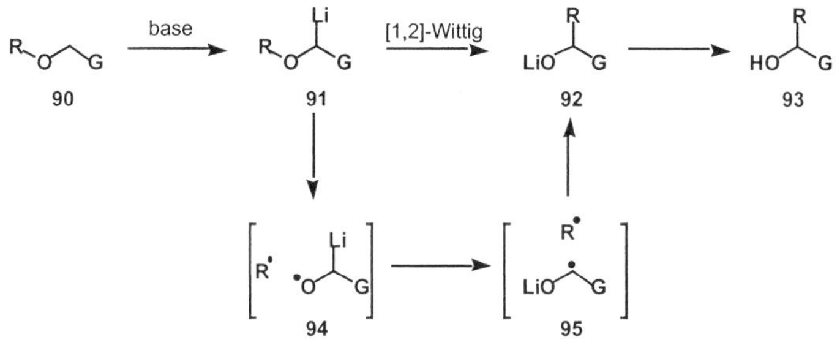

For this reason best results from [1,2]-Wittig rearrangements are obtained when the migrating group R and the organolithium substituent G are radical stabilising.[65] If G is vinyl or phenyl then the migrating group may be benzylic, secondary or primary (but not methyl). As G gets less radical stabilising, so the scope for variation of R decreases. For example, benzyloxymethyllithium **96** (G = H) does not undergo [1,2]-Wittig rearrangement,[71] although its homologue **97** does.[72]

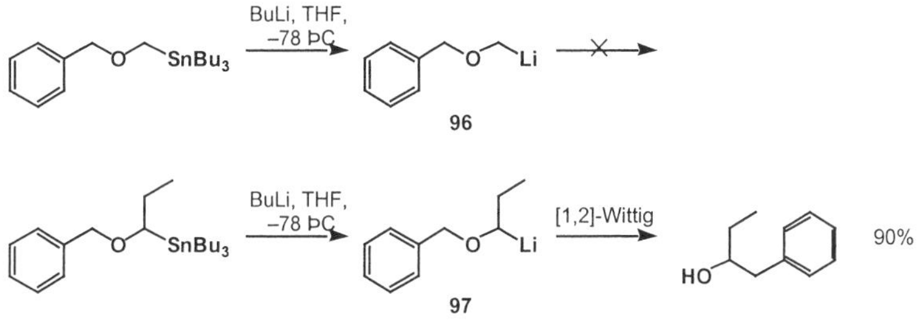

Successful [1,2]-Wittig rearrangements can be summarised in a chart thus:[72]

[1,2]-Wittig rearrangements with...		organolithium substituent G =			
		H	*n*-alkyl	vinyl	Ph
migrating group R =	Me	×	×	×	×
	Et	×$^{N}_{U}$	×	D4	D4
	i-Pr	×	×	D4	D4
	Bn	×	D4	D4	D4

Certain side reactions plague [1,2]-Wittig rearrangements; the most common are elimination, such as the formation of **98**,[73,74] and (with allyl ethers) competing [1,4]-rearrangement (mechanism unknown) to give compounds such as **99**.[75] A type of [1,4]-Wittig rearrangement has been observed on addition of *t*-BuLi to the benzyl styryl ether **100**.[76,77]

t-BuMe$_2$SiO(CH$_2$)$_8$ ... O ... Ph —BuLi→ t-BuMe$_2$SiO ... Li ... Ph → t-BuMe$_2$SiO ... **98** 90%

—i-PrLi→ ... Li —"[1,4]-Wittig"→ ... CHO **99** 29% + HO ... 23% (product of [1,2]-Wittig)

100 —t-BuLi→ t-Bu ... Li ... O ... Ph —"[1,4]-Wittig"→ t-Bu ... Ph ... OH

8.3.2 Stereospecificity

There are two aspects to the stereospecificity of the [1,2]-Wittig rearrangement: the stereochemical fate of the starting organolithium centre and the stereochemical fate of the migrating group. Schöllkopf showed[78,79] that the chiral benzylic groups of **101** and **102** migrate with predominant retention of stereochemistry, though without complete stereospecificity.

101 —BuLi, THF→ —[1,2]-Wittig→ —[O]→ 80% ee (90:10 retention:inversion)

102 —BuLi, THF→ —[1,2]-Wittig→ —[O]→ 20% ee (60:40 retention:inversion)

By contrast, [1,2]-Wittig rearrangement proceeds with *inversion* at the organolithium centre:[80-82] rearrangement of the equatorial organolithium **103**, for example, gives the alcohol **104**,[80] and rearrangement of axial organolithium **105** gives axial alcohol **106**.[81,82]

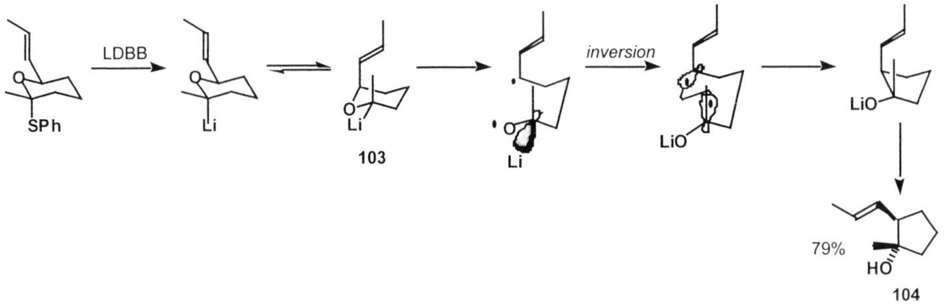

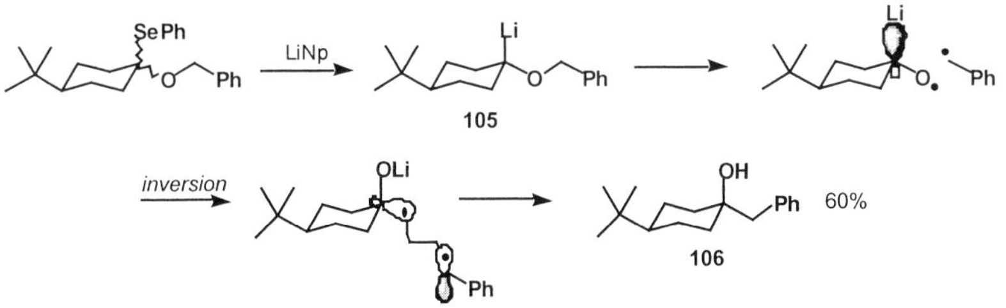

LDBB
SPh
Li
103
Li
inversion
LiO
LiO
79%
HO
104

SePh
Ph
LiNp
Li
105
Ph
Li
Ph
inversion
OLi
Ph
OH
Ph 60%
106

In the acyclic series, stannane (*R*)-**107** of 88% ee rearranges, on transmetallation with alkyllithiums, to the alcohol (*R*)-**108** of 42% ee, a reaction demonstrating 74% invertive stereospecificity.[72,83]

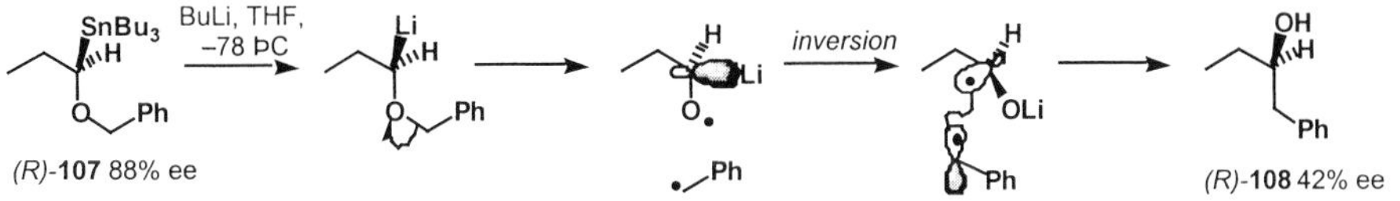

$SnBu_3$
BuLi, THF, –78 °C
(*R*)-**107** 88% ee
inversion
OLi
OH
(*R*)-**108** 42% ee

The stereochemical fate of the enantiomerically pure diastereoisomers **109a** and **109b** confirms this view of the rearrangement.[72] Recombination is evidently slower for the radical pair arising from **109a** than for the pair arising from **109b**, giving time for **112** to racemise, yielding a greater proportion of **110c**.

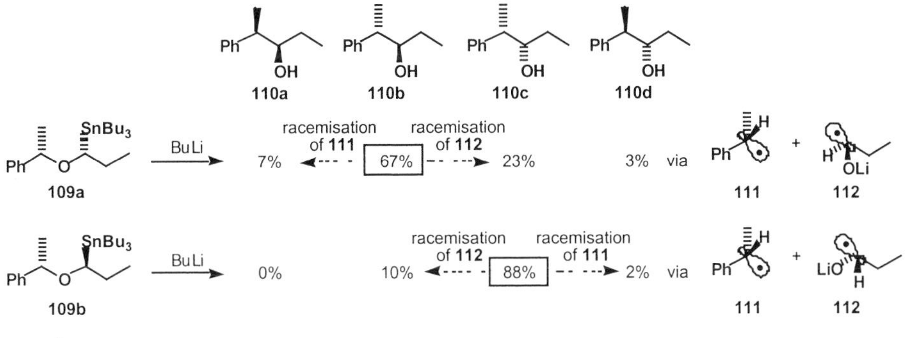

110a **110b** **110c** **110d**

109a BuLi: 7% (racemisation of **111**) ← 67% → 23% (racemisation of **112**); 3% via **111** + **112**

109b BuLi: 0%; 10% (racemisation of **112**) ← 88% → 2% (racemisation of **111**) via **111** + **112**

The first *asymmetric* [1,2]-Wittig rearrangement was not published until 1999, when Tomooka and Nakai showed that deprotonation and rearrangement of **113** using the complex of *t*-BuLi with the chiral bis-oxazoline **114** was able to generate **115** in 60% ee.[84] A similar rearrangement remarkably yielded the tertiary alcohol **116** in 65% ee. Stereochemistry in these rearrangements appears to be independent of the initial stereochemistry of the organolithium, since rearrangement of the racemic deuterated ether *d*-**113** gave deuterated **115** in the same ee as the undeuterated material.

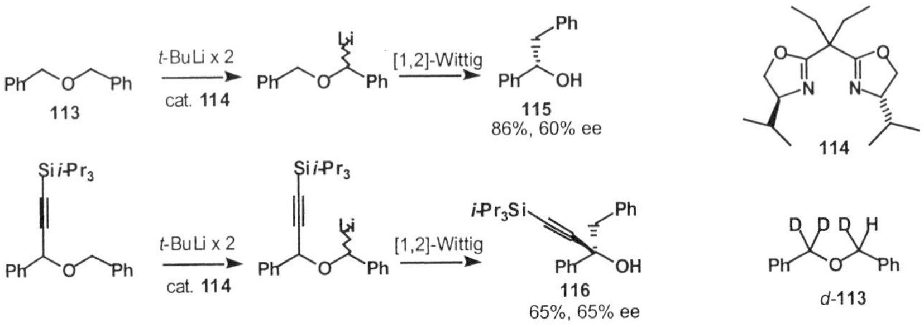

t-BuLi x 2, cat. **114**; [1,2]-Wittig

113 → **115** 86%, 60% ee

114

116 65%, 65% ee

d-**113**

8.3.3 [1,2]-Wittig rearrangements in synthesis

In a general sense, [1,2]-Wittig rearrangements have only a limited application in synthesis because yields and selectivities are frequently moderate at best. For example, the valuable stereoselective conversion of **117** to **118** (which also works with higher homologues) proceeds in only 14% yield.[85,86]

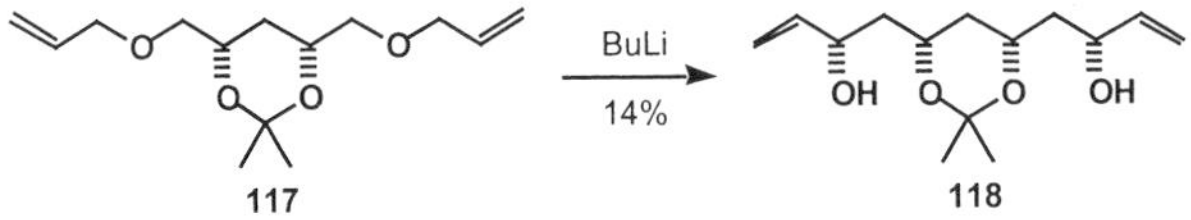

117 —BuLi, 14%→ **118**

Somewhat better yields are obtained in the construction of the taxane skeleton **120** from **119**.[87]

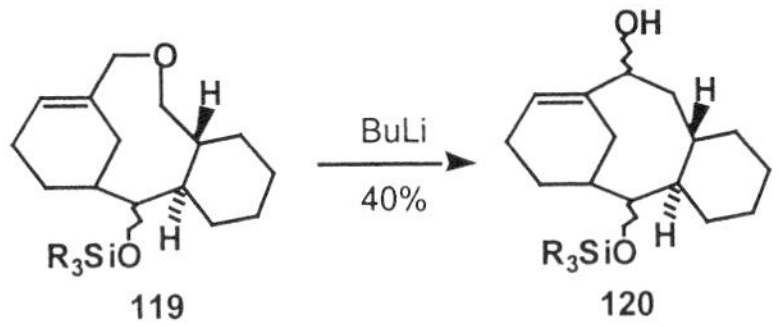

119 —BuLi, 40%→ **120**

Binaphthol ethers **121** undergo diastereoselective rearrangement to give alcohols **122**.[88]

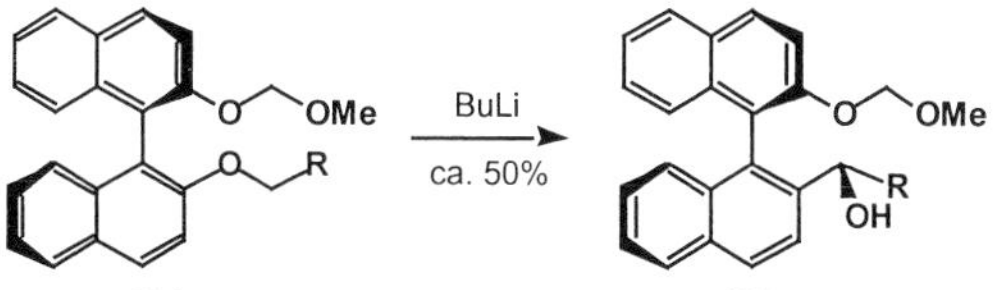

121 122

The stability of the radicals involved ensure that rearrangements of glucosides involving migration of the anomeric centre are particularly effective.[89] For example, **123** gives **124** in 40% yield and as a single diastereoisomer.

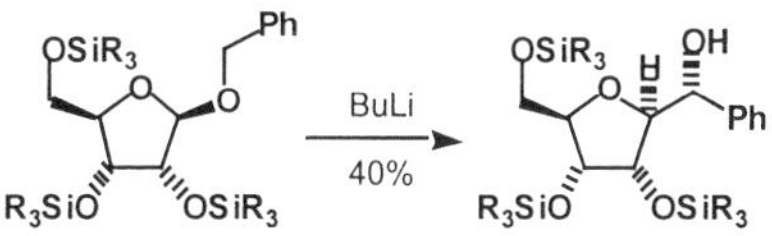

123 124

8.4 [2,3]-Wittig Rearrangements[90-92]

The first [2,3]-Wittig rearrangement, of **125** to give **126**, was observed in 1960 in the context of mechanistic studies on the [1,2]-Wittig rearrangement.[93] [1,2]-Wittig products can often contaminate those of [2,3]-Wittig rearrangements, with the ratio depending on temperature. For example, treatment of **127** with BuLi gives mainly [2,3]-Wittig product **128**, with the proportion of [1,2]-Wittig product increasing at a higher temperature.[94] The mechanisms of the [1,2]-Wittig and [2,3]-Wittig rearrangements are quite different: as described above (section 8.3), [1,2]-Wittig rearrangement takes place via a radical dissociation-recombination mechanism, while the [2,3]-Wittig rearrangement is a suprafacial six-electron pericyclic process. [1,2]-Wittig rearrangement by-products are a problem particularly with tertiary allylic systems.[90]

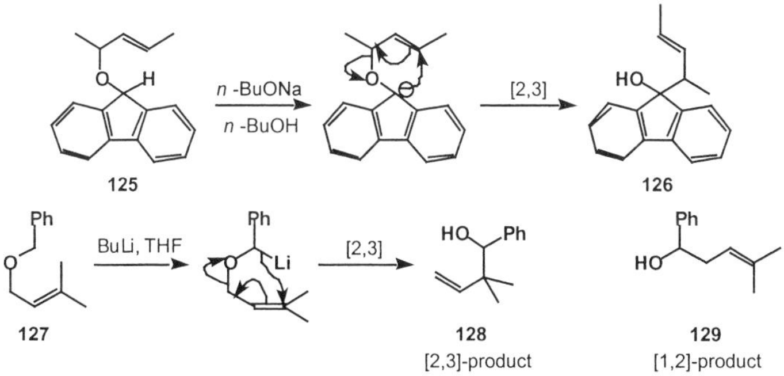

125 126

127 128 [2,3]-product 129 [1,2]-product

–25 ÞC: 7:1
+23 ÞC: 6:1

Reliable conditions for [2,3]-Wittig rearrangements were established by Still[95] and by Nakai[96] around 1980. To avoid competing reactions, the rearrangements must be carried out at low temperatures (–85 to –60 ˚C) and the major limitation on the application of the [2,3]-Wittig rearrangement is therefore the availability of substrates which can be converted to "carbanions" (in the context of this chapter, organolithiums) under these conditions.

In some cases, even reactions which appear perfectly set up for a [1,2]-Wittig rearrangement lead instead to [2,3]-Wittig rearrangement.[97,98] So, for example, **130** gives (especially at very low temperature, which disfavours homolysis to the radical intermediates in the [1,2]-Wittig) **132** by [2,3]-Wittig rearrangement even though a non-aromatic intermediate **131** is formed.[97]

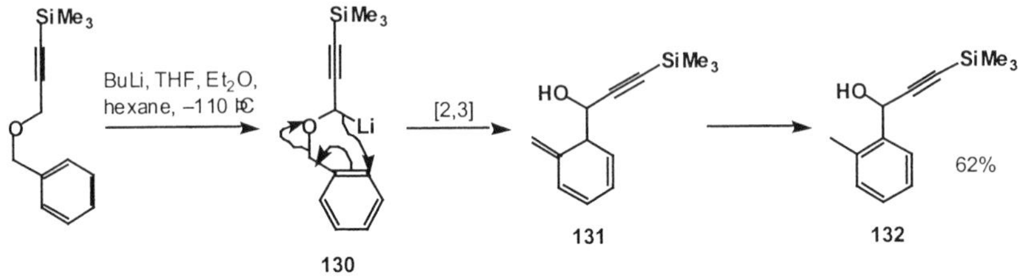

In the naphthalene series, a dearomatised intermediate **134** can itself be isolated from the rearrangement of **133**:[99]

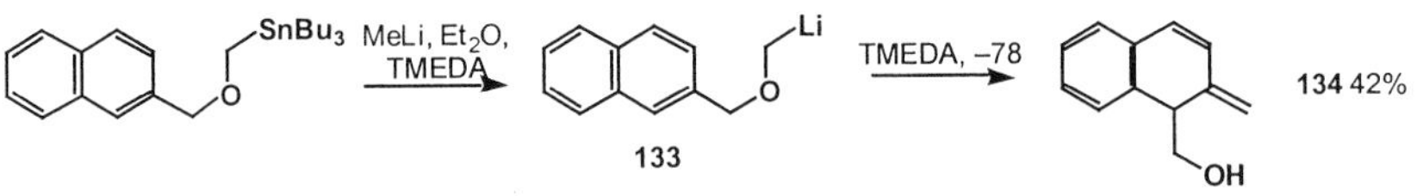

8.4.1 Regioselectivity

The three most important methods for initiating a [2,3]-Wittig rearrangement are (a) deprotonation, (b) tin-lithium exchange and (c) reductive lithiation. Regiochemistry is not an issue when methods (b) and (c) are used with a stannane[95] or a sulfide[100] starting material. **135** and **137** respectively give **136** and **138** cleanly and in good yield.

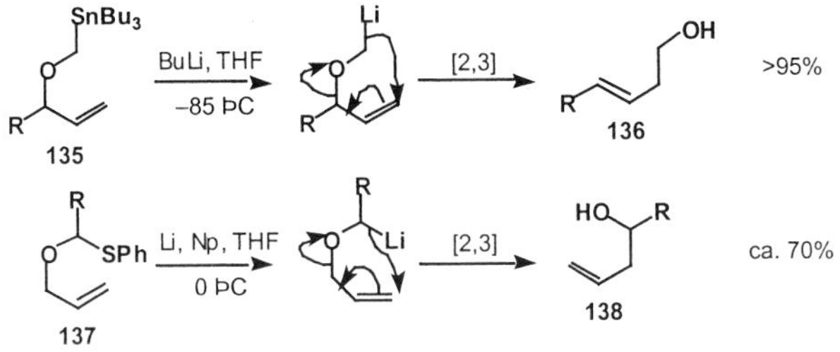

Deprotonation requires some degree of organolithium stabilisation, though too much stabilisation leads to slow rearrangement. Deprotonation works well with allyl benzyl ethers and allyl propargyl ethers (for example **139** → **140**,[101,102] **141** → **142**[103] and **143** → **144**:[94] the propargylic or benzylic position is deprotonated).

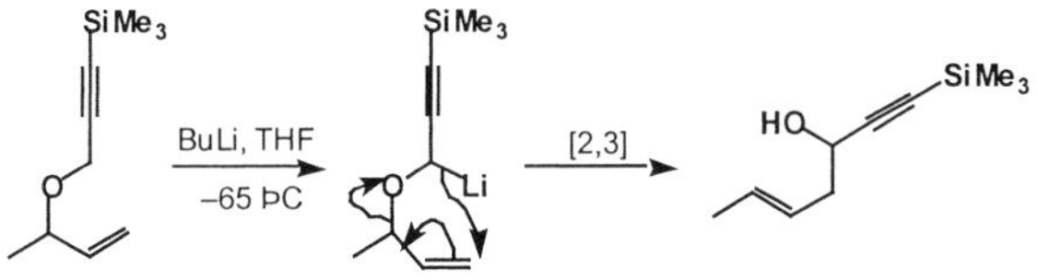

139 140 74%

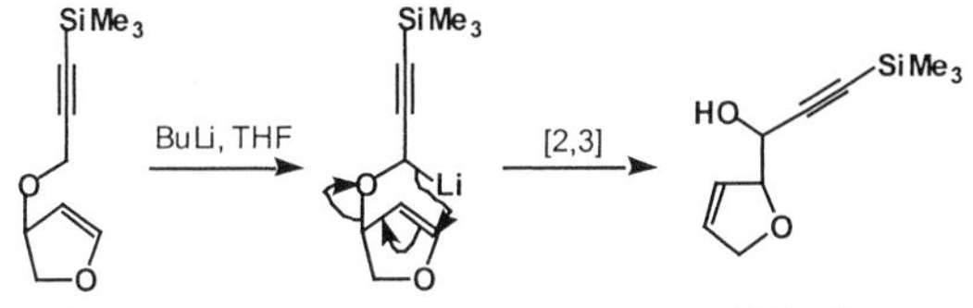

141 142 73%

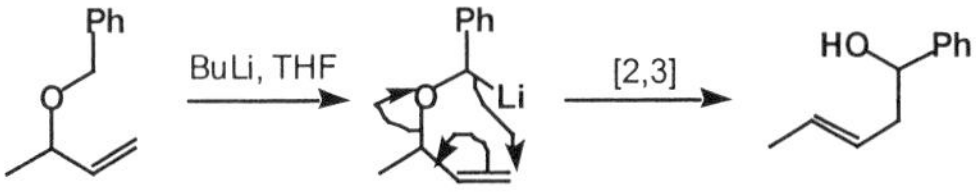

143 144

Regioselectivity may be an issue with unsymmetrical bis-allyl ethers, but provided one allyl group carries an α- or a γ-substituent, lithiation takes place at the less substituted allyl group. For example, **145** gives only **146**.[96] Regioselectivity can also be controlled by introducing a weakly anion stabilising group (MeS[104] or $SiMe_3$[105]) γ to the position to be deprotonated. For example, **147** rearranges selectively to give **148**.

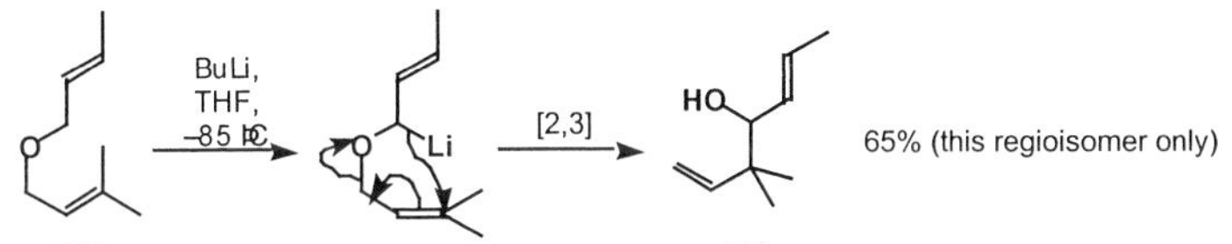

145 146

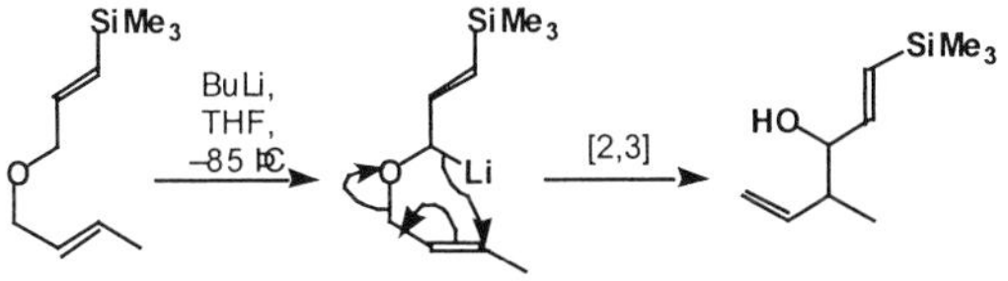

147 148

[2,3]-Wittig rearrangement of cyclic bis-allyl ethers or allyl propargyl ethers is an important method for carbocyclic ring synthesis by ring expansion or contraction, and in cyclic systems, the "rule" that the less substituted allyl group is lithiated breaks down. For example, lithiation of **149** with LiTMP gives **150** by lithiation at the prenyl-type methylene group.[106]

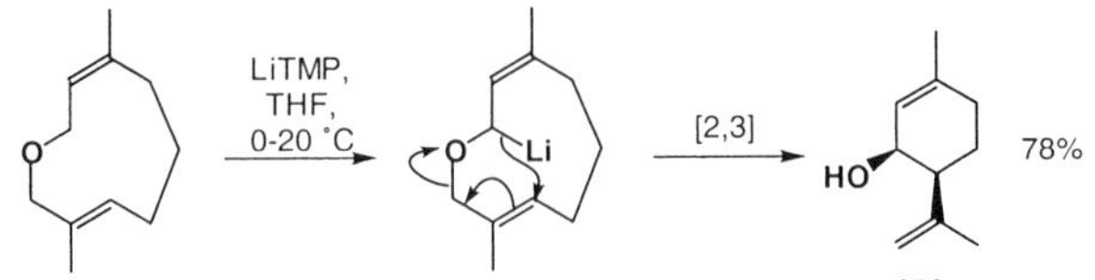

LiTMP, THF, 0-20 °C

[2,3]

78%

149

150

8.4.2 Diastereoselectivity[92]

8.4.2.1 Double bond geometry

The [2,3]-Wittig rearrangement is proposed to proceed via an envelope-like transition state approximating to **151**,[107] in which both R and G (the anion-stabilising group) adopt an *exo* (pseudoequatorial) arrangement. With pseudoequatorial R, the product is bound to contain a *trans* double bond, and *trans*-selectivity is indeed a general feature of the [2,3]-Wittig rearrangement.

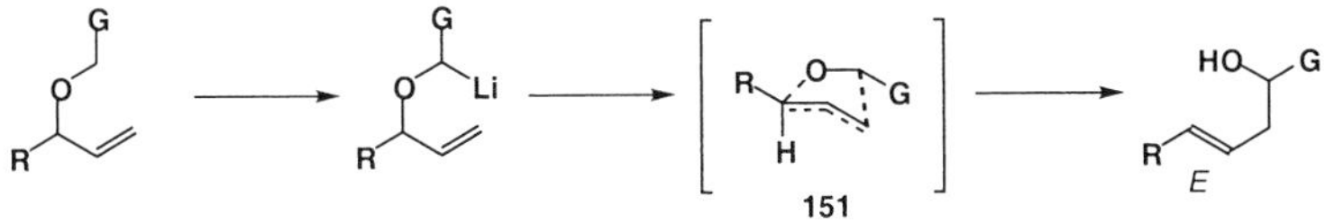

151

Two classes of organolithiums provide exceptions. If the allyl group bears a bulky β-substituent (a Me_3Si group, for example, as in **152**) interactions with the "R" group lead to *cis* selectivity.[108] The geometrical selectivity in the rearrangement of **152** to **153** was exploited in a subsequent pheromone synthesis.[109] The other exceptions are rearrangements where G = H, almost all of which are initiated by tin-lithium exchange of compounds such as **154**.[95] Rearrangement of **155** gives *Z* product **156** (the β-methyl group assists, but is not the sole reason for, the *Z*-selectivity).[110]

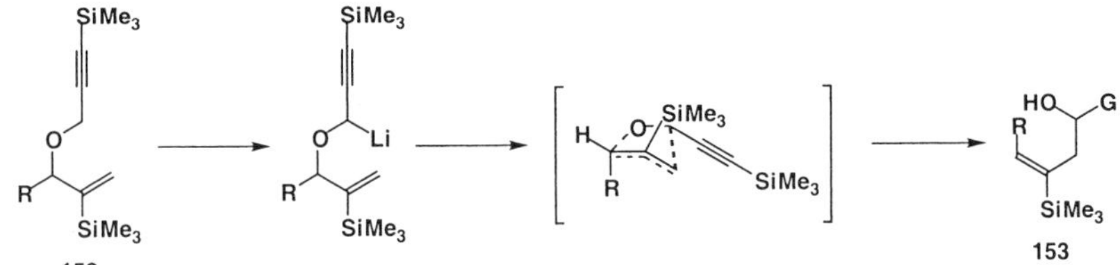

152

153

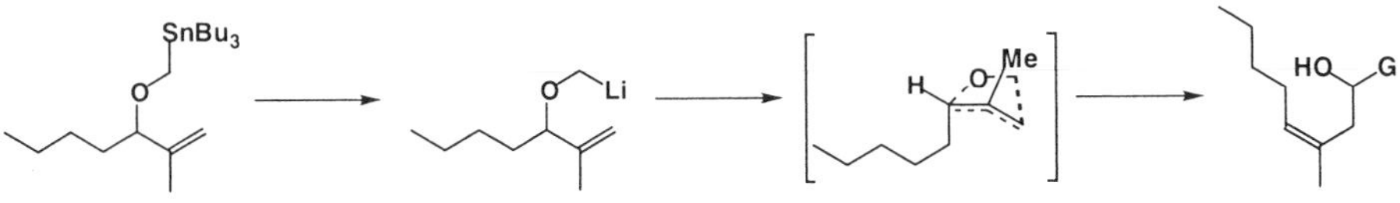

154

155

156

8.4.2.2 Syn/Anti relative stereochemistry

When the allylic system carries a terminal substituent, the [2,3]-Wittig rearrangement has the potential to control *syn/anti* relative stereochemistry. As suggested by the transition state **151**, which diastereoisomer is formed is dependent on the double bond geometry.[107] The degree of stereospecificity of the reaction (i.e. the extent to which double bond geometry translates directly into three dimensional relative stereochemistry) depends on the group G;[90] the best stereospecificity is provided by certain propargyl substituents. *E* and *Z*-**157**, for example, reliably give >99% of *anti*-**158** and *syn*-**158** respectively.[101,102]

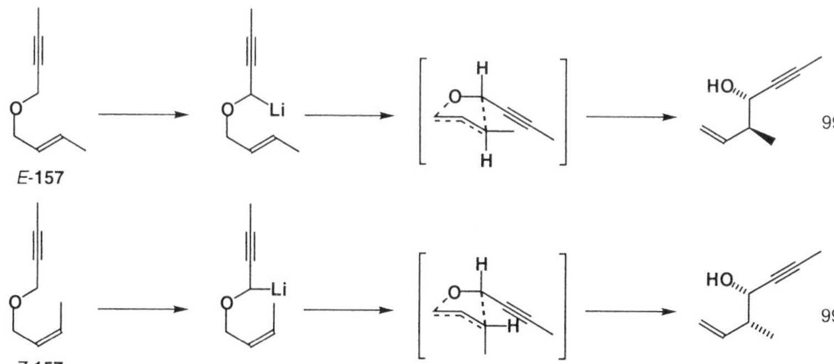

E-157 → 99% *anti*-**158**

Z-157 → 99% *syn*-**158**

But even quite small changes in the starting materials can change the situation completely.[90] For example,**159** gives *anti*-**160** irrespective of the starting enol ether geometry.[111]

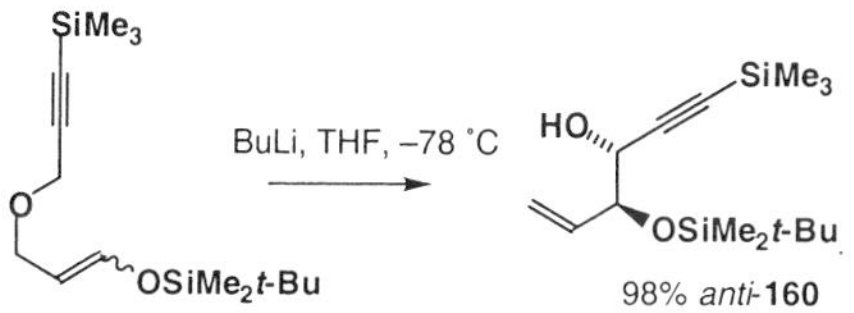

159 —BuLi, THF, –78 °C→ 98% *anti*-**160**, 79% yield

8.4.3 Stereospecificity and enantioselectivity[112]

Absolute stereochemistry may be controlled in [2,3]-Wittig rearrangements in a number of ways: by stereospecific rearrangement of a chiral starting material; by stereoselective rearrangement of a starting material bearing a chiral auxiliary, and by stereospecific rearrangement of a configurationally defined organolithium.

8.4.3.1 Stereospecific rearrangements of chiral allyl ethers

In a cyclic system, the stereochemistry of the product of a [2,3]-Wittig rearrangement may be ensured by the rearrangement's suprafacial nature. [2,3]-Wittig rearrangement of lithiomethyl ethers derived from stannanes provides a valuable method for stereospecific hydroxymethylation. For example, **161** gives **162** stereoselectively,[113,114] and the rearrangement of **163** provides a synthetic precursor **164** to a vitamin D metabolite.[115-118]

t-BuMe$_2$SiO 161 SnBu$_3$ BuLi, −78 °C t-BuMe$_2$SiO Li [2,3] HO t-BuMe$_2$SiO 162

Bu$_3$Sn 163 BuLi, −78 °C Li [2,3] OH 164

In acyclic systems, appropriate choices of allylic geometry and anion stabilising group similarly allow stereospecific rearrangement to take place, and such rearrangements may be rationalised by transition states related to **151**. The stereospecific rearrangement of **165** yields a compound **166** which may be readily converted to the insect pheromone **167**.[119] Other rearrangements of chiral allyl propargyl and allyl benzyl ethers are similar.[120,121]

SiMe$_3$ 165 → SiMe$_3$ Li → [Me$_3$Si O Me Me H] → Me$_3$Si OH 166 → OH 167 98% ee

Rearrangements of acyclic stannylmethyl ethers such as **168** may follow the stereospecificity evident with their cyclic counterparts.[95]

SnBu$_3$ 168 BuLi, −78 °C Li [2,3] OH

Even alkene starting materials without a terminal substituent may rearrange stereospecifically, and **169** gives products **171** with 60-96% ee.[122] The same transition state model **170** reliably predicts the sense of the chirality transfer.

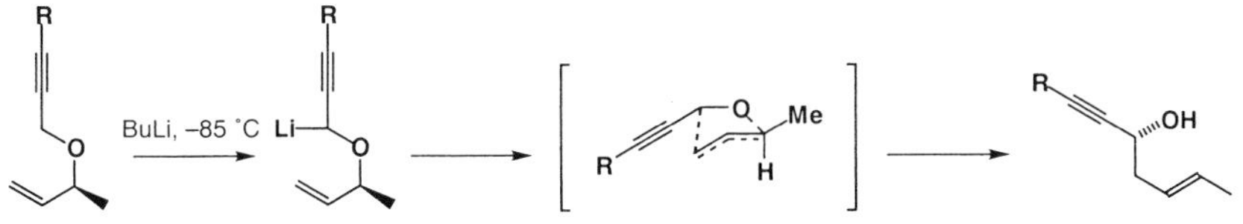

Glucose,[123] and terpene derived ketones,[124] have been used as chiral auxiliaries for the formation of new stereogenic centres by a similar method:[123]

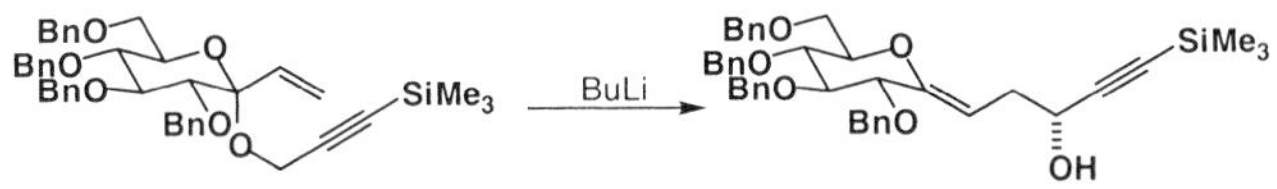

8.4.3.2 Stereoselective rearrangements with chiral auxiliaries

Chiral anion stabilising groups such as oxazolines and amides lead to good stereoselectivity in [2,3]-Wittig rearrangements,[90] but given that the starting materials are not organolithiums these rearrangements lie outside of the scope of this book. Chiral phosphonamides are effective in controlling asymmetric [2,3]-Wittig rearrangements.[125] Chiral benzyl groups, with asymmetry provided by a tricarbonylchromium complex, promote stereoselective rearrangement remarkably well:[126]

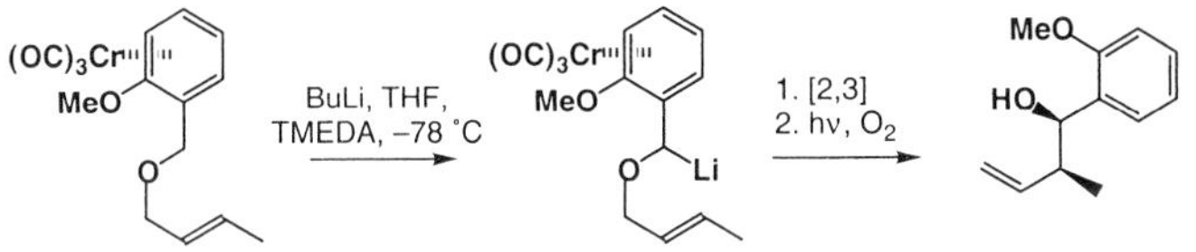

Chirality elsewhere in the starting material can lead to stereoselective rearrangement, often with very high levels of diastereoselectivity, as in this rearrangement of a lithiomethyl ether **172**.[127]

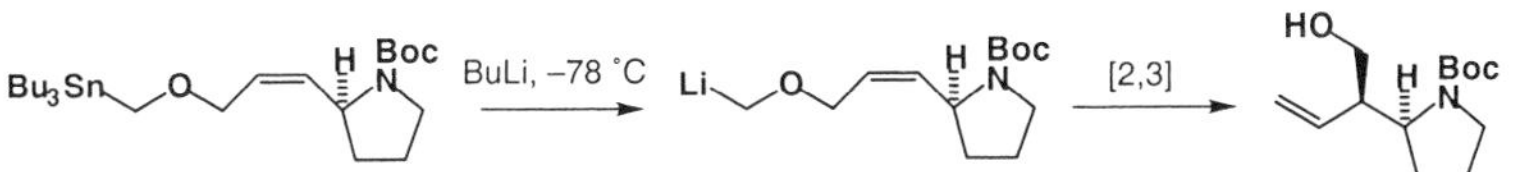

8.4.3.3 Stereospecific rearrangements of chiral organolithiums

The [2,3]-Wittig rearrangement proceeds with *inversion* at a configurationally defined organolithium centre.[80,128,129] The first evidence followed theoretical work[130] predicting that this should be the case. Cohen used the known axial preference of the formation of organolithiums by reduction of 2-alkylthiopyrans **173** to produce the organolithium **174** diastereoselectively, the vinylic and lithio substituents lying *trans*. After 12 h at –78 °C, a mixture of [1,2]- and [2,3]-Wittig products were produced, both stereoselectively; the [2,3]-

product **175** results from inversion at one reacting centre and retention at the other. Since the migration must be suprafacial across the allylic system, the conclusion must be that the reaction proceeds with inversion at the organolithium centre.[80]

LDBB
SPh
173
Li
174
O
Li
Li
inversion
OH
175 45%

More definitive evidence for inversion comes from an example lacking the diastereoselective bias present in **174**.[129] Stannane **176** was made by an asymmetric acylstannane reduction of known enantioselectivity.[131] Transmetallation of **176** (88% ee) gives **177** which rearranges to **178** (88% ee). The stereochemistry of the product **178**, which was proved by conversion to the naturally occurring ant pheromone **179** of known absolute configuration, demonstrates that the rearrangement proceeds with inversion of stereochemistry, the only assumption being retention of stereochemistry in the tin-lithium exchange step (section 5.2.1). Invertive rearrangement is also observed in the formation of **181** from **180**.[132]

$SnBu_3$
H
O
176 62% ee
BuLi, THF, −78 °C
Li
H
O
177
inversion
H
OH
178 62% ee
H_2, Pd
OH
179 62% ee
Li
H
O
H
Me

R
$SnBu_3$
O
180 95% ee
BuLi, THF, −78 °C
HO
181 95% ee
+ some [1,2]-rearrangement

Enantioselective deprotonation to yield a configurationally defined organolithium offers great potential for future asymmetric [2,3]-Wittig rearrangements. The first steps in this area were taken by Nakai and Tomooka, who treated **182** with *t*-BuLi in the presence of the bis-oxazoline ligand **114**. The product **183** is formed in 89% ee.[133]

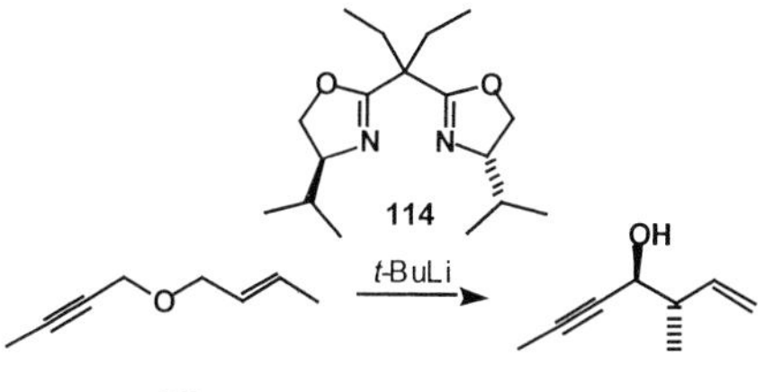

114

t-BuLi

182 183 89% ee

A comparable [1,2]-Wittig rearrangement demonstrably proceeds by asymmetric rearrangement and not asymmetric deprotonation, but in the case of related asymmetric [2,3]-Wittig rearrangements of **184** and *d*-**184**, different ees are obtained with racemic deuterated starting materials, suggesting that asymmetric deprotonation to form a configurationally stable organolithium intermediate plays a role in controlling the overall stereochemistry of the reaction.[133]

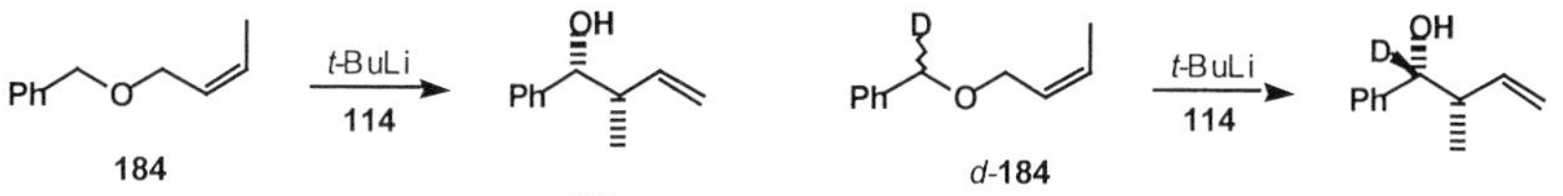

184 48% ee *d*-184 14% ee

A configurationally defined intermediate may be behind the stereoselective rearrangement of **185** to **187** promoted by the chiral base **186**.[134]

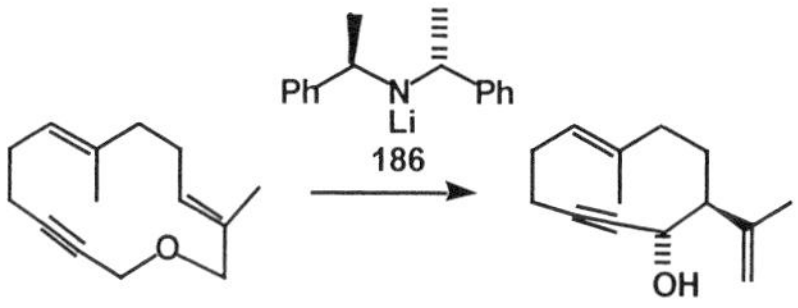

185 186 187 70% ee

8.4.4 [2,3]-Aza-Wittig rearrangements[135]

The nitrogen analogues of the [1,2]-Wittig and [2,3]-Wittig rearrangements are known as aza-Wittig rearrangements, and while many examples of [1,2]-aza-Wittig are known,[135] it is [2,3]-aza-Wittig rearrangements which have demonstrated the most synthetic potential. Synthetic application of the [2,3]-aza-Wittig rearrangement started in 1972, when Durst showed that the vinyl-substituted β-lactam **188** undergoes ring-expansion to the seven-membered lactam **189** in quantitative yield.[136] With phenyl-substituted lactam **190**, [1,2]-aza-Wittig rearrangement to **191** occurs.

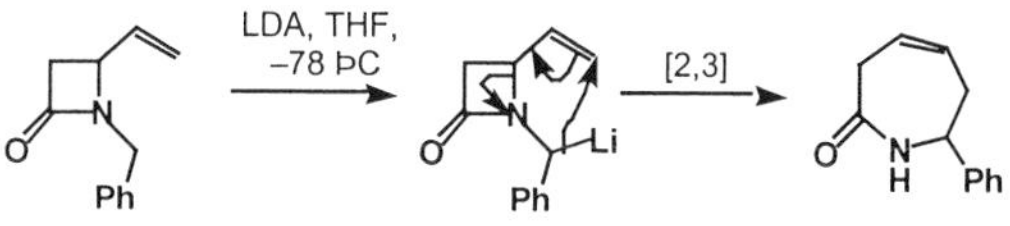

LDA, THF, –78 ÞC [2,3] 100%

188 189

Ph
LDA, THF, −78 °C
93%
190 191

The use of small rings to drive the [2,3]-aza-Wittig reaction was later revisited by Somfai, who demonstrated the use of vinylaziridines in the synthesis of six-membered lactams.[137-140]

[2,3]-Aza-Wittig rearrangement in the acyclic series are harder to control.[141] The most reliable turn out to be those of allyl amides such as **192** in which the allyl group bears a β-silyl substituent, whose function is to stabilise the anionic transition state of the reaction.[142,143]

LDA; THF, HMPA, −78 - −40 °C
84%
192

The rearrangement is stereospecific, with the major diastereoisomer of the produt being determined by the double bond geometry of the starting material.[144,145] Like the oxa analogue, [2,3]-aza-Wittig rearrangements proceed with inversion at the lithium-bearing centre.[146]

References

1. Shapiro, R. H. *Org. Reac.* **1975,** *23*, 405.
2. Adlington, R. M.; Barrett, A. G. M. *Acc. Chem. Res.* **1983,** *16*, 55.
3. Chamberlin, A. R.; Bloom, S. H. *Org. Reac.* **1990,** *391*, 1.
4. Shapiro, R. H.; Heath, M. J. *J. Am. Chem. Soc.* **1967,** *89*, 5734.
5. Bamford, W. R.; Stevens, T. S. *J. Chem. Soc.* **1952**, 4735.
6. Shapiro, R. H.; Lipton, M. F.; Kolonko, K. J.; Buswell, R. L.; Capuano, L. A. *Tetrahedron Lett.* **1975**, 1811.
7. Adlington, R. M.; Barrett, A. G. M. *J. Chem. Soc., Perkin Trans. 1* **1981**, 2848.
8. Adlington, R. M.; Barrett, A. G. M. *J. Chem. Soc., Chem. Commun.* **1979**, 1122.
9. Chamberlin, A. R.; Stemke, J. E.; Bond, F. T. *J. Org. Chem.* **1978,** *43*, 147.
10. Stemke, J. E.; Bond, F. T. *Tetrahedron Lett.* **1975**, 1815.
11. Stemke, J. E.; Chamberlin, A. R.; Bond, F. T. *Tetrahedron Lett.* **1976**, 2947.
12. Corey, E. J.; Lee, J.; Roberts, B. E. *Tetrahedron Lett.* **1997,** *38*, 8915.
13. Corey, E. J.; Dittami, J. P. *J. Am. Chem. Soc.* **1985,** *107*, 256.
14. Corey, E. J. *Tetrahedron Lett.* **1997,** *38*, 8919.
15. Evans, D. A.; Nelson, J. V. *J. Am. Chem. Soc.* **1980,** *102*, 774.
16. Maruoka, K.; Oishi, M.; Yamamoto, H. *J. Am. Chem. Soc.* **1996,** *118*, 2289.
17. Augeri, D. J.; Chamberlin, A. R. *Tetrahedron Lett.* **1994,** *35*, 5599.
18. Kim, S.; Oh, D. H.; Yoon, J.-Y.; Cheong, J. H. *J. Am. Chem. Soc.* **1999,** *121*, 5330.
19. Brook, A. G. *Acc. Chem. Res.* **1974,** *7*, 77.
20. Jankowski, P.; Raubo, P.; Wicha, J. *Synlett* **1994**, 985.

21. Moser, W. H. *Tetrahedron* **2001,** *57*, 2065.
22. Speier, J. L. *J. Am. Chem. Soc.* **1952,** *74*, 1003.
23. West, R.; Lowe, R.; Stewart, H. F.; Wright, A. *J. Am. Chem. Soc.* **1971,** *93*, 282.
24. Brook, A. G. *J. Am. Chem. Soc.* **1958,** *80*, 1886.
25. Brook, A. G.; Warner, C. M.; McGriskin, M. E. *J. Am. Chem. Soc.* **1959,** *81*, 981.
26. Brook, A. G.; Schwartz, N. V. *J. Am. Chem. Soc.* **1960,** *82*, 2435.
27. Brook, A. G.; Iachia, B. *J. Am. Chem. Soc.* **1961,** *83*, 827.
28. Lautens, M.; Delanghe, P. H.; Goh, J. B.; Zhang, C. H. *J. Org. Chem.* **1995,** *60*, 4213.
29. Lindermann, R. J.; Ghannam, A. *J. Am. Chem. Soc.* **1990,** *112*, 2396.
30. Jiang, X.; Bailey, W. F. *Organometallics* **1995,** *14*, 5704.
31. Fleming, I.; Roberts, R. S.; Smith, S. C. *J. Chem. Soc., Perkin Trans. 1* **1998**, 1215.
32. Naganuma, K.; Kawashima, T. *Chem. Lett.* **1999**, 1139.
33. Hudrlik, P. F.; Hudrlik, A. M.; Kulkarni, A. K. *J. Am. Chem. Soc.* **1982,** *104*, 6809.
34. Wilson, S. R.; Hague, M. J.; Misra, R. N. *J. Org. Chem.* **1982,** *47*, 747.
35. Wright, A.; West, R. *J. Am. Chem. Soc.* **1974,** *96*, 3227.
36. Kuwajima, I.; Kato, M.; Mori, A. *Tetrahedron Lett.* **1980,** *21*, 2745.
37. Reich, H. J.; Olson, R. E.; Clark, M. C. *J. Am. Chem. Soc.* **1980,** *102*, 1423.
38. Cunico, R. F.; Nair, S. K. *Synth. Commun.* **1996,** *26*, 803.
39. Bienz, S.; Enev, V.; Huber, P. *Tetrahedron Lett.* **1994,** *35*, 1161.
40. Takeda, K.; Tanaka, T. *Synlett* **1999,** *705.*
41. Koreeda, M.; Koo, S. *Tetrahedron Lett.* **1990,** *31*, 831.
42. Takeda, K.; Ubayama, H.; Sano, A.; Yoshii, E.; Koizumi, T. *Tetrahedron Lett.* **1998,** *39*, 5243.
43. Takeda, K.; Fujisawa, M.; Makino, T.; Yoshii, E.; Yamaguchi, K. *J. Am. Chem. Soc.* **1993,** *114*, 9351.
44. Takeda, K.; Nakayama, I.; Yoshii, E. *Synlett* **1994**, 178.
45. Takeda, K.; Nakajima, A.; Yoshii, E. *Synlett* **1996**, 255.
46. Takeda, K.; Takeda, M.; Nakajima, A.; Yoshii, E. *J. Am. Chem. Soc.* **1995,** *117*, 6400.
47. Takeda, K.; Nakajima, A.; Yoshii, E. *Synlett* **1996**, 753.
48. Takeda, K.; Nakajima, A.; Takeda, M.; Yasushi, O.; Sato, T.; Yoshii, E.; Koizume, T.; Shiro, M. *J. Am. Chem. Soc.* **1998,** *120*, 4947.
49. Takeda, K.; Nakajima, A.; Takeda, M.; Yoshii, E. *Org. Synth.* **1999,** *76*, 199.
50. Takeda, K.; Nakane, D.; Takeda, M. *Org. Lett.* **2000,** *2*, 1903.
51. Shinokubo, H.; Miura, K.; Oshima, K.; Utimoto, K. *Tetrahedron Lett.* **1993,** *34*, 1951.
52. Shinokubo, H.; Miura, K.; Oshima, K.; Utimoto, K. *Tetrahedron* **1996,** *52*, 503.
53. Smith, A. B.; Boldi, A. M. *J. Am. Chem. Soc.* **1997,** *119*, 6925.
54. Smith, A. B.; Lin, Q.; Nakayama, K.; Boldi, A. M.; Brook, C. S.; McBriar, M. D.; Moser, W. H.; Sobukawa, M.; Zhuang, L. *Tetrahedron Lett.* **1997,** *38*, 8671.
55. Smith, A. B.; Pitram, S. M. *Org. Lett.* **1999,** *1*, 2001.
56. Fischer, M. R.; Kirschning, A.; Michel, T.; Schaumann, E. *Angew. Chem., Int. Ed. Engl.* **1994,** *33*, 217.
57. Brauer, N.; Michel, T.; Schaumann, E. *Tetrahedron* **1998,** *54*, 11481.
58. Brauer, N.; Dreeben, S.; Schaumann, E. *Tetrahedron Lett.* **1999,** *40*, 2921.

59. Moser, W. H.; Endsley, K. E.; Colyer, J. T. *Org. Lett.* **2000,** *2*, 717.
60. Comanita, B. M.; Woo, S.; Fallis, A. G. *Tetrahedron Lett.* **1999,** *40*, 5283.
61. Clayden, J.; Julia, M. *Synlett* **1995**, 103.
62. Kim, K. D.; Magriotis, P. A. *Tetrahedron Lett.* **1990,** *31*, 6137.
63. Marumoto, S.; Kuwajima, I. *J. Am. Chem. Soc.* **1993,** *115*, 9021.
64. Tomooka, K.; Yamamoto, K.; Nakai, T. *Liebigs Ann. Chem.* **1997**, 1275.
65. Schöllkopf, U. *Angew. Chem., Int. Ed. Engl.* **1979,** *18*, 763.
66. Marshall, J. A. In *Comprehensive Organic Synthesis;* Trost, B. M.; Fleming, I. Eds., 1990; Vol. 3; pp. 975.
67. Wittig, G.; Löhmann, L. *Liebigs Ann. Chem.* **1942,** *550*, 260.
68. Lansbury, P. T.; Pattison, V. A.; Sidler, J. D.; Bieber, J. B. *J. Am. Chem. Soc.* **1966,** *88*, 78.
69. Schäfer, H.; Schöllkopf, U.; Walter, D. *Tetrahedron Lett.* **1968**, 2809.
70. Garst, J. F.; Smith, C. D. *J. Am. Chem. Soc.* **1976,** *98*, 1526.
71. Still, W. C. *J. Am. Chem. Soc.* **1978,** *100*, 1481.
72. Tomooka, K.; Igarishi, T.; Nakai, T. *Tetrahedron* **1994,** *50*, 5927.
73. Maercker, A. *Angew. Chem., Int. Ed. Engl.* **1987,** *26*, 972.
74. Matsushita, M.; Nagaoka, Y.; Hioki, H.; Fukuyama, Y.; Kodama, M. *Chem. Lett.* **1996**, 1039.
75. Felkin, H.; Tambuté, A. *Tetrahedron Lett.* **1969**, 821.
76. Wei, X.; Johnson, P.; Taylor, R. J. K. *J. Chem. Soc., Perkin Trans. 1* **2000**, 1109.
77. Wei, X.; Taylor, R. J. K. *Tetrahedron Lett.* **1996,** *37*, 4209.
78. Schöllkopf, U.; Fabian, W. *Liebigs Ann. Chem.* **1961,** *642*, 1.
79. Schöllkopf, U.; Schäfer, H. *Liebigs Ann. Chem.* **1963,** *663*, 22.
80. Verner, E. J.; Cohen, T. *J. Am. Chem. Soc.* **1992,** *114*, 375.
81. Hoffmann, R. W.; Brückner, R. *Chem. Ber.* **1992,** *125*, 1957.
82. Hoffmann, R. W.; Rückert, T.; Brückner, R. *Tetrahedron Lett.* **1993,** *34*, 297.
83. Tomooka, K.; Igarishi, T.; Nakai, T. *Tetrahedron Lett.* **1993,** *34*, 8139.
84. Tomooka, K.; Yamamoto, K.; Nakai, T. *Angew. Chem., Int. Ed. Engl.* **1999,** *38*, 3741.
85. Schreiber, S. J.; Goulet, M. T. *Tetrahedron Lett.* **1987,** *28*, 1043.
86. Schreiber, S. J.; Goulet, M. T.; Schulte, G. *J. Am. Chem. Soc.* **1987,** *109*, 4718.
87. Yadav, J. S.; Ravishankar, R. *Tetrahedron Lett.* **1991,** *32*, 2629.
88. Kiyooka, S.; Tsutsui, T.; Kira, T. *Tetrahedron Lett.* **1996,** *37*, 8903.
89. Tomooka, K.; Yamamoto, H.; Nakai, T. *J. Am. Chem. Soc.* **1996,** *118*, 3317.
90. Nakai, T.; Mikami, K. *Org. Reac.* **1994,** *46*, 105.
91. Nakai, T.; Mikami, K. *Chem. Rev.* **1986,** *86*, 885.
92. Mikami, K.; Nakai, T. *Synthesis* **1991**, 594.
93. Cast, J.; Stevens, T. S.; Holmes, J. *J. Chem. Soc.* **1960**, 3521.
94. Rautestrauch, V. *J. Chem. Soc., Chem. Commun.* **1970**, 4.
95. Still, W. C.; Mitra, A. *J. Am. Chem. Soc.* **1978,** *100*, 1927.
96. Nakai, T.; Mikami, K.; Taya, S.; Fujita, Y. *J. Am. Chem. Soc.* **1981,** *103*, 6492.
97. Tomooka, K.; Harada, M.; Hanji, T.; Nakai, T. *Chem. Lett.* **2000**, 1394.

98. Garbi, A.; Allain, L.; Chokrki, F.; Crousse, B.; Bonnet-Delpon, D.; Nakai, T.; Bégué, J.-P. *Org. Lett.* **2001,** *3*, 2529.
99. Clayden, J.; Kenworthy, M. N. *Org. Lett.*, in press.
100. Broka, C. A.; Shen, T. *J. Am. Chem. Soc.* **1989,** *111*, 2981.
101. Mikami, K.; Azuma, K.; Nakai, T. *Chem. Lett.* **1983,** 1379.
102. Mikami, K.; Azuma, K.; Nakai, T. *Tetrahedron* **1984,** *40*, 2303.
103. Tulshian, D. B.; Fraser-Reid, B. *J. Org. Chem.* **1984,** *49*, 518.
104. Wada, M.; Fukui, A.; Nakamura, H.; Takei, H. *Chem. Lett.* **1977**, 557.
105. Mikami, K.; Kishi, N.; Nakai, T. *Chem. Lett.* **1989**, 1683.
106. Marshall, J. A.; Lebreton, J. *J. Org. Chem.* **1988,** *53*, 4108.
107. Mikami, K.; Kimura, Y.; Kishi, N.; Nakai, T. *J. Org. Chem.* **1983,** *48*, 279.
108. Mikami, K.; Kishi, N.; Nakai, T. *Chem. Lett.* **1982**, 1643.
109. Kaye, A. D.; Pattenden, G.; Roberts, R. M. *Tetrahedron Lett.* **1986,** *27*, 2033.
110. Still, W. C.; MacDonald, J. H.; Collum, D. B.; Mitra, A. *Tetrahedron Lett.* **1979**, 593.
111. Nakai, E.; Nakai, T. *Tetrahedron Lett.* **1988,** *29*, 5409.
112. Nakai, T.; Tomooka, K. *Pure Appl. Chem.* **1997,** *69*, 595.
113. Paquette, L. A.; Sugimura, T. *J. Am. Chem. Soc.* **1986,** *108*, 3017.
114. Sugimura, T.; Paquette, L. A. *J. Am. Chem. Soc.* **1987,** *109*, 3017.
115. Trost, B. M.; Mao, M. K.-T.; Balkovec, J. M.; Buhlmayer, P. *J. Am. Chem. Soc.* **1986,** *108*, 4965.
116. Castedo, L.; Mascareñas, J. L.; Mourino, A. *Tetrahedron Lett.* **1987,** *28*, 2099.
117. Crimmins, M. T.; Gould, L. D. *J. Am. Chem. Soc.* **1987,** *109*, 6199.
118. Eguchi, S.; Ebihara, K.; Morisaki, M. *Chem. Pharm. Bull.* **1988,** *36*, 4638.
119. Sayo, N.; Azuma, K.; Mikami, K.; Nakai, T. *Tetrahedron Lett.* **1984,** *25*, 565.
120. Sayo, N.; Kitahara, E.; Nakai, T. *Chem. Lett.* **1984**, 259.
121. Tsai, D. J.-S.; Midland, M. M. *J. Org. Chem.* **1984,** *49*, 1842.
122. Sayo, N.; Shirai, F.; Nakai, T. *Chem. Lett.* **1984**, 255.
123. Tomooka, K.; Nakamura, Y.; Nakai, T. *Synlett* **1995**, 321.
124. Keegan, D. S.; Midland, M. M.; Werley, R. T.; McLoughlin, J. I. *J. Org. Chem.* **1991,** *56*, 1185.
125. Denmark, S. E.; Miller, P. C. *Tetrahedron Lett.* **1995,** *36*, 6631.
126. Uemura, M.; Nishimura, H.; Minami, T.; Hayashi, Y. *J. Am. Chem. Soc.* **1991,** *113*, 5402.
127. Priepke, H.; Bruckner, R.; Harms, K. *Chem. Ber.* **1990,** *123*, 555.
128. Hoffmann, R. W.; Brückner, R. *Angew. Chem., Int. Ed. Engl.* **1992,** *31*, 647.
129. Tomooka, K.; Igarishi, T.; Watanabe, M.; Nakai, T. *Tetrahedron Lett.* **1992,** *33*, 5795.
130. Wu, Y. D.; Houk, K. N.; Marshall, J. A. *J. Org. Chem.* **1990,** *55*, 1421.
131. Chang, P. C.-M.; Chong, J. M. *J. Org. Chem.* **1988,** *53*, 5584.
132. Tomooka, K.; Komine, N.; Nakai, T. *Synlett* **1997**, 1045.
133. Tomooka, K.; Komine, N.; Nakai, T. *Tetrahedron Lett.* **1998,** *39*, 5513.
134. Marshall, J. A.; Lebreton, J. *J. Am. Chem. Soc.* **1988,** *110*, 2925.
135. Vogel, C. *Synlett* **1997**, 497.
136. Durst, T.; van den Elzen, R.; LeBelle, M. J. *J. Am. Chem. Soc.* **1972,** *94*, 9261.

137. Åhman, J.; Somfai, P. *J. Am. Chem. Soc.* **1994,** *116*, 9781.
138. Åhman, J.; Somfai, P.; Tanner, D. *J. Chem. Soc., Chem. Commun.* **1994**, 2785.
139. Åhman, J.; Somfai, P. *Tetrahedron Lett.* **1996,** *37*, 2495.
140. Lindström, U. M.; Somfai, P. *Synthesis* **1998**, 109.
141. Coldham, I.; Middleton, M. L.; Taylor, P. L. *J. Chem. Soc., Perkin Trans. 1* **1998**, 2817.
142. Anderson, J. C.; Flaherty, A.; Swarbrick, M. E. *J. Org. Chem.* **2000,** *65*, 9152.
143. Anderson, J. C.; Smith, S. C.; Swarbrick, M. E. *J. Chem. Soc., Perkin Trans. 1* **1997**, 1517.
144. Anderson, J. C.; Dupau, P.; Siddons, D. C.; Smith, S. C.; Swarbrick, M. E. *Tetrahedron Lett.* **1998,** *39*, 2649.
145. Anderson, J. C.; Flaherty, A. *J. Chem. Soc., Perkin Trans. 1* **2001**, 267.
146. Gawley, R. E.; Zhang, Q.; Campagna, S. *J. Am. Chem. Soc.* **1995,** *117*, 11817.

CHAPTER 9

Organolithiums in Synthesis

Organolithiums are so widespread in synthesis that a comprehensive survey is of course impossible. This chapter highlights nine syntheses of biologically important compounds whose key steps rely upon an aspect of selective organolithium chemistry.

9.1 Ochratoxin: ortholithiation and anionic *ortho*-Fries rearrangement[1]

Ochratoxins A **1** and B **2** are metabolites of *Aspergillus ochraceus* and *Penicillium viridicatum* whose presence in agricultural products poses a health hazard to both animals and humans. Snieckus' synthesis[1] uses a series of directed ortholithiations to control the 1,2,3,4(,5)-polysubstitution pattern, introducing the two carbonyl substituents (an amide and a lactone) as tertiary amides to capitalise on their powerful lithiation chemistry. The routes to both ochratoxins are more or less identical, and only ochratoxin B is described here. Yields in the route to ochratoxin A are on average significantly higher.

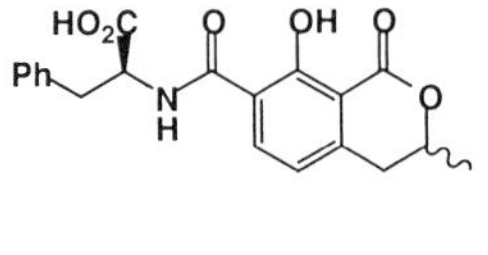

ochratoxin A **1**

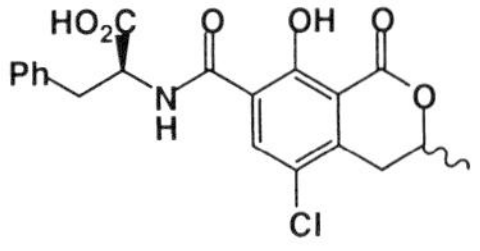

ochratoxin B **2**

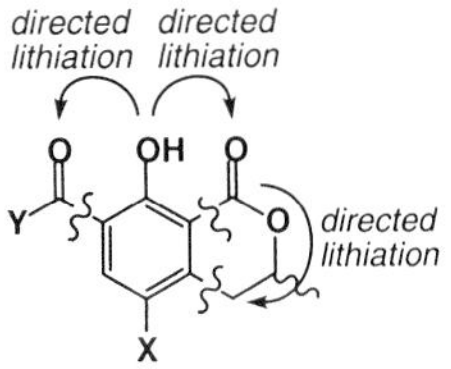

The starting point for the metallation sequence was *p*-chlorophenol, whose directing power was first maximised by conversion to the carbamate **3**. Ortholithiation and reaction with *N,N*-diethylcarbamoyl chloride gave the amide **4**. The second *ortho* carbonyl substituent was then introduced simply by allowing the lithiated carbamate **5** to undergo an anionic *ortho*-Fries rearrangement to the bis-amide **6**. The phenol was protected as its methyl ether **7**.

The alkyl substituent *meta* to the methoxy substituent was easily introduced into the symmetrical diamide **7** by yet another ortholithiation. Allyl electrophiles react poorly with aryllithiums, so the ortholithiated amide **8** was first transmetallated to the Grignard reagent before allylation (allyl bromide) to give **9**.

A combined lactonisation and deprotection to give **10** was achieved by refluxing **9** in 6N HCl. Peptide coupling with phenylalanine affords a mixture of diastereoisomers of the target compound **1**.

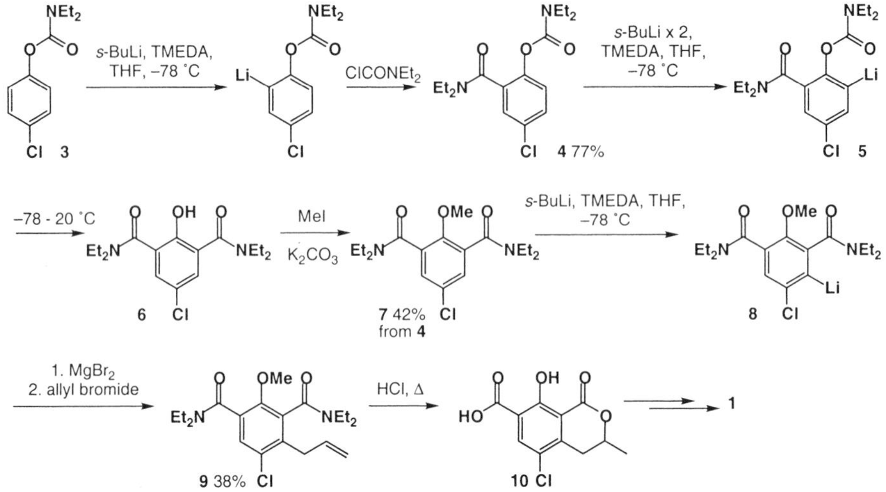

[Further background: section 2.3]

9.2 Corydalic acid methyl ester: lateral lithiation[2]

Corydalic acid methyl ester **11** , isolated from *Corydalis incisa*, is derived from a proposed biosynthetic intermediate in the route to the tetrahydroprotoberberine alkaloids. The 1,2,3,4-tetrasubstituted ring of **11** demands control by an ortholithiation strategy, and the synthetic route proposed by Clark and Jahangir[2] employs a lateral lithiation of **12** and addition to an imine as the key disconnection at the centre of the molecule.

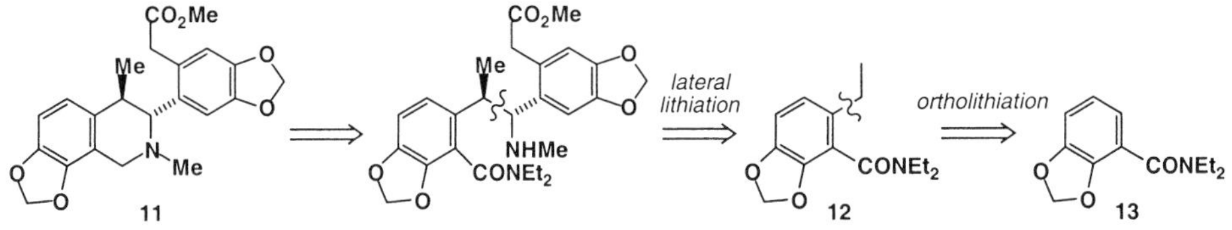

The amide **13** may be made by ortholithiation of benzodioxolane **14**, though a higher-yielding preparation starts from 1,2-dihydroxybenzoic acid **15**. Ortholithiation of **13**, directed by the tertiary amide group, is straightforward, and gives the alkylated amide **12**.

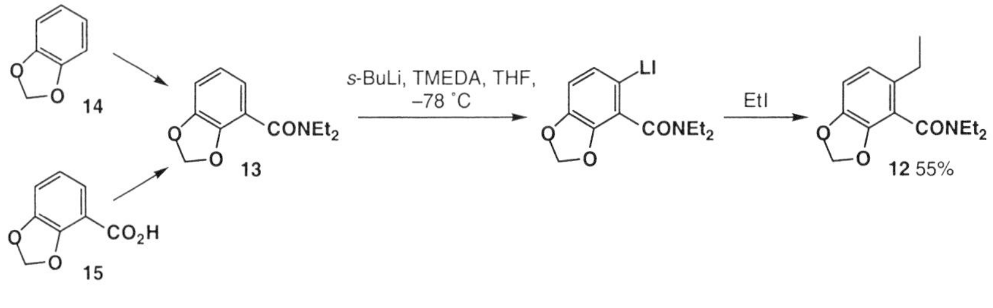

The directing effect of the amide group can then be used a second time in the lateral lithiation of **12** to give an organolithium **16** which adds to the imine **17** in a stereoselective manner, probably under thermodynamic control (imine additions of laterally lithiated amides appear to be reversible). Warming the reaction mixture to room temperature leads to a mixture of **18** and some of the (ultimately required) cyclised product **19**. The uncyclised product could readily be cyclised to **19** by treatment with *t*-BuLi at –30 °C.

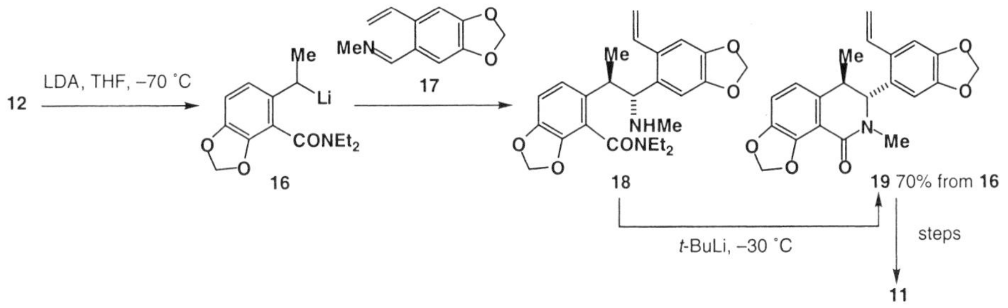

12 —LDA, THF, –70 °C→ 16 —17→ 18 —*t*-BuLi, –30 °C→ 19 70% from 16 —steps→ 11

Conversion to the target molecule **11** was achieved by oxidation ($TlNO_3$ then $KMnO_4$) of the vinyl group and reduction of the lactam.

[Further background: sections 2.3 and 2.4]

9.3 Fredericamycin A: *ortho*, lateral and α-lithiation[3]

Fredericamycin has a remarkable structure: two aromatic systems join at a chiral spiro centre in a compound with exceptional anticancer activity. The first total synthesis[3] drew heavily on directed metallation chemistry for the synthesis of the lower portion of the molecule.

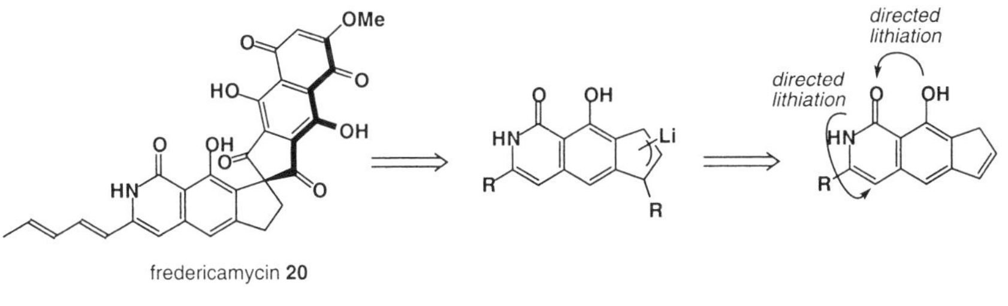

fredericamycin **20**

directed lithiation *directed lithiation*

The starting material was the dihydroxyindane **21**, differentially protected, with a phenolic MOM ether sited appropriately for directed metallation of the aromatic ring. Annelation of the pyridone ring was achieved by an ortholithiation, introducing an amide and giving **22**. A second ortholithiation of **22** is now directed by the more powerful amide group, giving **23**. Lateral metallation and reaction with the nitrile **24** gave the pyridone **25**.

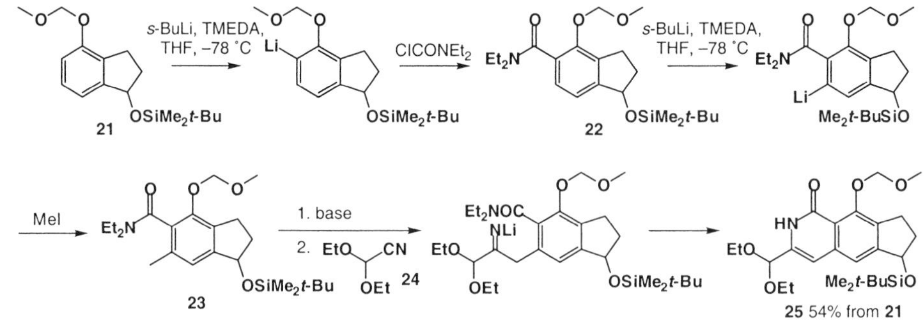

After some attempts to functionalise **25** directly, it turned out that the best way to take this compound through to fredericamycin was to promote an elimination to the indene **26**. TBAF deprotection and selenoxide elimination achieved this, and O-protection of the pyridone gave **27**. Lithiation gives an allyllithium **28** which is silylated regioselectively at the terminus further from the MOM group, giving **29**. A further lithiation of the indenylsilane gives an allyllithium **30** whose steric bias results in regioselective attack with >85% of the anhydride **31** attacking the required position closer to the MOM group. Without the silylation step, incorrect regiochemistry results. Desilylation to give **32** occurs on work-up. The spiro connection is made by reduction of the lactone, aldol condensation and then oxidation to a diketone. Deprotection and Wittig olefination completes the route to fredericamycin **20**.

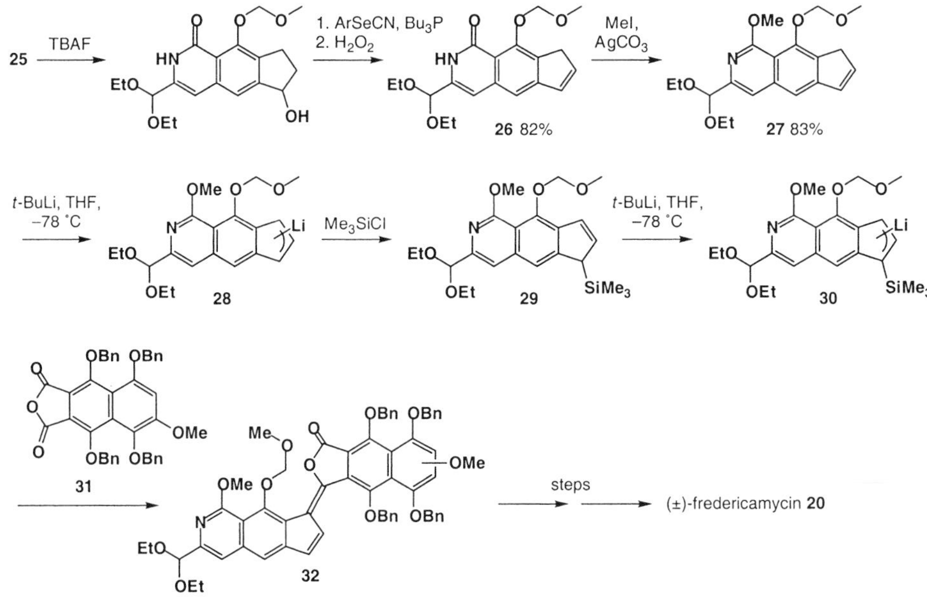

[Further background: sections 2.2-2.4]

9.4 (±)-Atpenin B: lithiation of an aromatic heterocycle[4]

atpenin B **33**

directed lithiation / directed lithiation/ rearrangement / directed lithiation / directed lithiation

The fully substituted pyridine atpenin B **33**, an antibiotic produced by *Penicillium*, was first made[4] in 1994 by a series of directed metallations which started with 2-chloropyridine **34** and introduced the substituents stepwise around the ring. 2-Chloropyridine can be lithiated (*ortho*-directed by the chlorine) by either LDA or by PhLi, and the organolithium **35** was oxidised to **36** with trimethylborate and then peracetic acid. *O*-methylation and substitution of chloride by methoxide gave 2,3-dimethoxypyridine **37**.

34 —LDA or PhLi→ **35** —1. $B(OMe)_3$ 2. $MeCO_3H$→ **36** 65% —1. MeI 2. MeONa→ **37** 58%

The second metallation in the sequence, which transformed **37** into **39**, required excess *n*-BuLi (a common characteristic of the lithiation of molecules bearing more than one directing group). Oxidation of the organolithium **38**, as before, gave 4-pyridone **39**. The new oxygen substituent of **39** has itself to act as a directing group, and for this reason **39** was converted to the carbamate **40**.

37 —BuLi x 2.2→ **38** —1. $B(OMe)_3$ 2. $MeCO_3H$→ **39** 64% —$AgCO_3$, $ClCONi\text{-}Pr_2$→ **40** 71%

Lithiation gave **41** which was brominated with cyanogen bromide to yield the 5-bromopyridine **42**. In fact, the next substituent to be introduced must be the hydroxyl group at C-6, in order to reserve the introduction of the acyl group at C5 until the end of the sequence. Treatment of the C-5 brominated **42** with LDA in the presence of catalytic bromine sets up an equilibrium between **43** and **44** (a "halogen-dance") in which **44** (with the lithium *ortho* to the carbamate) predominates. Hydrolysis gives the C-6 brominated **45**. Regioselectivity is then correct for a bromine-lithium exchange of **45** to give **46** and oxidation again using the boronate–peracetic acid method yields **47**.

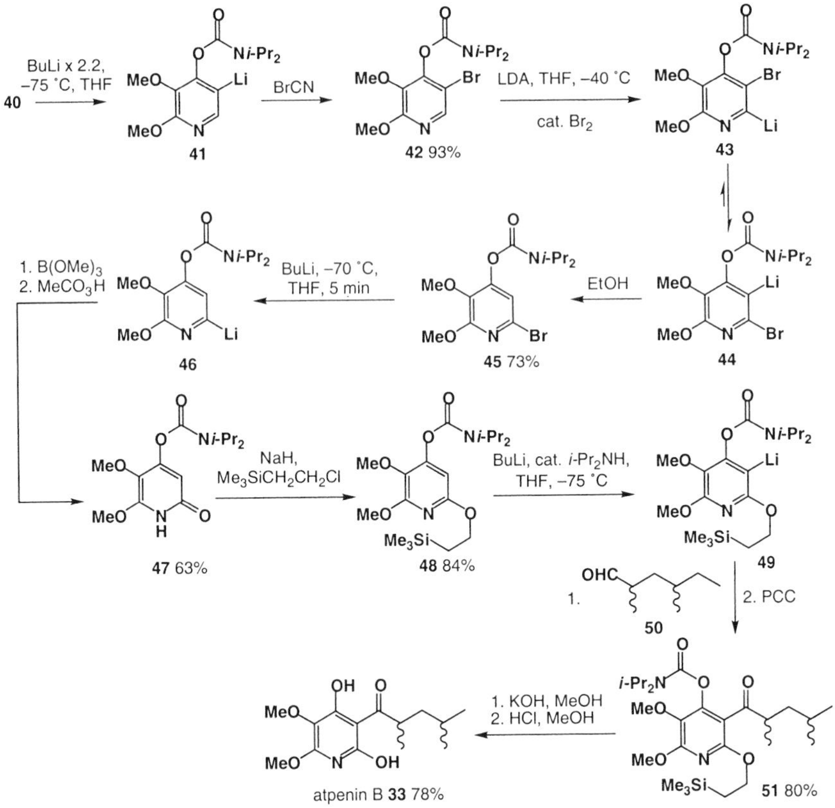

O-Protection of this 2-pyridone was followed by lithiation to give **49** with BuLi and *i*-Pr_2NH followed by reaction with a diastereoisomeric mixture of the aldehyde **50**. Oxidation gave a diastereoisomeric mixture of the ketones **51** whose deprotection yields a stereosiomeric mixture of **33**.

[Further background: sections 2.3.2 and 2.3.3]

9.5 Flurbiprofen: metallation with LiCKOR superbases[5]

Flurbiprofen **52** (Froben®, Cebutid®) is a non-steroidal anti-inflammatory whose structure makes it an ideal target for synthesis using a combination of the powerful deprotonating ability of the superbases and the weakly directing effects of a fluoro or an aryl substituent.[5]

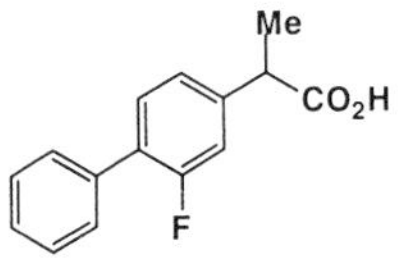

flurbiprofen **52**

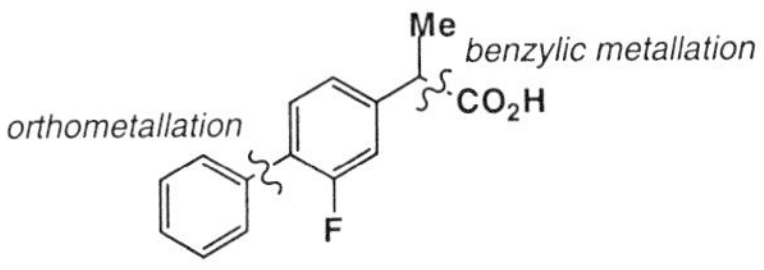

Deprotonation of 3-fluorotoluene **53** with *n*-BuLi–KO*t*-Bu or, better, *t*-BuLi–KO*t*-Bu follows the selectivity expected with these superbases and leads to metallation at the least hindered position *ortho* to the fluoro substituent. Trapping the metallated intermediate **54** with dimethylfluoroboronate gave, after hydrolysis, a boronic acid whose Suzuki arylation provided the biaryl **55**. These four steps may be carried out in a single pot, giving **55** in 79% overall yield from **53**.

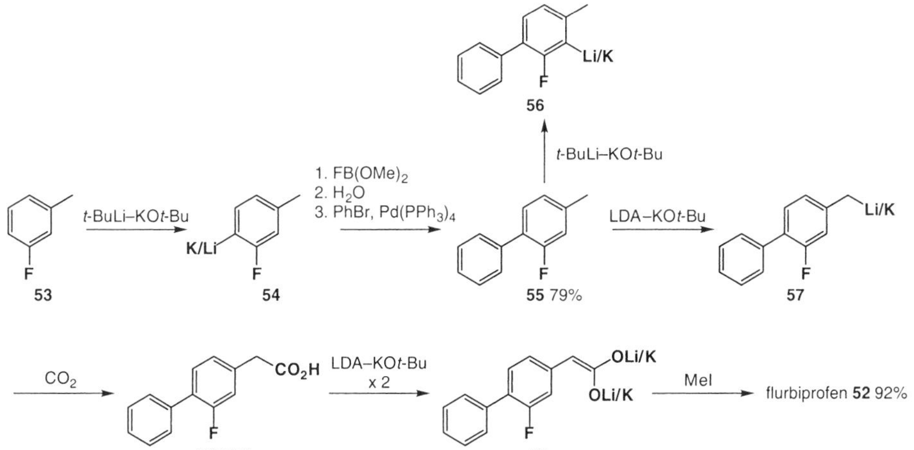

While metallation of **55** with a LiCKOR reagent such as BuLi–KO*t*-Bu again leads to orthometallation and **56**, reversible metallation of **55** by LDA-KO*t*-Bu generates the more stable benzylically metallated species **57**. Carboxylation and alkylation (2 equiv. LDA-KO*t*-Bu then MeI) of **57** gives flurbiprofen **52**.

[Further background: section 2.6]

9.6 California Red Scale Pheromone: α- and reductive lithiation[6]

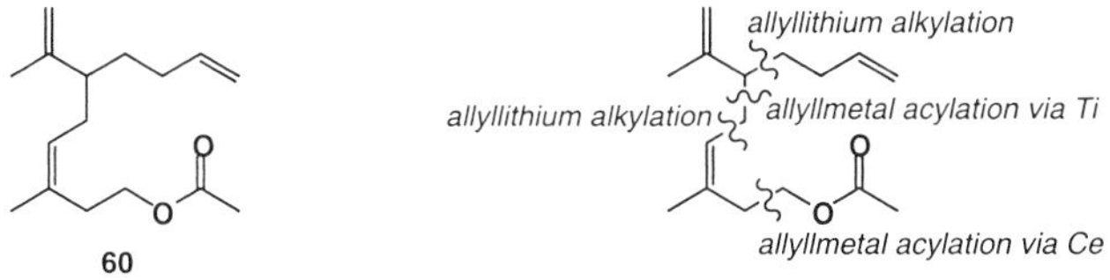

California Red Scale is a worldwide citrus pest which can be controlled by means of the pheromone **60**. Cohen used reductive lithiation to generate versatile allylic nucleophiles applicable to this type of target.[6] The allyl sulfide **61** is lithiated by BuLi and reacts with butenyl bromide α to sulfur to yield **63**. Reductive lithiation (Li, DBB) of the product yields allyllithium **64**. A regioselective reaction of this nucleophile with formaldehyde at the more substituted terminus is ensured by transmetallation to the allyl titanium **65**, which gives **66** after treatment with formaldehyde and bromination.

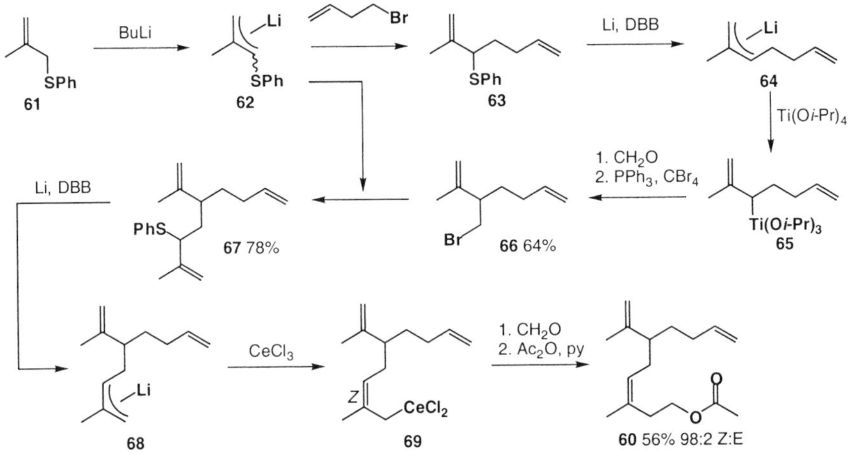

A second molecule of the lithiated sulfide **62** is then used to introduce the next allylic unit, and reductive lithiation of the product **67** gives allyllithium **68**. On this occasion, reaction at the less substituted terminus of the allyl system is required, and this can be achieved by transmetallation (at low temperature, to preserve the *cis* configuration) to the allylcerium **69**. Reaction with formaldehyde followed by acetylation gives the pheromone **60** contaminated by less than 2% of the undesired *E* isomer.

[Further background: section 4.4.1]

9.7 C1-C9 of the Bryostatins: diastereoselective bromine-lithium exchange[7]

Bromine-lithium exchange of the dibromide **70** in a Trapp solvent mixture is diastereoselective, and leads predominantly to carbenoid **71a**, with about 25% of diastereoisomer **71b**.[7] Stereochemical purity improves with time because the minor stereoisomer **71b** is selectively converted to the bicyclo[3.1.0]hexane **72** at –110 °C The remaining diastereoisomerically pure **71a** reacts stereospecifically with electrophiles, such as the borate ester **73**, which give the bromoboronate **74**. Hoffmann realised the potential of this new method for 1,3-stereocontrol by employing the boronate **74** in a synthesis of the C1-C9 fragment of the important anti-cancer compound bryostatin. Nucleophilic substitution with the dianion of *t*-butyl acetoacetate gives **75** which undergoes stereospecific oxidation and stereoselective reduction to **76**. Acetonide protection gives **77**, a known intermediate for the synthesis of the bryostatins.

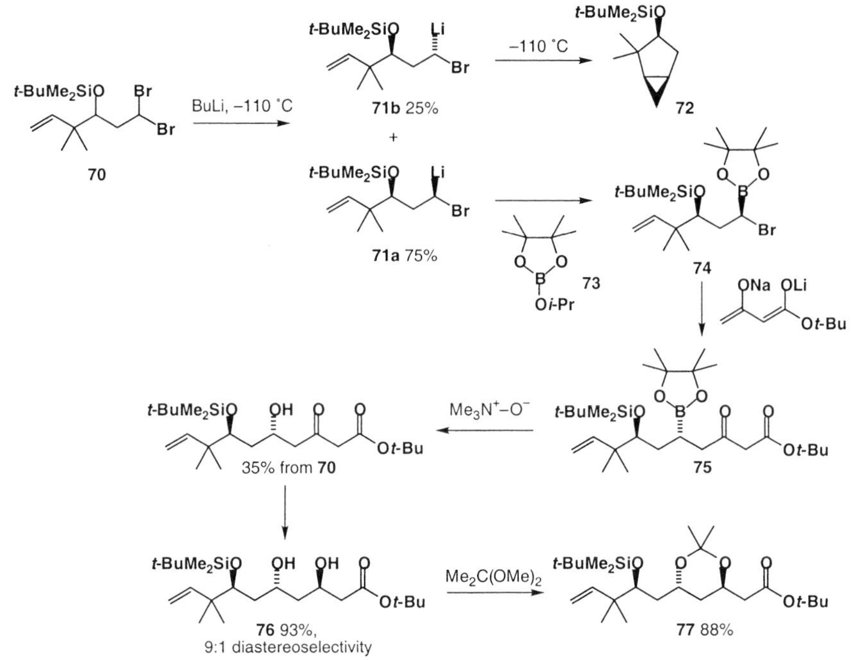

Dibromide **70** is made from the diol **78**, so an asymmetric route to **77** can be envisaged using the Sharpless ADH reaction to generate enantiomerically pure **78**.

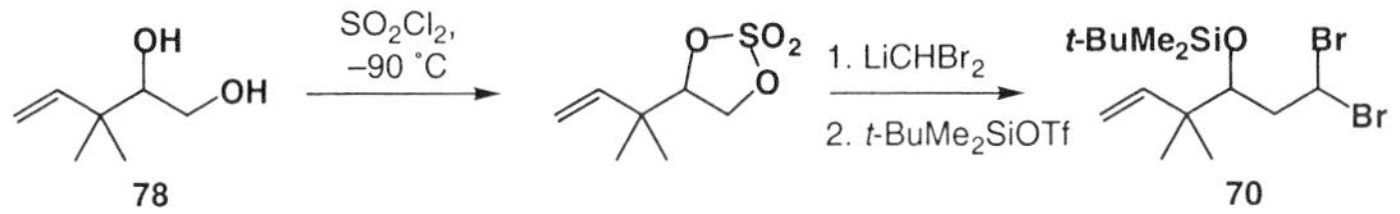

[Further background: section 3.1.7]

9.8 (*S*)-1-Methyldodecyl acetate, a Drosophila pheromone: (–)-sparteine assisted enantioselective lithiation[8]

alkylation of a configurationally stable and enantiomerically pure organolithium

The simple compound **79** is an aggregation pheromone of the fruit fly *Drosophila mulleri*. Hoppe proposed to synthesise it by asymmetric deprotonation and retentive methylation of a carbamate derivative of *n*-dodecanol.[8] Given that the carbamate group (required for directed α-lithiation) must, at the end of the synthesis, be removed from the molecule, the best choice is the tetramethyloxazolidine-derived **81**, readily formed by acylation with the carbamoyl chloride **80**. Lithiation of **81** with excess *s*-BuLi-(–)-sparteine gives, by asymmetric deprotonation, a configurationally stable complex of organolithium **82** with (–)-sparteine.

Alkylation by methyl iodide gives the carbamate **83** with greater than 98% ee. Cleavage of the carbamate is achieved with first acid (to give **84**) then base, via intramolecular attack on the carbonyl group. Acetylation of the alcohol **85** gives the target pheromone **79**.

80 → ROH, NaH → **81** 95% → s-BuLi, (–)-sparteine, Et_2O, –78 °C → **82** → MeI → **83** 69% → $MeSO_3H$, MeOH → **84** → $Ba(OH)_2$, MeOH → **85** 95% → Ac_2O, Et_3N, DMAP → **79** 95%; 98.4% ee

[Further background: section 5.4]

9.9 (–)-Paroxetine: (–)-sparteine-promoted asymmetric lithiation and substitution[9]

As described in section 5.4, (–)-sparteine-directed lithiation of **86** yields diastereoisomerically pure complexes **87**. The same reaction of the vinylogous Boc-protected amines **89** is also possible, giving **90**, and the organolithiums **87** and **90** react stereospecifically with Michael acceptors to yield compounds such as **88** and **91**.[10]

86 → n-BuLi, (–)-sparteine → **87** → **88**

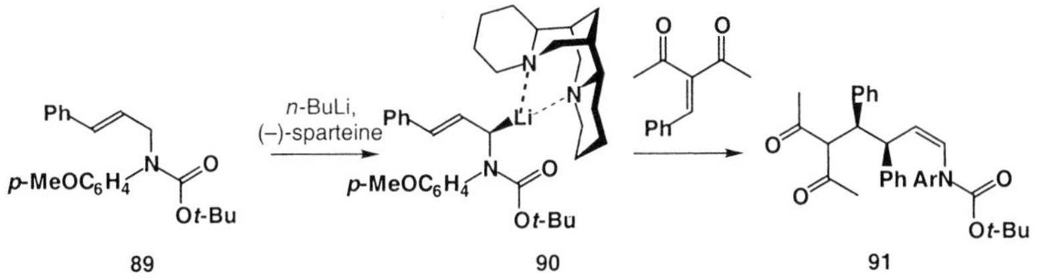

Beak used this method in a synthesis of (–)-paroxetine **97** (Paxil or Seroxat), a selective serotonin reuptake inhibitor.[9] Lithiation of **92** with *n*-BuLi-(–)-sparteine and addition of the product **93** to the nitroalkene **94** yields the protected *Z*-enamine **95**, as usual for reactions of lithiated allylamides, in >94% ee. Hydrolysis and reduction of the product, followed by mesylation and cyclisation gave the *trans*-substituted piperidine **96**. Displacement of the hydroxyl group by sesamol yielded (–)-paroxetine **97**.

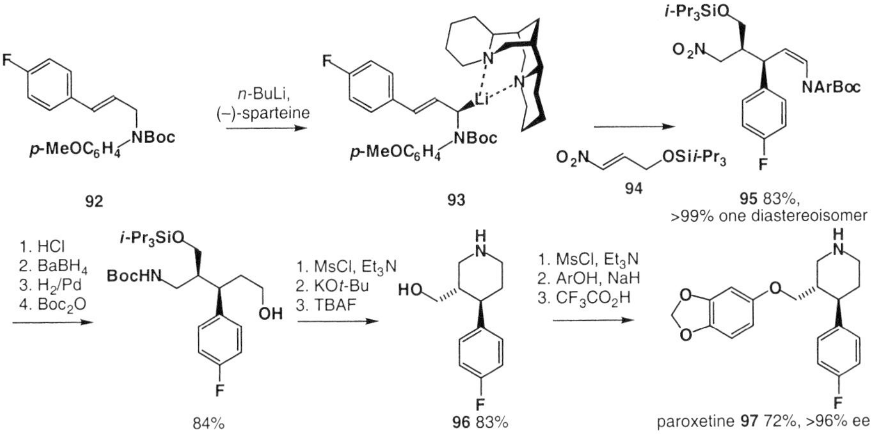

[Further background: section 5.4 and chapter 6]

References

1. Sibi, M. P.; Chattopadhyay, S.; Dankwardt, J. W.; Snieckus, V. *J. Am. Chem. Soc.* **1985,** *107*, 6312.
2. Clark, R. D.; Jahangir *J. Org. Chem.* **1989,** *54*, 1174.
3. Kelly, T. R.; Ohashi, N.; Armstrong-Chong, R. J.; Bell, S. H. *J. Am. Chem. Soc.* **1986,** *108*, 7100.
4. Trécourt, F.; Mallet, M.; Mongin, O.; Quéguiner, G. *J. Org. Chem.* **1994,** *59*, 6173.
5. Schlosser, M. *Chem. Eur. J.* **1998,** *4*, 1969.
6. Cohen, T.; Bhupathy, M. *Acc. Chem. Res.* **1989,** *22*, 152.
7. Hoffmann, R. W.; Stiasny, H.-C. *Tetrahedron Lett.* **1995,** *36*, 4595.
8. Hintze, F.; Hoppe, D. *Synthesis* **1992**, 1216.
9. Johnson, T. A.; Curtis, M. D.; Beak, P. *J. Am. Chem. Soc.* **2001,** *123*, 1004.
10. Curtis, M. D.; Beak, P. *J. Org. Chem.* **1999,** *64*, 2996.

Index